*Edited by*
*Srinivasan Chandrasekaran*

**Click Reactions in Organic Synthesis**

*Edited by Srinivasan Chandrasekaran*

# Click Reactions in Organic Synthesis

WILEY-VCH
Verlag GmbH & Co. KGaA

**Editor**

***Prof. Srinivasan Chandrasekaran***
Indian Institute of Science
Department of Organic Chemistry
C. V. Raman Avenue
560 012 Bangalore
India



**Library of Congress Card No.:** applied for

**British Library Cataloguing-in-Publication Data**
A catalogue record for this book is available from the British Library.

**Bibliographic information published by the Deutsche Nationalbibliothek**
The Deutsche Nationalbibliothek lists this publication in the Deutsche Nationalbibliografie; detailed bibliographic data are available on the Internet at <http://dnb.d-nb.de>.



**Print ISBN:** 978-3-527-33916-7
**ePDF ISBN:** 978-3-527-69415-0
**ePub ISBN:** 978-3-527-69418-1
**Mobi ISBN:** 978-3-527-69416-7
**oBook ISBN:** 978-3-527-69417-4

**Cover Design** Formgeber, Mannheim, Germany
**Typesetting** SPi Global, Chennai, India
**Printing and Binding** Markono Print Media Pte Ltd, Singapore

Printed on acid-free paper

# Contents

# List of Contributors

*Franck Amblard*
Emory University School of Medicine
Center for AIDS Research
and Department of Pediatrics
Atlanta, GA 30322
USA

*Kengadarane Anebouselvy*
University of Hyderabad, Central University (P.O.)
Catalysis Laboratory
School of Chemistry
Prof. CR Rao Road
Gachibowli
500 046 Hyderabad
India

*Maria José Arévalo*
Universidad de Extremadura
Escuela Politécnica
Departamento de Química Orgánica e Inorgánica
Avenida de la Universidad
s/n, 10004 Cáceres
Spain

*Kalpattu Kuppusamy Balasubramanian*
INSA Senior Scientist
Department of Biotechnology
Indian Institute of Technology Madras
Chennai 600036
India

*Sebastien Boucle*
Emory University School of Medicine
Center for AIDS Research
and Department of Pediatrics
Atlanta, GA 30322
USA

*Poonam Chauhan*
Indian Institute of Technology
Department of Chemistry
208016 Kanpur
India

*Gaojian Chen*
Soochow University
Center for Soft Condensed Matter Physics and Interdisciplinary Research
1 ShiZi Street
Jiangsu Province
Suzhou 215006
P. R. China

*Kui Chen*
Soochow University
Center for Soft Condensed
Matter Physics and
Interdisciplinary Research
1 ShiZi Street
Jiangsu Province
Suzhou 215006
P. R. China

*Maria Victoria Gil*
Universidad de Extremadura
Departamento de Química
Orgánica e Inorgánica
Facultad de Ciencias
Avda. de Elvas
s/n, 06071 Badajoz
Spain

*Balasubramanian Gopalan*
Orchid Chemicals and
Pharmaceuticals Limited
Drug Discovery Research
Plot #476/14
Old Mahabalipuram Road
Sholinganallur
600119 Chennai
India

*Freek A. B. M. Hoogstede*
Radboud University
Institute for Molecules and
Materials
Department of Synthetic Organic
Chemistry
Heyendaalseweg 135
6525 AJ Nijmegen
The Netherlands

*Ahmed Khalil*
Emory University School
of Medicine
Center for AIDS Research
and Department of Pediatrics
Atlanta, GA 30322
USA

*Óscar López*
Universidad de Sevilla
Departamento de Química
Orgánica
Facultad de Químicas
c/Profesor García González
s/n, 41012 Seville
Spain

*Joydeb Mandal*
Indian Institute of Science
Department of Inorganic and
Physical Chemistry
560012 Bangalore
India

*Amrita Mishra*
Banaras Hindu University
Department of Chemistry
Faculty of Science
221005 Varanasi, UP
India

*Kunj B. Mishra*
Banaras Hindu University
Department of Chemistry
Faculty of Science
221005 Varanasi, UP
India

*Dario Pasini*
University of Pavia
Department of Chemistry and
INSTM Research Unit
Viale Taramelli
10-27100 Pavia
Italy

*Dhevalapally B. Ramachary*
University of Hyderabad
Central University (P.O.)
Catalysis Laboratory
School of Chemistry
Prof. CR Rao Road
Gachibowli
500 046 Hyderabad
India

*S. Ramakrishnan*
Indian Institute of Science
Department of Inorganic and
Physical Chemistry
560012 Bangalore
India

*Ramesh Ramapanicker*
Indian Institute of Technology
Department of Chemistry
208016 Kanpur
India

*Floris P. J. T. Rutjes*
Radboud University
Institute for Molecules and
Materials
Department of Synthetic Organic
Chemistry
Heyendaalseweg 135
6525 AJ Nijmegen
The Netherlands

*Ozkan Sari*
Emory University School
of Medicine
Center for AIDS Research
and Department of Pediatrics
Atlanta, GA 30322
USA

*Raymond F. Schinazi*
Emory University School
of Medicine
Center for AIDS Research
and Department of Pediatrics
Veterans Affairs Medical Center
Atlanta, GA 30322
USA

*Anoop S. Singh*
Banaras Hindu University
Department of Chemistry
Faculty of Science
221005 Varanasi, UP
India

*Vinod K. Tiwari*
Banaras Hindu University
Institute of Science
Department of Chemistry
Centre of Advanced Study
221005 Varanasi, UP
India

*Weidong Zhang*
Soochow University
Center for Soft Condensed
Matter Physics and
Interdisciplinary Research
1 ShiZi Street
Jiangsu Province
Suzhou 215006
P. R. China

# Preface

Organic chemists are generally familiar with "Name Reactions" and many of these have had profound influence over the way we practice organic chemistry. However, there are a few reactions without a name that changed the course of history (science) and are unparalleled in terms of their impact over a wide range of scientific disciplines. "Olefin metathesis" is one such reaction that led to the award of Nobel Prize to the pioneers who contributed to the development of this reaction. In a similar vein, we have a group of reactions without a proper name that have taken the scientific community by storm in less than fifteen years, that is, "Click Reactions."

Click chemistry is a chemical concept enunciated by Barry Sharpless, Scripps Research Institute, USA, in 2001, which highlights the importance of using a set of powerful, highly reliable, selective reactions under simple reaction conditions to join small molecular units together quickly for the rapid synthesis of new compounds via heteroatom links and create molecular diversity. Several types of reactions have been identified that fulfill the criteria- thermodynamically favored reactions that lead specifically to one product such as nucleophilic ring opening reactions of epoxides and aziridines, nonaldol type carbonyl reactions, additions to carbon carbon multiple bonds, Michael additions, and cycloaddition reactions. The best-known click reactions are the copper-catalyzed reaction of azides and alkynes or the so-called CuAAC reaction and the thiol-ene reaction.

After the advent of click chemistry, the synthesis of molecules with outstanding, multifaceted properties has become very popular and easier simply by conjugating two or more molecules that possess remarkable individual properties. Over the past fifteen years, developments in the area of click chemistry have been extraordinary, and these reactions have been explored to their limits in various fields.

Who would have imagined that this simple set of reactions referred to as "click reactions" would revolutionize the approach to science cutting across disciplines such as drug discovery, polymer synthesis, materials science, chemical biology, supramolecular chemistry, and cosmetic chemistry in a short span of time. The tenth anniversary of the discovery of this concept of click reactions was celebrated with a lot of fanfare in the scientific community in 2011.

A number of review articles appeared in leading scientific journals covering the impact of this simple but elegant set of reactions in the design and synthesis of novel molecular architectures. While there are a couple of books that have been published dealing with click chemistry and its application to biology and material science, no book has been published so far that highlights the "click concept" and its far-reaching implications in various facets of organic synthesis. Hopefully, this book would fulfill this need and would also serve as a ready reckoner for accessing all information in pursuit of newer vistas in scientific research.

This book *Click Reactions in Organic Synthesis* covers ten different topics that would illustrate the scope of click reactions in various facets of organic synthesis. Leading experts who are active in this field have contributed to this venture. In the introductory chapter, Ramapanicker and Chauhan provide the mechanistic and synthetic perspectives of click reactions. Gopalan and Balasubramanian discuss the applications of click reactions in the synthesis of pharmaceuticals and drug discovery/development in Chapter 2. In Chapter 3, Gil and coworkers present the perspectives of green chemical synthesis. Ramachary and Anebouselvy have focused the attention on metal-free click reactions in organic synthesis in Chapter 4. Rutjes and Hoogstede have provided details on the use of click chemistry to peptide synthesis in Chapter 5. In Chapter 6, work related to the application of click chemistry to the synthesis of carbohydrate derivatives is discussed by Tiwari and coworkers. Synthesis of polymers and modifications using CuAAC click reactions are enunciated by Ramakrishnan and Mandal in Chapter 7. Chemistry related to thiol-ene (click) reactions in polymer synthesis and modifications are presented by Chen and coworkers in Chapter 8. Pasini highlights the importance of click chemistry in the synthesis of macrocycles in Chapter 9. In the last chapter, Schinazi and coworkers discuss the importance of click chemistry in DNA synthesis and modifications. I feel honored by the excellent contributions that the authors have delivered, and I owe my special thanks to all these scientists. I would also like to thank publishing editors /staff of Wiley-VCH for their fruitful collaboration.

Finally, I would like to place on record my sincere thanks to the pioneer of this area of work, Professor Barry Sharpless on the occasion of the 15th anniversary of "Click Reactions."

*Bangalore* *Srinivasan Chandrasekaran*
*April 2, 2016*

# 1
# Click Chemistry: Mechanistic and Synthetic Perspectives

*Ramesh Ramapanicker and Poonam Chauhan*

Search for reactions that can be used to link two or more diversely functionalized molecules with minimum effort and without the formation of side products has become increasingly important in the past 15 years. As organic molecules started to find their place as easily tunable and functional materials, the requirement of new conjugation reactions that can be used effectively by nonsynthetic organic chemists became unavoidable. Such a reaction should be easier to carry out, yield high selectivity, should be compatible with water and other protic solvents, and should lead to quantitative conversions. Click chemistry is a collection of such reactions that has evolved as an efficient tool for ligation, which gained quick acceptance in biotechnology, material and polymer science, medicinal chemistry, and so on. Among all the click reactions, copper-catalyzed 1,3-dipolar Huisgen cycloaddition (HDC) between a terminal alkyne and an azide is *the jewel in the crown*. Owing to its remarkable functional group tolerance, researchers can fearlessly introduce easily functionalizable groups such as hydroxyl, carboxyl, and amino groups into conjugate molecules using this reaction.

The concept of click chemistry was first introduced by Sharpless and coworkers in 2001 at the Scripps Research Institute [1]. Click chemistry is not limited to a set of organic reactions, but is a synthetic philosophy inspired by nature in terms of their efficiency, selectivity, and simplicity. Any reaction that can produce conjugate molecules efficiently from smaller units under simpler reaction conditions can be considered as a click reaction. The catchy term *click* refers to reactions that are modular in approach, efficient, selective, versatile in nature, give single product (high yielding), and can be performed in benign and easily removable solvents without the need for chromatographic purification. There are various reactions with different mechanisms that can be considered as click reactions, provided they follow a simple common reaction trajectory [1].

Sharpless first introduced the concept of click chemistry to provide an effective conjugation technique in drug discovery [2], but the concept and methodology were widely accepted, and click chemistry found its applications in almost all facets of research and technology, which employ organic molecules, such as

*Click Reactions in Organic Synthesis*, First Edition. Edited by Srinivasan Chandrasekaran.

polymer science [3], nanoscience [4], bioconjugation [5], and development of sensors [6] .

In this chapter, we have provided a detailed account of various click reactions with emphasis on their mechanisms and synthetic details. The discussions are based on the following classification of click reactions.

## 1.1 Cycloaddition Click Reactions

### 1.1.1 Azide–Alkyne Huisgen 1,3-Dipolar Cycloaddition

The classical HDC reaction between an alkyne and an azide is the most discussed among click reactions. Both alkynes and azides are unreactive under physiological conditions and undergo a cycloaddition reaction only at elevated temperatures (Scheme 1.1) [7, 8]. Although both alkynes and azide functions can easily be introduced on to the substrates, the cycloaddition reaction is highly exothermic ($\Delta H^0$ is between −50 and −65 kcal/mol) and has a high activation barrier of 25–26 kcal/mol (for methyl azide and propyne). Hence, the uncatalyzed reaction is generally slow and is not regioselective [9]. The difference between HOMO-LUMO energy levels of both azide and alkyne are comparable, thus both dipole HOMO and dipole LUMO pathways can operate in this reaction leading to a mixture of 1,4 and 1,5-triazole regioisomers. It is, however, observed that the use of electron-deficient terminal alkynes can impart 1,4-regioselectivity to a reasonable extent. These factors limit the use of uncatalyzed Huisgen cycloaddition as an effective conjugation technique.

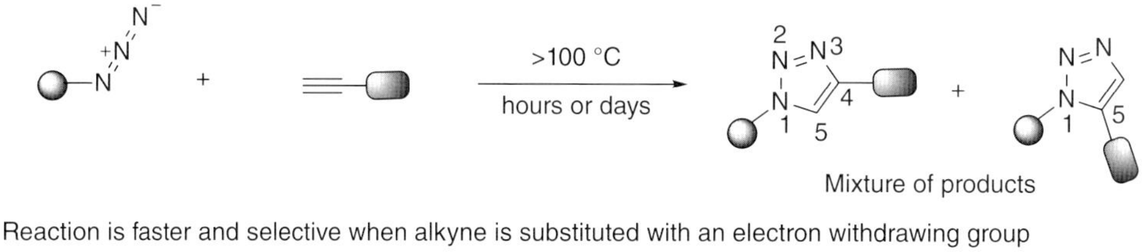


**Scheme 1.1** Huisgen 1,3-dipolar cycloaddition between alkynes and azides.

### 1.1.2 Copper-Catalyzed Azide–Alkyne Cycloaddition (CuAAC) Click Reaction

Sharpless [9] and Meldal [10] independently reported a Cu(I)-catalyzed version of the cycloaddition reaction between azides and terminal alkynes, which is $10^7$ times faster than the uncatalyzed reaction. The interaction between Cu(I) and terminal alkynes makes the latter a better 1,3-dipolarophile, enhancing its reaction with azides. The Cu(I)-catalyzed reaction is highly regioselective and only the 1,4-adducts are formed. The Cu(I)-catalyzed reactions can be carried out at room temperature and at a much faster rate.

Sharpless reported the possibility of using *in situ* generated copper(I), obtained through the reduction of copper sulfate pentahydrate ($CuSO_4{\cdot}5H_2O$) with ascorbic acid, as an efficient catalyst for carrying out azide–alkyne conjugation reactions in solutions [9]. The reactions worked well when a mixture of water and an alcohol is used as the solvent. The solvent mixture allowed effective dissolution of the metal salt and the organic components needed to be conjugated. Meldal and coworkers reported a very practical application of azide–alkyne cycloaddition catalyzed with cuprous iodide in conjugating peptides through side chains or the backbone in solid phase [10]. Both reactions were selective for the formation of 1,4-disubstituted 1,2,3-triazoles and together revolutionized the concept of click reactions (Scheme 1.2).

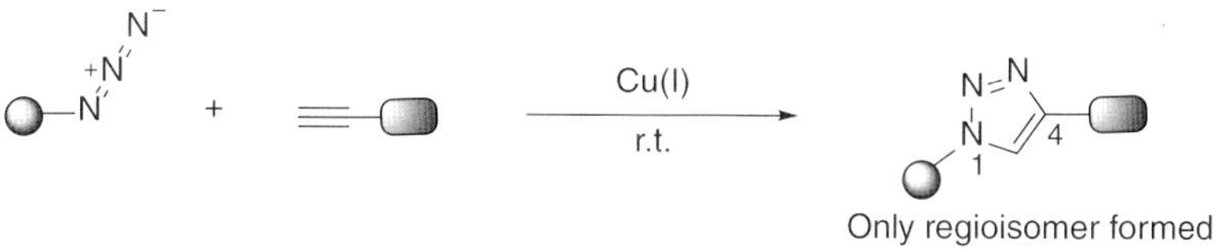


**Scheme 1.2** CuAAC click reaction.

In addition to being a stable linker, the triazole group has certain other advantages. On comparison with an amide bond, which was otherwise the most common linkage used, a triazole group exhibits certain interesting and unique properties. Unlike an amide bond, triazoles are not susceptible to hydrolytic cleavage. They cannot be reduced or oxidized under normal conditions. A triazole linkage, with an extra atom in its backbone, places the carbon atoms linked to 1- and 4-positions at a distance of 5.0 Å, while an amide linkage places the carbon atoms only at 3.8 Å apart from each other. The nitrogen atoms at 2- and 3-positions of the triazole have weak hydrogen-bond-accepting properties. The inherent dipole moment in a triazole ring leads to polarization of the $C_5$–H bonds, making them hydrogen bond donors and enabling C–H···X hydrogen bonds, similarly to an amide bond [11]. These properties also enabled Cu(I)-catalyzed triazole formation to gain attention as an effective conjugation method.

Conjugation of functional molecules through triazoles received immediate attention especially in drug discovery. Linhardt *et al.* synthesized some sialic acid conjugates using copper catalyzed azide–alkyne cycloaddition (CuAAC), which are potential neuraminidase inhibitors with good $IC_{50}$ values (Figure 1.1) [12]. There are a large number of such examples of CuAAC being used effectively for assembling small molecular units to obtain more functional and useful molecules. An interesting example is the synthesis of the rigid macrocycle **C** (Figure 1.2) by Flood *et al.*, in which triazole units function as rigid structural units and provide acidic hydrogens to interact and detect chloride ions in organic solvents [13]. In

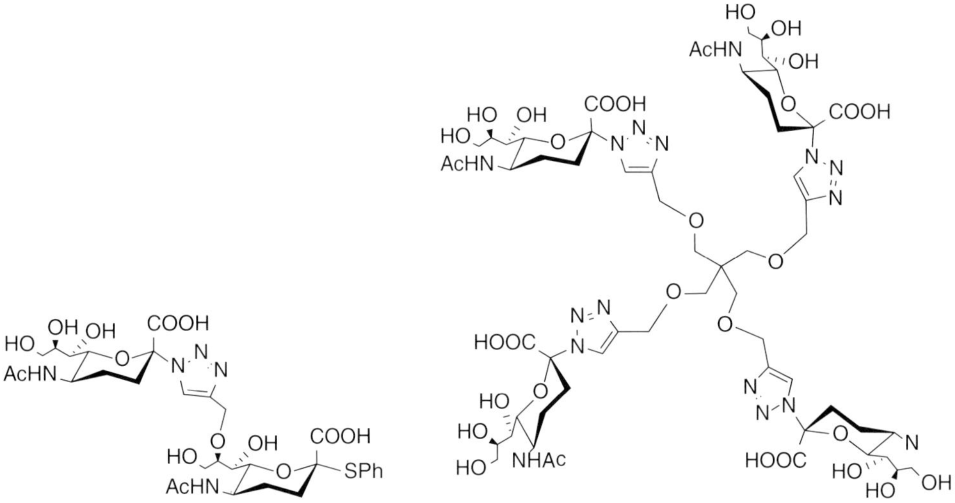

**A** ($IC_{50}$ value = 17 μM) **B** ($IC_{50}$ value = 20 μM)

**Figure 1.1** Sialic-acid-based neuraminidase inhibitors; a disaccharide mimic **A** and a dendrimer **B**.

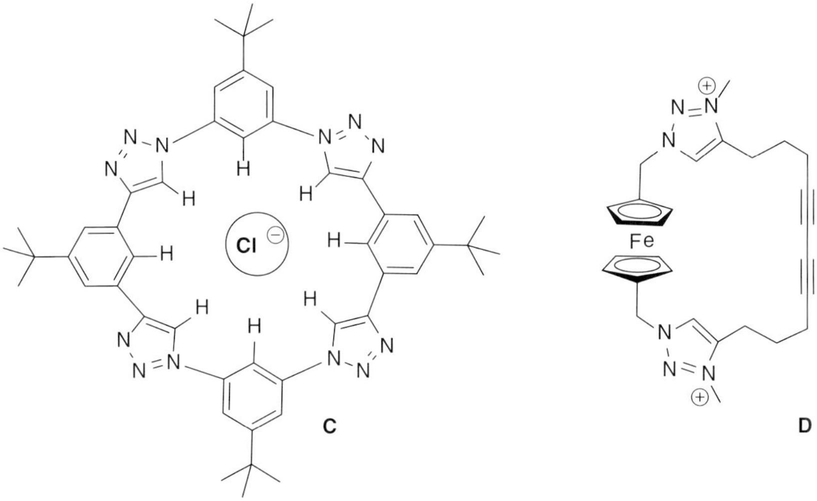

**C** **D**

**Figure 1.2** Triazole-containing macrocycles used for the detection of anions.

a similar attempt, Beer *et al.* have reported a ferrocene-containing bis(triazole) macrocycle **D** (Figure 1.2), in which they have increased the anion binding tendency of the C–H of triazole by converting triazole units to cationic triazolium moieties. Alkylation of a triazole increases its binding capability with anions such as chloride and benzoate ions even in polar organic solvents [14].

#### 1.1.2.1 Mechanism of CuAAC Click Reactions

A detailed mechanistic analysis of CuAAC was reported by Jan H. van Maarseveen and coworkers in 2006 [15]. The report was based on comprehensive kinetic studies and DFT calculations. Studies showed that the Cu-catalyzed cycloaddition reaction proceeds through a stepwise mechanism and the activation energy is 11 kcal/mol less than that of the uncatalyzed reaction, which has an activation energy of 26 kcal/mol. However, a concerted mechanism involving Cu–acetylene π-complex and the azide was calculated to have a higher activation energy of 27.8 kcal/mol. The reaction begins with the formation of a Cu–alkyne π complex, which then forms a copper acetylide after deprotonation of the alkyne proton. Coordination of copper with the alkyne makes the acetylenic proton more acidic, increasing its acidity by up to 9.7 pH units, which allows the deprotonation to occur in aqueous media even in the absence of a base. The copper acetylide exists in equilibrium between a monomer and a dimer. One of the Cu ions in the dimer coordinates with the azide nitrogen and activates it. This complex then cyclizes to give a metallacycle via a nucleophilic attack of the terminal nitrogen of the azide group on the internal carbon of the alkyne. The metallacycle then undergoes a ring contraction through a transannular interaction between the lone pair of electrons on the substituted nitrogen of the azide and the C=Cu bond. This relatively faster step yields a Cu triazolide, which undergoes protonation to liberate the 1,4-disubstituted triazole and regenerates the Cu(I) catalyst (Scheme 1.3) [16].

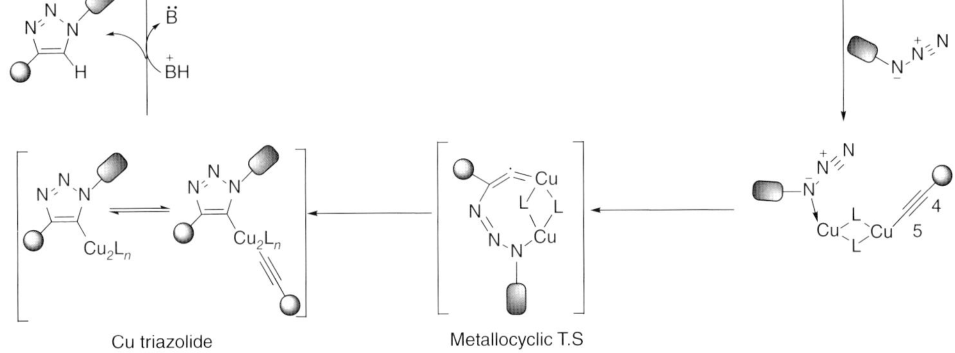


**Scheme 1.3** Mechanism of the CuAAC reaction as proposed by Jan H. van Maarseveen [15].

#### 1.1.2.2 Catalysts used for CuAAC Click Reactions

The success achieved in CuAAC click reactions prompted researchers to look for better and more stable catalysts to carry out the azide–alkyne cycloaddition to triazoles. However, despite several efforts, $Cu^{+1}$ is found to be the best catalyst. The unique activity of $Cu^{+1}$ over other metal ions is due to its ability to involve the terminal alkynes in both $\sigma$ and $\pi$ interactions and the possibility of immediate replacement of the ligands in its coordination sphere (generally in aqueous medium). However, $Cu^{+1}$ is thermodynamically unstable and oxidizes to $Cu^{+2}$ or disproportionates to a mixture of $Cu^{+2}$ and Cu under aerobic conditions. $Cu^{+2}$ is catalytically inactive, and its generation halts the reaction.

The thermodynamic instability of $Cu^{+1}$ places importance on its introduction to a reaction mixture. It is observed that $Cu^{+1}$ species are relatively stable in organic solvents and in the absence of water and oxygen. Cu(I) salts such as CuI, CuBr, and $CuOTf{\cdot}C_6H_6$ have been found to be efficient catalysts in organic solvents. The use of Cu(I) salts in organic solvents is generally carried out with the addition of a tertiary amine base such as diisopropylethylamine (DIPEA) or 2,6-lutidine [17]. This is attributed largely to the requirement of a base to deprotonate the Cu–alkyne $\pi$ complex, so as to generate the copper acetylide. It is also observed that amines and certain solvents such as acetonitrile [18] stabilize the Cu(I) species through coordination, preventing its degradation through oxidation or disproportionation.

When the reactions are carried out in aqueous media or in a mixture of water and an alcohol (most commonly *tert*-butanol), the degradation of Cu(I) salts is inevitable. It is found that the use of a Cu(II) salt such as $CuSO_4{\cdot}5H_2O$ along with reducing agents such as sodium ascorbate, hydrazine, or tris(2-carboxyethyl)phosphine (TCEP) generates $Cu^{+1}$ *in situ*. This method, where the active catalyst is generated in the reaction mixture via reduction of Cu(II) salts, works well in aqueous solutions and even in the presence of oxygen [19]. Additionally, the ability of water to act as a base allows these reactions to be carried out in the absence of an external base such as DIPEA. The continuous presence of a reducing agent such as sodium ascorbate ensures the regeneration of Cu(I) even if the active catalysts is quenched by air [19b]. It is also advantageous that carrying out reactions in aqueous media with *in situ* generation of Cu(I) species allows the use of substrates with unprotected amino and hydroxyl functions.

A third but less explored method for introduction of $Cu^{+1}$ into the reaction mixture is by vigorously shaking or by microwave irradiation of a solution containing metallic copper. While the amount of $Cu^{+1}$ ions produced in solution by vigorous shaking is quite less leading to extended reaction times (12–48 h) [20], microwave irradiation completes the reaction in 10–15 min at elevated temperatures [21]. An advantage of this method is the isolation of products with negligible copper contamination. Various other forms of copper such as Cu(I)-modified zeolites, copper oxide nanoparticles [22], or copper nanoparticles adsorbed on charcoal [23] have all been utilized successfully for CuAAC reactions.

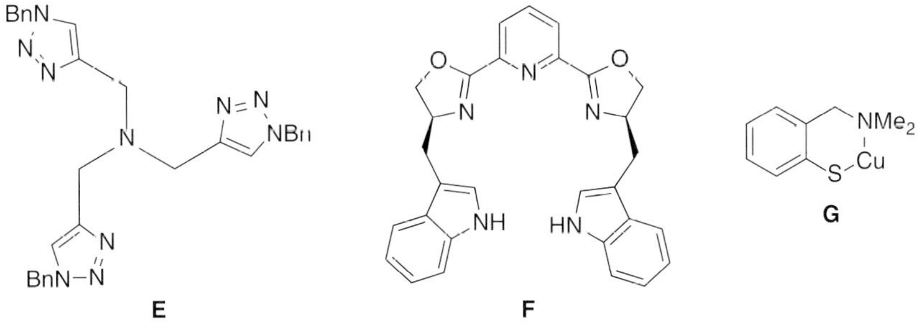


**Figure 1.3** Ligands used in CuAAC click reactions.

#### 1.1.2.3 Ligands used for CuAAC Click Reactions

Although CuAAC can be performed with $Cu^{+1}$ generated *in situ* or provided as a Cu(I) salt in the absence of any ligands, certain ligands such as those that can form heterocyclic chelates with $Cu^{+1}$ ions are shown to increase the rate of the reaction (Figure 1.3) [24]. The role of these ligands is assumed to be based on restraining $Cu^{+1}$ from interactions, which lead to its degradation. Tris-(benzyltriazolylmethyl)amine (TBTA, **E**), a tetradentate ligand, is shown to be very efficient in increasing the rate of CuAAC click reactions [25]. Owing to its tetradentate-binding ability, it completely surrounds the Cu(I) center and does not provide any free binding sites for destabilizing interactions. The tertiary amino group in TBTA can also act as the required base, when reactions are carried out in organic solvents. Certain ligands are known to reduce the minimum catalyst loading by almost 10 times with no increase in reaction time [26]. Some common nitrogen-based ligands used in facilitating CuAAC are shown in Figure 1.3. Other than those based on nitrogen, ligands containing oxygen, phosphorous [27], carbon [28], and sulfur [29] as donor atoms are also reported.

### 1.1.3 Ruthenium-Catalyzed Azide–Alkyne Cycloaddition (RuAAC) Click Reactions

Among various other metal ions studied for catalyzing HDC between azides and alkynes, Ru(II) catalysts were found to be the most notable. The catalytic activity and regioselectivity of the reaction were found to be dependent on the ligand environment of the Ru center. Unlike the Cu(I)-catalyzed reactions, azide–alkyne cycloaddition reactions catalyzed by ruthenium complexes showed a preference for the formation of 1,5-disubstituted triazoles to the formation of 1,4-disubstituted triazoles (Scheme 1.4). Out of the various ruthenium complexes studied for catalysis of this cycloaddition reaction, the most successful catalysts are $C_p{}^*RuCl$, $C_p{}^*RuCl(PPh_3)_2$, $C_p{}^*RuCl(COD)$, and $C_p{}^*RuCl(NBD)$. The reactions are performed with 1–2 mol% of the catalyst in THF/dioxane or in any nonprotic solvent at temperatures ranging from ambient to 80 °C. Another salient feature of Ru-catalyzed reactions is the possibility to use internal alkynes for the reaction to obtain 1,4,5-trisubstituted triazoles as the products in good

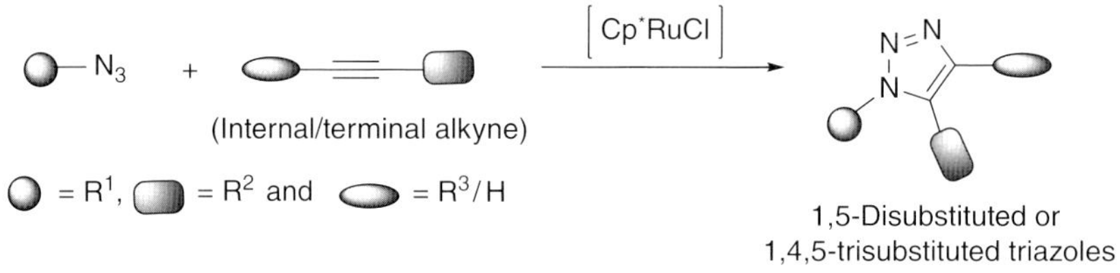


**Scheme 1.4** Formation of 1,5-disubstituted or 1,4,5-trisubstituted triazoles via Ru-catalyzed 1,3-dipolar cycloaddition reaction between azides and alkynes.

yields (Scheme 1.4) [30]. Other than Cu and Ru, attempts have been made to use other metals such as Ni, Fe, Sm, Ce, and Zn also as catalysts for HDC reactions, but none of them gave satisfying results to be used widely [31].

Unlike CuAAC reactions, the Ru-catalyzed version of HDC reactions was more dependent on the steric details of the azides than those of the alkyne components. Primary and secondary azides in the presence of catalytic amount of Ru complexes react with a wide range of terminal alkynes, but tertiary azides seem to be less reactive [24]. Electronic and steric properties of the alkynes too play a crucial role in these reactions, but not as much as those of the azides. Alkynes having H-bond donor groups such as propargyl alcohols and propargyl amines show high regioselectivity even for unsymmetrical alkynes. Strong H-bond between OH or $NH_2$ of the alkyne and Cl on the Ru complex is the driving force for the reaction. The new bond is always formed between β carbon of alkyne and terminal nitrogen of the azides.

#### 1.1.3.1 Mechanism of RuAAC Click Reactions

Mechanistic insights into ruthenium-catalyzed azide–alkyne cycloaddition (RuAAC) reactions were provided by Fokin and coworkers in 2008, based on DFT calculations [32]. The mechanism is proposed to have two important steps. After the initial coordination of the alkyne and azide onto ruthenium, an irreversible oxidative coupling takes place, which also involves the formation of a C–N bond by the nucleophilic attack of the electronegative carbon of the activated alkyne on the terminal electrophilic nitrogen of the coordinated azide, forming a six-membered ruthenacycle intermediate. This cyclic intermediate then undergoes a rate-determining reductive elimination to give a triazolyl complex, which liberates a 1,5-disubstituted triazole product through ligand exchange (Scheme 1.5).

### 1.1.4 Strain-Promoted Azide–Alkyne Cycloaddition (SPAAC) Reactions

Apart from the applications in synthesizing drug molecules with a triazole linkage, azide–alkyne cycloaddition reactions have also been used for various biological applications such as site-specific protein/viruses modifications and functionalization of cell surfaces. Use of transition-metal-catalyzed reactions for

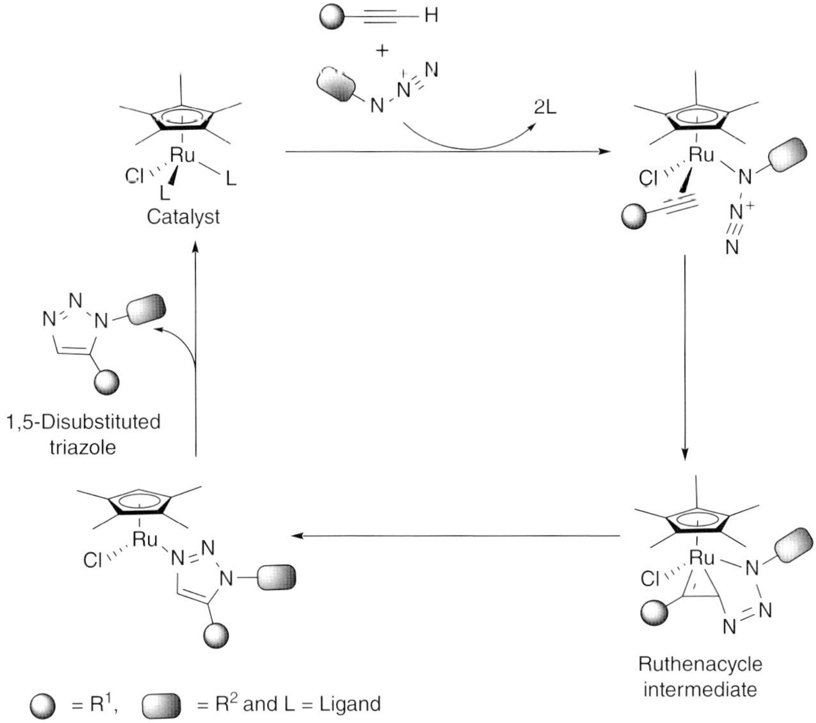


**Scheme 1.5** Proposed mechanism of RuAAC click reactions.

such applications is not advisable as metal salts could be detrimental to living cells. Copper salts are known to degrade oligonucleotide strands, and copper is cytotoxic at higher concentrations. This has placed an importance on the search for click reactions that can be carried out without the use of metal catalysts.

Use of electron-deficient alkynes is a possible option for increasing the rate of an uncatalyzed azide–alkyne cycloaddition reaction. However, this requirement places a serious restriction on the nature of functionalities that can be incorporated on the alkyne. A rapid cycloaddition reaction between neat cyclooctyne, the smallest stable cycloalkyne, and phenyl azide to give a triazole product in high yields was reported as early as in 1961. The release of substantial ring strain of nearly 18 kcal/mol in the cyclooctyne was the driving force for this reaction. Bertozzi and coworkers explored the possibility of using this strain-promoted azide–alkyne cycloaddition (SPAAC) reaction as a click reaction for bioconjugation [33]. They introduced electron-withdrawing groups (EWGs) on to the cyclooctyne system to increase its reactivity toward cycloaddition reactions further. Mono- and difluorinated cyclooctyne derivatives were prepared, which have lower energy LUMO providing an increased second-order rate constants for cycloaddition reactions (Scheme 1.6) [34].

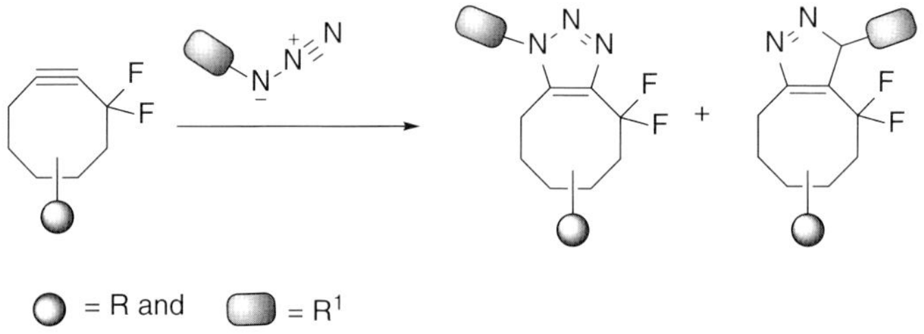


**Scheme 1.6** An example for SPAAC click reaction.

Boons and coworkers employed a different strategy and introduced benzyl groups adjacent to the alkyne function, thereby increasing the ring strain in the cyclooctyne molecule [35]. They succeeded in using these benzyl derivatives for reaction with azides and employed this strategy in visualizing labeled glycoconjugates metabolically in living cells.

Various other cycloaddition reactions including Diels–Alder reaction and hetero-Diels–Alder reactions have been employed as click reactions. However, all such reactions have found no or limited applications as conjugation methods.

### 1.1.5
### Organocatalytic Triazole Formation

As illustrated earlier, 1,2,3-triazoles have the potential to be very useful pharmacophores. This has placed some importance in methods leading to their formation even if they are not be used as click reactions for conjugation. One such approach that has gained recent attention is the synthesis of 1,2,3-triazoles through organocatalytic cycloaddition reactions. Compared to metal catalysts, organocatalysts are eco-friendly, insensitive to oxygen and water, and are easily available. Enolates and enamines can be easily produced by condensation of amine and aldehyde, so they have been explored by many researchers as dipolarophiles [36]. A representative example is the proline-catalyzed reaction between Hagemann's ester and tosyl azide to give fused triazoles, reported by Ramachary *et al.* [34b] (Scheme 1.7).

In general, such reactions begin with the *in situ* formation of an enamine by reaction between a carbonyl compound and a secondary amine, which also acts as the catalyst (Scheme 1.8). The enamine thus generated act as a dipolarophile, which reacts with the azide. The cycloaddition reaction between the dipolarophile and azide leads to the formation of a five-membered triazoline intermediate in equilibrium with other intermediates H and I. The protonated secondary ammonium ion I undergoes an elimination leading to the formation of a 1,2,3-triazole (Scheme 1.8). Other than enamines, enolates, peptidyl phosphoranes, vinyl sulfones, and iminolates are some examples that are frequently utilized as dipolarophiles. The dipolarophiles generated are categorized into two types: activated dipolarophiles (e.g., Hagemann's esters and β-ketoesters) and

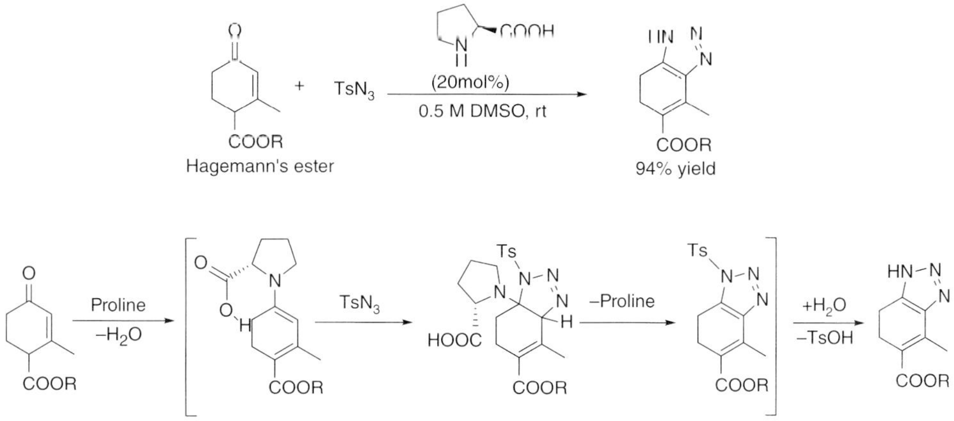


**Scheme 1.7** Proline-catalyzed synthesis of fused triazole from Hagemann's ester and tosyl azide.

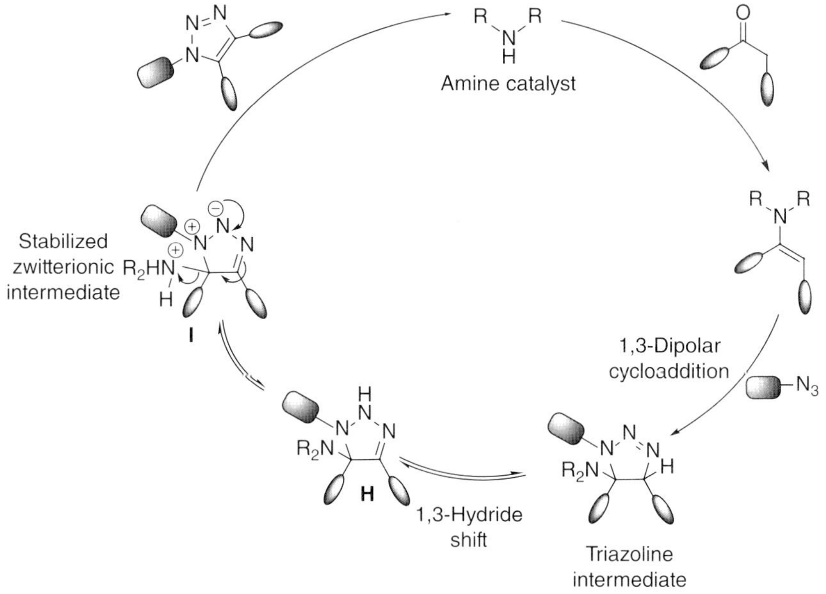


**Scheme 1.8** Mechanism for amine-catalyzed 1,3-dipolar cycloaddition between aldehydes and azides.

unactivated dipolarophiles (e.g., alkyl or allyl ketones and aldehydes). Among amine catalysts that have been found to catalyze these reactions, secondary amines (such as pyrrolidine, morpholine, and diethylamine) and amino acids (such as proline) are the most effective [37].

## 1.2 Thiol-Based Click Reactions

Thiols react with a wide range of substrates and with a number of different functional groups. Generally, the reactions are high yielding and easy to follow. Many of these reactions can be carried out under benign conditions and have been utilized for routine organic synthesis, polymerization, and surface functionalization. The wide range of reactivity of thiols makes them very good conjugation tools, but at the same time, this makes them very susceptible to many side reactions. Thiols, especially the low-molecular-weight molecules are foul-smelling and have low self-stability. Proper selection of substrates and careful handling can circumvent most of the disadvantages, leaving thiols as efficient members in the toolbox of click chemistry.

The history of sulfur-based cross-linking began as early as in 1839, when Goodyear used elementary sulfur to cross-link unsaturated polymers, and the technique is known as *vulcanization* [38]. Ever since then, sulfur and thiols are seen as easily available conjugation tools. The reactions of thiols can broadly be classified as radical reactions and nucleophilic reactions. Radical reactions in particular make them selective toward certain groups under specific reactions conditions, tolerating a large number of other functional groups.

### 1.2.1 Radical Click Reactions of Thiols

The reactions of thiols toward alkenes and alkynes proceed quite smoothly in the presence of light or a radical initiator. The reactions do not need any transition metals as catalysts and are highly preferred as a conjugation method for the preparation of functional molecules for biological applications. Thioether bonds generated in such reactions are stable to strong acids, strong bases, and reducing conditions. These reactions have found their applicability in tailoring solid surfaces with specific properties, immobilization of macromolecules such as proteins, and surface engineering and patterning.

#### 1.2.1.1 Thiol–Ene Radical Click Reaction

The reaction of thiols with alkenes was first introduced by Posner in 1905 [39]. The reactions can be initiated either by using a radical initiator or directly by irradiating thiols with a UV source, preferably at 254 nm. Irradiation of thiols promotes homolysis of the S–H bond resulting in the formation of a thiyl radical [40]. The self-initiation of thiols leading to radical reactions on irradiation with UV light of low wavelengths was first reported by Cramer and coworkers. The reaction of thiyl radicals with alkenes is regioselective and tolerates a wide variety of functional groups. The reaction conditions are mild and are compatible with water and oxygen. These characteristics along with the self-initiation properties of thiols have provided the thiol–ene reaction the status of being a very useful click reaction. The reactions are often termed as *hydrothiolation* of an alkene (Scheme 1.9).

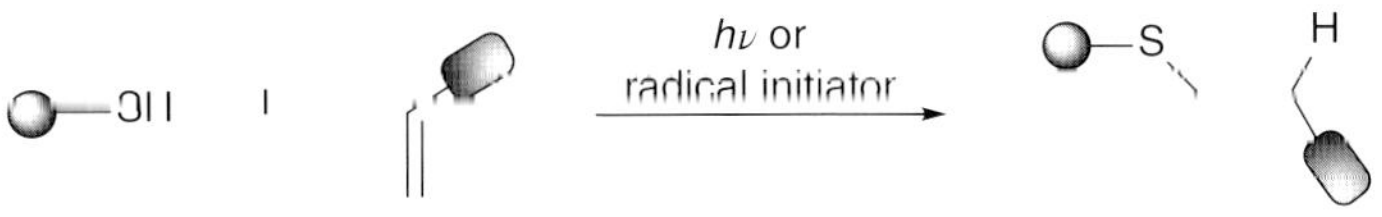


**Scheme 1.9** The hydrothiolation of a C=C bond in the presence of hν or a radical initiator.

The thiol–ene radical reactions can sometime lead to the formation of unwanted by-products through radical recombination reactions. The addition of a thiyl radical to olefins is reversible, until the free radical product formed abstracts a hydrogen radical from another thiol giving a thioether product and propagating the radical reaction [41].

Although it is advantageous to initiate these reactions through direct irradiation of thiols, the reaction rates are often quite slow, resulting from a slow rate of radical formation. The rates can be improved considerably by using a combination light and a photoinitiator. Initial studies and applications were based on the use of hydrogen-abstracting initiators such as benzophenone. It was later found that the reaction rates increase tremendously on the use of Norrish type-1 photoinitiators such as dimethoxyphenylacetophenone (DMPA, for ultraviolet initiation) and phosphine oxide (for visible light initiation).

The mechanism of the reaction has three steps, similarly to all radical reactions. In the initiation step, the thiol or the photoinitiator is irradiated with a light of suitable frequency to generate a thiyl radical. The thiyl radical thus generated undergoes an anti-Markovnikov addition to the alkene to generate a carbon radical as intermediate. The carbon radical then reacts with another thiol molecule forming a thioether and another thiyl radical, and the reaction propagates to complete the radical cycle. The reaction terminates through recombination of thiyl radicals and carbon radicals with each other or between themselves (Scheme 1.10).

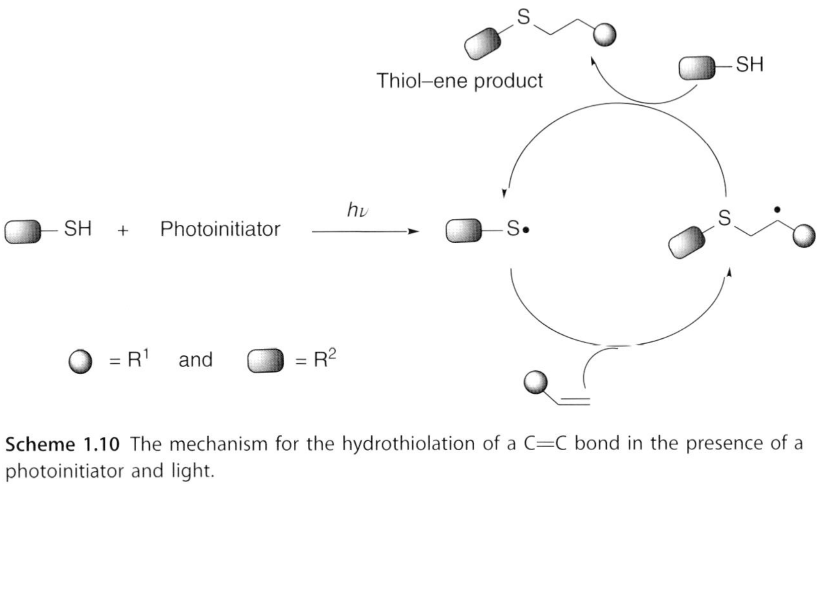


**Scheme 1.10** The mechanism for the hydrothiolation of a C=C bond in the presence of a photoinitiator and light.

The formation of the carbon sulfur bond follows an anti-Markovnikov regioselectivity, which ensures the formation of the most stable carbon radical [42]. There are several reports establishing a general trend for the reaction of thiols with alkenes. Comprehensive reports in this regard were published by Hoyle *et al.* [43], where they compared the reaction of three families of thiols, namely alkyl-3-mercaptopropionates, alkyl thioglycolates, and alkyl thiols, with various alkenes. The reactivity order provided by them is as follows: norbornene > vinyl ethers > propenyl > alkenes ≈ allyltriazines ≈ allyl isocyanurates > acrylates > N-substituted maleimides > acrylonitrile ≈ methacrylates > styrene > conjugated dienes.

A general observation is that the reactivity of an alkene decreases with decrease in electron density of the double bond. Norbornene has an unusually high reactivity owing to a distorted double bond, and addition of thiyl radical leads to a decrease in ring strain. Conjugated olefins such as methacrylates, styrene, and 1,3-dienes have very low reactivity, which is attributed to a very low rate of abstraction of protons by the corresponding carbon radicals from thiol molecules. The decreased reactivity of such carbon radicals are a result of their increased stability achieved through conjugation. It is observed that terminal double bonds are more reactive to hydrothiolation than internal double bonds. Hoyle and coworkers have shown that 1-hexene is 8 times more reactive than *trans*-2-hexene and 18 times more reactive than *trans*-3-hexene [41a].

Among the various families of thiol that have been studied, propionates and glycolates are more reactive than alkyl thiols. This difference in reactivity is proposed to be resulting from the weakening of S–H bond through H-bonding with the carbonyl of the ester function and from polar effects [44].

#### 1.2.1.2 Thiol–Yne Radical Click Reaction

Thiol–yne radical reactions follow a similar initiation step to that of thiol–ene reactions. After the initial addition of a thiol to the alkyne and formation of a vinyl radical with a β-thioether function, a hydrogen abstraction from another thiol molecule generates a new thiyl radical. Subsequent addition of a thiyl radical on the vinyl thioether forms another carbon radical, which abstracts a hydrogen from another thiol molecule to give a 1,2-dithioether, and the thiyl radical generated reenters the chain process (Scheme 1.11). The addition of the first thiol to the alkyne is the rate-limiting step, and the second thiol addition to the intermediate thiol–alkene is a faster step. Studies revealed that the second addition is approximately three times faster than the first addition [45].

In short, the thiol–yne radical click reaction is the formation of a 1,2-dithioether through double addition of thiols on to an alkyne. The reaction has largely been used to generate multifunctional polymer structures. Repetitive thiol–yne reactions are used to form multifunctional molecules, which are further used to make dendrimers [42] or hyperbranched polymers [46].

It has been found that the reactions of thiols with internal alkynes are slower than those with terminal alkynes. These reactions are generally very sensitive to steric crowding. Sulfanyl and related radicals are electron-deficient in nature and

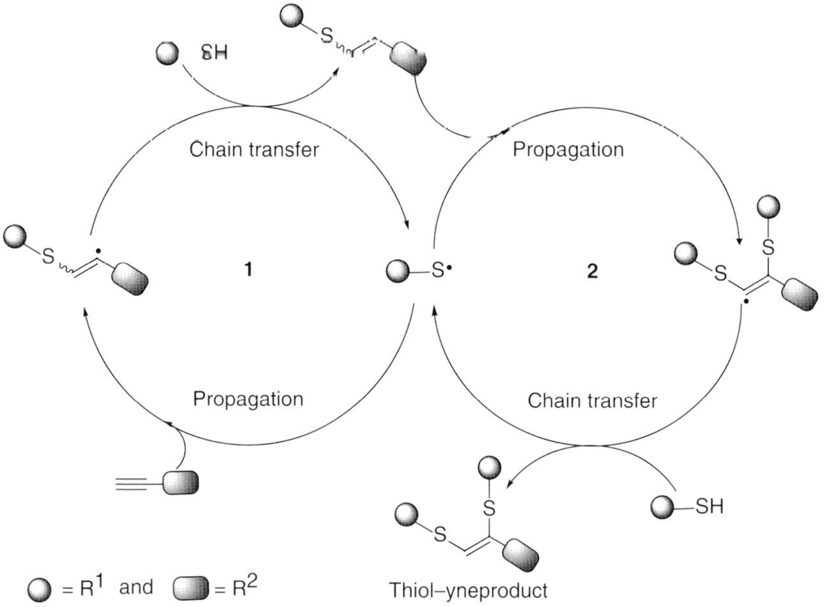


**Scheme 1.11** The reaction mechanism of thiol–yne addition reaction.

are more prone to react with electron-rich alkynes. It is, however, possible to effect these reactions on strained internal alkynes, which are not necessarily electron-rich. In one such example, rapid reaction of thiols with cyclooctyne in the absence of radical initiation is reported. The driving force for this reaction comes from the release of strain in the cyclooctyne system [47].

An elegant example for the use of thiol–yne click reactions to form highly functionalized dendrimers was reported by Stenzel coworkers [42]. They used tripropargyl ester of trimesic acid as the core of a dendrimer, which was functionalized with 1-thioglycerol molecules to get 12 hydroxyl functions attached to the dendritic core. The hydroxyl groups were esterified with anhydride-bearing alkyne groups and were further functionalized with thiol–yne click reactions. The method was repeated to get a dendrimer with as many as 192 hydroxyl groups (Scheme 1.12).

### 1.2.2 Nucleophilic Addition Click Reactions of Thiols

Thiols and thiolate anions are very good nucleophiles. Various click reactions have been developed based on the nucleophilic attack of thiols on to the electrophilic substrates such as epoxides, isocyanates, halides, and Michael acceptors. The reactions are generally initiated by bases, which are added in catalytic amounts or are produced in catalytic amounts by photolatent bases that act as photoinitiators for these reactions [48]. The rates of these reactions are dependent on the substrates

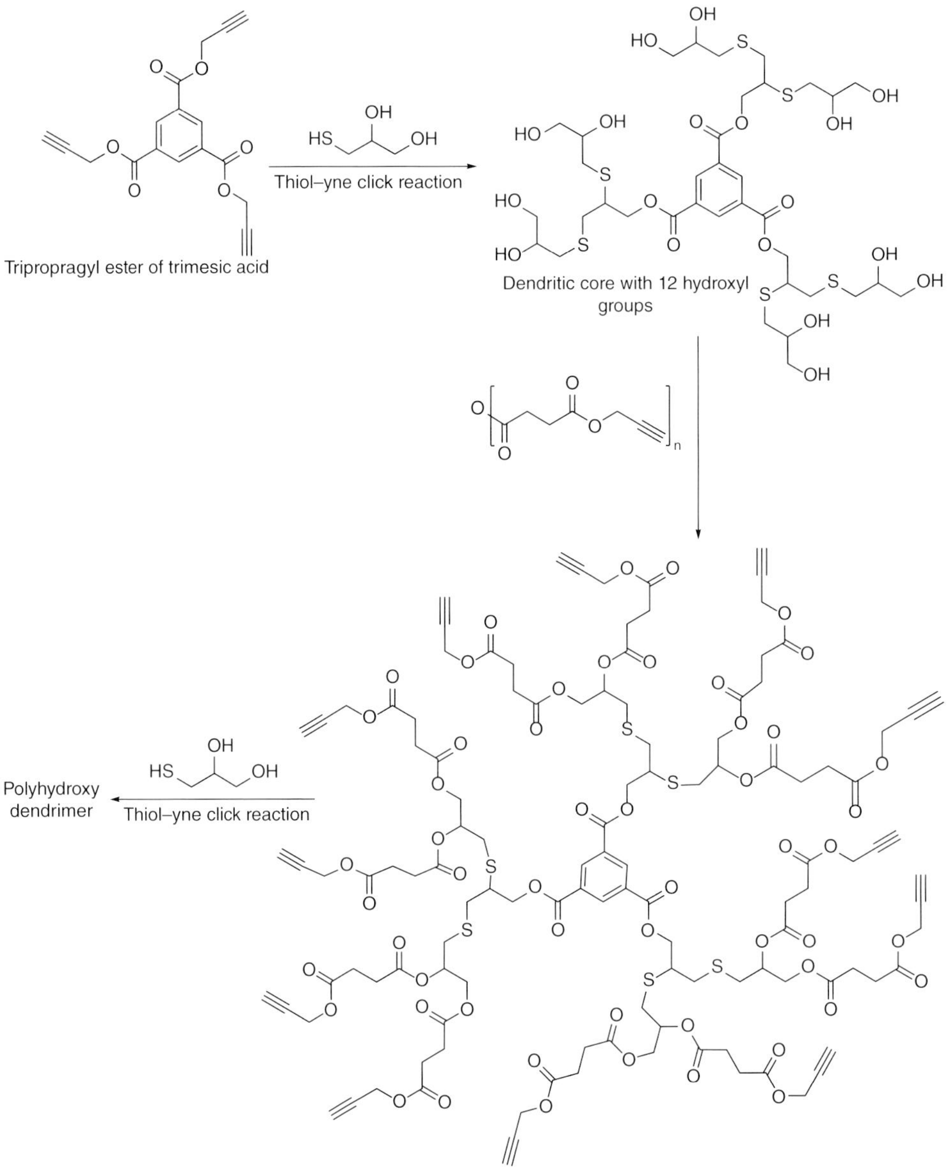


**Scheme 1.12** Multistep thiol–yne mediated synthesis of a highly functional dendrimer.

and their inherent susceptibility to attack by thiols and thiolate ions. This section discusses the various click reactions developed based on this concept.

#### 1.2.2.1 Thiol–Epoxide Click Reactions

Epoxides are strained compounds that undergo ring opening reactions in the presence of nucleophiles. These reactions are carried out either in acidic medium, where the epoxide is protonated making it more electrophilic, or using strong anionic nucleophiles. Thiols, which are considerably more acidic (p$K_a$ of RSH ~5–10 and p$K_a$ of PhSH is 6.4) than water (p$K_a$ = 15.7) and alcohols (p$K_a$ ~ 17) are readily deprotonated to give thiolate ions in the presence of very dilute basic solutions [49]. In general, tertiary amines are used as bases for the generation of thiolate ions. Thiolate ions react immediately and effectively with epoxides following an $S_N2$ reaction pathway and yielding alkoxide anions with a β-thioether substituent. The nucleophilic attack usually happens on the less substituted carbon of the epoxide. The alkoxide ions are protonated either from a protonated base that was used to initiate the reaction or from a molecule of thiol, which generates another thiolate anion and propagates the reaction (Scheme 1.13) [50].

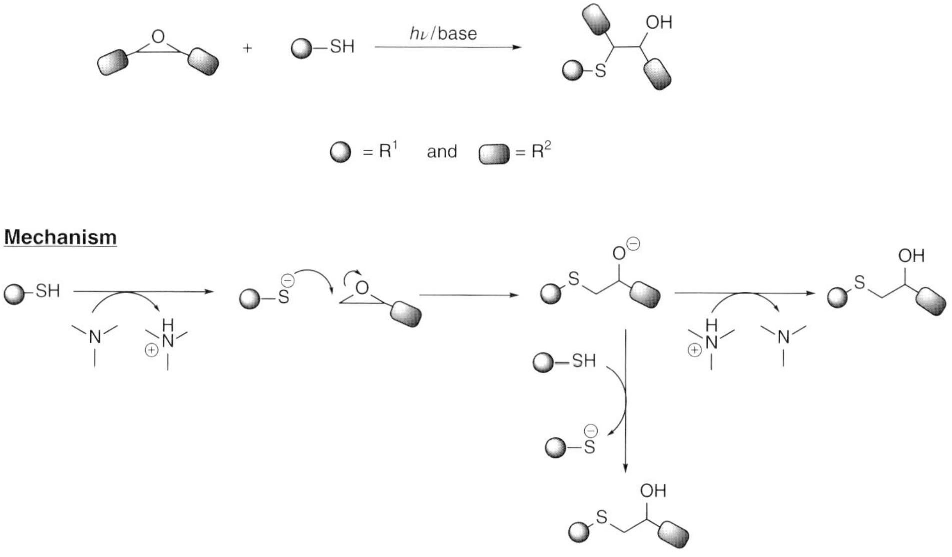


**Scheme 1.13** Base catalyzed thiol–epoxy ring-opening click reaction.

Fringuelli and coworkers have explored the use of InCl3 (Lewis acid), TsOH (Brønsted acid), n-Bu3P (Lewis base), K2CO3 (Brønsted base), and so on, as catalysts in solvent-less conditions for thiol–epoxide click reactions [51].

#### 1.2.2.2 Thiol–Isocyanate Click Reactions

Isocyanates are very reactive compounds, which react readily with alcohols, amines, water, and thiols or thiolate ions. Reaction of isocyanates with thiols gives thiocarbamates in very high to quantitative yields. This reaction has found its place in organic and polymer chemistry and satisfies all the criteria required to term it as a click reaction. The reaction has a very good potential to be used for modular approaches in surface engineering [52].

Polyurethanes are extremely versatile polymeric materials due to their elasticity, responsive nature toward impact, stretchability, and other possible physical manipulations. They have been utilized in various fields to make optical devices, adhesives, coatings, and also in many biomedical applications. Thiol–isocyanate click reactions have been used to generate polythiourethanes, a sulfur analog of polyurethane, in a very efficient way and in high yields [50].

In the presence of catalytic amount of a base (such as triethylamine ($NEt_3$)), thiols are deprotonated to thiolate anions, which react with isocyanates forming thiourethanes. The thiol–isocyanate reaction is fast and proceeds readily without any side product even in the presence of water, alcohol, or amines. The most common base used for this reaction is 1,5-diazabicyclo(4.3.0)non-5-ene (*DBN*). The reaction has the potential to be used more often as a very effective conjugation method (Scheme 1.14).

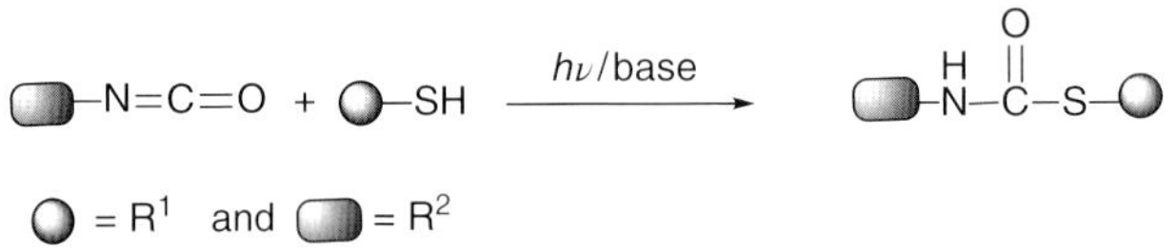


Mechanism

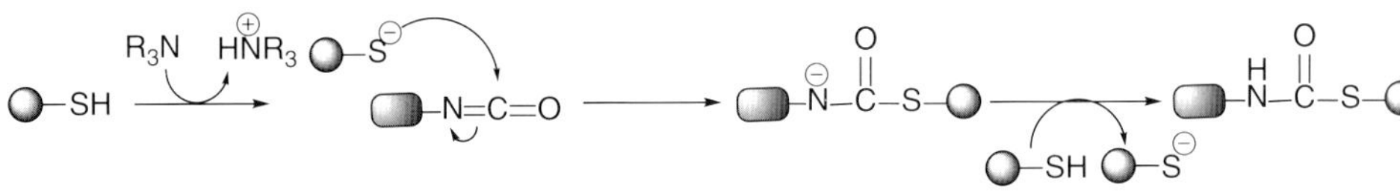


**Scheme 1.14** Tertiary-amine-catalyzed thiol–isocyanate click reaction.

### 1.2.2.3 Thiol–Michael Addition Click Reactions

Hydrothiolation of C=C double bond can be performed in the presence of mild bases or using nucleophilic catalysis. Unlike thiol–ene radical reaction (which can proceed with almost all olefins), thiol Michael addition reactions require activated carbon–carbon double bonds, which are in conjugation with an EWG. In the presence of trialkylamine bases such as $NEt_3$, the reaction proceeds smoothly to give the addition products in very high yields. On deprotonation of thiols by base such as $NEt_3$, thiolate anions are formed along with triethylammonium cations. Since thiolate anion is a strong nucleophile, it attacks at the electrophilic β-carbon of the electron-deficient olefin and generates a carbon-centered anion as intermediate. This anion is a strong base thus abstracts proton from either a thiol or an ammonium cation and ultimately forms a thioether product regioselectively

(Scheme 1.15). The final abstraction of proton is thermodynamically controlled and is a fast step. However, the attack of thiolate anion on the Michael acceptor is kinetically controlled. The overall rate and yield of these reactions can be altered by changing various factors such as solvent polarity, pH, strength of the base (catalyst), and nature of EWGs on the C=C bond [53–55].

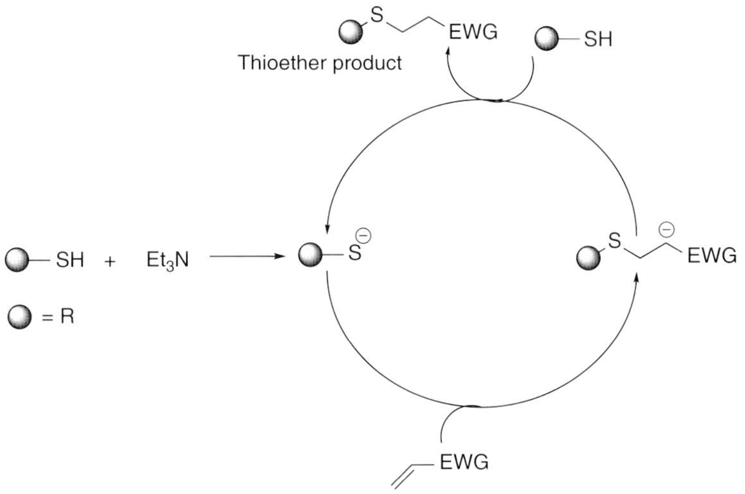


**Scheme 1.15** The base-catalyzed mechanism for the hydrothiolation of an activated C=C bond.

Apart from base catalysis, Michael addition of thiols can also be performed using nucleophilic catalysis. Primary and secondary amines and certain phosphines are the most commonly used catalysts. Nucleophile mediated thiol–Michael addition reactions have extensively been studied. The nucleophiles attack the Michael acceptors to generate a carbanion, which abstracts protons from thiols to generate thiolate anions, which in turn propagate the reaction (Scheme 1.16) [56]. Nucleophilicity of the catalyst plays a crucial role in the kinetics of the nucleophile-based thiol–Michael addition reactions, as stronger the nucleophile, more easily the thiolate anion will be generated.

Chan *et al.* studied bulk reaction of hexanethiol (5 mmol) with hexyl acrylate (5 mmol) in the presence of 0.43 M hexylamine ($pK_a = 10.56$), *n*-dipropylamine ($pK_a = 11$), and $NEt_3$ ($pK_a = 10.75$) under ambient condition for 500 s. After 500 s, reaction with hexylamine showed approximately 95% conversion, *n*-dipropylamine showed approximately 60% conversion, and $NEt_3$ showed less than 1% conversion. These amines have almost the same $pK_a$, but there is a huge difference in their kinetic profiles. Apart from these amines, they also studied various weak nucleophiles, which have varying basicity, such as pyridine ($pK_a = 5.14$), aniline ($pK_a = 9.34$), and 1,8-bis(dimethylamino)naphthalene (a proton sponge with $pK_a = 12.1$), and found that they yield less than 1% conversion. Based on these observations, the catalysis was attributed to the kinetic

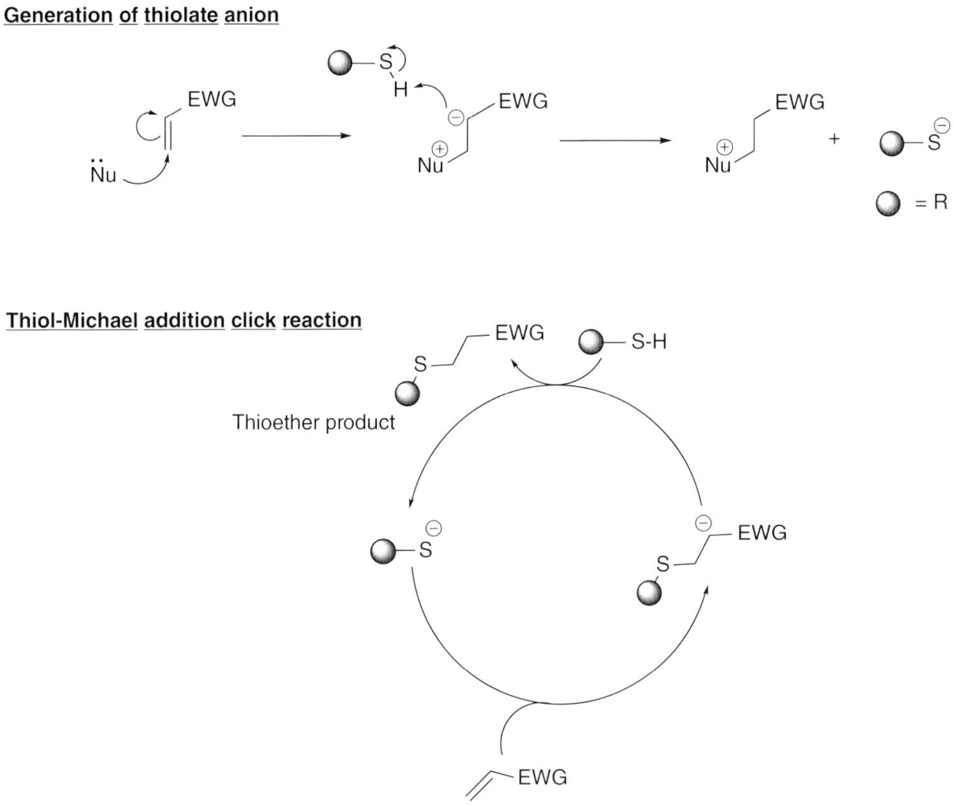


Scheme 1.16 Nucleophilic catalysis of thiol–Michael addition reactions.

profile of the catalysts in nucleophilic reactions rather than to their basicity [56, 57]. Stewart *et al.* have reported nucleophilic catalysis of thiol–Michael addition reactions using phosphines [55a].

Owing to the low $pK_a$ values of thiols, thiol–Michael addition reactions can be performed under ambient conditions in water or other protic solvents. However, both base- and nucleophile-catalyzed thiol–Michael addition reaction are affected to some extent by the presence of external protic species, especially the nucleophile-mediated pathway is more affected due to low catalytic concentration of nucleophiles compared to bases in these reactions.

#### 1.2.2.4 Thiol–Halogen Nucleophilic Substitution Reaction

Another example of thiol click reaction is the rapid and efficient substitution of leaving group bearing substrates by thiols, a soft nucleophile [58]. These reactions proceed better in the presence of mild organic bases such as trialkylamines. The halide salts formed during this displacement reactions can be removed easily as precipitates in a very simple and effective manner.

Displacement of bromine by various thiols by $S_N2$ nucleophilic substitution is definitely one of the best examples of click reaction. These reactions proceed well even in the presence of other nucleophiles such as alcohols and amines, owing to the increased nucleophilicity of thiols and thiolates. 2-Mercaptoethanol and other aliphatic thiol–dialcohol molecules when added to polymers, end-functionalized by halogen atoms, result in selective thiol end-functionalization of polymer chains.

## 1.3 Miscellaneous Click Reactions

Other than the click reactions mentioned earlier, there are few others that can be included in this list [1].

1) Nucleophilic ring-opening reactions of epoxides, aziridines, aziridinium ions, episulfonium ions, and cyclic sulfates have been included by Sharpless under the category of click reactions. They are termed as *spring-loaded reactions* owing to their increased reactivity resulting from the strain in the ring systems.
2) Nonaldol carbonyl-type reactions such as formation of hydrazones, oximes, amides, ureas, and isoureas are also effective click reactions. A recent report demonstrates the use of oxime-based click reactions for the formation of hydrogels [59]. An eight-armed aminooxy poly(ethylene glycol) was reacted with glutaraldehyde to form oxime-linked hydrogel efficiently. This hydrogel has tunable mechanical properties and can be used to support cell adhesion. Staudinger reaction has also been used by many researchers for bioorthogonal ligation and for synthesis of radiopharmaceuticals (Scheme 1.17) [60–62].

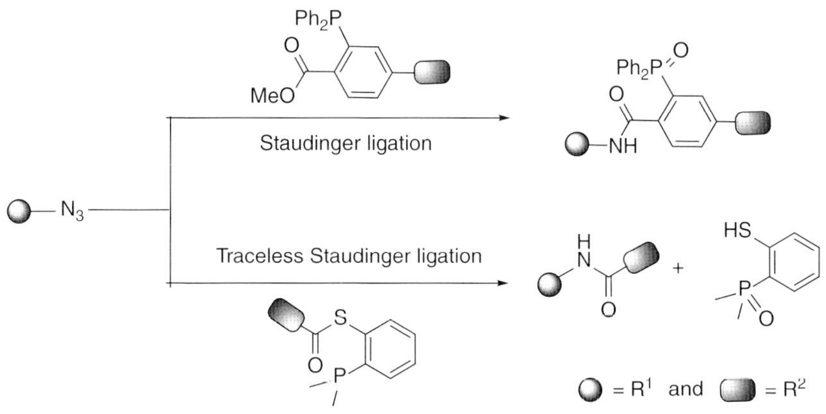


**Scheme 1.17** Staudinger and traceless Staudinger ligation click reactions.

Another report incorporating novel hydrazide/hydrazone click reaction was by Benny *et al.* in 2011. This report emphasizes on the high potential of these reactions for labeling biomolecules with $^{99m}Tc(CO)_3$. The hydrazone moiety is stable at physiological pH and unstable under strongly acidic and basic conditions, so can be used efficiently for drug delivery applications [63].

3) Addition to carbon–carbon multiple bonds leading to the formation of three-membered rings (epoxidations, aziridinations), dihydroxylations, nitrosyl–halide addition, sulfenyl–halide addition, and a few Michael additions are also grouped under click reactions [2].

## References


1. Kolb, H.C., Finn, M.G., and Sharpless, K.B. (2001) *Angew. Chem. Int. Ed.*, **40**, 2004.
2. (a) Adolfsson, H. *et al.* (1999) *Tetrahedron Lett.*, **40**, 3991–3994; (b) Kolb, H.C. *et al.* (1994) *Chem. Rev.*, **94**, 2483–2547; (c) Gontcharov, A.V. *et al.* (1999) *Org. Lett.*, **1**, 783–786; (d) Kuhle, E. (1970) *Synthesis (Mass.)*, **11**, 563–586.
3. Li, H., Aneja, R., and Chaiken, I. (2013) *Molecules*, **18**, 9797–9817.
4. Hoogenboom, R. (2010) *Angew. Chem. Int. Ed.*, **49**, 3415–3417.
5. (a) Tong, R., Tang, L., Ma, L., Tu, C., Baumgartner, R., and Cheng, J. (2014) *Chem. Soc. Rev.*, **43**, 6982–7012; (b) Fuaad, A.A.H., Azmi, F., Skwarczynski, M., and Toth, I. (2013) *Molecules*, **18**, 13148–13174.
6. Zhang, X. and Zhang, Y. (2013) *Molecules*, **18**, 7145–7159.
7. (a) Huisgen, R. (1961) *Proc. Chem. Soc., London*, 357–396; (b) Huisgen, R. (1963) *Angew. Chem.*, **75**, 604–637.
8. (a) Huisgen, R. (1963) *Angew. Chem. Int. Ed. Engl.*, **2**, 565–598; (b) Huisgen, R. (1963) *Angew. Chem.*, **75**, 742–754.
9. Rostovtsev, V.V., Green, L.G., Fokin, V.V., and Sharpless, K.B. (2002) *Angew. Chem.*, **114**, 2708–2711.
10. Tornøe, C.W., Christensen, C., and Meldal, M. (2002) *J. Org. Chem.*, **67**, 3057–3064.
11. (a) Krivopalov, V.P. and Shkurko, O.P. (2005) *Russ. Chem. Rev.*, **74**, 339–379; (b) Horne, W.S., Yadav, M.K., Scout, C.D., and Ghadiri, M.R. (2004) *J. Am. Chem. Soc.*, **126**, 15366–15367; (c) Palmer, M.H., Findlay, R.H., and Gaskell, A.J. (1974) *J. Chem. Soc., Perkin Trans. 2*, **4**, 420–428.
12. Weiwer, M., Chen, C.-C., Kemp, M.M., and Linhardt, R.J. (2009) *Eur. J. Org. Chem.*, **16**, 2611.
13. (a) Li, Y. and Flood, A.H. (2008) *Angew. Chem., Int. Ed.*, **47**, 2649–2652; (b) Lee, S., Hua, S., Park, H. and Flood, A.H. (2010) *Org. Lett.*, **12** (9), 2100–2102.
14. White, N.G. and Beer, P.D. (2012) *Beilstein J. Org. Chem.*, **8**, 246–252.
15. (a) Bock, V.D., Hiemstra, H., and Maarseveen, J.H. (2006) *Eur. J. Org. Chem.*, **2006** (1), 51–68; (b) Hein, C.D., Liu, X.-M., and Wang, D. (2008) *Pharm. Res.*, **25** (10), 2216–2230.
16. Hein, J.E. and Fokin, V.V. (2010) *Chem. Soc. Rev.*, **39**, 1302–1315.
17. Horne, W.S., Stout, C.D., and Ghadiri, M.R. (2003) *J. Am. Chem. Soc.*, **125**, 9372–9376.
18. Kolthoff, I.M. and Coetzeem, J. (1957) *J. Am. Chem. Soc.*, **79**, 1852.
19. (a) Schindler, S. (2000) *Eur. J. Inorg. Chem.*, **2000** (11), 2311–2326; (b) Díaz, D.D., Punna, S., Holzer, P., McPherson, A.K., Sharpless, K.B., Fokin, V.V., and Finn, M.G. (2004) *J. Polym. Sci., Part A: Polym. Chem.*, **42**, 4392–4403.
20. (a) Himo, F., Lovell, T., Hilgraf, R., Rostovtsev, V.V., Noodleman, L., Sharpless, K.B., and Fokin, V.V. (2005) *J. Am. Chem. Soc.*, **127**, 210–216; (b) Lee, L.V., Mitchell, M.L., Huang, S.-J., Fokin, V.V., Sharpless, K.B., and Wong, C.-H. (2003) *J. Am. Chem. Soc.*, **125**, 9588–9589.

21. Appukkuttan, P., Dehaen, W., Fokin, V.V., and Lycken, E.V. (2004) *Org. Lett.*, **6**, 4223–4225.
22. Molteni, G., Bianchi, C.L., Marinoni, G., Santo, N., and Ponti, A. (2006) *New J. Chem.*, **30**, 1137–1139.
23. Lipshutz, B.H. and Taft, B.R. (2006) *Angew. Chem. Int. Ed.*, **45**, 8235–8238.
24. González, S.D. (2011) *Catal. Sci. Technol.*, **1**, 166–178.
25. Chan, T.R., Hilgraf, R., Sharpless, K.B., and Fokin, V.V. (2004) *Org. Lett.*, **6**, 2853–2855.
26. (a) Deiters, A., Cropp, T.A., Mukherji, M., Chin, J.W., Anderson, J.C., and Schultz, P.G. (2003) *J. Am. Chem. Soc.*, **125**, 11782–11783; (b) Zhan, W.-H., Barnhill, H.N., Sivakumar, K., Tian, H., and Wang, Q. (2005) *Tetrahedron Lett.*, **46**, 1691–1695; (c) Link, J.A. and Tirrell, D.A. (2003) *J. Am. Chem. Soc.*, **125**, 11164–11165.
27. Balderas, F.P., Munoz, M.O., Sanfrutos, J.M., Mateo, F.H., Flores, F.G.C., and González, F.S. (2003) *Org. Lett.*, **5**, 1951–1954.
28. Gonzalez, S.D., Correa, A., Cavallo, L., and Nolan, S.P. (2006) *Chem. Eur. J.*, **12**, 7558–7564.
29. Wang, F., Fu, H., Jiang, Y.Y., and Zhao, Y.F. (2008) *Green Chem.*, **10**, 452–456.
30. (a) Zhang, L., Chen, X.G., Xue, P., Sun, H.H.Y., Williams, I.D., Sharpless, K.B., Fokin, V.V., and Jia, G. (2005) *J. Am. Chem. Soc.*, **127**, 15998–15999; (b) Majireck, M.M. and Weinreb, S.M. (2006) *J. Org. Chem.*, **71**, 8680–8683.
31. (a) Hong, L., Lin, W., Zhang, F., Liu, R., and Zhou, X. (2013) *Chem. Commun.*, **49**, 5589–5591; (b) Smith, C.D. and Greaney, M.F. (2013) *Org. Lett.*, **15**, 4826; (c) Wang, Y.C., Xie, Y.Y., Qu, H.E., Wang, H.S., Pan, Y.M., and Huang, F.P. (2014) *J. Org. Chem.*, **79**, 4463–4469.
32. (a) Boren, B.C., Narayan, S., Rasmussen, L.K., Zhang, L., Zhao, H., Lin, Z., Jia, G., and Fokin, V.V. (2008) *J. Am. Chem. Soc.*, **130**, 8923–8930; (b) Rasmussen, L.K., Boren, B.C., and Fokin, V.V. (2007) *Org. Lett.*, **9**, 5337–5339.
33. (a) Baskin, J.M., Prescher, J.A., Laughlin, S.T., Agard, N.J., Chang, P.V., Miller, I.A., Lo, A., Codelli, J.A., and Bertozzi, C.R. (2007) *Proc. Natl. Acad. Sci. U.S.A.*, **104** (43), 16793–16797; (b) Jewett, J.C., Sletten, E.M., and Bertozzi, C.R. (2008) *Org. Lett.*, **10** (14), 3097–3099; (c) Jewett, J.C., Sletten, E.M., and Bertozzi, C.R. (2010) *J. Am. Chem. Soc.*, **132**, 3688–3690; (d) Gordon, C.G., Mackey, J.L., Jewett, J.C., Sletten, E.M., Houk, K.N., and Bertozzi, C.R. (2012) *J. Am. Chem. Soc.*, **134**, 9199–9208.
34. Ess, D.H., Jones, G.O., and Houk, K.N. (2008) *Org. Lett.*, **10**, 1633–1636.
35. (a) Ning, X., Guo, J., Wolfert, M.A., and Boons, G.-J. (2008) *Angew. Chem. Int. Ed.*, **47**, 2253–2255; (b) Becer, C.R., Hoogenboom, R., and Schubert, U.S. (2009) *Angew. Chem. Int. Ed.*, **48**, 4900–4908.
36. (a) John, J., Thomas, J., and Dehaen, W. (2015) *Chem. Commun.*, **51**, 10797–10806; (b) Ramachary, D.B., Ramakumar, K., and Narayana, V.V. (2008) *Chem. Eur. J.*, **14**, 9143–9147.
37. Lima, C.G.S., Ali, A., Berkel, S.S.V., Westermann, B., and Paixa, M.W. (2015) *Chem. Commun.*, **51**, 10784–10796.
38. Stevens, M.P. (1999) *Polymer Chemistry*, 3rd edn, Oxford University Press, Oxford and New York, pp. 265–267.
39. Posner, T. (1905) *Ber. Dtsch. Chem. Ges.*, **38**, 646–657.
40. Cramer, N.B., Scott, J.P., and Bowman, C.N. (2002) *Macromolecules*, **35**, 5361.
41. Dondoni, A. (2008) *Angew. Chem. Int. Ed.*, **47**, 8995–8997.
42. Chen, G., Kumar, J., Gregory, A., and Stenzel, M.H. (2009) *Chem. Commun.*, 6291–6293.
43. (a) Hoyle, C.E., Le, T.Y., and Roper, T. (2004) *J. Polym. Sci., Part A: Polym. Chem.*, **42**, 5301–5338; (b) Morgan, C.R., Magnotta, F., and Ketley, A.D. (1977) *J. Polym. Sci., Polym. Chem. Ed.*, **15**, 627–645.
44. Stock, L.M. (1972) *J. Chem. Educ.*, **49**, 400–404.
45. Fairbanks, B.D., Scott, T.F., Kloxin, C.J., Anseth, K.S., and Bowman, C.N. (2008) *Macromolecules*, **42**, 211–217.
46. Konkolewicz, D., Weale, A.G., and Perrier, S. (2009) *J. Am. Chem. Soc.*, **131**, 18075–18077.
47. (a) Agard, N.J., Prescher, J.A., and Bertozzi, C.R. (2004) *J. Am. Chem. Soc.*, **126**, 15046–15047; (b) Johnson, J.A.,

Baskin, J.M., Bertozzi, C.R., Koberstein, J.T., and Turro, N.J. (2008) *Chem. Commun.*, 3064–3066; (c) Laughlin, S.T., Baskin, J.M., Amacher, S.L., and Bertozzi, C.R. (2008) *Science*, **320**, 664–667; (d) DeForest, C.A., Polizzotti, B.D., and Anseth, K.S. (2009) *Nat. Mater.*, **8** (8), 659–664.

48. Dietliker, K., Husler, R., Birbaum, J.-L., Ilg, S., Villeneuve, S., Studer, K., Jung, T., Benkhoff, J., Kura, H., Matsumoto, A., and Oda, H. (2007) *Prog. Org. Coat.*, **58**, 146–157.
49. (a) Gough, J.D., Gargano, J.M., Donofrio, A.E., and Lees, W.J. (2003) *Biochemistry*, **42**, 11787–11797; (b) Hupe, D.J. and Jencks, W.P. (1977) *J. Am. Chem. Soc.*, **99**, 451–464.
50. Hoyle, C.E., Lowe, A.B., and Bowman, C.N. (2010) *Chem. Soc. Rev.*, **39**, 1355–1387.
51. (a) Fringuelli, F., Pizzo, F., Tortoioli, T., and Vaccaro, L. (2004) *Targets Heterocycl. Syst.*, **8**, 146–161; (b) Fringuelli, F., Pizzo, F., Tortoioli, T., and Vaccaro, L. (2003) *Tetrahedron Lett.*, **44**, 6785–6787.
52. Hensarling, R.M., Rahane, S.B., LeBlanc, A.P., Sparks, B.J., White, E.M., Locklin, J., and Patton, D.L. (2011) *Polym. Chem.*, **2**, 88–90.
53. (a) Nair, D.P. and Podgórski, M. (2014) *Chem. Mater.*, **26**, 724–744; (b) Xi, W.X., Wang, C., Kloxin, C.J., and Bowman, C.N. (2012) *ACS Macro Lett.*, **1** (7), 811–814.
54. Mather, B.D., Viswanathan, K., Miller, K.M., and Long, T.E. (2006) *Prog. Polym. Sci.*, **31**, 487.
55. (a) White, T.J., Natarajan, L.V., Tondiglia, V.P., Lloyd, P.F., Bunning, T.J., and Guymon, C.A. (2007) *Macromolecules*, **40**, 1121; (b) Kilambi, H., Reddy, S.K., Beckel, E.R., Stansbury, J.W., and Bowman, C.N. (2007) *Chem. Mater.*, **29**, 641; (c) Kilambi, H., Stansbury, J.W., and Bowman, C.N. (2008) *J. Polym. Sci., Part A: Polym. Chem.*, **46**, 3452; (d) Friedman, M., Cavins, J.F., and Wall, J.S. (1965) *J. Am. Chem. Soc.*, **87**, 3672; (e) Hued, C.D. and Gershbein, L.L. (1947) *J. Am. Chem. Soc.*, **69**, 2328–2335.
56. Chan, J.W., Shin, J., Hoyle, C.E., Bowman, C.N., and Lowe, A.B. (2010) *Macromolecules*, **43**, 4937–4942.
57. (a) Stewart, I.C., Bergam, R.G., and Toste, F.D. (2003) *J. Am. Chem. Soc.*, **125**, 8696; (b) Lowe, A.B. (2010) *Polym. Chem.*, **1**, 17–36.
58. (a) Samaroo, D., Vinodu, M., Chen, X., and Drain, C.M. (2007) *J. Comb. Chem.*, **9**, 998–1011; (b) Rosen, B.M., Lligadas, G., Hahn, C., and Percec, V. (2009) *J. Polym. Sci., Part A: Polym. Chem.*, **47**, 3940; (c) Becer, C.R., Babiuch, K., Pilz, D., Hornig, S., Heinze, T., Gottschaldt, M., and Schubert, U.S. (2009) *Macromolecules*, **42**, 2387–2394.
59. Grover, G.N., Lam, J., Nguyen, T.H., Segura, T., and Maynard, H.D. (2012) *Biomacromolecules*, **13**, 3013–3017.
60. Zeng, D., Zeglis, B.M., Lewis, J.S., and Anderson, C.J. (2013) *J. Nucl. Med.*, **54**, 829–832.
61. Gaeta, A., Woodcraft, J., Plant, S., Goggi, J., Jones, P., Battle, M., Trigg, W., Luthra, S.K., and Glaser, M. (2010) *Bioorg. Med. Chem. Lett.*, **20**, 4649–4652.
62. Carroll, L., Boldon, S., Bejot, R., Moore, J.E., Declerck, J., and Gouverneur, V. (2011) *Org. Biomol. Chem.*, **9**, 136–140.
63. Ganguly, T., Benjamin, B.K., Bucar, D.-K., MacGillivray, L.R., Berkman, C.E., and Benny, P.D. (2011) *Chem. Commun.*, **47**, 12846–12848.

# 2 Applications of Click Chemistry in Drug Discovery and Development

*Balasubramanian Gopalan and Kalpattu Kuppusamy Balasubramanian*

## 2.1 Introduction

The thermally induced 1,3-dipolar cycloaddition of alkynes with azides to obtain 1,2,3-triazoles is well known and further developed by Huisgen [1]. A Cu(I)- and Ru(I)-catalyzed variation of this cycloaddition to a regioselective synthesis of 1,4- and 1,5-substituted (Scheme 2.1) and functionalized 1,2,3-triazoles, respectively (Scheme 2.2), termed as *click reaction*, was developed independently by Sharpless *et al.* [2] and Meldal *et al.* [3] as delineated.

In recent years, several groups [4–7] independently reported the use of small organic azides and carbonyl compounds to afford substituted 1,2,3-triazoles with high levels of regioselectivity including metal-free, one-pot strategy toward trisubstituted 1,2,3-triazoles [8, 9]. Concise New click chemistry methods for the synthesis of 1,2,3-triazoles have been reviewed [10, 11]. This chapter is divided into two parts. Part A deals with 1,2,3-triazoles that are under various stages of investigation in the field of drug discovery and development, while part B deals with the synthesis of 1,2,3-triazole-based drugs, which have already been in the market or have been approved.

## 2.2 Part A: Application of Click Chemistry to Drug Discovery and Development

Drug discovery and development involve syntheses of new chemical entities/new molecular entities from the basic understanding of different disease processes. Lead is an important part of drug discovery in a therapeutic biological target, defined as a substance or a compound that has demonstrated a validated biological activity in an *in vivo* model that has relevance to a human or animal disease or pathology and that it evokes sufficient interest to plan a major chemical program with the objective of finding a candidate for clinical evaluation. One of the least risks, lead-hunting strategies is creation of large chemical libraries of structurally diverse small molecules (molecular weight <500 Da), which are subsequently

*Click Reactions in Organic Synthesis*, First Edition. Edited by Srinivasan Chandrasekaran.

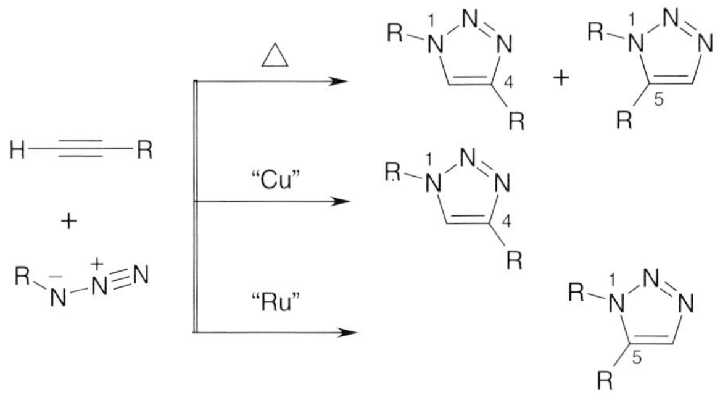


Scheme 2.1

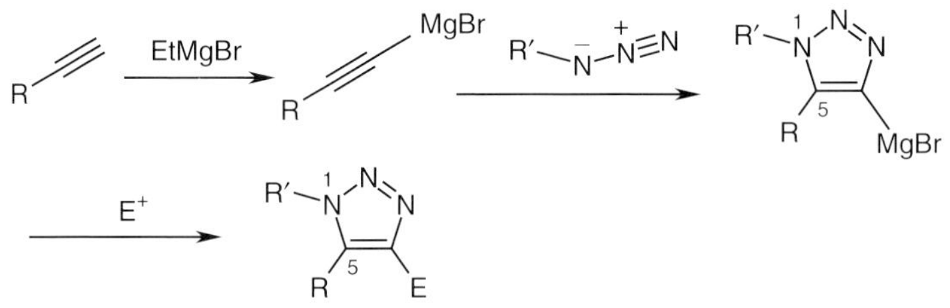


Scheme 2.2

screened for the desired activity. Click chemistry has been proven to be a powerful tool in the creation of libraries of small molecules in hit identification and lead generation process of the early-stage drug discovery as reported in a recent review [12].

Synthesis of biologically active compounds possessing the 1,2,3-triazole core through 1,3-dipolar cycloaddition reactions has been reviewed [13]. The triazole skeleton can be present as a single disubstituted ring, a linker between two molecules, or embedded in a polyheterocycle.

The 1,2,3-triazole core is stable against acidic and basic hydrolysis as well as oxidative and reductive conditions, reflecting a high aromatic stabilization and relative resistance to metabolic degradation [14]. At the same time, it has a high dipole moment of 5D [15] and participates actively in hydrogen bond formation as well as in dipole–dipole interaction [16]. These features together make the variously substituted triazole moiety structurally similar to the amide bond, mimicking a *Z* or an *E* amide bond, depending on its substitution pattern. In particular, many features underline the fact that the 1,4-disubstituted triazole moiety shows similarity with a *Z* amide bond: (i) the role of the carbonyl oxygen lone pair of the amide bond is played by the 3-nitrogen atom lone pair, (ii) the polarized C–H bond (in the 5-position) acts as a hydrogen bonding donor in the same way as the amide N–H bond, and (iii) the electrophilic and polarized C4 is electronically similar to the carbonyl carbon [17].

In addition to the potential of triazoles in mimicking the amide moiety and consequently the peptide bond, their capability to act as bioisosteres of the acyl phosphate and *trans*-olefinic moieties has also been predicted [18]. The *trans*-olefinic moiety of resveratrol **1** (an anticancer and anti-inflammatory drug) is replaced with a triazole nucleus in compounds **2** and **3** as delineated in Figure 2.1.

This modification did not alter the spatial positioning and the gap between the phenolic hydroxyl groups crucial for the biological activity of the phytoalexin. Five lead compounds were identified from 72 compounds synthesized and screened.

Another interesting and important use of triazoles was in the generation of non-classical bioisosteres of potentially labile groups (ester, amide) with heterocycles capable of mimicking the same electrostatic potential maps [19] (Figure 2.2).

Analogously, the triazole derivatives **5** and **6** have been shown to act as a nonclassical bioisosteres of the ester group as in arecoline **4**, a muscarinic agonist [20, 21].

Among the best known examples of triazole-containing drugs, an eminent role is played by tazobactam **7**, a commercially available beta-lactamase inhibitor in combination with broad-spectrum antibiotics. Indeed, biological studies performed on tazobactam and related triazole-containing compounds **8** and **9** showed that the compounds of this class are potent beta-lactamase inhibitors with higher potency than that of sulbactam and clavulanic acid, underlying a pivotal role by the triazole ring in terms of potency [22, 23](Figure 2.3).

Triazoles have also been used to improve pharmacokinetic properties of the desired drugs. Cephalosporins with good oral bioavailability were obtained by linking the triazole moiety to the cephalosporin core as in compound **9**, shown earlier [24].

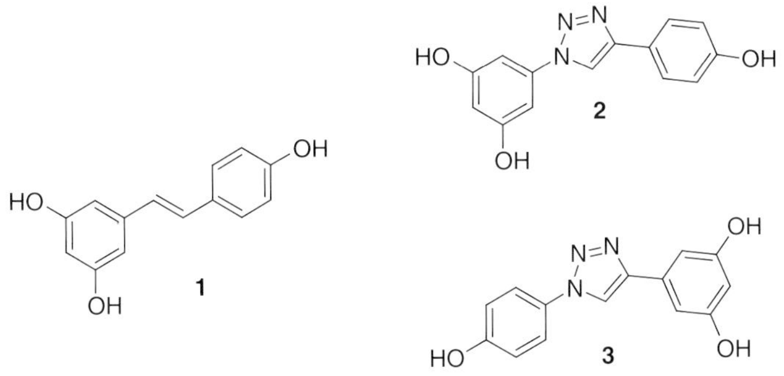


**Figure 2.1** Triazole bioisostere of Resveratrol.

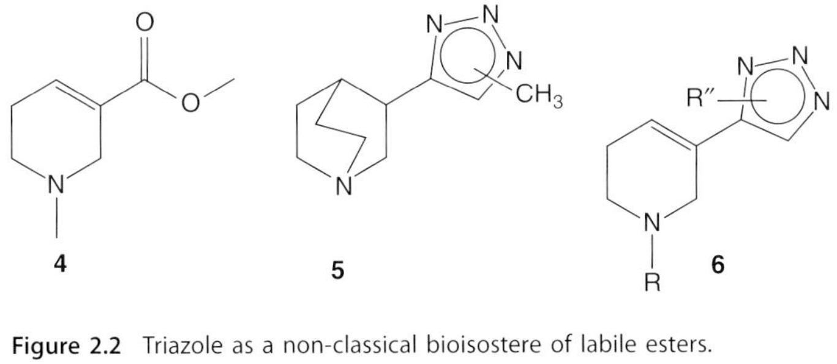


**Figure 2.2** Triazole as a non-classical bioisostere of labile esters.

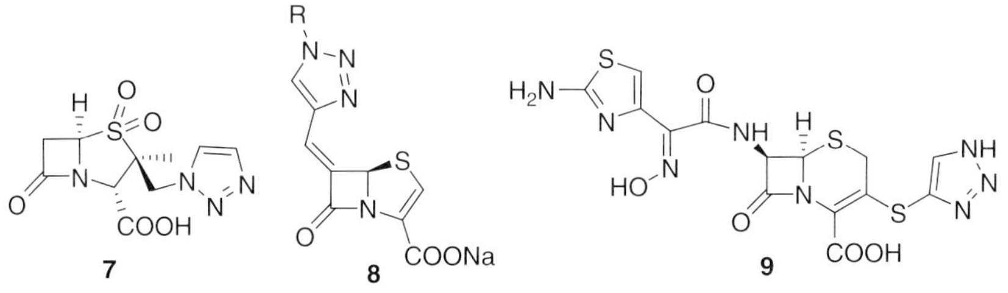


Figure 2.3 Role of triazole in improving intrinsic potency and bioavailability.

The most innovative application of click reaction in drug discovery is kinetically controlled target-guided synthesis, also known as *in situ click chemistry*. In this method, a target (enzyme, protein) is incubated with several building blocks (selected inhibitors modified with azide and alkyne group) and the target facilitates the formation of an irreversible bonding between two selected building blocks (via click reaction), resulting in the formation of a more potent inhibitor that deactivates the target that recruited them. In order to address target selectivity, a futuristic approach was carried out in which an engineered enzyme at room temperature could bring the azide and alkyne together by using itself as their reaction vessel [25]. By this strategy, only compounds that fit correctly into the active site of the enzyme should react to give new potential ligands (Figure 2.4).

This strategy was used to identify extremely potent inhibitors of acetyl cholinesterase (AchE), the key enzyme that hydrolyzes acetylcholine at the synaptic cleft and constitutes a pharmacological target for many clinically relevant conditions including Alzheimer's disease and muscular dystrophy. Reactions between compounds 10 and 11 leading to the triazoles 12 were observed only in the presence of engineered enzyme, thus revealing the rate accelerating enzyme -template effect. The significant finding of this study was the *in situ* discovery of

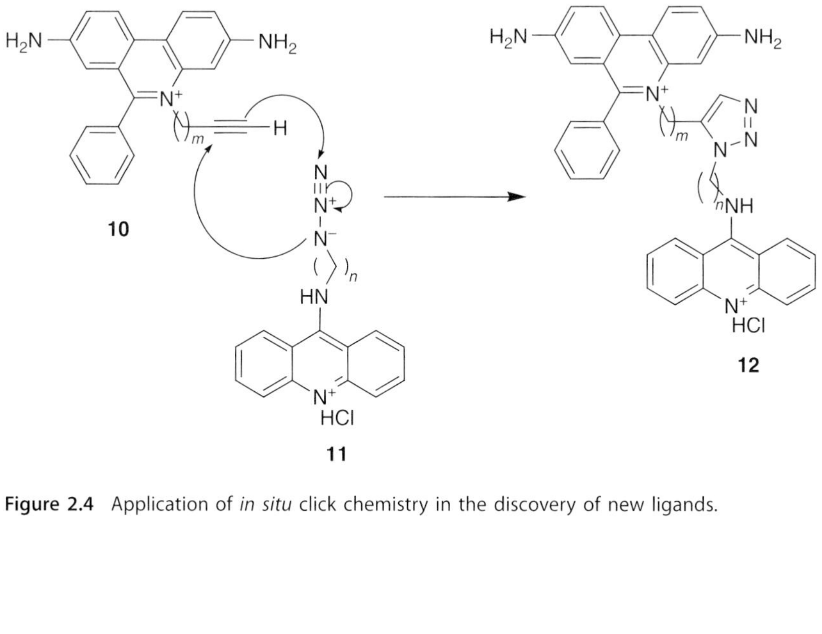


Figure 2.4 Application of *in situ* click chemistry in the discovery of new ligands.

the most potent noncovalent inhibitors of Ach-E ever discovered with dissociation constants in the femtomolar range ($K_d$ 33–100 fM) [26, 27]. The application of this enzyme catalysis was also demonstrated for carbonic anhydrase (CA) inhibitors [28], underlining the versatility of this strategy with other enzymes.

The enzyme-directed synthesis is an emerging fragment-based lead discovery approach in which the biological target is able to assemble its own multidentate ligands from a pool of building blocks. For the first time, the use of human AchE as an enzyme template for the design and synthesis of new potent huprine-based inhibitors has been reported [29]. The huprine moiety would be a suitable catalytic site binder for *in situ* click chemistry [30]. Huprine derivatives are the best AChE active-site ligands reported to date, especially huprines X and Y (Figures 2.5 and 2.6) that exhibit high affinity toward AChE acylation site [31] and can be readily functionalized at position 9 or 12 [32].

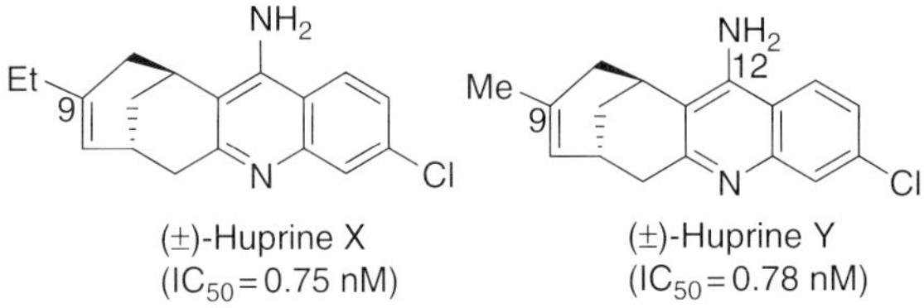


**13** **14**

**Figure 2.5** Enzyme- directed fragment- based lead discovery of Huprin-based inhibitors.

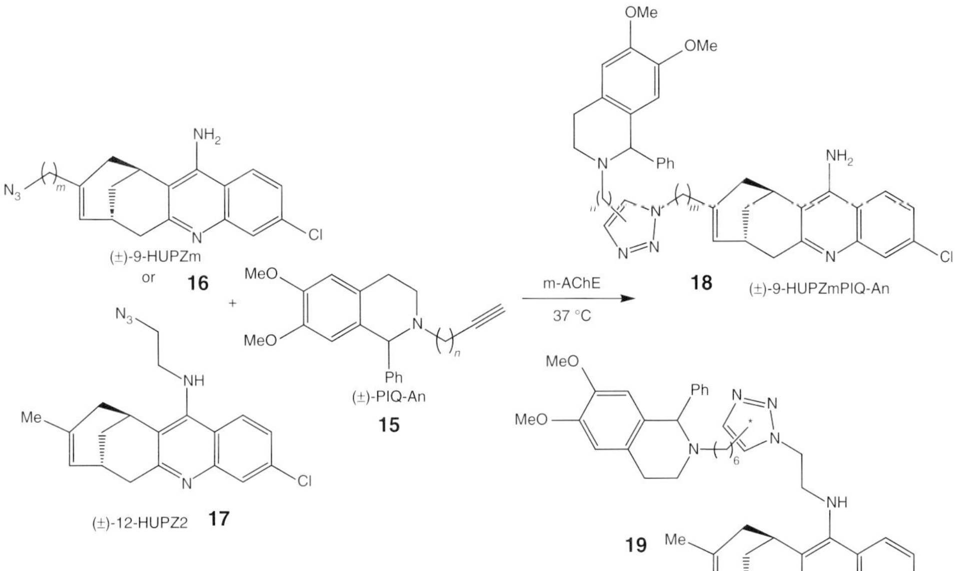


**Figure 2.6** Huprin moiety as a catalytic site binder for *in situ* click chemistry.

**Table 2.1** m-AChE-directed synthesis of huprine-based heterodimers.

| Entry | Ligand | *M* | *n* | Click *in situ* | $IC_{50}$ [a] (nM) |
|---|---|---|---|---|---|
| 1 | (±)-12-HUPZ2PIQ-A6 | n.a[b] | 6 | — | 7.0 ± 1.0 |
| 2 | (±)-9-HUPZ4PIQ-A1 | 4 | 1 | — | 1.5 ± 0.3 |
| 3 | (±)-9-HUPZ4PIQ-A2 | 4 | 2 | Anti[c] | 0.4 ± 0.1 |
| 4 | (±)-9-HUPZ4PIQ-A3 | 4 | 3 | Syn/anti[d] | 1.2 ± 0.3 |
| 5 | (±)-9-HUPZ4PIQ-A4 | 4 | 4 | — | 3.1 ± 0.6 |
| 6 | (±)-9-HUPZ2PIQ-A4 | 2 | 4 | — | 13.0 ± 2.5 |

a) Corresponding to the anti-isomer.
b) Not applicable.
c) Detected after a 2-day incubation period.
d) Detected after a 7-day incubation period.

Phenyl tetrahydroisoquinoline scaffold was selected as it was proven to be a convenient peripheral site binder [27]. There is a growing interest in developing multitarget-directed drugs and particularly new potent dual-binding site, AChE inhibitors are able to exert a dual action [33] (inhibition of cholinesterase activity and inhibition of AChE-mediated Aβ deposition) [34]. The results of the m-AChE-directed syntheses of heterodimeric huprine-based inhibitors are summarized (*vide* Table 2.1).

*In situ* click chemistry, sustained by docking experiments, is a flexible tool capable of highlighting different click chemistry sites within an enzyme, depending on the combination of building blocks selected.

In a recent review article [35], several methods of syntheses of 1,2,3-triazoles using click chemistry and its applications aimed at finding new lead candidates against pathologies such as cancer, AIDS, and Alzheimer's disease have been elegantly described.

### 2.2.1
### Carbonic Anhydrase Inhibitors

CAs are ubiquitous metalloenzymes that catalyze the reversible synthesis of bicarbonate ion and proton from carbon dioxide and water. This family includes 15 isoenzymes (I–XIV, V further subclassified as VA and VB) endowed with different molecular features, cellular localization, distribution in organs and tissues, and kinetic properties. It has been demonstrated that the dysfunction of these enzymes is related to different pathologies such as cancer (CAs VII, IX, XII), neurodegeneration (CA VIII), obesity (CAs VA and VB), and sterility (CA XIII) [36]. Based on the knowledge that coumarin derivatives are CA inhibitors [37], a library of sulfocoumarin derivatives **21** bearing a triazole moiety was synthesized (Scheme 2.3). These compounds **21** (R = Ph) were found to

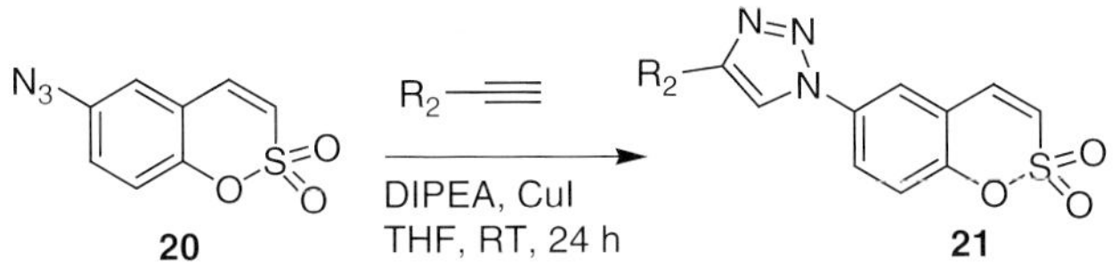


Scheme 2.3

be active against human CAs IX and XII with $IC_{50}$ values in the nanomolar range (29–32 nm) but showing much weaker inhibition (6.86–7.76) μM against the ubiquitous off-target cytosolic isoforms (CAs I and II). These compounds are hydrolyzed by the esterase CA activity to 2-hydroxyphenyl-vinylsulfonic acids, which thereafter bind to the enzyme in a region rarely occupied by other inhibitors [37].

On the basis of the interesting results just discussed, the isomeric 5-substituted 1,2,3-triazoles 22 using a ruthenium-based catalyst were synthesized (Scheme 2.4) [38]. This new series of compounds (**22**, R = Ph) resulted in activity against human CAs IX (67.1 nM) and XII (9.5 nM) with $K_i$ values in the nanomolar range but were inactive against CAs I (>10 μM) and CAs II (>10 μM).

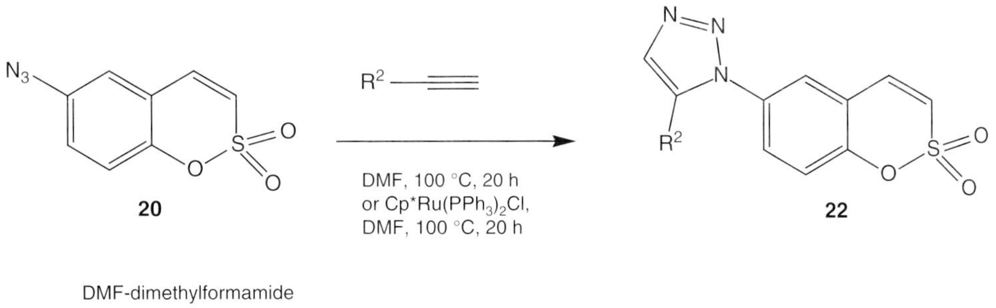


Scheme 2.4

As the human CAs IX and XII are validated, antitumor targets, such prodrug, isoform-selective inhibitors as the sulfocoumarins reported earlier, may be useful for identifying suitable drug candidates for clinical trials.

Bioisosterism and scaffold hopping are two widely used approaches in medicinal chemistry for lead optimization. The study [39] highlights the physicochemical properties of the 4-hydroxy-1,2,3-triazole scaffold, a less investigated heterocyclic system. The aim is to use this system to modulate the acidic moieties present in lead compounds and to region-direct substituents in set directions through targeted substitution on the three nitrogen atoms of the 1,2,3-triazole ring and to enhance the binding affinity to the target.

### 2.2.2 Targeting Onchocerca Volvulus Chitinase-1 (OvCHT1) using the Hydroxytriazole Moiety within a Scaffold Hopping Approach

Onchocerciasis is a neglected tropical disease caused by the parasitic worm *Onchocerca volvulus* and is the world's second leading infectious cause of blindness in sub-Saharan Africa and South America [40]. Additionally, the possible emergence of ivermectin-resistant *O. volvulus* accelerates the need to develop new innovative therapeutic strategies in this field, since the broad-spectrum antiparasitic drug ivermectin is the product currently most widely used to treat the disease [41]. *O. volvulus* chitinase-1 (OvCHT1) is a larval-stage-specific chitinase that was recently identified as a potential biological target affecting nematode development [42]. As a result of a blind screening program, Closantel was discovered to be a potent and selective inhibitor of OvCHT1. Starting from this lead, a series of closantel analogs were prepared to afford chitinase inhibitors with $IC_{50}$ values in low nanomolar range [43]. Of these, compound MDG-1-33A (Table 2.2) was found to inhibit OvCHT1 with an $IC_{50}$ value of 1.98 μM [43]. Using a scaffold hopping approach, the iodophenolic core, found to be essential for chitinase inhibition, in compound MDG-1-33A, was replaced by the hydroxyl triazole moiety to obtain derivatives **24**–**27**.

The results of these scaffold hopping procedures afforded three promising MDG-1-33A variants **26**–**28** that inhibit OvCHT1 chitinase in the nanomolar range.

Click chemistry is frequently employed in drug discovery and greatly helps advance research programs in the pharmaceutical industry. It facilitates library synthesis to support medicinal chemistry optimization, helps identify the targets and off-targets of drug candidates, and can facilitate the determination of drug efficacy in clinical trials. In the past decade, a large number of patent applications covering the various types and utilities of click chemistry have been filed. The first patent analysis of click chemistry applications to drug discovery and chemical biology has recently been reviewed [44]. In this review, a branched hydroxamate histone deacetylase (HDAC) inhibitor **29** demonstrated potent growth inhibition ($GI_{50} < 1$ μM) against three tumor cell lines: adenocarcinomic human alveolar basal epithelial cells (A549), human liver carcinoma cell line (HepG2), breast cancer cell line (MDA-MB-231) [45].

The conjugation of two antimicrobial agents, quinolones and aminoglycosides, as in compound **30**, was made in order to improve activity and overcome bacterial resistance [46] (Figure 2.7). Some of these conjugates exhibit high potency against both Gram-negative and Gram-positive bacteria including MRSA.

Thus, click chemistry has impacted drug discovery from early exploratory research through the clinical stages of development. It helps to identify the targets of hits from phenotypic screens and determine target engagement by drug candidates.

**Table 2.2** $IC_{50}$ of OvCHT1 inhibition.

| Compounds | Structure | $IC_{50}$ (μM)[a)] |
|---|---|---|
| | 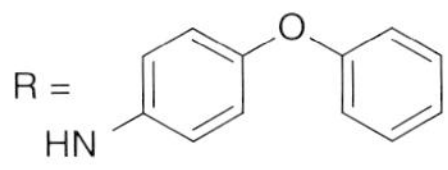 | |
| **23** (MDG-1-33A) | 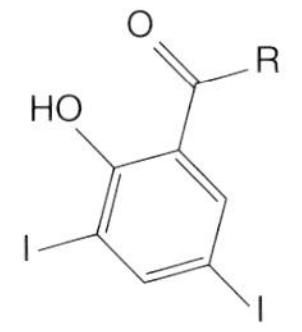 | 1.98 [31] |
| **24** | 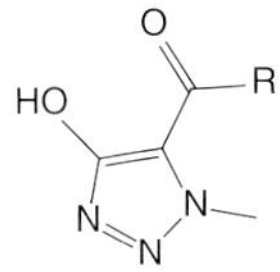 | 4.69 ± 0.29 |
| **25** | 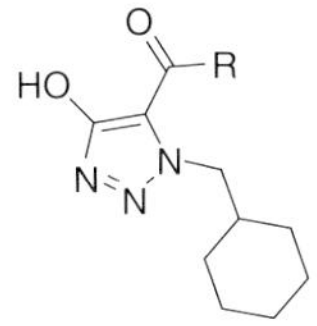 | 5.11 ± 0.06 |
| **26** | 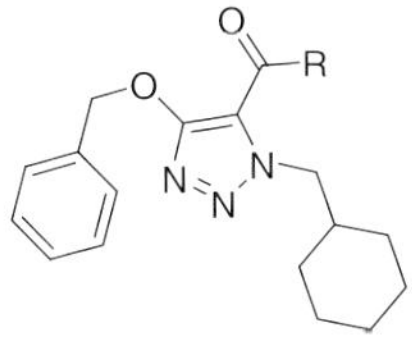 | 0.52 ± 0.03 |
| **27** | 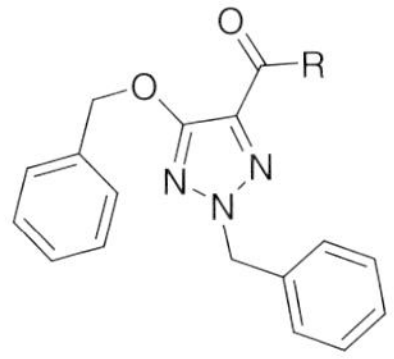 | 0.41 ± 0.01 |
| **28** | 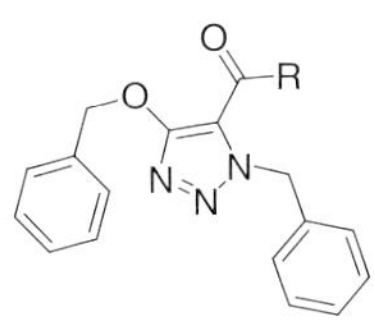 | 0.41 ± 0.04 |

a) *In vitro* OvCHT1 inhibition activity was measured by fluorescence-based assay. Each dataset represents mean ± SD of three independent assays.

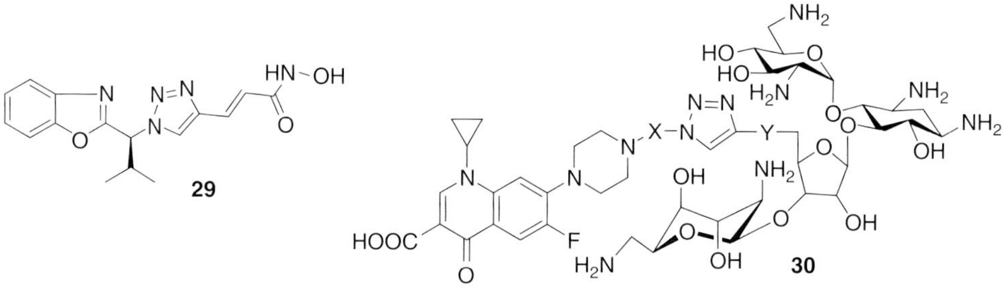


**Figure 2.7** Conjugation strategy for combating bacterial resistance.

An exhaustive review [47] on click chemistry in Drug Development and Diverse Chemical Biology applications and another review [48] pertaining to the biological properties of 1*H*-1,2,3- and 2*H*-1,2,3-triazoles have recently been published. There have also been many review articles that summarize the impact of click chemistry in drug discovery particularly in the early stages [49, 50].

## 2.2.3
## 1,2,3-Triaole-Derived Anticancer Agents

Cancer, also called *malignant neoplasm*, is caused by the disorder of cell proliferation mechanism and has become a major health burden in developed and developing countries. Although several anticancer agents are in clinical use and have significant curative effects, it is still necessary to screen new molecules with different modes of actions. Enzymes, expressed at an estimated 18–29% within eukaryotic genomes [47], and enzyme inhibitors play a significant role in cancer therapy.

A recent review [51] focuses mainly on the applications of click reaction in the synthesis of agents with mechanism-based anticancer activity, which are divided into four groups: Topoisomerase II inhibitors, HDAC inhibitors, protein tyrosine kinase (PTK) inhibitors, and antimicrotubule agents.

### 2.2.3.1 Topoisomerase II Inhibitors

Topoisomerase enzymes (topoisomerases I and II) are enzymes that control the changes in DNA structure by breaking and rejoining the phosphodiester backbone of DNA strand during the normal cell cycles. Topoisomerases have become popular targets for cancer chemotherapy drugs.

Podophyllotoxin, **31**, is an antimicrotubule agent acting at the colchicine-binding site on tubulin [52]. Due to its severe toxicity, it cannot be used clinically, but its semisynthetic variants, etoposide and teniposide, had been used as effective anticancer drugs for the treatment of lung cancer, Kaposi's sarcoma, and lymphoma. A series of podophyllotoxin derivatives, the majority of which proved to be more potent than etoposide, **32**, have been synthesized and their activity investigated. The selected compound **33**, prepared employing click chemistry approach, exhibited significant anticancer activity (Figure 2.8), which revealed

**Table 2.3** $IC_{50}$ values (μM).

| Entry | $R_1$ | $R_2$ | PC-3 | HeP-2 | HCT-15 | MCF-7 |
|---|---|---|---|---|---|---|
| 1 | $CH_3$ | Ethyl | 0.03 | 0.06 | 0.4 | 0.4 |
| 2 | $CH_3$ | Propyl | 5.1 | 1.6 | 0.03 | 1.4 |
| 3 | $CH_3$ | $CH_2OH$ | 0.06 | 0.06 | 15 | 0.01 |
| 4 | H | Ethyl | 0.06 | 0.05 | 1.9 | 0.04 |
| 5 | H | Propyl | 0.2 | 2.9 | 0.8 | 5.7 |
| 6 | H | $CH_2OH$ | 8.2 | 6.7 | 18 | 0.6 |

**Table 2.4** Cytotoxicity data.

| Cytotoxicity | $IC_{50}$ (μM) |
|---|---|
| HL-60 | < 0.01 |
| A-549 | 0.14 |
| HeLa | 0.51 |
| HCT-8 | 0.05 |

that the presence of methyl, ethyl, and hydroxyl groups in triazole moiety could increase the anticancer activity [53], in agreement with docking experiments (Tables 2.3). The carbamate analogs of 4β-(1,2,3-triazol-1-yl) podophyllotoxin were evaluated for their cytotoxic effects against human cancer cell lines HL-60, A-549, HeLa, and HCT-8. Among all these, the triazolylmethyl carbamate **34** (Figure 2.9) showed the most potent cytotoxicities, which induced cell cycle arrest in the G2/M phase accompanied by apoptosis and inhibited the formation of microtubules in A-549 cells. This compound can also cause the inhibition of DNA topoisomerase II [54] (Tables 2.4).

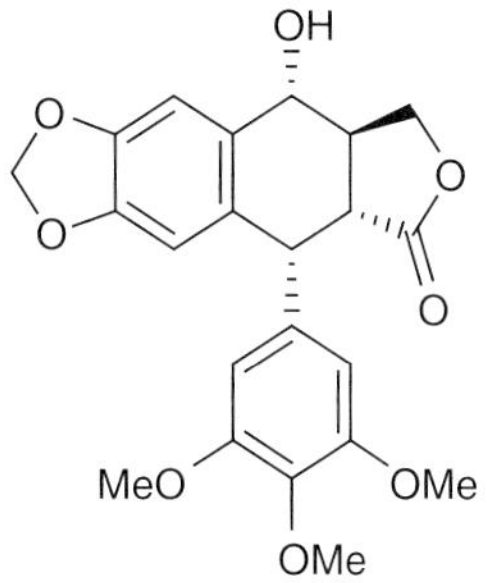


Podophyllotixin

**31**

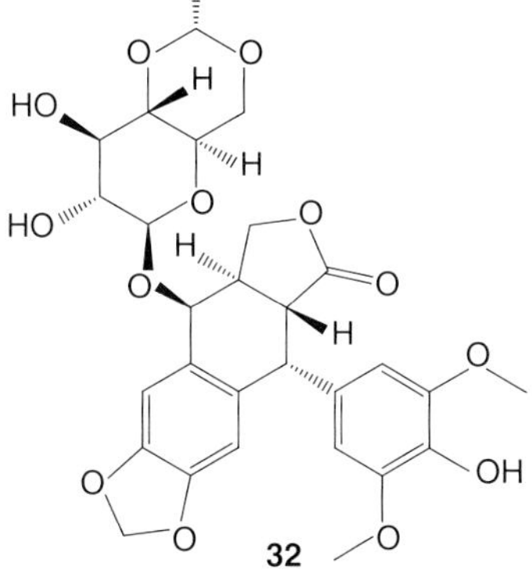


**32**

Etoposide

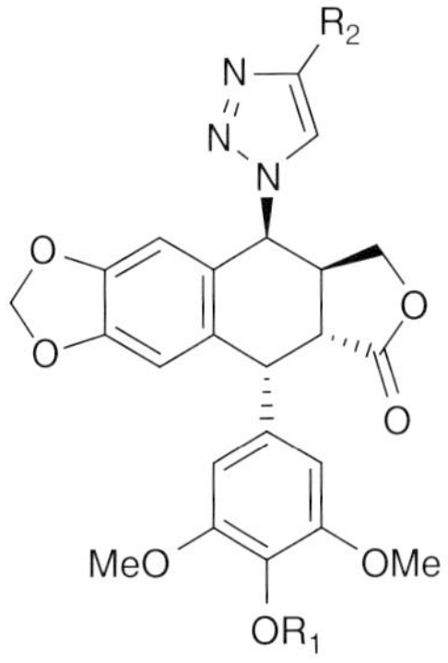


33

Figure 2.8 Triazole variant of podophyllotoxin as antimicrotubule agents.

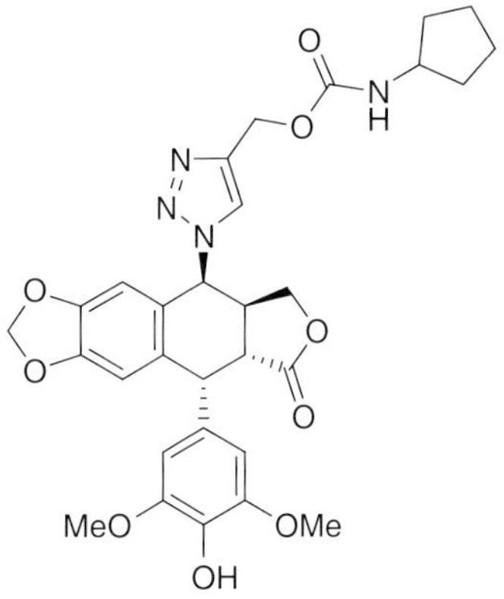


34

Figure 2.9 Triazolylmethyl analogue of podophyllotoxin for improved anticancer activity.

Most of the compounds exhibited improved anticancer activity in human cancer cell lines compared to the marketed drug etoposide **32.**

#### 2.2.3.2 Histone Deacetylase Inhibitors

HDACs are a family of enzymes that catalyze the deacetylation of lysine side chain in chromatin. These enzymes are involved in a wide range of biological processes such as cell differentiation, proliferation, angiogenesis, and apoptosis. HDAC inhibitors have been clinically validated as a therapeutic strategy for cancer treatment. This topic has been covered by critical reviews [55–58]. The classic pharmacophore for HDAC inhibitors consists of three distinct structural motifs: the zinc-binding group, a hydrophobic linker, and a recognition cap moiety.

A new chemical scaffold of HDAC inhibitors through click chemistry was reported [59] in which the precursors corresponding to the "cap" moiety contained an azido group and the zinc-chelating functionality precursors

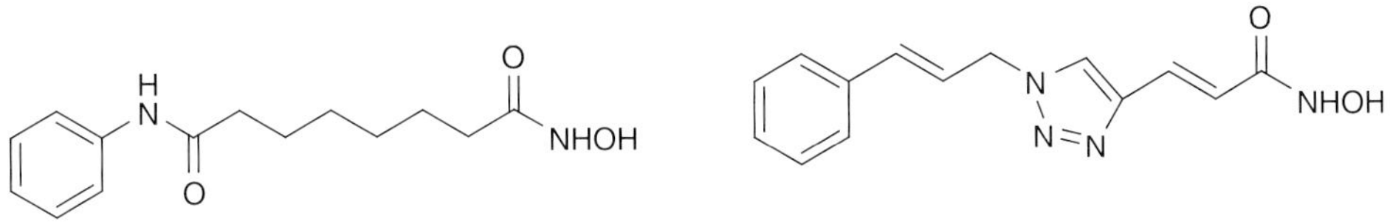


**35** SAHA **36** (NSC746457)

**Figure 2.10** Traizole NSC746457 as a novel HDAC-1 inhibitor.

contained an alkyne group. Among the compounds synthesized, NSC746457, 36, (Figure 2.10) could inhibit HDAC 1 with an $IC_{50}$ value of $104 \pm 30$ nM and proved to be quite potent in the cancer cell line screening with $GI_{50}$ values ranging from 3.92 μM to 10 nM. It was also discovered that the triazole ring was of suitable size to fit into the narrow active pocket of HDAC protein and that it was correctly positioned to form a π–π interaction with two reserved phenylalanine residues in the docking structure of NSC746457 with HDAC 2 indicating that the triazole ring contributed to the binding affinity. Further optimization using click-chemistry-based approach including the replacement of the trans-styryl moiety with a 2-substituted benzo-heteroaromatic ring and introduction of a substituent onto the central methylene carbon resulted in the isopropyl derivative **37** and tertiary butyl variant **38** exhibiting excellent potency [60] (Figure 2.11) against HDAC enzymes. In *in vitro* safety tests, compound **39** (Figure 2.12) was far less toxic to nontransformed cells than tumor cells exhibiting a 10-fold potency compared to the marketed drug, suberoylanilide hydroxamic acid (SAHA) [61] (Table 2.5).

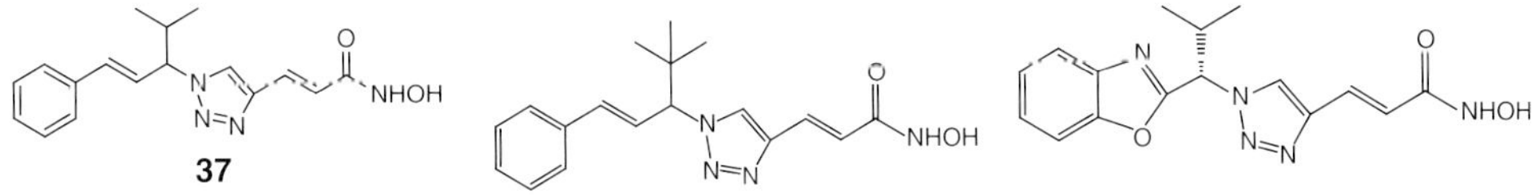


**37** **38** **39**

**Figure 2.11** Optimisation of NSC746457 as potent HDAC inhibitors.

**Table 2.5** $IC_{50}$ values (nM).

| | $IC_{50}$ (nM) | |
|---|---|---|
| TSAHA | HDACs | 67 |
| **36** | HDAC1 | $104 \pm 30$ |
| **37** | HDACs | 22 |
| **38** | HDACs | 18 |
| **39** | HDACs | 7 |

The current data on the chiral lead **39** provide strong evidence to warrant future *in vivo* preclinical studies.

***HDAC8 Inhibitors*** Using click chemistry, two HDAC8-selective inhibitors **40** and **41** (Figure 2.7) were reported [62] in which the zinc-binding group that coordinates with the active-site zinc ion was linked to the capping structure, which interacts with the residues on the rim of the active site via a triazole moiety in the presence of Cu(I). Compound **41** showed more potent activity compared to the known HDAC8 inhibitor PCI-34058 ($IC_{50} = 0.31\,\mu M$) with an $IC_{50}$ value of $0.070\,\mu M$. It was also found that these inhibitors caused selective acetylation of cohesion in cells and exerted growth inhibitory effects on T-cell lymphoma and neuroblastoma cells ($GI_{50} = 3-80\,\mu M$), which suggest that HDAC8 inhibitors have the potential to be anticancer agents. Based on these results, a series of derivatives of compound **41** were designed and synthesized as HDAC8-selective inhibitors. Among these, triazole **42** (Figure 2.12), showed the most potent activity with an $IC_{50}$ value of $0.053\,\mu M$ (Table 2.6).

These findings are useful for further development of HDAC8-selective inhibitors and should pave way for the development of new anticancer drugs [63].

#### 2.2.3.3 Protein Tyrosine Kinase Inhibitors

PTKs catalyze the phosphorylation of phenolic group of tyrosine residue in many substrate proteins by the transfer of *c*-phosphate moiety of ATP. PTKs play a crucial role in the signal transduction pathways. Aberrant regulation of kinase activity has been implicated in many diseases including cancer.

Src family kinases, one genre of PTKs, play an important role in the regulation of normal cellular signal transduction pathways, such as cell division, growth factor signaling, differentiation, survival, adhesion, migration, and invasion [64]. Src mutations and/or overexpression had been correlated with tumor growth,

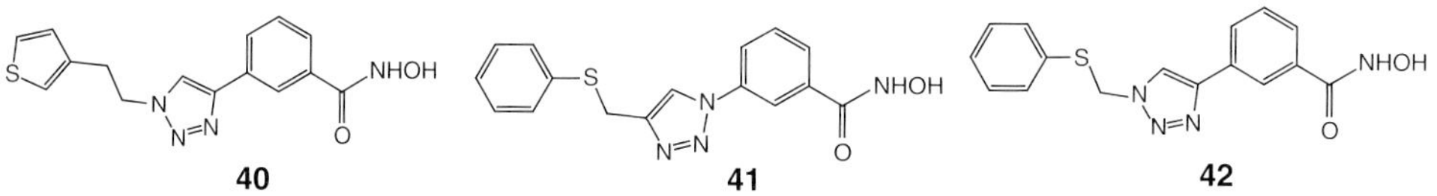


**40** **41** **42**

**Figure 2.12** Triazole moiety as selective HDAC-8 inhibitor.

**Table 2.6** $IC_{50}$ values (μM).

| Compounds | HDACs | HDAC1 | HDAC3 | HDAC8 | HDAC4 | HDAC6 |
|---|---|---|---|---|---|---|
| **PCI-34058** | >100 | >100 | >100 | 0.31 | >100 | 9.6 |
| **40** | 44 | >100 | >100 | 0.1 | >100 | 1.1 |
| **41** | 54 | 38 | >100 | 0.070 | 44 | 2.4 |

metastasis, and angiogenesis [65, 66]. Src kinase has been implicated in the development of a variety of cancers [67, 68].

The cellular form of Abelson (Abl) tyrosine kinase is a close relative of the Src family of tyrosine kinases. Abl tyrosine kinase is directly implicated in chronic myelogenous leukemia. The rapid synthesis of Abl tyrosine kinase inhibitors incorporating the head portion of imatinib and a 1,2,3-triazole ring with a suitable spacer attached to the N at position 1 using click chemistry has been reported [69]. Among these compounds, compounds **43** and **44** showed better inhibition against Src kinase in enzymatic screening, with a potency (Figure 2.13) comparable to the drug imatinib **45** (Table 2.7).

The click chemistry strategy employed should provide a useful tool in the fields of kinase research and drug discovery.

#### 2.2.3.4 Antimicrotubule Agents

The tubulin system plays a key role during mitosis and disturbing its dynamic equilibrium can prevent cell division and induce apoptosis. Microtubules, dynamic protein polymers composed of α-tubulin and β-tubulin heterodimers, are major components of the cytoskeleton with an important role in variety of cellular functions. They have become well-established cellular targets for

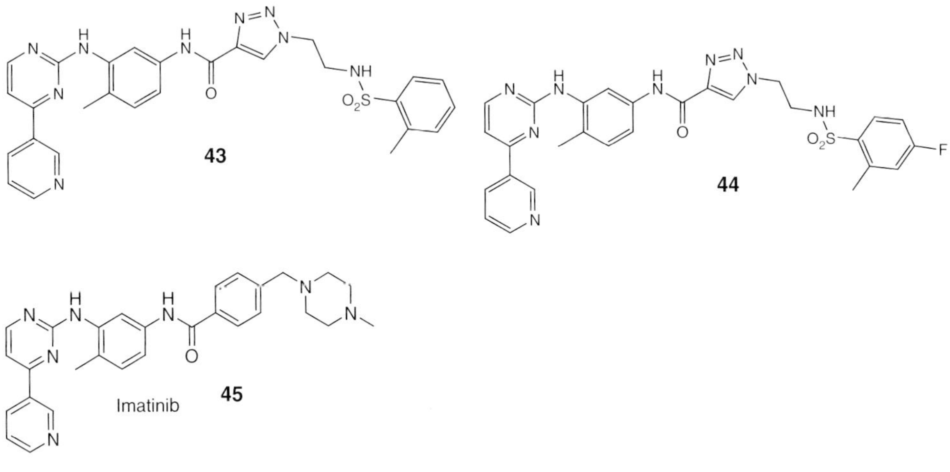


**Figure 2.13** Triazole linked Ematinib analog as potent Src Kinase inhibitor.

**Table 2.7** $IC_{50}$ values (μM).

| | $IC_{50}$ (μM) | |
|---|---|---|
| **Compounds** | **Abl** | **Scr** |
| **43** | 0.7 | 26.6 |
| **44** | 1.12 | 17.9 |
| Imatinib | 0.8 | — |

anticancer drugs [70, 71]. Searching for novel microtubule-binding agents that have enhanced tumor specificity, reduced neurotoxicity, and insensitivity to chemoresistance mechanisms is the main objective [72]. A series of *N*-((1-benzyl-1*H*-1,2,3-triazol-4-yl)methyl)arylamides as a new scaffold were synthesized for rapid access to microtubule agents by click reaction. Among these inhibitors, compound **46b** led by the initial hit compound **46a** showed potent activity (Figure 2.14) against MCF-7 human breast tumor cells [73] (Table 2.8).

Development of combinatorial library of compounds and extensive SAR studies were made possible by click chemistry approach. These findings will facilitate the design and optimization of potent, cell-permeable antimicrotubule agents.

#### 2.2.3.5 HSP 90 Inhibitors

Heat shock proteins (HSPs) play a key role in cell protection, acting as a protective factor against the misfolding of essential proteins involved in maintaining cell functionality. HSP90 is a protein highly conserved from bacteria to mammals, and it is documented to interact with more than 200 different "client" proteins involved in signal transduction, protein trafficking, receptor maturation, and innate and adaptive immunity. Consequently, HSP90 has emerged as an interesting molecular target for developing new anticancer agents against several solid and hematological malignancies as well as toward leukemia stem cells [74].

Based on a recent report that points out how substituted triazoleamido carboxylates can be employed in HSP90 binding [75], 5-amido-4-carboxytriazole **47** was prepared [75] and tested for its binding assay with HSP90 (Figure 2.15).

The substituted amido-triazole carboxylates can be used as stereodefined scaffolds for the preparation of triazole-containing peptidomimetics or as generic scaffolds for the preparation of diverse trisubstituted amino triazoles.

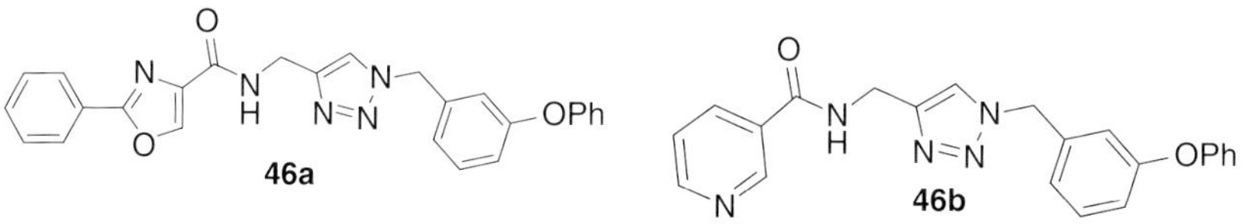


**Figure 2.14** Substituted triazoles as novel microtubule agents.

**Table 2.8** $IC_{50}$ values (nM).

| | $IC_{50}$ (nM) |
|---|---|
| Compounds | MCF-7 |
| **46a** | 560 |
| **46b** | 46 |

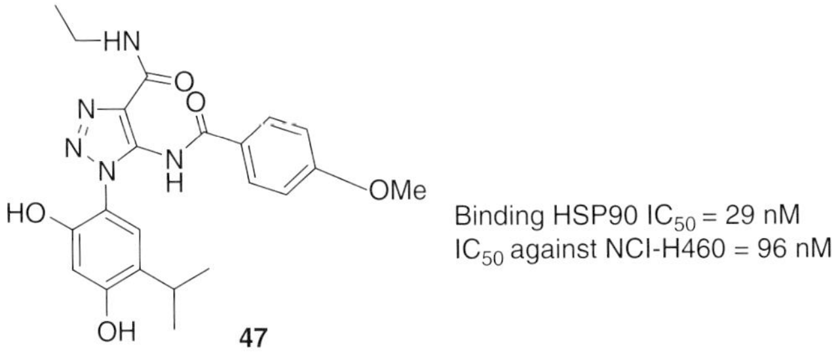


**Figure 2.15** Functionalised triazoles as HSP-90 inhibitors.

#### 2.2.3.6 Autophagy-Dependent Apoptosis in Cancer Therapy

Autophagy is a catabolic process of organelle protein turnover through lysosomal degradation necessary for balancing sources of energy that ultimately ensure cellular survival during stress [76]. Autophagy is considered as a new developing target for cancer therapy [77].

From a small library of 1,4,5-trisubstituted 1*H*-1,2,3-triazole compounds reported previously [78], compound 52, identified as a lead, was found to induce autophagy and apoptosis in the treated animal group. Its combination with *N*-acetyl cysteine and chloroquine significantly inhibited tumor regression, suggesting that ROS generation and autophagy are the key mediators of compound-**52**-induced cancer cell death *in vivo* [79] (Scheme 2.5).

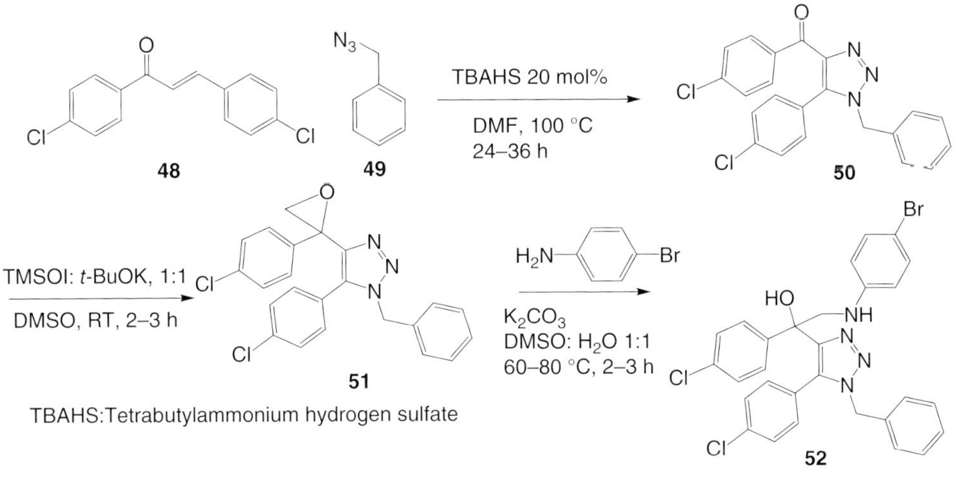


Scheme 2.5

The biological data suggest that compound **52** inhibits breast cancer cell growth both *in vitro* and *in vivo* through induction of autophagy-mediated apoptosis and will help to develop a promising therapeutic strategy for the treatment of breast cancer.

#### 2.2.3.7 Anticancer Activity of 4β-Triazole-Podophyllotoxin

A series of novel 4β-triazole-podophyllotoxin glycosides were synthesized [80] utilizing click reaction and evaluated for their cytotoxic activity against a panel of five human cancer cell lines (HL-60, SMMC-7721, A-540, MCF-7, SW480). Most of these compounds showed weak cytotoxicity. The galactoside-derived triazole derivative **53** showed the highest activity with $IC_{50}$ values ranging from 2.85 to 7.28 μM (Table 2.3), which is more potent than the control drugs etoposide **33** and cisplatin against four of five cancer cell lines (except against HL-60) tested (Figure 2.16 and Table 2.9).

In addition, chemical stability investigation indicates that the conjugated sugar residue seems to improve the stability of podophyllotoxin residues under the physiological conditions.

#### 2.2.3.8 1,2,3-Triazole-Substituted Oleanolic Acid Derivatives as Anticancer Agents

A series of novel oleanolic acid coupled 1,2,3-triazole derivatives have been designed and synthesized through click chemistry [81]. The antiproliferative evaluation indicated that some compounds exhibited excellent anticancer activity against five cancer cell lines (HeLa, HepG2, HCT116, A375-S2, and HTT1080). Among these derivatives, compound **54** (Figure 2.17) showed the best inhibitory

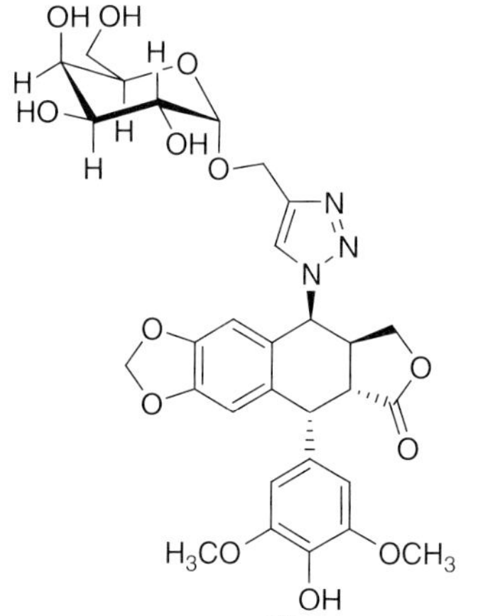

**53**

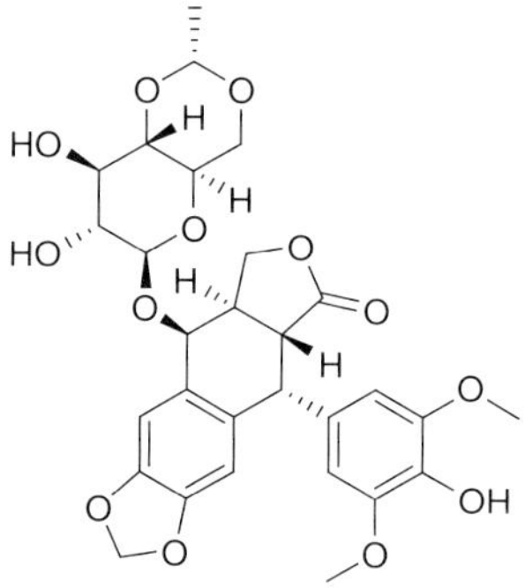

**32** Etoposide

**Figure 2.16** Galactoside-derived triazoles as novel anticancer agents.

**Table 2.9** *In vitro* anticancer activity ($IC_{50}$ μM) of compound **52**.

| Compounds | IC50 (μM) | | | | |
|---|---|---|---|---|---|
| | HL-60 | SMMC-7721 | A-549 | MCF-7 | SW480 |
| **53** | 2.85 | 3.99 | 4.07 | 7.28 | 5.52 |
| **32** (Etoposide) | 0.31 | 8.12 | 11.92 | 32.82 | 17.11 |
| Cisplatin | 1.17 | 6.43 | 9.24 | 15.86 | 13.42 |

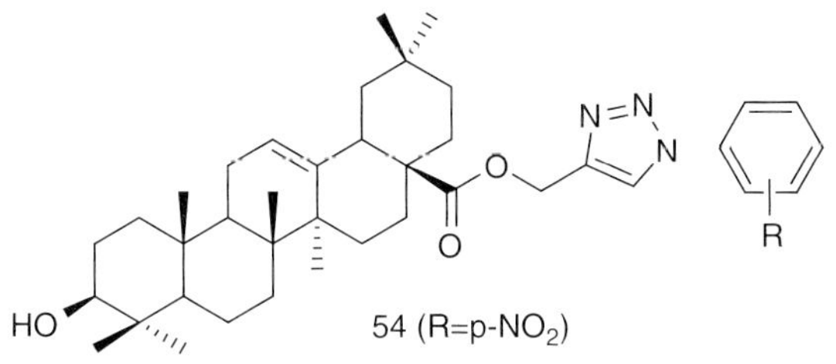
54 (R=p-NO$_2$)

**Figure 2.17** Oleanolic acid-coupled triazoles as novel anticancer agents.

**Table 2.10** $IC_{50}$ value of compound **54** against human cancer cell lines.

| Human cancer cell lines | 54 ($IC_{50}$ in μM) | 5-Fluorouracil ($IC_{50}$ in μM) |
|---|---|---|
| HeLa | 10.85 | 26.18 |
| HepG2 | 24.15 | 67.64 |
| HCT116 | 12.28 | 35.16 |
| A375-S2 | 4.97 | 90.74 |
| HT1080 | 3.51 | 25.46 |

activity against HTT1080 (human fibrosarcoma) cells, and it also significantly induced apoptosis (Table 2.10).

From the biological studies, compound **54** is found to be a potent apoptosis inducer in HT1080 cells and thus can serve as a promising lead candidate for further studies.

There are several recent reports in the literature on 1,2,3-triazole-linked substituted dihydrofurocoumarin hybrids [82], carbanucleoside [83], berberine derivatives [84], dehydroabietic acid variants [85], carnosic acid, and carnosol analogs [86] showing anticancer activity *in vitro* and are in early stages of drug discovery.

#### 2.2.3.9 Anti-Infective Agents

Anti-infectives or anti-infective agents are drugs that either kill an infectious agent or inhibit it from spreading. Antibacterials, antivirals, antifungals, and antiparasitic agents are broadly categorized under anti-infective agents. Antibacterial agents can benefit cancer patients in several ways, for example, by (i) killing oncogenic-related microorganisms, (ii) protecting them from recurring immune-suppression-induced infection, and (iii) their direct anti-proliferative/cytotoxic effects

Recent advances on diverse and potent biological profiles of 1,2,3-triazolo-nucleosides along with emerging application of click chemistry in their synthesis and their perspectives have been reviewed [87]. In the article, the work is primarily addressed to antiviral, antimicrobial, and anticancer potency of these important structural motifs in which the 1,2,3-triazole ring acts as a nucleobase surrogate or is linked to a nucleobase or a sugar/sugar-mimicking moiety.

#### 2.2.3.10 1,2,3-Triazole Nucleoside

Many glycoconjugates are involved in important biological functions, including on the cell surface such as recognition of host compounds, immunological responses, inflammation, cell–cell recognition, bacterial and viral infection, cell communication, metastasis, and many important functions inside the cells [88]. From this series of nucleoside analogs in which the nucleobase is replaced with 1,2,3-triazole scaffold, emerged the TSAO analogs **55–58**, which showed potent and selective activity against human immunodeficiency virus type-1 (HIV-1). Structure–activity relationship (SAR) studies with the TSAO class of compounds revealed that stringent requirements existed with regard to the structural determinants for optimum anti-HIV-1 activity in cell culture. The sugar part of the TSAO molecules played a crucial role in the interaction with their target enzyme. The presence of *t*-butyldimethylsilyl (TBDMS) groups at both the C-2 and C-5 positions of the ribose and the unique 3-spiro-5-(4-amino-1,2-oxathiole-2,2-dioxide) moiety of the nucleoside in the D-ribo configuration were prerequisites for the anti-HIV-1 activity [89].

The prototype compound of this family was the thymine derivative TSAO-T (Figure 2.18). TSAO derivatives were targeted at the HIV-1-encoded reverse transcriptase (RT) with which they interacted at a nonsubstrate binding site. Within the non-nucleoside reverse transcriptase inhibitors (NNRTIs), the TSAO nucleosides occupy a unique position in that they interfered at the interface between the P51 and P66 subunits of RT. A variety of 1,2,3-triazole TSAO derivatives (Figure 2.18) substituted at the 5-position of the triazole moiety were evaluated for their inhibitory effects against HIV-1- and HIV-2-induced cytopathicity in CEM and MT-4 cell cultures. The most active TSAO derivatives were those that contained either *N*-ethyl (**55**) *N*-cyclopropyl carbamoyl (**56**)

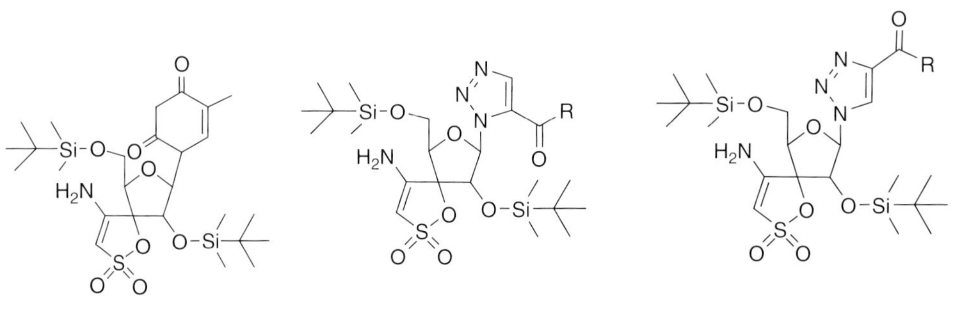


**55** (R = $NHCH_2CH_3$), **56** (R = 59 NH-cyclopropyl), **57** (R = azetidinyl), **58** (R = $N(CH_3)_2$)

**Figure 2.18** 1,2,3-Triazole variants of TSAO nucleosides.

function or an azetidine carbamoyl (**57**) function (Figure 2.18) at the 5-position of the triazole moiety. The values were, as a rule, slightly lower in CEM cells and were comparable with the inhibitory activity of the prototype TSAO-T and its dimethyl carbamoyl derivative (**58**). A bulky substituent on the nitrogen of the carbamoyl moiety resulted in a more pronounced loss of antiviral activity in MT-4 cells. The loss was less pronounced in CEM cells. None of the tested compounds showed any measurable anti-HIV-2 activity in MT-4 or CEM cells. Furthermore, differences in activity among TSAO analogs with substitution patterns on C-4 and C-5 were found. When the unsubstituted and substituted carbamoyl functions were introduced at C-4 (**59**) instead of C-5 of the triazole moiety, the antiviral activities were decreased by 10-fold [89] (Table 2.11).

Since the polymerase activity of HIV-1 RT is entirely dependent on the heterodimeric structure of the enzyme, RT dimerization represents a target for the development of a new therapeutic class of HIV inhibitors. Therefore, the ability of TSAO derivatives to inhibit RT dimerization was also evaluated. Good correlation between the ability of TSAO derivatives to inhibit RT dimerization and the enzyme's polymerase activity was observed [90].

#### 2.2.3.11 1,2,3-Triazole Carbonucleosides

Carbonucleosides are compounds in which the oxygen atom of the furanose ring is replaced with a methylene group. These compounds have high metabolic stability compared to nucleosides because they are unaffected by phosphorylase and hydrolase enzymes that cleave the glycosidic bond of natural nucleosides [91]. For this reason, 1,2,3-triazole carbocyclic nucleosides are important compounds and have led to interesting biological agents [92–96].

A significant amount of synthetic efforts have been directed toward finding structural analogs of Neplanocin A (NPA, **60**) (Figure 2.12), a carbocyclic nucleoside as potential antiviral and antitumor agent [97, 98]. A practical and convenient methodology for the synthesis of target 1,2,3-triazole carbocyclic nucleoside **61**, structurally related to NPA is reported (Figure 2.19) [99].

Ring-closing metathesis (RCM) reaction, which is one of the most powerful methods for the formation of small-sized rings or cyclopentenol ring [100, 101], was successfully applied in the synthesis of chiral cyclopentenol, as the key intermediate of carbocyclic nucleosides. Interestingly, the 1,2,3-triazole NPA analog

**Table 2.11** $EC_{50}$ values (μM).

| Compounds | EC 50 (μM) HIV-1 | |
|---|---|---|
| | MT-4 | CEM |
| **55** | 0.17 | 0.13 |
| **56** | 0.19 | 0.08 |
| **57** | 0.14 | 0.08 |
| **58** | 0.06 | 0.12 |
| TSAO-T | 0.06 | 0.06 |

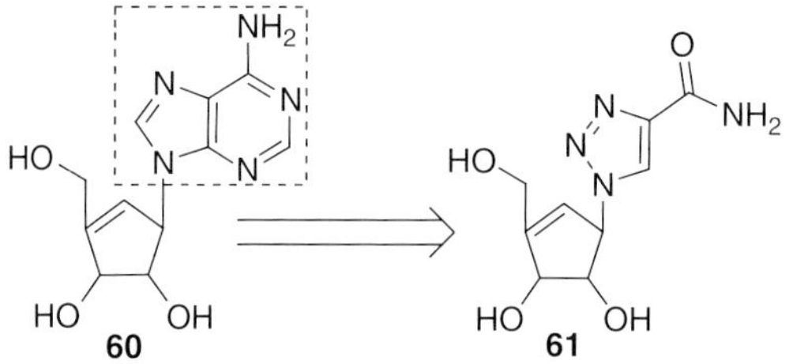

**60** Neplanocin A, **61**

**Figure 2.19** 1,2,3-Triazole carbanucleoside analogue of Neplanocin A.

**61** (Figure 2.19) exhibited the most potent antiviral activity among the evaluated five-membered ring heterocycles (1,2,4-triazole and imidazole) carbocyclic nucleosides (Figure 2.20 and Table 2.12).

Triazolyl nucleoside derivatives can be regarded as a promising class for further drug development. The syntheses of diverse types of triazolyl nucleoside derivatives and their screening may provide a new perspective and advancement of research on novel classes for more potent candidates to treat infections or life-threatening diseases. The growing applications of click chemistry in the area of drug discovery may witness the emerging of new 1,2,3-triazole-bearing nucleosides in the market [87].

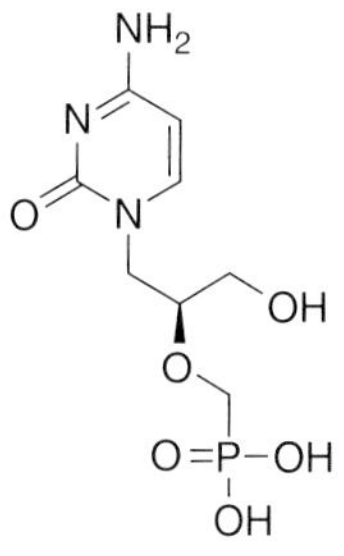

**62** Cidofovir

**Figure 2.20** Cidofovir as a reference standard for antiviral agent.

**Table 2.12** $EC_{50}$ values (μM).

| | EC 50 (μM) | |
|---|---|---|
| | 61 | Cidofovir 62 |
| Vaccinia | 0.4, SI > 750 | 6 |
| Cowpox | 39, SI > 2.1 | 15 |
| SARSCoV | 47 | — |

SI – selectivity index.

#### 2.2.3.12 β-Lactamase Inhibitors as Antibacterial Agents

Boronic acid transition-state inhibitors (BAT-SIs) represent one of the most promising classes of β-lactamase inhibitors. A small library of 1-amido-2-triazolyl-ethaneboronic acids as new class of BAT-SIs have been synthesized through click chemistry and screened against two clinically relevant class C β-lactamases from *Enterobacter* spp. and *Pseudomonas aeruginosa* [102]. Among these compounds, compound **65** showed a good activity (4 nM) with a significant reduction of bacterial resistance to the combination with cefotaxime (Scheme 2.6 and Table 2.13).

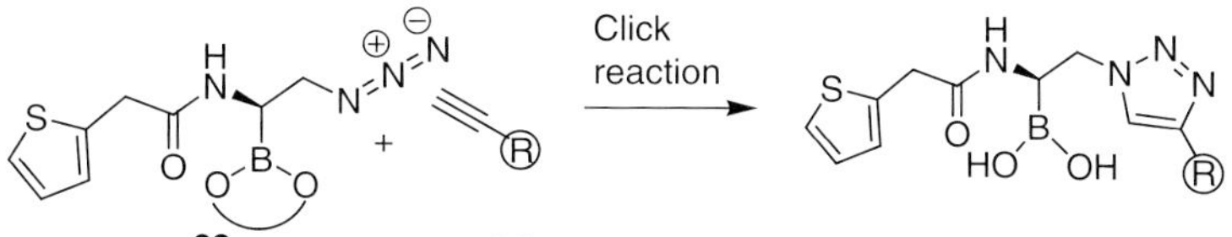


**63** **64** **65**

**65** (R = COOH); **66** (R = *m*-$C_6H_4$-COOH)

Scheme 2.6

The design, synthesis, and biological properties of a new class of chiral boronic acids that effectively target two representative β-lactamases found in nosocomial

**Table 2.13** $K_i$ (μM) value of compound **46** and reference compounds **67** and **68**.

| Compounds | Structure | $K_i$ (μM) vs. PDC-3 | vs. P99 |
|---|---|---|---|
| **65** | 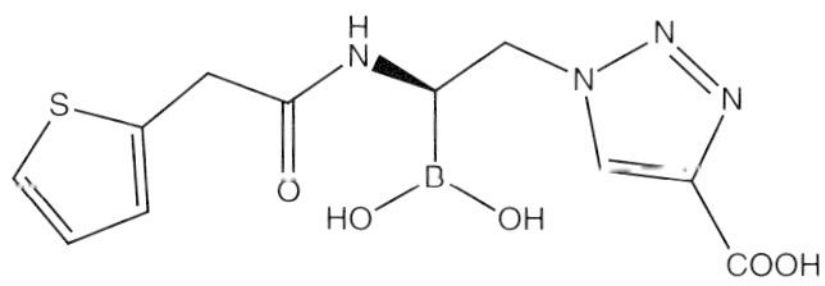 | 0.004 ± 0.001 | 0.023 ± 0.002 |
| **66** | 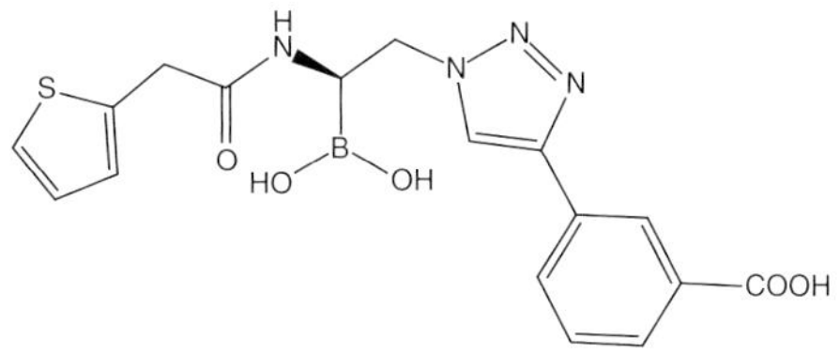 | 0.008 ± 0.001 | 0.014 ± 0.001 |
| Reference compound **67** | 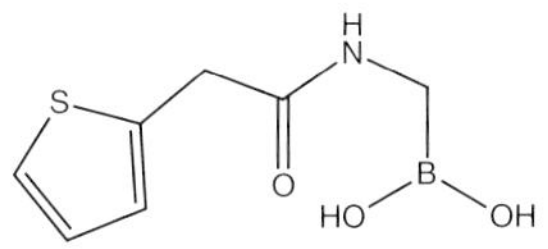 | 0.411 ± 0.001 | 0.451 ± 0.002 |
| Reference compound **68** | 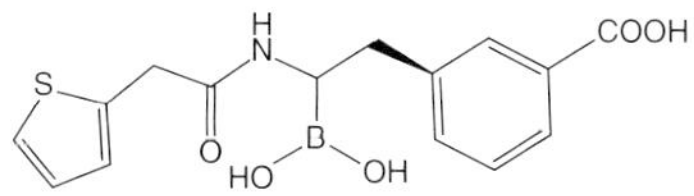 | 0.013 ± 0.006 | 0.015 ± 0.005 |

pathogens have been reported. The improved synthetic accessibility of these boronic acids through click chemistry opens up the possibilities of obtaining a large number of compounds with a wider range of substituents that could lead to a better understanding of the β-lactamase binding site and help to locate it. Extension of the studies could lead to other microbiological targets of importance and diverse β-lactamases.

Several articles have appeared in recent literature on 1,2,3-triazole-based scaffolds as antimicrobial agents that are in hit stages of discovery [103–106].

Tuberculosis (TB) is a bacterial infection caused by *Mycobacterium tuberculosis* (Mtb). Mtb is an obligate, intracellular, nonmotile bacillus that primarily infects humans. The bacterium is also known for its lipid-rich cell wall that is impermeable to most dyes. Mtb also divides at an incredibly slow pace taking 15–20 h. Mycobacteria can be classified into nonpathogenic organism, such as *Mycobacterium smegmatis,* which is fast growing and most often used as a laboratory model for Mtb research and pathogenic organisms, which cause diseases in humans and animals such as Mtb and *Mycobacterium bovis.* A review on triazoles as a promising antitubercular agent has appeared recently in literature [107].

#### 2.2.3.13 1,2,3-Triazole-Linked Carbazoles as Antitubercular Agents

A recent report [108] describes the synthesis of a small library of novel substituted 1,2,3-triazolylmethylcarbazoles, **69** and **70** rationally designed by incorporating two pharmacophoric heterocyclic rings, namely carbazole and triazole in one structure. Evaluation of these compounds for antimycobacterial activity against Mtb $H_{37}Rv$ (ATCC27294) led to two compounds with MIC of 6.25 μg/ml (standards isoniazid: 0.1 μg/ml; ethambutol: 3.13 μg/ml; and pyrazinamide: 50 μg/ml) as most promising antitubercular agents with low cytotoxicity (Figure 2.21).

The results of this study demonstrate the potential utility of scaffold hopping approach in designing new antitubercular analogs of carbazole with appended 1,2,3-triazole fragment as potent antitubercular agents for further optimization.

#### 2.2.3.14 1,4-Diaryl-Substituted 1,2,3-Traizoles as Antimycobacterial (Mtb) Agents

Three series of 1,4-substituted 1,2,3-triazoles have been synthesized (Scheme 2.7) as small libraries and evaluated for their as antitubercular activity [109].

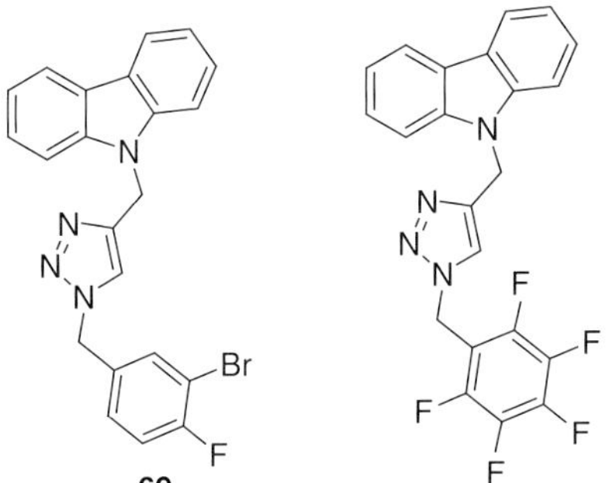

**69** **70**

**Figure 2.21** 1,2,3-Triazole linked carbazole analogues as anti tubercular agents.

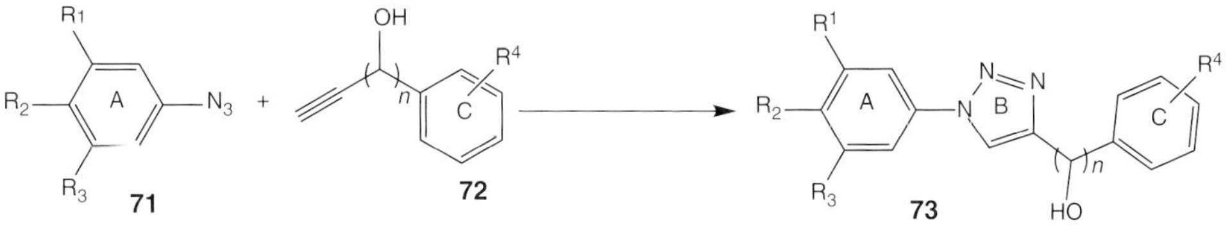


Scheme 2.7

Among the first series (Figure 2.22), compounds **74** and **75** showed good activity. The aryl substituent on the alcohol seemed to determine the activity. The compounds bearing a nitro functionality are the most potent since Mtb can reductively activate such molecules [110, 111].

The nitro aromatic compounds have generally been associated with a high frequency of mutation in Mtb due to the ease of inactivation of the various processes required for nitro reduction [110] making compounds **76–78** forming the second series (Figure 2.23) particularly attractive for follow-up since these are unlikely prodrugs and might inhibit a specific target. From this preliminary study, it was seen that compounds consisting of three aromatic rings were a better class of potential antitubercular compounds. Additionally, these compounds containing

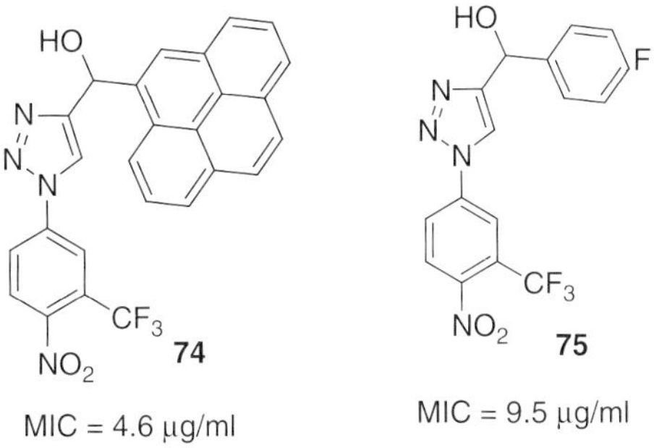


**Figure 2.22** Diaryltriazole carbinols as antitubercular agents.

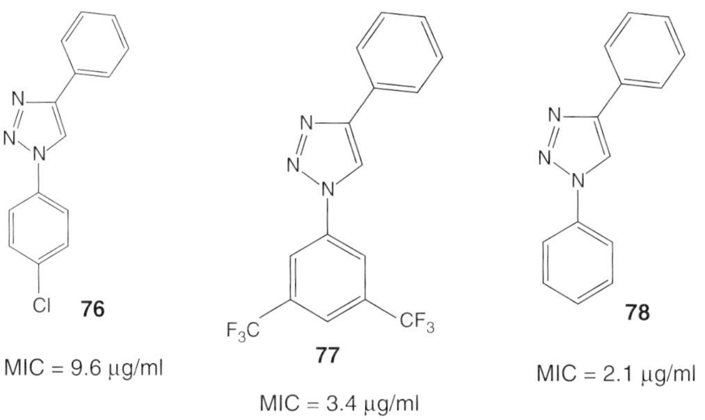


**Figure 2.23** Diaryltriazoles as follow up antitubercular agents.

an *N*-aryltriazole moiety bearing a 3,5-trifluoromethyl or unsubstituted showed the most promise.

Interestingly, amino substitution on this series compound **79** (Figure 2.24) was not tolerated at all, with no bactericidal properties observed. Another surprising result reported was the addition of amide groups to the phenyl ring of triazoles **80** resulted in compounds with vastly decreased MIC values or total abolition of antimicrobial properties.

From these studies, it is clear that simple 1,4-diaryl-substituted 1,2,3-traizolemolecular scaffold is an excellent starting point for the development of more potent antitubercular agents and mode of action studies.

#### 2.2.3.15 1,2,3-Triazole-Adamantylacetamide Hybrids as Antitubercular Agents

A series of novel 1,2,3-triazole-adamantylacetamide hybrids designed by combining bioactive fragments from antitubercular I-A09 and substituted 1-adamantyl urea were synthesized [112] using click chemistry strategy. Evaluation of all new compounds for *in vitro* antitubercular activity against Mtb H37Rv (ATCC27294) resulted in the hybrid **84** (Figure 2.25) as the most promising lead.

The results demonstrate the potential utility of new hybrid analogs of admantane **85** with appended 1,2,3-triazole fragment as potent antitubercular agents for further optimization.

Several articles have appeared in recent literature on 1,2,3-triazole-based scaffolds as antitubercular agents that are in very early discovery stages [113–115].

#### 2.2.3.16 Non-Nucleoside HIV Integrase Inhibitors

As a first line of design, L-708906 (compound **86**) was modified to structure (I) (Figure 2.26), which did not show the expected activity [116]. Hence, it was further modified to structure (II), which led to a series of 5-chloro-2-hydroxy-3-triazolylbenzoic acid bearing compounds as potential HIV-integrase inhibitors [117]. The modification comprised replacement of N-5 position of structure I

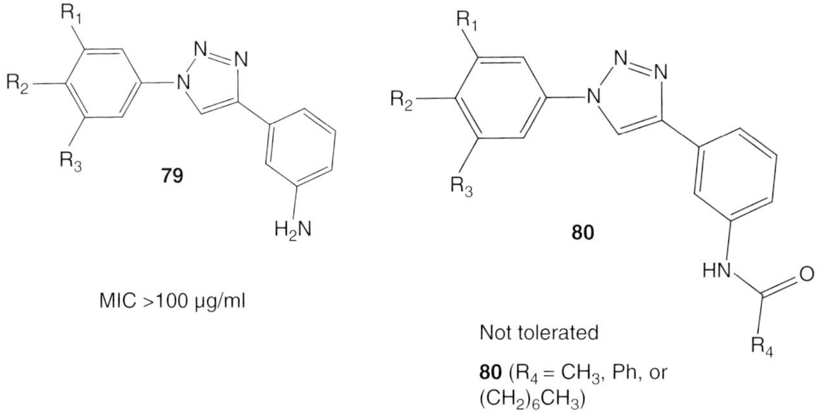


**Figure 2.24** Evaluation of aminoaryl and its acyl derivatives as antitubercular agents.

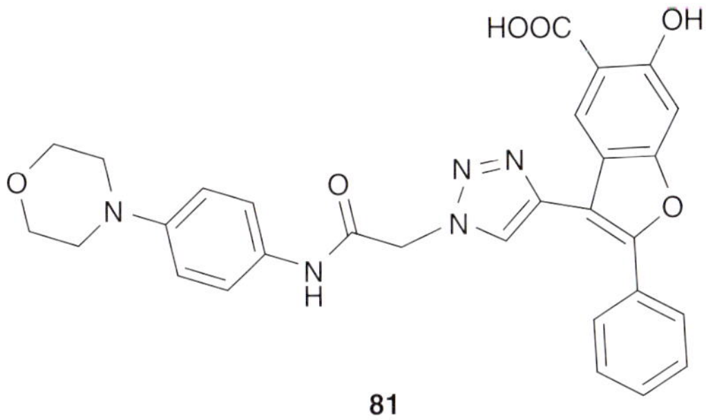


**81**

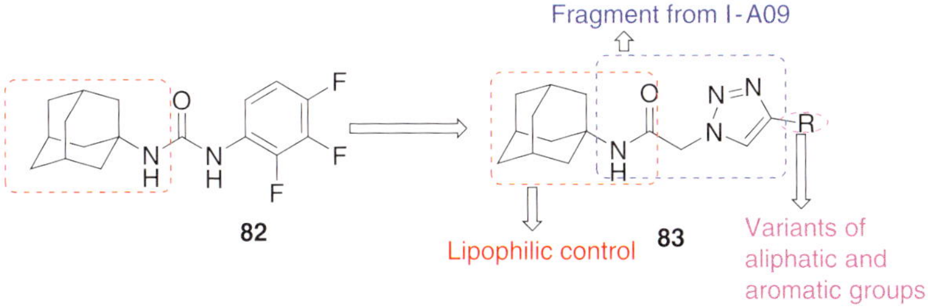


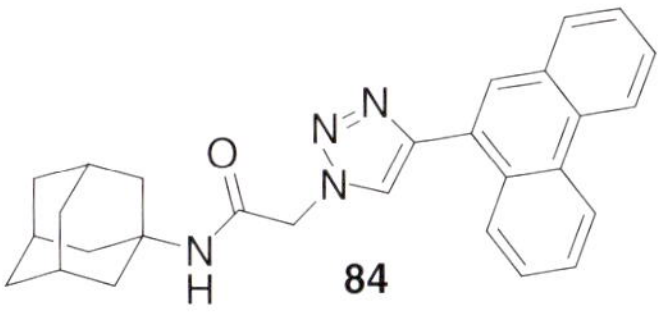


(MIC=3.12 µg/ml; SI>15)

**Figure 2.25** 1,2,3-Triazole-admantyl acetamide hybrids as antitubercular agents.

pharmacophore by an aromatic carbon atom and the fused pyridine ring by a triazolyl group. The triazole, in this series of compounds, played a dual role of a linker and a metal ion binder. Chlorine atom introduced on C-5 of benzoic acid increased the acidic property of the phenolic OH at C-2 position favoring metal chelation.

The compounds of this series were found to be more active than L-708906, **86**. Among which, compounds **87–89** (Figure 2.26) exhibited potent HIV-integrase inhibitory activity (Scheme 2.8).

**Scheme 2.8**

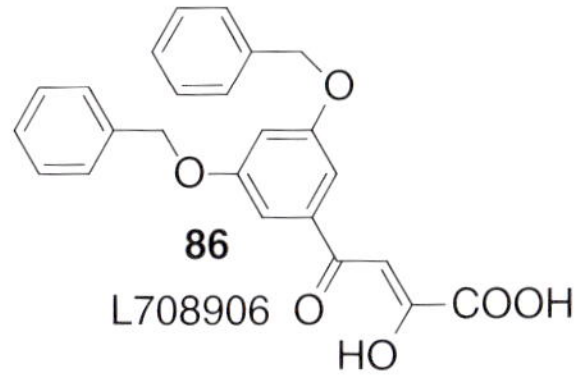


**86**
L708906

$IC_{50}$ = 28 µg/ml

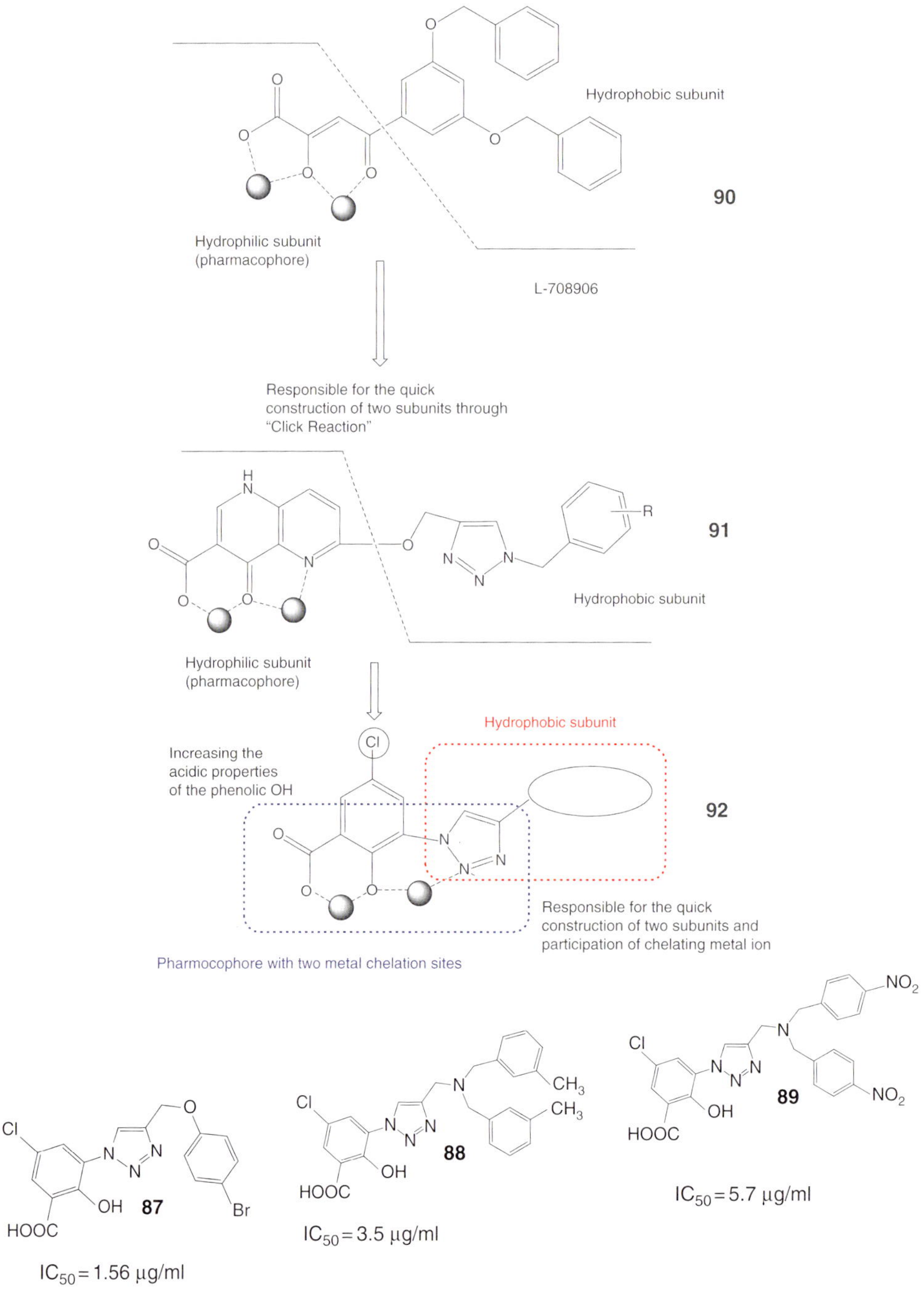


**Figure 2.26** Improved variants of L-708906 as non-nucleoside HIV integrase inhibitors.

It is evident from the results that the click chemistry reaction methodology has been quite useful in the design of HIV-integrase inhibitors producing a promising new starting point for further elaboration and optimization

#### 2.2.3.17 Miscellaneous Therapeutic Segments: 1,2,3-Triazole-Linked Dopamine D3 Receptor (D3R)-*Selective Therapeutic Agents*

D3 receptor (D3R)-selective ligands may be used to treat a variety of neuropsychiatric disorders associated with aberrant dopamine signaling including schizophrenia [118], Parkinson's disease and associated dyskinesias [106, 107, 119, 120], and drug addiction [121–123].

A common molecular template used in the development of D3R-selective antagonists and partial agonists incorporates a butylamide linker between two pharmacophores, a phenylpiperazine moiety, and an extended aryl ring system (Figure 2.27). Bioisosteric replacement of the amide linkage by 1,2,3-triazole moiety resulted in a library of compounds, among which compound **93** showed a greater affinity for the D3R [124] (Figure 2.27 and Table 2.14).

In the present series, indolyl-triazole **93** has the best overall profile in terms of D3R affinity, selectivity, lipophilicity, and metabolic stability.

Reviews on psychopharmacological [125], antineuropathic, anticonvulsant [126], and antidiabetic [127] activities of substituted triazole analogs have appeared.

#### 2.2.3.18 Peptidomimetics: 1,2,3-Triazole as a Disulfide Bond Mimetic

The design of disulfide bond mimetics is an important strategy in optimizing cysteine-rich peptides in drug development. A very interesting article [128] on 1,2,3-triazole as a disulfide bond mimetic of the drug lead conotoxin MrIa

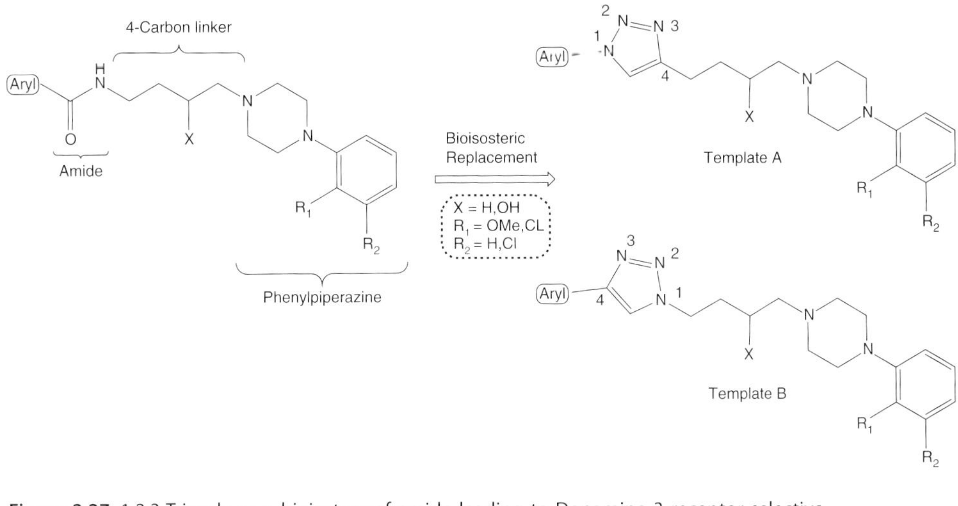


**Figure 2.27** 1,2,3-Triazole as a bioisotere of amide leading to Dopamine 3 receptor selective ligands.

**Table 2.14** $K_i$ values ±SEM.

| Compounds | D3R $K_i$ ± SEM |
|---|---|
| 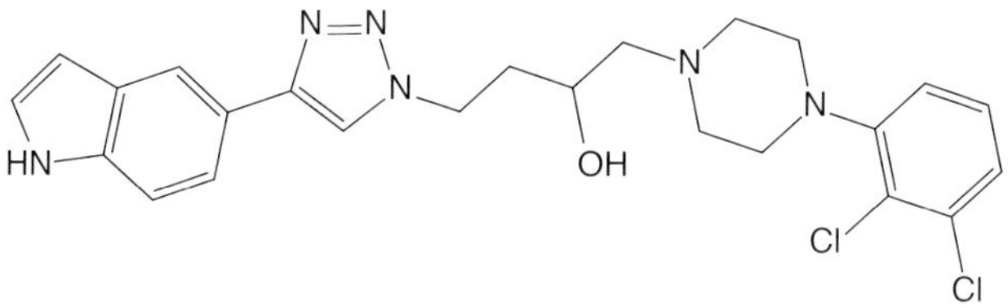 **93** | 5.85 ± 0.780 |
| 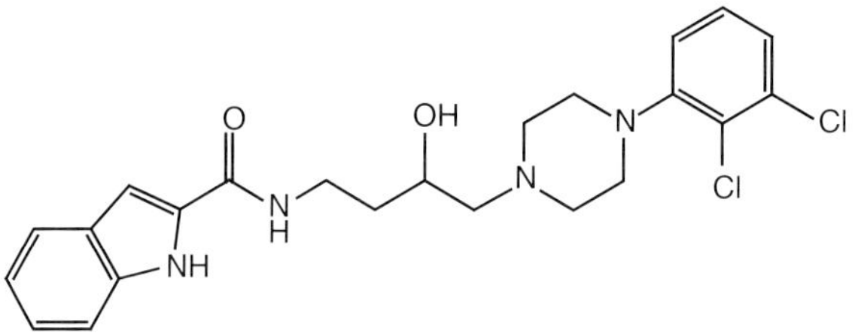 **94** PG-648 | 1.88 ± 0.112 |

in which one disulfide bond is selectively replaced by 1,4-disubstituted 1,2,3-triazole bridge is described. The mimetics with a triazole replacing *Cys*4 − *Cys*13 disulfide bond retained tertiary structure and full *in vitro* and *in vivo* activity as norepinephrine reuptake inhibitors.

## 2.3 Part B: Synthesis of Triazole-Based Drugs Currently in use

Click chemistry is a very powerful tool in drug discovery, because it is very efficient in the creation of compound libraries through diversity. Although 1,2,3-triazole motif is widespread, to date, only four drugs (Figure 2.28) have emerged out in the market. Some of them are in different stages of clinical trials, and many of them are in discovery stages in several therapeutic segments especially in the area of oncology and anti-infectives (Figure 2.29). The syntheses of these triazole-derived pharmaceuticals are described as follows:

Synthesis of all these compounds would be delineated.

### 2.3.1 Tazobactam

Tazobactam, a penicillanic acid sulfone derivative, is a drug that inhibits the action of bacterial beta-lactamases especially those belonging to the SHV-1 and

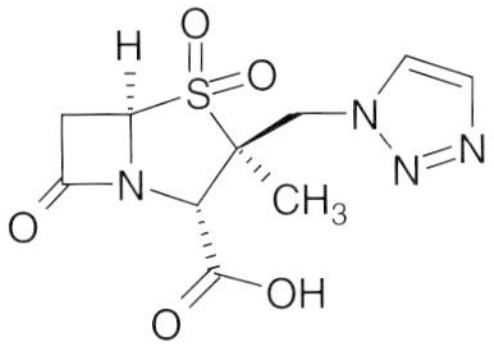

Tazobactam

**7**

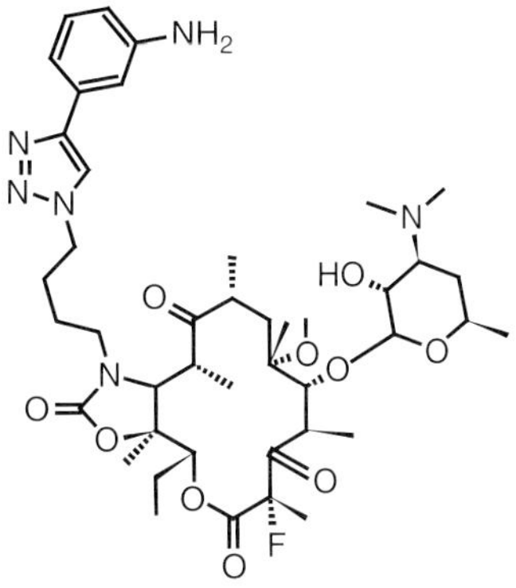

Solithromycin

**95**

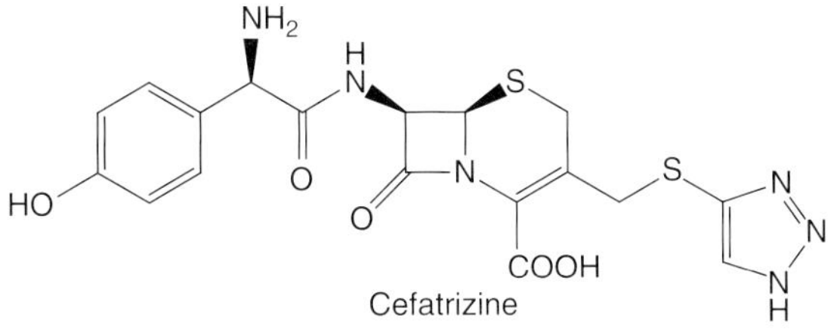

Cefatrizine

**96**

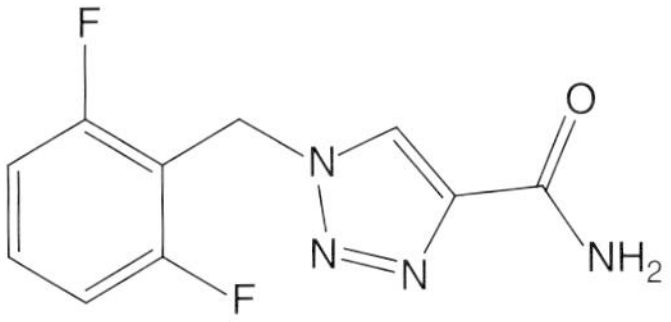

Rufinamide

**97**

**Figure 2.28** 1,2,3-Triazole- based drugs currently in use.

TEM groups. Tazobactam is beta-lactamase inhibitor that helps piperacillin work better. Recently, its combination with ceftolozane and metronidazole was found to be a safe and effective new antimicrobial combination approved for the treatment of complicated gastrointestinal and urinary-tract infection [129].

#### 2.3.1.1 Synthesis of tazobactam from intermediate 102

The synthesis of tazobactam was first reported by Taiho Pharmaceutical Company Ltd, Japan [130]. Starting with the benzhydryl ester of penicillanic acid 102 (Scheme 2.9), tazobactam has been synthesized [131a] in six steps in an overall yield of 40% as outlined in Scheme 2.10. The impurities that are formed in this synthesis are listed in (Figure 2.30).

#### 2.3.1.2 Other reports on Tazobactam synthesis

There are other reports available in the literature for the synthesis of tazobactam [131b, c, 132] (Figure 2.30).

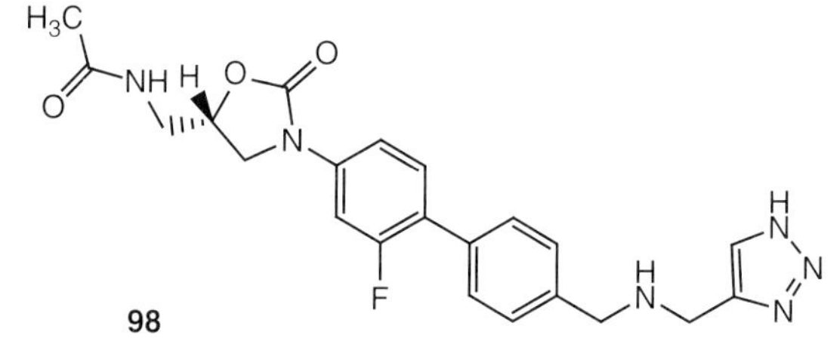

98

Radezolid (Phase-3,QIDP)

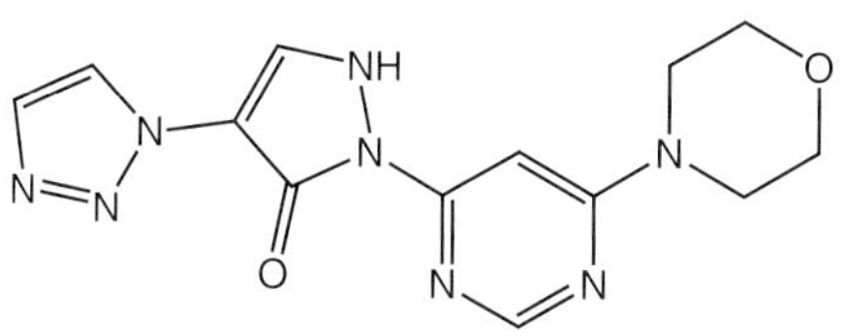

Molidustat (Phase-2b)

99

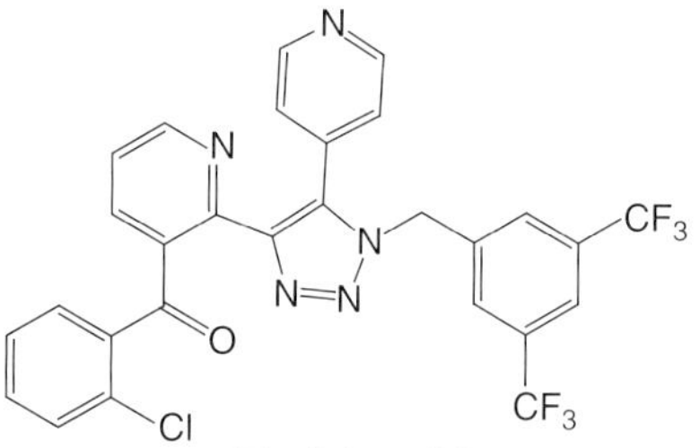

Tradipitant (Phase-2)

100

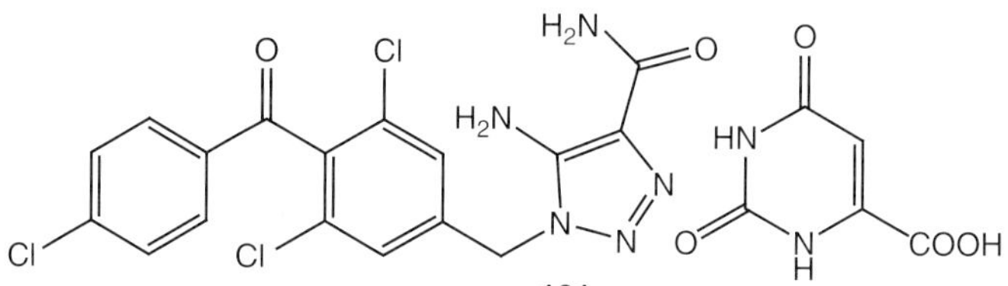

101

Carboxyamido triazole orotate
(Phase-3,discontinued)

**Figure 2.29** 1,2,3-Triazole-based pharmaceuticals in clinical trials.

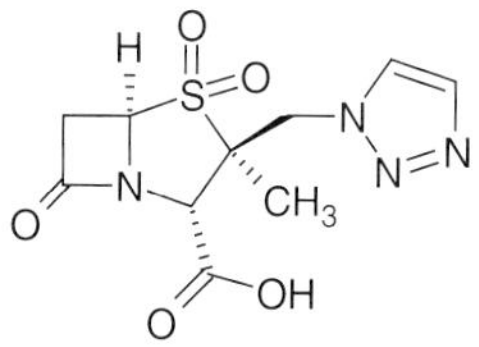

7

Tazobactam

## 2.3.2 Solithromycin

Solithromycin, a fluoroketolide, was first reported by Optimer Pharmaceuticals and is currently developed by Cempra. This antibiotic is active against clinical

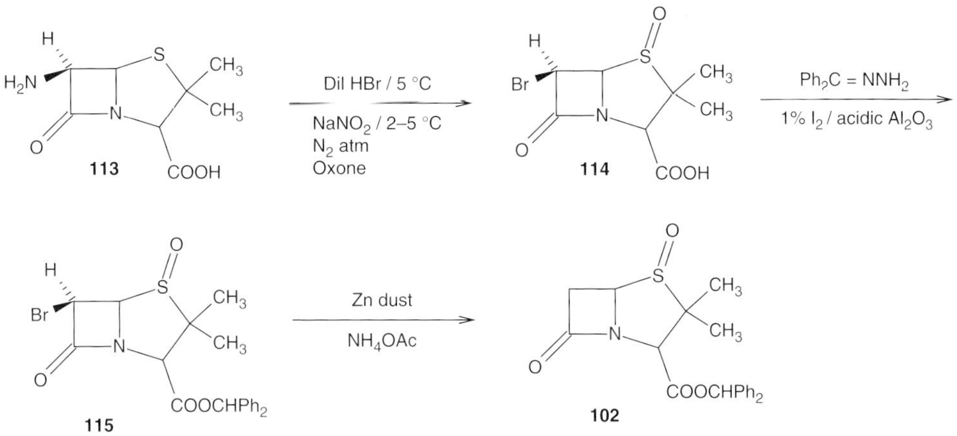


Scheme 2.9

isolates of Legionella Pneumophila Sero group. Legionellosis is a major public concern in industrialized countries. Manifestation of the disease ranges from a mild respiratorial illness (Pontiac fever) to a severe and rapidly fatal pneumonia. Solithromycin has eightfold higher activity than that of azithromycin.

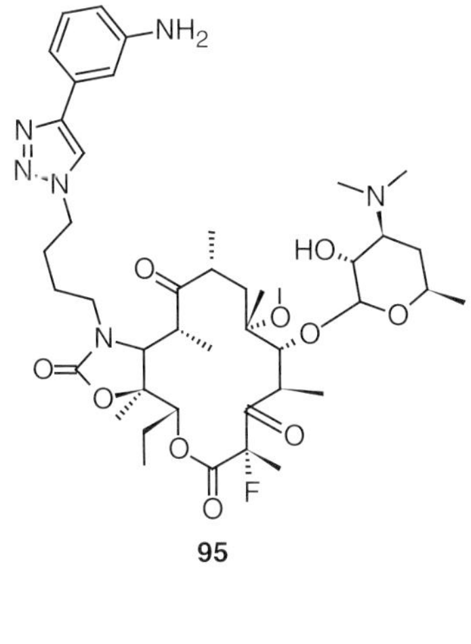


95

#### 2.3.2.1 **Synthesis of Solithromycin**

Solithromycin has been synthesized [133] in eight steps starting from clarithromycin as outlined in Schemes 2.11 and 2.12.

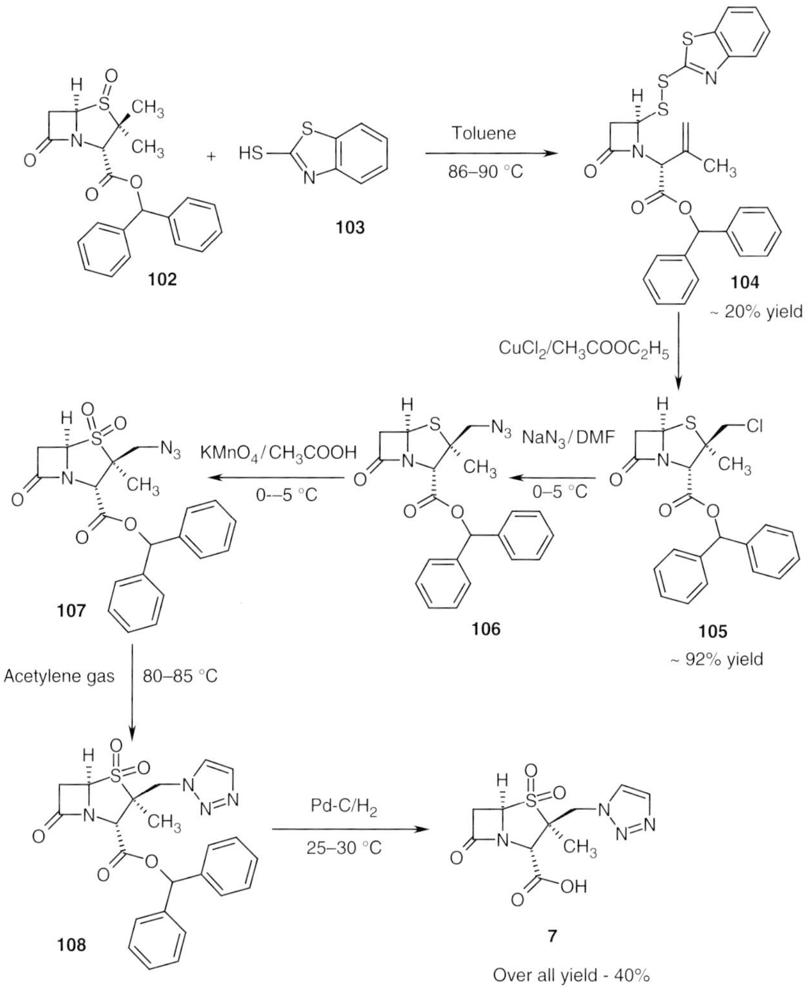


Scheme 2.10

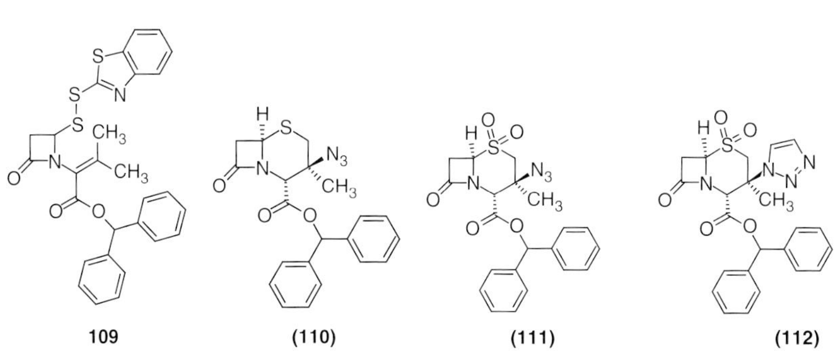


**Figure 2.30** Impurities formed during the synthesis of tazobactam [131a].

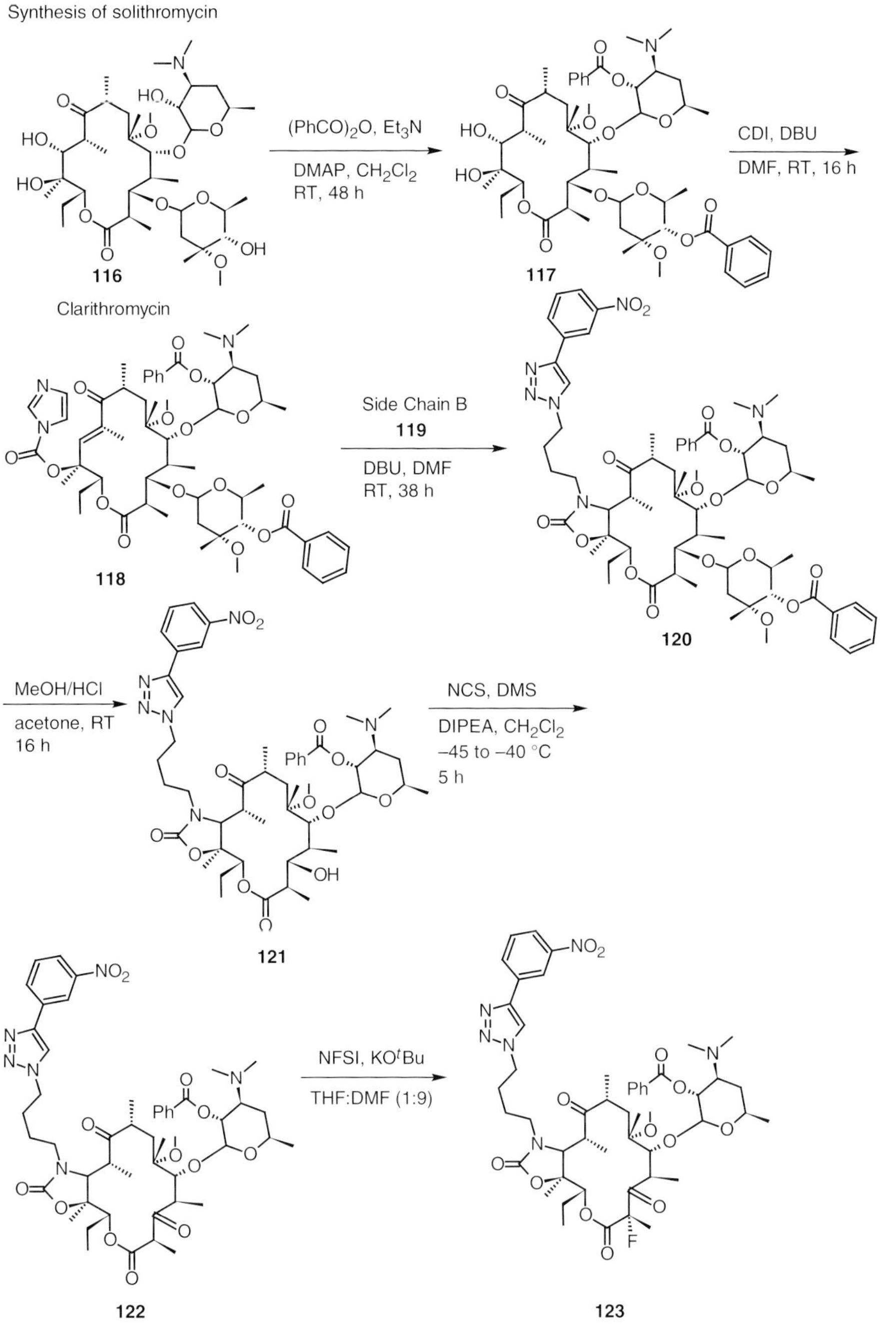
Synthesis of solithromycin
116
Clarithromycin
$(PhCO)_2O$, $Et_3N$
DMAP, $CH_2Cl_2$
RT, 48 h
117
CDI, DBU
DMF, RT, 16 h
118
Side Chain B
119
DBU, DMF
RT, 38 h
120
MeOH/HCl
acetone, RT
16 h
121
NCS, DMS
DIPEA, $CH_2Cl_2$
–45 to –40 °C
5 h
122
NFSI, KO$^t$Bu
THF:DMF (1:9)
123

Scheme 2.11

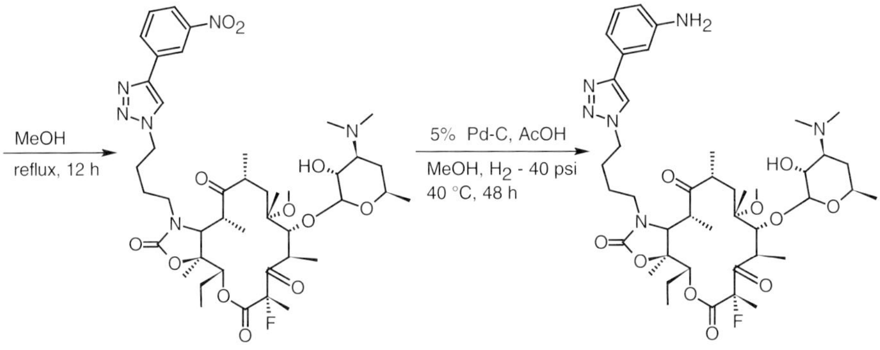


124

95

Scheme 2.11 (*Continued*)

Synthesis of side chain B of solithromycin

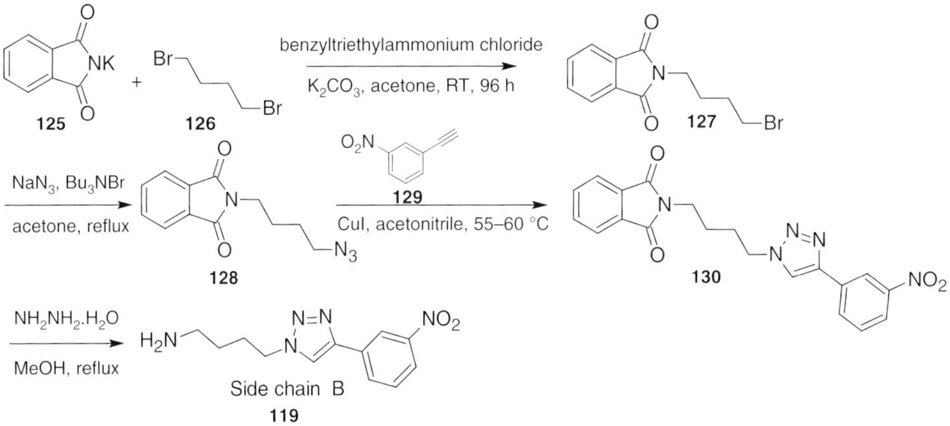


125 126 127 128 129 130 119

Scheme 2.12

2.3.3
**Cefatrizine**

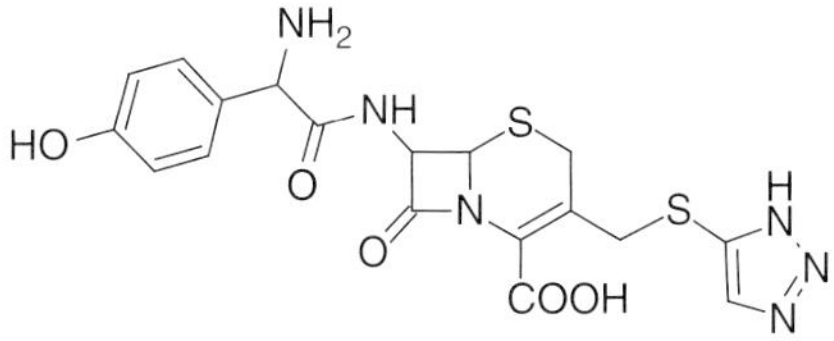

Cefatrizine
**96**

Cefatrizine is a second-generation semisynthetic cephalosporin antibiotic introduced in 1991 [134]. Cephalosporins are derivatives of 7-aminocephalosporic acid and are closely related to penicillins in structure. Cefatrizine is used to treat a large variety of bacterial infections, such as respiratory tract, ear, skin, and urinary tract infections (Scheme 2.13).

Synthesis of cefatrizine

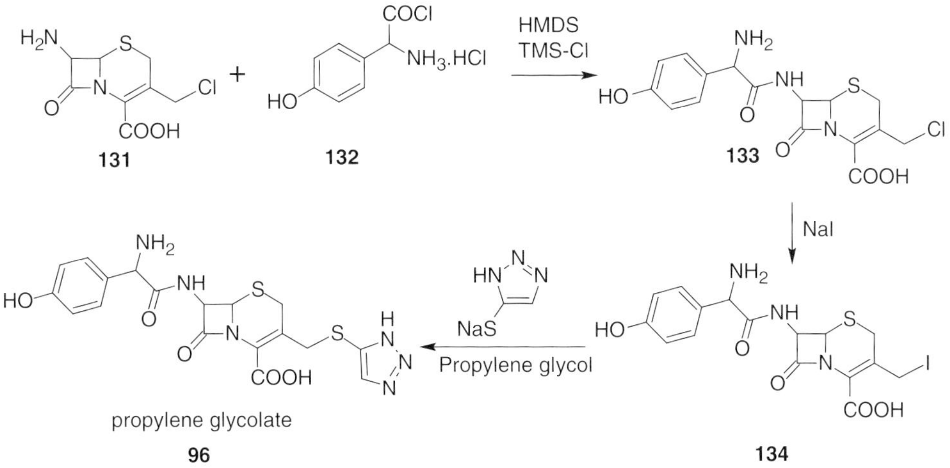


HMDS: Hexamethyldisilazane; TMS-Cl: Trimethylsilyl chloride

Scheme 2.13

### 2.3.4 Radezolid

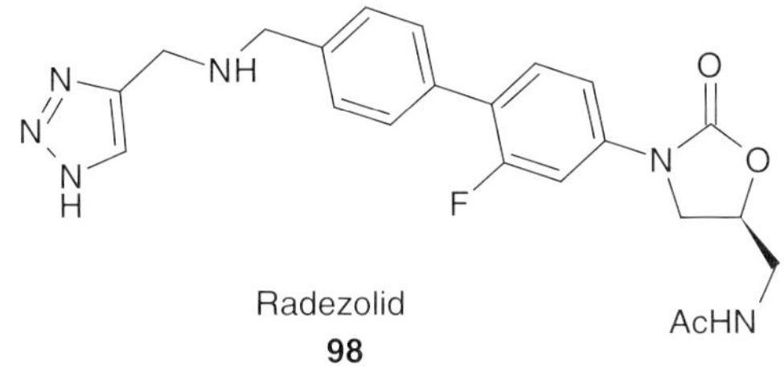

Radezolid
**98**

Radezolid is an oxazolidinone antibiotic used for treating acute bacterial skin and skin structure infections (ABSSSI) and community-acquired bacterial pneumonia (CABP) [135]. It has been given QIDP (Qualified Infectious Disease Product) status by USFDA (Scheme 2.14).

Synthesis of radezolid

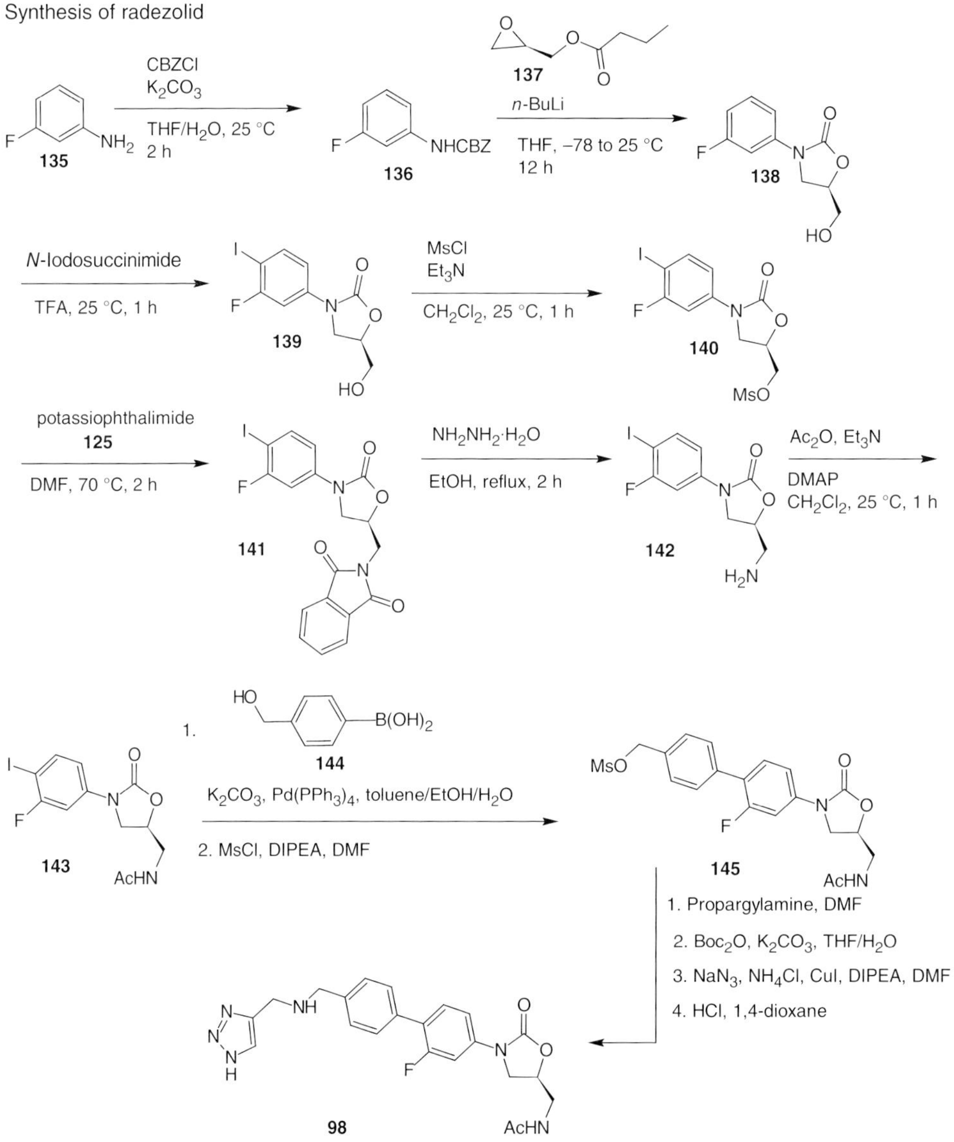


$Boc_2O$: Di-*tert*-butyl carbonate; CBZ: Carboxybenzl; DIPEA: Diisopropylethylamine; DMAP: Dimethylaminopyridine; DMF: Dimethylformamide; TFA: Trifluoroacetic acid; THF: Tetrahydrofuran

**Scheme 2.14**

### 2.3.5
### Molidustat

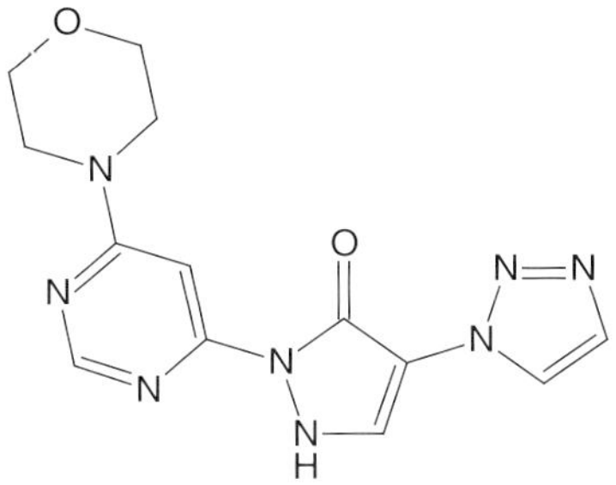


Molidustat
**99**

Molidustat (BAY 85-3934) is a novel inhibitor of hypoxia-inducible factor (HIF) prolyl hydroxylase (PH), which stimulates erythropoietin (EPO) production and the formation of red blood cells. Phase I data have shown that inhibition of HIF-PH by Molidustat results in an increase in endogenous production of EPO [136]. Molidustat is currently under clinical trials at Bayer for the treatment of patients suffering from renal anemia due to chronic kidney disease.

#### 2.3.5.1 Synthesis of Molidustat

Molidustat can be synthesized [137, 138] as described in Scheme 2.8. The compound **148** in the scheme is prepared by alkylation of triazole. Alternatively, they can also be prepared by a (3+2) cycloaddition of compounds **146** and **147** to a cycloadduct, followed by a retro-Diels–Alder reaction of the adduct to provide the desired compound **148** along with compound **149** (Scheme 2.15).

### 2.3.6
### Tradipitant

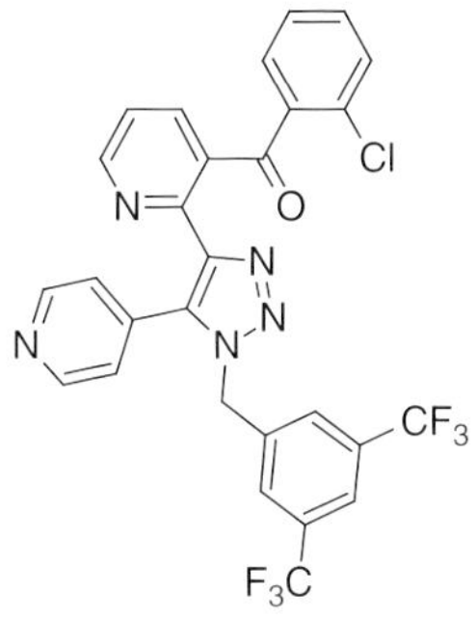


Tradipitant
**100**

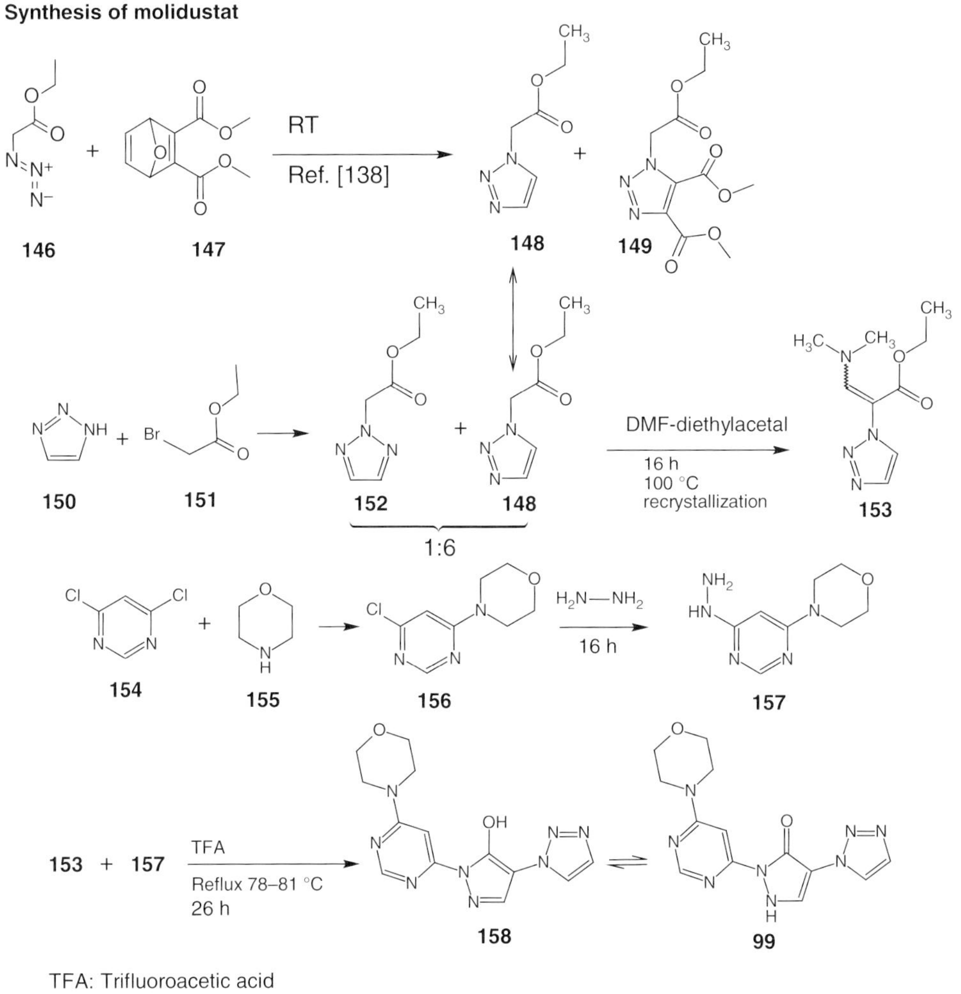


Scheme 2.15

Tradipitant [139] is safe and effective as monotherapy for chronic pruritus in patients with atopic dermatitis. It is a second-generation neurokinin-1 receptor antagonist used as a therapy for alcoholism.

Tradipitant has been synthesized [139] in three steps starting from intermediate as outlined in Scheme 2.9. Compound **161**, under directed-*ortho* metalation conditions, undergoes lithiation on the pyridine at a position ortho to the sulfonyl group in preference to ortho hydrogen on the phenyl sulfone moiety of compound **161** to provide compound **162,** which further reacts with 4-acetyl pyridine to give compound **163**. Compound **163** exists as a mixture of tautomers and geometrical isomers that undergo a (3+2) cycloaddition reaction with compound **160** to give tradipitant (Scheme 2.16).

Synthesis of Tradipitant

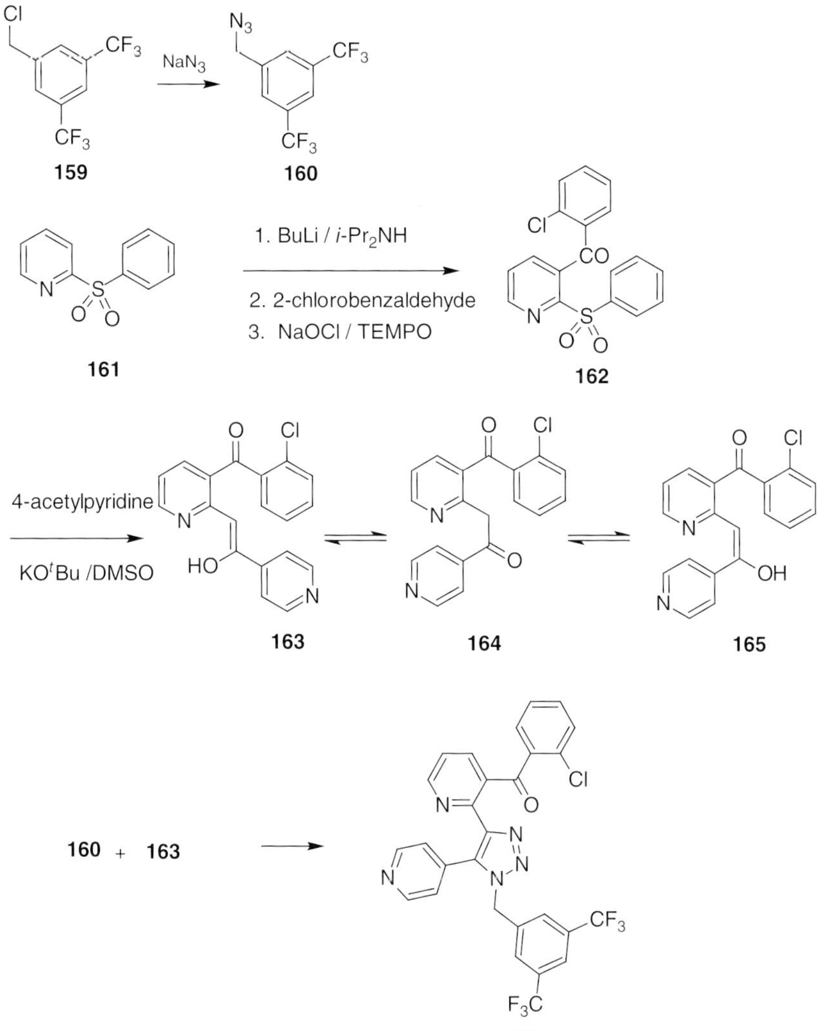


DMSO: Dimethylsulfoxide; TEMPO: 2,2,6,6-Tetramethylpiperidinyloxyl radical

Scheme 2.16

### 2.3.7 Carboxyamidotriazole

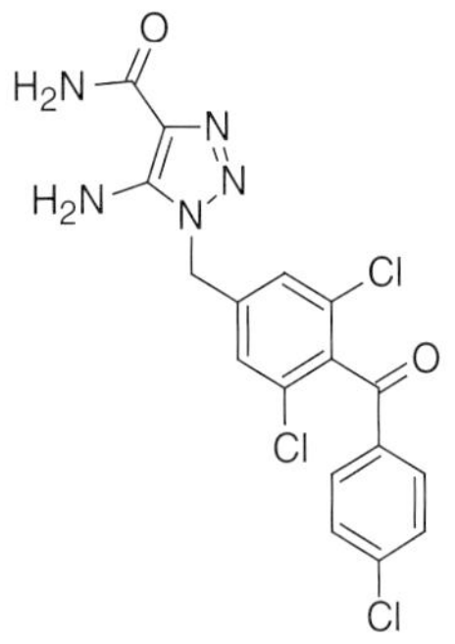

Carboxyamidotriazole
**101**

Carboxyamidotriazole, an inhibitor of calcium-mediated signal transduction, is one of the first cytostatic signal inhibitory anticancer drugs discovered [140, 141]. It was tested in patients suffering from solid cancers in Phase I, Phase II, and Phase III trials in National Cancer Institute. However, it was discontinued because of its lack of efficacy in human trials (Scheme 2.17).

### 2.3.8 Rufinamide

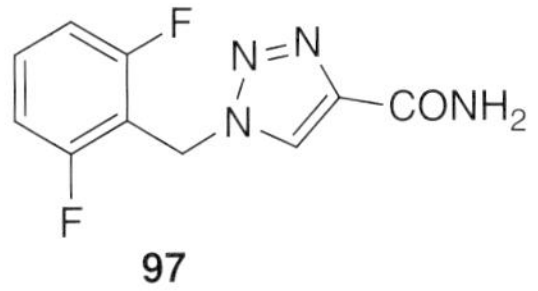

**97**

Rufinamide, a triazole derivative that is structurally distinct from currently marketed antiepileptic drugs, is a sodium channel blocker in development for the adjunctive treatment in patients with partial seizures and more effective for generalized seizures in children with Lennox–Gastaut syndrome, a severe form of childhood epilepsy [142] (Scheme 2.18 and Table 2.15).

#### 2.3.8.1 Rufinamide – Novartis Process

HCl is a side product of this reaction. In organic solvents, this polymerizes chloroacrylonitrile. In water, a two-phase system is observed where the HCl is dissolved in the water layer [143a].

Synthesis of carboxyamidotriazole

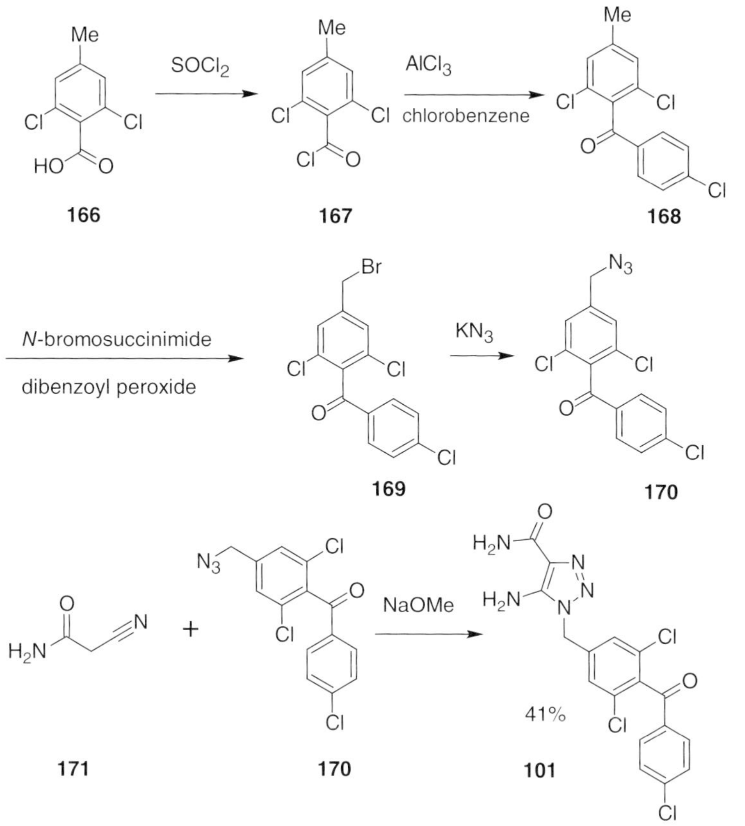

Scheme 2.17

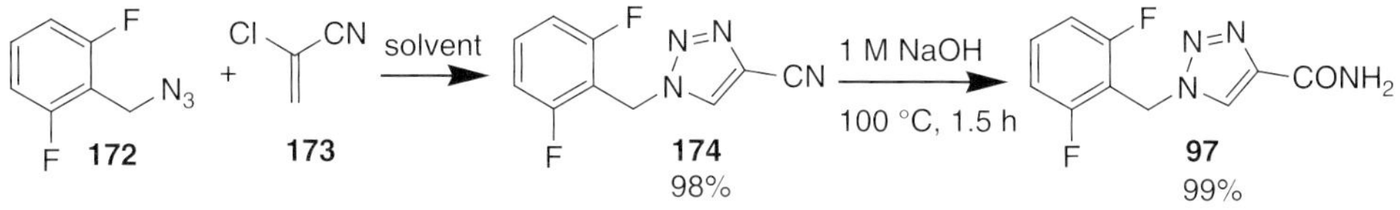

Scheme 2.18

Table 2.15 Effect of solvents on yield.

| Solvent | Temperature (°C) | Yield (%) |
|---|---|---|
| Neat | 80 | 72 |
| *n*-Heptane | 80 | 46 |
| Ethanol | 77 | 40 |
| Water | 80 | 98 |

### 2.3.8.2 An Efficient Synthesis of Rufinamide

A new solventless and metal-free synthesis of Rufinamide and analogs [143b, 144] is discussed (Schemes 2.19 and 2.20).

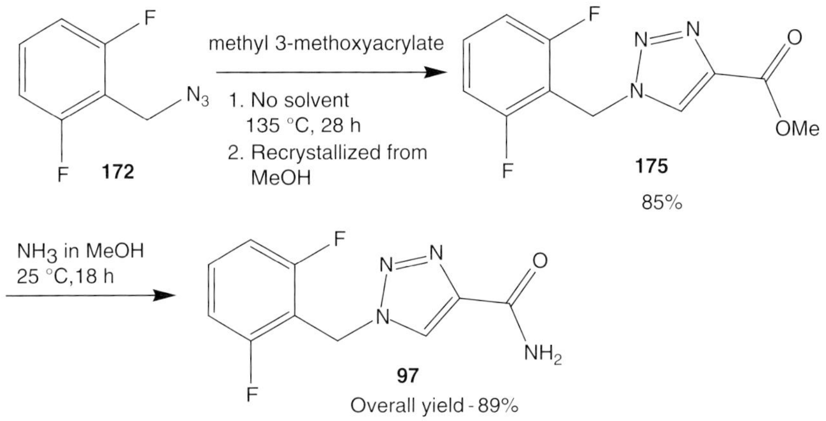


Scheme 2.19

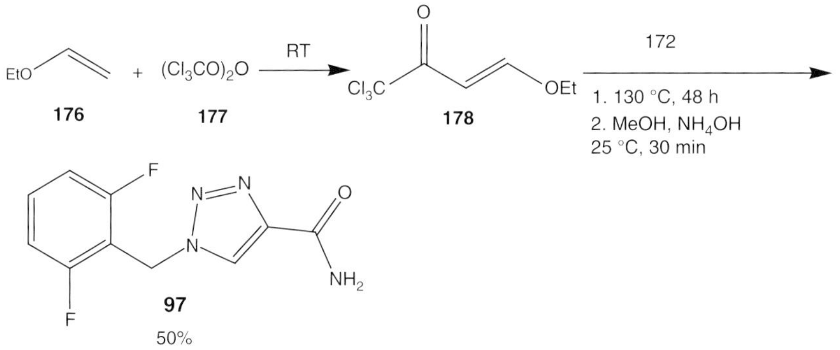


Scheme 2.20

Recently, a straightforward and versatile approach to the synthesis of Rufinamide analogs from alkyl halides via a one-pot three-component reaction has been reported [145].

Continuous-flow processing offers unprecedented opportunities to accelerate, integrate, simplify scale up, and automize chemical reactions in combination with an inherently safer and greener nature over traditional batch-based synthesis.

### 2.3.8.3 Continuous-Flow Total Synthesis of Rufinamide

Continuous-flow synthesis provides the opportunity to synthesize and consume the energetic organoazides without accumulation thereof [146]. The route is convergent and features copper tubing reactor–catalyzed cycloaddition reaction (Scheme 2.21).

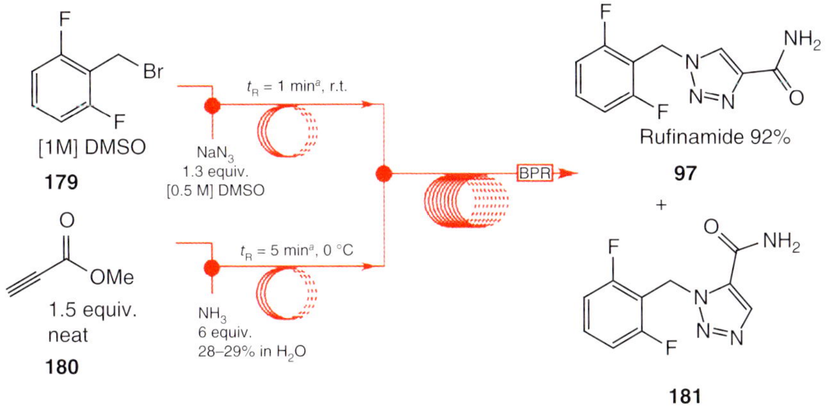


Scheme 2.21

In each of the three steps, safety hazards were very significantly minimized because organoazide intermediate was neither accumulated nor isolated. The otherwise costly and unstable propiolamide was prepared in line and could be employed without concerns of polymerization or storage and also without the need for further functional group manipulation. The overall average residence time was 11 min, and rufinamide was obtained in 92% overall yield and high selectivity.

Pressure-accelerated azide–alkyne cycloaddition for the synthesis of rufinamide precursor has been reported [147] and schematically represented as shown as follows (Scheme 2.22).

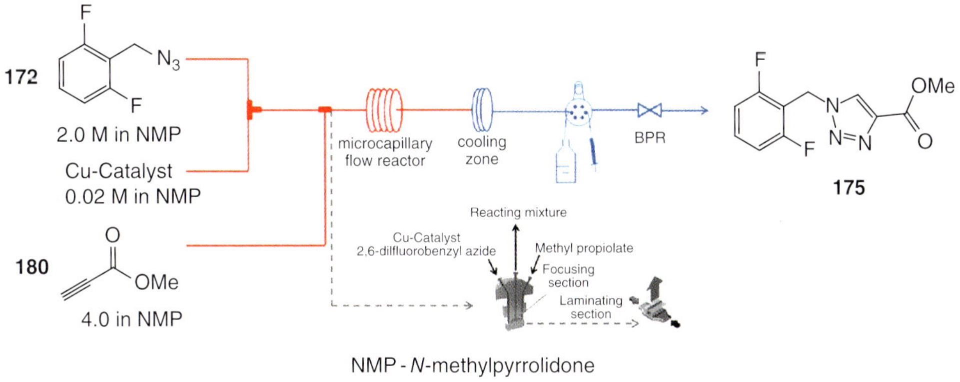


Scheme 2.22

Novel process window principles were applied to the 1,3 dipolar cycloaddition to yield the rufinamide precursor. The conditions for the activation of the reaction and the regioselectivity toward the 1,4-cycloadduct were optimized.

Life-cycle assessment of multistep rufinamide synthesis from isolated reactions in a batch to a continuous microreactor network is reported [148]. A continuous solvent- and catalyst-free flow process utilizing relatively inexpensive and green dipolarophile, (*E*)-methyl 3-methoxyacrylate, was shown. Here, benzyl chloride, instead of very reactive benzyl bromide, was employed for azide formation, which was produced utilizing benzyl alcohol and hydrogen chloride.

Flow chemistry [149] as a versatile tool for the synthesis of triazoles has been reviewed. Catalyst-free continuous-flow reactions between azides and further dipolarophiles complement the copper-catalyzed transformations and have been proven to be convenient for the synthesis of diversely substituted triazoles. The safety aspects associated with the handling of azides have been recognized as one of the most important driving forces behind the advent of continuous-flow processes for triazole synthesis. Better heat-transfer characteristics and increased mixing efficiency can be accounted for an improved safety profile as well as for the elimination of common by-products such as diacetylenes and bis-triazoles.

Although, click chemistry has several limitations, the importance of click chemistry in medicinal chemistry is undeniable. Click chemistry has proven to be a powerful tool in the pharmaceutical field, in particular, in the early discovery phase. However, given the fact that 1,2,3-triazole may be not a suitable moiety in the design of drugs, other cores exhibiting more biocompatibility and that are similar to 1,2,3-triazole can be designed based on the concept of click chemistry discovery.

## References


1. (a) Huisgen, R. (1963) *Angew. Chem. Int. Ed. Engl.*, **2**, 565; (b) Huisgen, R. (1963) *Angew. Chem. Int. Ed. Engl.*, **2**, 633; (c) Huisgen, R. (1984) *1,3-Dipolar Cycloaddition Chemistry*, vol. 1, Chapter 1, Wiley-Interscience, New York, pp. 1–176.
2. (a) Kolb, H.C., Finn, M.G., and Sharpless, K.B. (2001) *Angew. Chem. Int. Ed.*, **40**, 2004–2021; (b) Rostovtsev, V.V., Green, L.G., Fokin, V.V., and Sharpless, K.B. (2002) *Angew. Chem. Int. Ed.*, **41**, 2596; (c) Krasinski, A., Fokin, V.V., and Sharpless, K.B. (2004) *Org. Lett.*, **6**, 1237; (d) Zhang, L., Chen, X., Sun, H.H.Y., Williams, I.D., Sharpless, K.B., Fokin, V.V., and Jia, G. (2005) *J. Am. Chem. Soc.*, **127**, 15998.
3. Tornoe, C.W., Christensen, C., and Meldal, M. (2002) *J. Org. Chem.*, **67**, 3057.
4. (a) Ramachary, D.B., Ramakumar, K., and Narayanan, V.V. (2008) *Chem. Eur. J.*, **14**, 9143; (b) Ramachary, D.B. and Shashank, A.B. (2013) *Chem. Eur. J.*, **19**, 13175; (c) Ramachary, D.B., Shashank, A.B., and Karthik, S. (2014) *Angew. Chem. Int. Ed.*, **53**, 10420; (d) Shashank, A.B., Karthik, S., Madhavachary, R., and Ramachary, D.B. (2014) *Chem. Eur. J.*, **20**, 16877.
5. Belkheira, M., El Abed, D., Pons, J.M., and Bressy, C. (2011) *Chem. Eur. J.*, **17**, 12917.
6. (a) Danence, L.J.T., Gao, Y., Li, M., Huang, Y., and Wang, J. (2011) *Chem. Eur. J.*, **17**, 3584; (b) Wang, L., Peng, S.Y., Danence, L.J.T., Gao, Y., and Wang, J. (2012) *Chem. Eur. J.*, **18**, 6088; (c) Li, W., Jia, Q.F., Du, Z.Y., and Wang, J. (2013) *Chem. Commun.*, **49**, 10187; (d) Yeung, D.K.J., Gao, T., Huang, J.Y., San, S.F., Guo, H.B., and Wang, J. (2013) *Green Chem.*, **15**, 2384; (e) Wang, L., Huang, J.Y., Gong, X.J., and Wang, J. (2013) *Chem. Eur. J.*, **19**, 7555;

(f) Li, W., Du, Z.Y., Huang, J.Y., Jia, Q.F., Zhang, K., and Wang, J. (2013) *Green Chem.*, **16**, 3003; (g) Li, W. and Wang, J. (2014) *Angew. Chem. Int. Ed.*, **53**, 14186; (h) Li, W., Du, Z.Y., Zhang, K., and Wang, J. (2015) *Green Chem.*, **17**, 781.

7. (a) Seus, S., Goncalvas, L.C., Deobald, A.M., Savegnago, L., Alves, D., and Paixao, M.W. (2012) *Tetrahedron*, **68**, 10456; (b) Seus, N., Goldani, B., Lenardo, E.J., Savegnago, L., Paixao, M.W., and Alves, D. (2014) *Eur. J. Org. Chem.*, **2014**, 1059; (c) Thomas, J., John, J., Parekh, N., and Dehaen, W. (2014) *Angew. Chem. Int. Ed.*, **53**, 10155; (d) Ramasastry, S.S.V. (2014) *Angew. Chem. Int. Ed.*, **53**, 14310.
8. Wei, F., Li, H., Song, C., Ma, Y., Zhou, L., Tung, C.-H., and Xu, Z. (2015) *Org. Lett.*, **17**, 2860.
9. Jia, Q., Yang, G., Chen, L., Du, Z., Wei, J., Zhong, Y., and Wang, J. (2015) *Eur. J. Org. Chem.*, **2015**, 3435.
10. Totobenazara, J. and Burke, A.J. (2015) *Tetrahedron Lett.*, **56**, 2853.
11. Musumeci, F., Schenone, S., Desogus, A., Nieddu, W., Deodato, D., and Botta, L. (2015) *Curr. Med. Chem.*, **22**, 1.
12. Wang, X., Huang, B., Liu, X., and Zhan, P. (2015) *Drug Discovery Today*. doi: 10.1016/j.drudis.2015.08.004
13. Lauria, A., Delisi, R., Mingoia, F., Terenzi, A., Martorana, A., Barone, G., and Almerico, A.M. (2014) *Eur. J. Org Chem.*, **2014** (16), 3289–3306.
14. Ferreira, S.B., Sodero, A.C., Cardoso, M.F., Lima, E.S., Kaiser, C.R., Silva, F.P., and Ferreira, V.P. (2010) *J. Med. Chem.*, **53**, 2364–2375.
15. Bourne, Y., Kolb, H.C., Radic, Z., Sharpless, K.B., Taylor, P., and Marchot, P. (2004) *Proc. Natl. Acad. Sci. U.S.A.*, **101**, 1449–1454.
16. Whiting, M., Muldoon, J., Lin, Y.C., Silverman, S.M., Lindstrom, W., Olsen, A.J., Kolb, H.C., Finn, M.G., Sharpless, K.B., Elder, J.H., and Fokin, V.V. (2006) *Angew. Chem. Int. Ed.*, **45** (9), 1435–1439.
17. Tron, G.C., Pirali, T., and Billington, R.A. (2008) *Med. Res. Rev.*, **28**, 278–308.
18. Pagliai, F., Pirali, T., Del Grosso, E., Di Brisco, R., Tron, G.C., Sorba, G., and Genazzani, A.A. (2006) *J. Med. Chem.*, **49**, 467–470.
19. Pirali, T., Gatti, S., Di Brisco, R., Tacchi, S., Zaninetti, R., Brunelli, E., Massarotti, A., Sorba, G., Canonico, P., Moro, L., Genazzani, A.A., Tron, G.C., and Billington, R.A. (2007) *ChemMedChem*, **2**, 437–440.
20. Wadsworth, H.J., Jenkins, S.M., Orlek, B.S., Cassidy Clark, M.S.G., Brown, F., Riley, G.J., Graves, D., Hawkins, J., and Naylor, C.B. (1992) *J. Med. Chem.*, **35**, 1280–1290.
21. Moltzen, E.K., Pedersen, H., Boegesoe, K.P., Meier, E., Frederiksen, K., Sanchez, C., and Lemboel, H.L. (1994) *J. Med. Chem.*, **37**, 4085–4099.
22. Bennett, I.S., Brooks, G., Broom, N.P., Calvert, S.H., Coleman, K., and Francois, I. (1991) *J. Antibiot.*, **44**, 969–977.
23. Bennett, S., Broom, N.P., Bruton, G., Calvert, S., Clarke, B.P., Coleman, K., Edmondson, R., Edwards, P., Jones, D., Osborne, N.F., and Walker, G. (1991) *J. Antibiot.*, **44**, 331–337.
24. Kume, M., Kubota, T., Kimura, Y., Nakashimizu, H., Motokawa, K., and Nakano, M. (1993) *J. Antibiot.*, **46**, 177–192.
25. Lewis, W.G., Green, L., Grynszpan, G., Radiac, Z., Carlier, P.R., Taylor, P., Finn, M.G., and Sharpless, K.B. (2002) *Angew. Chem.*, **114**, 1095–1099.
26. Manetssch, R., Krasinski, A., Radie, Z., Raushel, J., Taylor, P., Sharpless, K.B., and Kolb, H.C. (2004) *J. Am. Chem. Soc.*, **126**, 12809–12818.
27. Krasinski, A., Radie, Z., Manetsch, R., Raushel, J., Taylor, P., Sharpless, K.B., and Kolb, H.C. (2005) *J. Am. Chem. Soc.*, **127**, 6686–6692.
28. Macharla, V.P., Colasson, B., Lee, L.V., Roper, S., Sharpless, K.B., Wong, C.H., and Kolb, H.C. (2004) *Angew. Chem. Int. Ed.*, **44**, 116–120.
29. Oueis, E., Santoni, G., Ronco, C., Syzgantseva, O., Tognetti, V., Joubert, L., Romieu, A., Weik, M., Jean, L., Sabot, C., Nachon, F., and Renard, P.Y. (2014) *Org. Biomol. Chem.*, **12**, 156–161.

30. Ronco, C., Carletti, E., Colletier, J.P., Weik, M., Nachon, F., Jean, L., and Renard, P.Y. (2012) *ChemMedChem*, **7**, 400–405.
31. (a) Camps, P., El Achab, R., Gorbig, D.M., Morral, J., Munoz-Torrero, D., Badia, A., Banos, J.E., Vivas, N.M., Barril, X., Orozio, M., and Luque, J. (1999) *J. Med. Chem.*, **42**, 3227–3242; (b) Camps, P., El Achab, R., Morral, J., Munoz-Torrero, D., Badia, A., Banos, J.E., Vivas, N.M., Barril, X., Orozio, M., and Luque, J. (2000) *J. Med. Chem.*, **43**, 4657–4666.
32. (a) Ronco, C., Sorin, G., Nachon, F., Foucault, R., Jean, L., Romieu, A., and Renard, P.Y. (2009) *Bioorg. Med. Chem.*, **17** (13), 4523–4536; (b) Nachon, F., Carletti, E., Ronco, C., Trovaslet, M., Nicolet, Y., Jean, L., and Renard, P.Y. (2013) *Biochem. J*, **453** (3), 393–399.
33. Cavalli, A., Bolognesi, M.L., Minarini, A., Rosini, M., Tumiatti, V., Recanatini, M., and Melchiorre, C. (2008) *J. Med. Chem.*, **51** (3), 347–372.
34. Viayna, E., Gómez, T., Galdeano, C., Ramírez, L., Ratia, M., Badia, A., Clos, M.V., Verdaguer, E., Junyent, F., Camins, A., Pallàs, M., Bartolini, M., Mancini, F., Andrisano, V., Arce, M.P., Rodriguez-Franco, M.I., Bidon-Chanal, A., Luque, F.J., Camps, P., and Muñoz-Torrero, D. (2010) *ChemMedChem*, **5**, 1855–1870.
35. Musumeci, F., Schenone, S., Desogus, A., Nieddu, E., Deodato, D., and Botta, L. (2015) *Curr. Med. Chem.*, **22** (17), 2022–2050.
36. Alterio, V., Di Fiore, A., D'Ambrosio, K., Supuran, C.T., and De Simone, G. (2012) *Chem. Rev.*, **112**, 4421–4468.
37. Tars, K., Vullo, D., Kazaks, A., Leitans, J., Lends, A., Grandane, A., Zalubovskis, R., Scozzafava, A., and Supuran, C.T. (2013) *J. Med. Chem.*, **56** (1), 293–300.
38. Grandane, A., Tanc, M., Zalubovskis, R., and Supuran, C.T. (2014) *Bioorg. Med. Chem. Lett.*, **24**, 1256–1260.
39. Pippione, A.C., Dosio, F., Ducime, A., Federico, A., Martina, K., Sainas, S., Frølund, B., Gooyit, M., Janda, K.D., Boschia, D., and Lollia, M.L. (2015) *Med. Chem. Commun.*, **6**, 1285–1292.
40. Allen, J.E., Adjei, O., Bain, O., Hoerauf, A., Hoffmann, W.H., Makepeace, B.L., Schulz-Key, H., Tanya, V.N., Trees, A.J., Wanji, S., and Taylor, D.W. (2008) *PLoS Negl. Trop. Dis.*, **2** (4), e217.
41. Lustigman, S., Prichard, R.K., Gazzinelli, A., Grant, W.N., Boatin, B.A., McCarthy, J.S., and Basáñez, M.G. (2007) *PLoS Negl. Trop. Dis.*, **6** (4), e1582.
42. Gloeckner, C., Garner, A.L., Mersha, F., Oksov, Y., Tricoche, N., Eubanks, L.M., Lustigman, S., Kaufmann, G.F., and Janda, K.D. (2010) *Proc. Natl. Acad. Sci. U.S.A.*, **107** (8), 3424–3429.
43. Gooyit, M., Tricoche, N., Lustigman, S., and Janda, K.D. (2014) *J. Med. Chem.*, **57** (13), 5792–5799.
44. Xu, H. and Jones, L.H. (2015) *Pharm. Pat. Anal.*, **4** (2), 109–119.
45. Accenda Tech Co., Ltd. (2011) Patent WO/2011/003365.
46. Technion Research & Development Foundation Ltd. (2010) Patent WO/2010/113151.
47. Thirumurugan, P., Matosiuk, D., and Jozwiak, K. (2013) *Chem. Rev.*, **113** (7), 4905–4979.
48. da Silva, F.C., Cardoso, M.F.C., Ferreira, P.G., and Ferreira, V.F. (2015) *Top. Heterocycl. Chem.*, **40**, 117–166.
49. Naik, P., Wagulde, S., Kshetri, S., Somani, R., and Kale, M. (2014) *Indo Am. J. Pharm. Res.*, **4** (8), 3435–3446.
50. Haider, S., Sarwar Alam, M., and Hamid, H. (2014) *Inflamm. Cell Signal.*, **1**, e95.
51. (a) Ma, N., Wang, Y., Zhao, B.X., Ye, W.C., and Jiang, S. (2015) *Drug Des., Dev. Ther.*, **9**, 1585–1599 and references cited therein; (b) Ahmad Dar, M., Shrivastava, S., and Firdoos Iqbal, P. (2015) *World J. Pharm. Res.*, **4** (1), 1949–1975 and references cited therein.
52. Jordan, A., Hadfield, J.A., Lawrence, N.J., and McGown, A.T. (1998) *Med. Res. Rev.*, **18** (4), 259–296.
53. Reddy, D.M., Srinivas, J., Chashoo, G., Saxena, A.K., and Sampath Kumar, H.M. (2011) *Eur. J. Med. Chem.*, **46** (6), 1983–1991.
54. Yang, J.S., Song, D., Ko, W.J., Kim, B., Kim, B.K., Park, S.K., Won, M., Lee, K.,

Lee, K., Kim, H.M., and Han, G. (2013) *Eur. J. Med. Chem.*, **63**, 621–628.
55. Bolden, J.E., Peart, M.J., and Johnstone, R.W. (2006) *Nat. Rev. Drug Discovery*, **5** (9), 769–784.
56. Emanuele, S., Lauricella, M., and Tesoriere, G. (2008) *Int. J. Oncol.*, **33** (4), 637–646.
57. Haberland, M., Montgomery, R.L., and Olson, E.N. (2009) *Nat. Rev. Genet.*, **10** (1), 32–42.
58. Mottamal, M., Zheng, S., Huang, T.L., and Wang, G. (2015) *Molecules*, **20** (3), 3898–3941.
59. Shen, J., Woodward, R., Kedenburg, J.P., Liu, X., Chen, M., Fang, L., Sun, D., and Wang, P.G. (2008) *J. Med. Chem.*, **51** (23), 7417–7427.
60. Hou, J., Feng, C., Li, Z., Fang, Q., Wang, H., Guoxian, G., Shi, Y., Liu, P., Xu, F., Zheng, Y., Shen, J., and Wang, P. (2011) *Eur. J. Med. Chem.*, **46** (8), 3192–3200.
61. Hou, J., Li, Z., Fang, Q., Feng, C., Zhang, H., Guo, W., Wang, H., Gu, G., Tian, Y., Liu, P., Liu, R., Lin, J., Shi, Y.K., Yin, Z., Shen, J., and Wang, P.G. (2012) *J. Med. Chem.*, **55** (7), 3066–3075.
62. Suzuki, T., Ota, Y., Ri, M., Bando, M., Gotoh, A., Itoh, Y., Tsumoto, H., Tatum, P.R., Mizukami, T., Nakagawa, H., Iida, S., Ueda, R., Shirahige, K., and Miyata, N. (2012) *J. Med. Chem.*, **55** (22), 9562–9575.
63. Suzuki, T., Muto, N., Bando, M., Itoh, Y., Masaki, A., Ri, M., Ota, Y., Nakagawa, H., Iida, S., Shirahige, K., and Miyata, N. (2014) *ChemMedChem*, **9** (3), 657–664.
64. Summy, J.M. and Gallick, G.E. (2003) *Cancer Metastasis Rev.*, **22** (4), 337–358.
65. Martin, G.S. (2001) *Nat. Rev. Mol. Cell Biol.*, **2** (6), 467–475.
66. Schlessinger, J. (2000) *Cell*, **100**, 293–296.
67. Frame, M.C. (2002) *Biochim. Biophys. Acta*, **1602** (2), 114–130.
68. Fizazi, K. (2007) *Ann. Oncol.*, **18** (11), 1765–1773.
69. Kalesh, K.A., Liu, K., and Yao, S.Q. (2009) *Org. Biomol. Chem.*, **7** (24), 5129–5136.
70. Perez, E.A. (2009) *Mol. Cancer Ther.*, **8** (8), 2086–2095.
71. Jordan, M.A. and Wilson, L. (2004) *Nat. Rev. Cancer*, **4** (4), 253–265.
72. Dumontet, C. and Jordan, M.A. (2010) *Nat. Rev. Drug Discovery*, **9** (10), 790–803.
73. Stefely, J.A., Palchaudhuri, R., Miller, P.A., Peterson, R.J., Moraski, G.C., Hergenrother, P.J., and Miller, M.J. (2010) *J. Med. Chem.*, **53** (8), 3389–3395.
74. Taddei, M., Ferrini, S., Giannotti, L., Corsi, M., Manetti, F., Giannini, G., Vesci, L., Milazzo, F.M., Alloatti, D., Guglielmi, M.B., Castorina, M., Cervoni, M.L., Barbarino, M., Foderà, R., Carollo, V., Pisano, C., Armaroli, S., and Cabri, W. (2014) *J. Med. Chem.*, **57** (6), 2258–2274.
75. Ferrini, S., Chandanshive, J.Z., Lena, S., Franchini, M.C., Giannini, G., Tafi, A., and Taddei, M. (2015) *J. Org. Chem.*, **80** (5), 2562–2572.
76. Mathew, R., Karantza-Wadsworth, V., and White, E. (2007) *Nat. Rev. Cancer*, **7** (12), 961–967.
77. (a) Sui, X., Chen, R., Wang, Z., Huang, Z., Kong, N., Zhang, M., Han, W., Lou, F., Yang, J., Zhang, Q., Wang, X., He, C., and Pan, H. (2013) *Cell Death Dis.*, **4**, e838; (b) Yang, Z.J., Chee, C.E., Huang, S., and Sinicrope, F.A. (2011) *Mol. Cancer Ther.*, **10** (9), 1533–1541.
78. Ajay, A., Gupt, M.P., Devender, N., and Tripathi, R.P. (2012) *Mol. Diversity*, **16** (2), 335–350.
79. Hamidullah Khan, Saini, K.S., Ajay, A., Devender, N., Bhattacharjee, A., Das, S., Dwivedi, S., Gupt, M.P., Bora, H.K., Mitra, K., Tripathi, R.P., and Konwar, R. (2015) *Int. J. Biochem. Cell Biol.*, **65**, 275–287.
80. Zi, C.-T., Li, G.-T., Li, Y., Zhou, J., Ding, Z.-T., Jiang, Z.-H., and Hu, J.-M. (2015) *Nat. Prod. Bioprospect.*, **5** (2), 83–90.
81. Wei, G., Luan, W., Wang, S., Cui, S., Li, F., Liu, Y., Yang, L., and Cheng, M. (2015) *Org. Biomol. Chem.*, **13**, 1507–1514.
82. Lipeevaa, A.V., Pokrovsky, M.A., Baev, D.S., Shakirova, M.M., Bagryanskaya, I.Y., Tolstikova, T.G., Pokrovsky, A.G.,

and Shults, E.E. (2015) *Eur. J. Med. Chem.*, **100**, 119–128.
83. Romeo, R., Carnovale, C., Giofrè, S.V., Chiacchio, M.A., Garozzo, A., Amata, E., Romeo, G., and Chiacchio, U. (2015) *Beilstein J. Org. Chem.*, **11**, 328–334.
84. Jin, X., Yan, L., Li, H.J., Wang, R.L., Hu, Z.L., Jiang, Y.Y., Cao, Y.B., Yan, T.H., and Sun, Q.Y. (2015) *Anticancer Agents Med. Chem.*, **15** (1), 89–98.
85. Pertino, M.W., Verdugo, V., Theoduloz, C., and Schmeda-Hirschmann, G. (2014) *Molecules*, **19** (2), 2523–2535.
86. Pertino, M.W., Theoduloz, C., Butassi, E., Zacchino, S., and Schmeda-Hirschmann, G. (2015) *Molecules*, **20** (5), 8666–8686.
87. Malic, S.R. and Mesic, A. (2005) *Curr. Med. Chem.*, **22** (12), 1462–1499.
88. Gonzaga, D.T.G., Da Rocha, D.R., Da Silva, F.D., and Ferreira, V.F. (2013) *Curr. Top. Med. Chem.*, **13** (22), 2850–2865.
89. Velazquez, S., Alvarez, R., Prez, C., Gago, F., De Clercq, E., Balzarini, J., and Camarasa, M.-J. (1998) *Antiviral Chem. Chemother.*, **9**, 481–489.
90. Sluis-Cremer, N., Hamamouch, N., San Felix, A., Velazquez, S., Balzarini, J., and Camarasa, M.J. (2006) *J. Med. Chem.*, **49** (16), 4834–4841.
91. Zhu. X. F. *Nucleosides Nucleotides Nucleic Acids* **19**(3), 651-690.
92. Broggi, J., Kumamoto, H., Berteina-Raboin, S., Nolan, S.P., and Agrofoglio, L.A. (2009) *Eur. J. Org. Chem.*, **12**, 1880–1888.
93. Broggia, J., Jouberta, N., Díez-Gonzalezb, S., Berteina-Raboina, S., Zevacoc, T., Nolanb, S.P., and Agrofoglio, L.A. (2009) *Tetrahedron*, **65** (6), 1162–1170.
94. Broggi, J., Díez-González, S., Petersen, J.L., Berteina-Raboin, S., Nolan, S.P., and Agrofoglio, L.A. (2008) *Synthesis*, **2008** (1), 141–148.
95. Broggi, J., Joubert, N., Aucagne, V., Berteina-Raboin, S., Diez-Gonzalez, S., Nolan, S., Topalis, D., Deville-Bonne, D., Balzarini, J., Neyts, J., Andrei, G., Snoeck, R., and Agrofoglio, L.A. (2007) *Nucleosides Nucleotides Nucleic Acids*, **26** (10-12), 1391–1394.
96. Joubert, N., Schinazi, R.F., and Agrofoglio, L.A. (2005) *Tetrahedron*, **61** (49), 11744–11750.
97. De Clercq, E. (1985) *Antimicrob. Agents Chemother.*, **28** (1), 84–89.
98. Agrofoglio, L., Suhasa, E., Faresea, A., Condoma, R., Richard Challandb, S., Earlc, R.A., and Guedja, R. (2005) *Tetrahedron*, **50** (36), 11744–11750.
99. Cho, J.H., Bernard, D.L., Sidwell, R.W., Kern, E.R., and Chu, C.K. (2006) *J. Med. Chem.*, **49** (3), 1140–1148.
100. Deiters, A. and Martin, S.F. (2004) *Chem. Rev.*, **104** (5), 2199–2238.
101. Yang, M., Ye, W., and Schneller, S.W. (2004) *J. Org. Chem.*, **69**, 3993–3996.
102. Caselli, E., Romagnoli, C., Vahabi, R., Taracila, M.A., Bonomo, R.A., and Prati, F. (2005) *J. Med. Chem.*, **58** (14), 5445–5458.
103. Jadhav, R.P., Raundal, H.N., Patil, A.A., and Bobade, V.D. (2015) *J. Saudi Chem. Soc.*, in press.
104. Khedar, P., Pericherla, K., Singh, R.P., Jha, P.N., and Kumar, A. (2015) *Med. Chem. Res.*, **24** (7), 3117–3126.
105. Satyanarayana Reddy, B., Srinivasa Reddy, B., Rao, C., and Subha, M.C.S. (2015) *World J. Pharm. Pharm. Sci.*, **4** (3), 1084–1090.
106. Abdel-Wahab, B.F., Mohamed, H.A., and Awad, G.E.A. (2015) *Eur. Chem. Bull.*, **4** (2), 106–109.
107. Keri, R.S., Patil, S.A., Budagumpi, S., and Nagaraja, B.M. (2015) *Chem. Biol. Drug Des.*, **86** (4), 410–423.
108. Surineni, G., Yogeeswari, P., Sriram, D., and Kantevari, S. (2015) *Med. Chem. Res.*, **24**, 1298–1309.
109. Altimari, J.M., Hockey, S.C., Boshoff, H.I., Sajid, A., and Henderson, L.C. (2015) *Chem. Med. Chem. Commun.*, **10**, 787–791.
110. Mukherjee, T. and Boshoff, H. (2011) *Future Med. Chem.*, **3**, 1427–1454.
111. Tawari, N.R. and Degani, M.S. (2010) *J. Comput. Chem.*, **31**, 739–751.
112. Addla, D., Jallapally, A., Gurram, D., Yogeeswari, P., Sriram, D., and Kantevari, S. (2014) *Bioorg. Med. Chem. Lett.*, **24** (8), 1974–1979.
113. Emmadi, N.R., Bingi, C., Kotapalli, S.S., Ummanni, R., Nanubolu, J.B., and

Atmakur, K. (2015) *Bioorg. Med. Chem. Lett.*, **25** (15), 2918–2922.

114. Sumana, P., Dayakara, C., Rajkumara, K., Yashwanthb, B., Yogeeswaric, P., Sriramc, D., Raob, J.V., and Rajua, B.C. (2015) *Bioorg. Med. Chem. Lett.*, **25** (11), 2390–2394.
115. Yempala, T., Sridevi, J.P., Yogeeswari, P., Sriram, D., and Kantevari, S. (2014) *Eur. J. Med. Chem.*, **71**, 160–167.
116. Zeng, J., Lü, X., Zeng, C., Hu, L., and Zhong, R. (2009) *Chin. J. Chem.*, **27** (5), 953–962.
117. Chen, J., Liu, C.-F., Yang, C.-W., Zeng, C.-C., Liu, W., and Hu, L.-M. (2015) *Med. Chem. Res.*, **24**, 2950–2959.
118. Gross, G. and Drescher, K. (2012) *Handbook of Experimental Pharmacology*, vol. 213, Springer, pp. 167–210.
119. Joyce, J.N. and Millan, M.J. (2007) *Curr. Opin. Pharmacol.*, **7** (1), 100–105.
120. Bézard, E., Ferry, S., Mach, U., Stark, H., Leriche, L., Boraud, T., Gross, C., and Sokoloff, P. (2003) *Nat. Med.*, **9**, 762–767.
121. Newman, A.H., Blaylock, B.L., Nader, M.A., Bergman, J., Sibley, D.R., and Skolnick, P. (2012) *Biochem. Pharmacol.*, **84** (7), 882–890.
122. Newman, A.H., Grundt, P., and Nader, M.A. (2005) *J. Med. Chem.*, **48**, 3663–3679.
123. Heidbreder, C.A. and Newman, A.H. (2010) *Ann. N.Y. Acad. Sci.*, **1187**, 4–34.
124. Keck, T.M., Banala, A.K., Slack, R.D., Burzynski, C., Bonifazi, A., Okunola-Bakare, O.M., Moore, M., Deschamps, J.R., Rais, R., Slusher, B.S., and Newman, A.H. (2015) *Bioorg. Med. Chem.*, **23** (14), 4000–4012.
125. Asif, M. (2014) *Knowl. Res.*, **1** (1), 68–78.
126. Asif, M. (2015) *Chem. Int.*, **1** (4), 174–183.
127. Asif, M. (2015) *Acta Velit*, **1** (2), 1–11.
128. Gori, A., Wang, C.-I.A., Harvey, P.J., Rosengren, K.J., Bhola, R.F., Gelmi, M.L., Longhi, R., Christie, M.J., Lewis, R.J., Alewood, P.F., and Brust, A. (2015) *Angew. Chem. Int. Ed.*, **54** (4), 1361–1364.
129. Skalweit, M.J. (2015) *Drug Des., Dev. Ther.*, **9**, 2919–2925.
130. Taiho Pharmaceutical CompanyLimited (1985) Patent US4507239.
131. (a) Narsihma Reddy, B., Narayan, G.K.A.S.S., Hunnur, R.K., Kaushik, V.K., Badarinath, G., Mamatha, B., and Aminul, I. (2012) *Phram. Lett.*, **4** (2), 674–682; (b) Maiti, S.N., Spevak, P., Reddy, A.V.N., Micetichl, R.G., Ishida, N., Miyake, Y., and Ogawa, K. (1994) *Eur. J. Med. Chem.*, **29**, 795–798; (c) T. P. C. Limited (1985) Patent US4562073.
132. Yang, L.I., Ligong, C., Guoyi, B., Wang, D., and Fenglei, W. (2002) *Trans. Tianjin Univ.*, **8**, 33.
133. (a) Cempra Pharmaceuticals, Inc. (2014) Patent WO2014145210. (b) Optimer Pharmaceuticals Inc (2005) Patent WO2004080391.
134. Biochimica O. P. O. S. SpA (1990) Patent US 4965355.
135. (a) Rib X. Pharmaceuticals Inc (2005) Patent WO2005019211. (b) Zhou, J., Bhattacharjee, A., Chen, S., Chen, Y., Duffy, E., Farmer, J., Goldberg, J., Hanselmann, R., Ippolito, J.A., Lou, R., Orbin, A., Oyelere, A., Salvino, J., Springer, D., Tran, J., Wang, D., Wu, Y., and Johnson, G. (2008) *Bioorg. Med. Chem. Lett.*, **18** (23), 6175; (c) For the preparation of Intermediate 143, Rib X Pharmaceuticals (2006) Patent WO 2006/133397.
136. http://www.investor.bayer.com http://www.medkoo.com/products/5657
137. Bayer Pharma Aktiengesellscaft (2015) Patent US20150087827.
138. (a) Rickborn, B. (1998) *Org. React.*, **52**; (b) Fisera, L. and Pavlovic, D. (1984) *Czech. Chem. Commun.*, **49**, 1990–2000.
139. (a) Eli Lilly and Company (2014) Patent US 8772496. (b) Eli Lilly and Company (2008) Patent WO2008079600. (c) Borghese, A. (2005) Patent WO2005042515. (d) Amegadzie, A.K. (2003) Patent WO2003091226.
140. Merck & Co., Inc. (1986) Patent US 4590201.
141. General Editor Ashton Acton, Q. (ed.) (2013) Clinical trials and studies,

Chapter 1, in *Macular Degeneration: New Insights for the Healthcare Professional*, p. 15.

142. (a) Sobera, L.A., Leeson, P.A., Rabasseda, X., and Castaner, J. (2000) *Drugs Future*, **25**, 1145; (b) Rogowski, M.A. (2000) *Epilepsy Res.*, **69**, 273; (c) Stafstrom, C.E. (2009) *Neuropsychiatr. Dis. Treat.*, **5**, 547.
143. (a) Novartis (2000) Patent US6156907. (b) Mudd, W.H. and Stevens, E.P. (2010) *Tetrahedron Lett.*, **51**, 3229 and references cited therein.
144. (a) Bonacorso, H.G., Moraes, M.C., Luz, F.M., Quintana, P.S., Zanatta, N., and Marin, M.A.P. (2015) *Tetrahedron Lett.*, **56**, 441; (b) Hojo, M., Masuda, R., Kokuryo, Y., Shioda, H., and Matsuo, S. (1976) *Chem. Lett.*, **5**, 499.
145. Gonzalez – Calderon, D., Aguirre-De Paz, J.G., Gonzalez – Gonzalez, C.A., Fuentes-Benites, A., and Gonzalez-Romero, C. (2015) *Tetrahedron Lett.*, **56**, 1713.
146. Zhang, P., Russel, M.G., and Jamison, T.F. (2014) *Org. Process Res. Dev.*, **18**, 1567.
147. Borukhova, S., Sieger, A.D., Noel, T., Wang, Q., Bush, M., and Hessel, V. (2015) *ChemSusChem*, **8**, 504.
148. Ott, D., Borukhova, S., and Hessel, V. (2015) *Green Chem.* doi: 10.1039/C5GC01932J
149. Otvos, S.B. and Fulop, F. (2015) *Catal. Sci. Technol.*, **5**, 4926–4941.

# 3
# Green Chemical Synthesis and Click Reactions

*Maria José Arévalo, Óscar López, and Maria Victoria Gil*

## 3.1
## Introduction

Novel green technologies play a key role in helping the society to move toward the elimination or appropriate handling of waste. In this sense, Click Chemistry presents features that can be considered within the "Twelve Principles of Green Chemistry" [1], including simple operation, mild reaction conditions, solvent-free conditions, readily available starting materials and reagents, high yields, high selectivity, and simple product isolation procedures.

Click chemistry has recently aroused great interest and emerged as one of the most powerful tools in drug discovery and chemical biology [2].

Among the different procedures that can be considered as click chemistry reactions, cycloadditions, and particularly the Cu(I)-catalyzed Huisgen 1,3-dipolar cycloaddition (CuAAc) between azides and terminal alkynes to form 1,2,3-triazoles, are the most widely used. This reaction has found applications across many diverse research areas. In the following sections, the potential of this click reaction and its applications are reviewed.

## 3.2
## Huisgen 1,3-Dipolar Cycloaddition

The classical thermal 1,3-dipolar cycloaddition that involves azides and alkynes (either terminal or internal) derivatives is known as the *Huisgen cycloaddition* and gives access to 1,2,3-triazoles as a mixture of 1,4- and 1,5-regioisomers (Scheme 3.1) [3]. Regiospecific couplings to furnish the 1,4-disubstituted 1,2,3-triazoles are promoted by copper(I) salts as faster procedures (more than 100 times faster).

The cycloaddition, in most cases, is carried out at room temperature in aqueous media and is compatible with an ample number of functional groups present in the molecules. Since its discovery, it has become one of the most applied organic reactions in this century in many areas of research.

*Click Reactions in Organic Synthesis*, First Edition. Edited by Srinivasan Chandrasekaran.

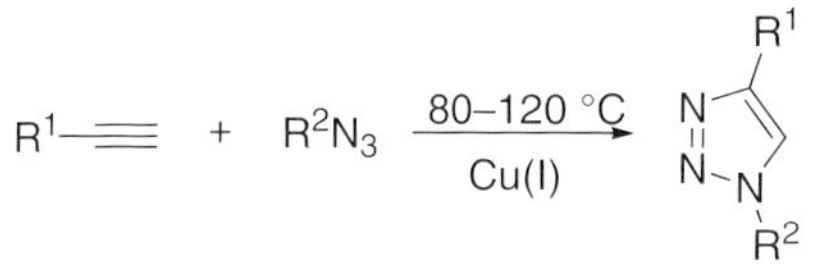

**Scheme 3.1** General 1,3-dipolar cycloaddition of azides and alkynes.

### 3.2.1 Green Perspectives on Reaction Conditions

The Huisgen cycloaddition may be performed under many different conditions, for example, organic and aqueous solvents and heterogeneous or homogeneous catalysis. Cu(I) ions are the most common catalysts used in this reaction. For an efficient cycloaddition, Cu(I) must be present at high levels. However, Cu(I) is thermodynamically unstable, as it is easily oxidized to Cu(II) and/or disproportionates to Cu(0) and Cu(II). The use of Cu(II) salts in the presence of a reducing agent such as sodium ascorbate is the procedure usually employed to form Cu(I) *in situ* and is very popular in the cycloaddition of terminal alkynes [4]. Some other conditions have been reported, trying to avoid the addition of a reducing agent in order to simplify the process.

#### 3.2.1.1 Copper(I) Catalysts

CuI, one of the most commonly used Cu(I) sources, is inefficient, when used alone. In most cases, it has been used as a combination of CuI/$NR_3$, in which the tertiary amine is an essential additive. During the reaction, CuI is included in a stable polymeric structure that then must be dissociated by amines to yield the "active" Cu(I) species for the formation of copper(I) acetylide, the key intermediate of this reaction. Sharpless reported the formation of diacetylenes, bis-triazoles, and 5-hydroxytriazoles as by-products if $NEt_3$ and DIPEA are used as additives. The catalytic system CuI/$NR_3$ is highly beneficial to be used in nonaqueous or nonprotonic solvents, which is of practical importance to a range of substrates that do not tolerate water or protonic solvents. On the other hand, the addition of HOAc promotes the CuI/$NR_3$-catalyzed cycloaddition favoring the hydrolysis of the cuprotriazole intermediate. In this context, the mixture of phenylethyne and benzyl azide in toluene reacts in 2.5 h to give the corresponding triazole in an 84% yield. The use of CuI, DIPEA, and HOAc leads to a 96% yield within 3 min [5]. Due to the excellent chelating properties of carboxylate anions with Cu(I), species such as benzoate, trifluoroacetate, and pivalate serve as ligands for Cu(I) ions [6].

Cu(I) acetate is a very efficient catalyst for the reaction of phenylethyne and benzylazide, giving the desired product in 98% yield, in 8 min. The catalyst forms $[(PhCtCCu)_2]_n$, while an equimolar amount of HOAc is produced, which promotes the reaction as mentioned earlier [7].

The readily available and inexpensive $Cu_2O$ seems to be quite robust and catalyzes aqueous azide–alkyne cycloaddition reactions efficiently, even with very low loadings (ppm levels). A range of functional groups on the acetylene including halo-, cyclopropyl-, ethoxy-, amide, ether, and double bonds were well tolerated. In addition to the tosyl azide, 4-chlorophenyl, 1,2-naphthyl, methyl, as well as bulky camphor sulfonyl azides, were all suitable to this $Cu_2O$-catalyzed procedure. Water plays an essential role in the increase of the reaction rate and appears to act as an activator for $Cu_2O$ [8].

#### 3.2.1.2 Copper(I) Complexes with Nitrogen- and Phosphorous-Donating Ligands

The catalytic activity of copper phosphine complexes has also been explored. Excellent yields are obtained using low catalyst loadings of $Cu(PPh_3)_2NO_3$ complex (0.5 mol%) at room temperature and under solventless conditions. The cycloaddition reaches completion within 40 min and further decrease of catalyst loading is possible [9]. Its activity has been compared with that of $CuOAc(PPh_3)_2$ and $C_3H_7COOCu(PPh_3)_2$, the latter showing higher activity in the reaction of benzyl azide and phenylacetylene. The presence of phosphine ligands and carboxylate ions makes the copper(I) ion more soluble in organic media and ensures a better accessibility of the transition metal for the reactant [10]. Other $PPh_3$-containing copper(I) complexes (1 mol%), such as $[Cu(phen)(PPh_3)_2]NO_3$, [Cu(phen)]Br, [Cu(phen)]Cl, have been evaluated as catalysts without any ligand assistance, in the absence of solvent. The $[Cu(phen)(PPh_3)_2]NO_3$ complex provides the best result in terms of reaction times [11].

$[CuBr(PPh_3)_3]$ complex shows catalytic activity in click reactions. Thus, although the catalyst is efficient in several solvents, reaction time is shorter in water or under solventless conditions. It is compatible with aryl, benzyl, and alkyl azides and alkynes with various substituents in terms of electronic and steric effects [12].

Phosphinite- and phosphonite-bearing complexes (Figure 3.1) displayed better catalytic activities, with complete conversions after 5–8 h under aqueous conditions [13].

A PTA–iminophosphorane Cu(I) complex (PTA = 1,3,5-triaza-7-phosphaadamantane) also catalyzes the CuAAc reaction in aqueous media, as an example, the reaction of (azidomethyl)(phenyl)sulfane ($PhSCH_2N_3$) and

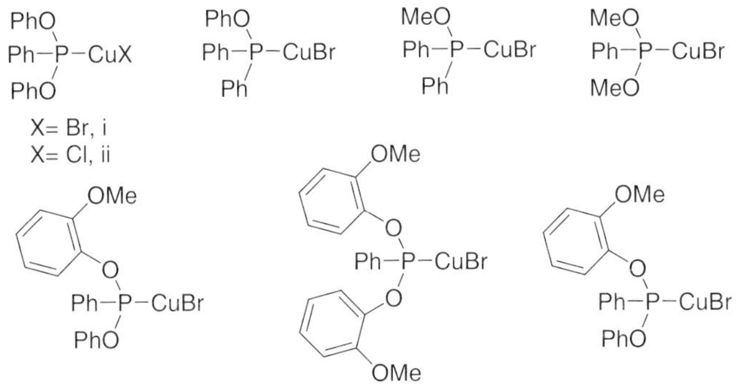


**Figure 3.1** Phosphinite and phosphonite copper complexes.

phenyl acetylene, catalyzed in water by a 0.5 mol% of catalyst, in the presence of 2,6-lutidine (10 mol%) can be highlighted [14].

Nitrogen-donating ligands are particularly promising as they form stable Cu(I) complexes. They stabilize the Cu(I) oxidation state and modulate the catalytic activity of the Cu center, allowing smoother reaction conditions and broader applicability.

Imidazoles carrying a long alkyl chain as a ligand for the copper(I)-catalyzed azide–alkyne cycloaddition are also very efficient. An alkyl chain can readily change its conformation in a host molecule to adjust its volume to the cavity size and may have an appreciable steric effect [15]. The copper(I) isonitrile complex prepared upon treatment of isonitriles with CuCl in THF in 95% yield as [CuLCl] units extended in a polymeric chain is stable in air or water for several months. This complex catalyzes the reaction of azides and acetylenes in water at room temperature in 0.5–5 mol% loading to give excellent yields in very short times. The catalyst may be recycled through five cycles [16–18].

Some of the most representative ligands are the commercially available sulfonated bathophenanthroline BPDS, the tris(benzimidazole) $(BimC_4A)_3$, and the tris-(triazole) TBTA ligands.

Analogs of TBTA that bear bulky *t*-butyl groups have also been found to be efficient ligands [17]. Tris(2-{4-[(dimethylamino)methyl]-1*H*-1,2,3-triazol-1-yl}ethyl)-amine (DTEA) outperforms TBTA in a number of reactions. The acceleratory effect of DTEA is tentatively attributed to its high basicity (four tertiary amino groups) and its ability to bind copper(I/II) in a five-membered ring chelation. Complete conversions within 1 h were recorded in $CH_3OH$, *i*PrOH, *t*BuOH, as well as in aprotic toluene, acetonitrile, and THF [18]. Tris-(triazolyl)methanol is an effective ligand for CuCl. The reaction between azides and alkynes with a 0.5 mol% of catalyst loading leads to completion in good yields in solvents such as water, $CH_2Cl_2$, THF or $CH_3CN$. This system also catalyzes the tandem processes involving an organyl bromide, sodium azide, and alkynes in acetonitrile/water, at 100 °C under microwave irradiation in a remarkably short time [19].

Highly reactive azides, such as picolyl, pyridine, bis(benzimidazole), or bis(triazole) azides, have been designed to assist the metal catalysis by the copper(I) chelation to form active complexes, acting both as reactant and as catalyst. They react faster than standard azides thanks to internal chelation of the copper catalyst, which is supposed to enhance the electrophilicity of the azido group and facilitate the formation of the metallacycle intermediate [13].

Hybrid ligands with hemilabile sulfur donor such as thioether-tethered 1,2,3-triazole-pyridyl ligands, 1-(2-picolyl)-4-(2-(methylthio)-pyridine)-1*H*-1,2,3-triazole, and 1-(4-picolyl)-4-(2-(methylthio)-pyridine)-1*H*-1,2,3-triazole form Cu complexes that with 0.5 mol% loading lead to near quantitative yield in the Huisgen cycloaddition in MeOH-$H_2O$ at 50 °C [20]. The copper(I) cation bonded to a monoanionic, *S,N*-bidentate-coordinating 2-aminoarenethiolate ligand, in a 1 mol% loading, also provides good yields. This catalyst is somewhat sensitive to atmospheric oxygen. The reaction does not work in water, but acetonitrile and $CH_2Cl_2$ are good solvents [21].

The catalytic activity of preformed copper(I) complexes bearing one or two α-diimine ligands has also been studied in the preparation of 1,2,3-triazoles from terminal alkynes and azides [22]. Bis(aryl)acenaphthenequinonediimine compounds (Ar-BIAN) are very efficient bidentate ligands for the formation of a wide range of robust homogeneous catalysts of the type $[Cu(Ar\text{-}BIAN)L_2]BF_4$, where L = PhCN, $PPh_3$ and $L_2$ = Ar-BIAN. Regarding solvents, these catalysts are active in nonaqueous solvents, such as THF and $CH_2Cl_2$, but also in the aqueous 1 : 1 *t*BuOH-$H_2O$ mixture [23].

*N*-heterocyclic carbenes (NHCs), as strong σ-donor ligands, are widely used. Although the majority of NHC ligands are imidazol-2-ylidenes or 1,2,4-triazol-5-ylidenes (normal NHC), some alternative NHC ligands (abnormal *N*-heterocyclic carbene, aNHC) are also known. Thus, $[Cu(Icy)_2]PF_6$ works at only 50 ppm loading, and even under neat conditions [24]. [CuX(SIMes)] (X = Cl, Br) functions particularly well under neat conditions, and the reaction can be carried out "on water" [25]. Its efficiency can be increased with external *N*-donor aromatic ligand(s). Thus, the order of different ligands according to the positive effect on the reaction rate is as follows: 4-dimethylaminopyridine (4-DMAP) > phenanthroline (Phen) > *N*-methylimidazole (NMI) > bathophenanthroline disulfonic sodium salt [26].

In bis-triazolylidene dicopper(I) complexes, two copper(I) ions are irreversibly bound in the same catalyst molecule, at an adequate distance to each other that facilitates the formation of a bridged μ-acetylide intermediate, thus allowing the use of aprotic solvents under homogeneous conditions with low catalyst loadings. Very good conversions can be reached when a 0.5–10 mol% load of catalyst is used [27, 28].

Copper complexes with *N,S*-heterocyclic carbenes also catalyze the Huisgen cycloaddition, leading to a quantitative yield in the reaction of benzylazide with phenylacetylene in acetonitrile with a precatalyst loading of 1% [29].

Regarding the aNHC ligands, the 1,3-bis(2,6-diisopropylphenyl)-2,4-diphenylimidazolium salt, as a aNHC ligand, forms a copper(I) chloro complex, which efficiently catalyzes reactions of azides, even those sterically hindered, with both terminal and internal alkynes, to give 1,4-substituted 1,2,3-triazoles in excellent yields at room temperature within short reaction time under solvent-free conditions with a loading of 0.005 mol% [30].

The catalytic properties in the CuAAc of several Cu(I) complexes have been tested. The halide-free cationic complex in a 0.1 mol% load allows a full conversion in 15 min, whereas iodide-containing complexes require 2–3 h. The latter mentioned ligand favors the reaction of internal or electron-rich alkynes with azides to form the corresponding triazole ring, as well as the reaction of bulky azides with acetylenes [31].

#### 3.2.1.3 Metalated Reagents as Catalysts

Azide-tagged copper(I) complexes of NHCs analogous to the well-known bis[2,6-diisopropylphenyl]imidazol-2-ylidene (IPr) and *N*,*N*′-bis[2,6-diisopropylphenyl] imidazolin-2-ylidene (SIPr) in the reaction with propargyl alcohols, yield triazole functionalized copper(I)–NHC complexes with 75% and 69% yield. This strategy opens the possibility of diversifying the substitution patterns of these copper(I) complexes (Scheme 3.2) [32].

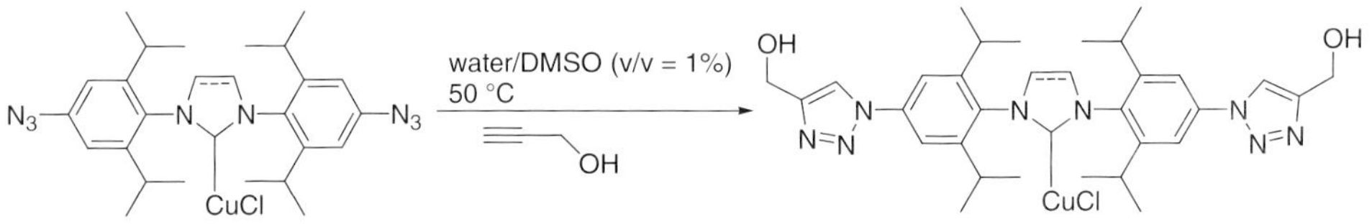


**Scheme 3.2** Reaction of N,N′-bis[2,6-diisopropylphenyl]imidazolin-2-ylidene with propargyl alcohol.

In a similar way, ruthenium(II)–azido complex reacts toward electron-deficient and strain-activated alkynes at room temperature. Nonactivated terminal and internal alkynes failed to react under such conditions [33].

On the other hand, metalated acetylides can promote the cycloaddition reaction and form a stable 5-metalated triazole ready to be functionalized at this position. Thus, stable, easily accessible, nontoxic, and diversely substituted 1-bismuth(III) acetylides undergo an efficient cycloaddition with organic azides in the presence of 5 mol% CuOTf in THF, under mild conditions to form stable 5-bismuth(III) triazolides in yields from 60% to 91%. In this study, it is hypothesized that there is a weak and reversible π interaction of the copper(I) catalyst with bismuth(III) acetylide [34, 35].

Furthermore, ruthenium bipyridyl and tripyridyl complexes bearing a terminal alkyne reacted with azides affording triazole complexes [36].

#### 3.2.1.4 Immobilized Copper Species

To improve reuse and recovery, copper species may be immobilized onto various supports although this system may suffer from leakage of catalysts, high reaction temperatures, low yields of low activity, and the need for additives [37].

Cuprous oxide on charcoal ($Cu_2O/C$, 5 mol%) in the presence of triethylamine catalyzes the formation of 1,4-disubstituted 1,2,3-triazoles in a $H_2O$/isopropanol mixture, in good to excellent yields (69–94%) [38].

Copper(I)–zeolite, CuI-USY, catalyzes Huisgen reaction in good yields, with the possibility of reusing the catalyst for three times without apparent loss of activity [39].

The polymer-stabilized monovalent copper Cu(I)–poly(2-aminobenzoic acid) shows a high catalytic activity toward azide–alkyne 1,3-dipolar cycloaddition reactions carried out at room temperature, in the presence of water as a solvent. Under aerated conditions, catalytic efficiency was proved for at least five cycles [40]. $Cu(OAc)_2$ immobilized on a polystyrene-anchored imidazole ligand, Cu-PsIm catalyzes (0.1 mol% loading) the reaction of phenyl azide and phenylacetylene at 60 °C to form the corresponding triazole in 97% yield, for 7 h [41].

MCM-41 has a high surface area (~1000 $m^2/g$), a large pore volume, and pore diameters that can be tuned between 20 and 100 Å. In click chemistry, the MCM-41-supported copper(II) allows the one-pot synthesis of 1,2,3-triazoles from alkyl bromides, sodium azide, and alkynes. The reaction may be carried out in water, at room temperature. Short reaction times give high yields (70–98%) [42].

Melamine-formaldehyde resin (MFR; melamine (1,3,5-triazine-2,4,6-triamine)) has the ability to host several cations through the presence of spaces and/or cavities between triazine cores and also multicoordinating N-atoms in its structure. Doped nanosized $Cu_2O$ on melamine-formaldehyde resin (nano-$Cu_2O$ MFR) is a highly efficient heterogeneous catalyst for the Huisgen 1,3-dipolar cycloaddition between different terminal alkynes and β-azido alcohols at room temperature using a 0.072 mol% of Cu. The catalyst can be recovered, recycled, and reused for many consecutive trials without significant loss of its activity [18].

Copper–aluminum hydrotalcite (3 : 1) (Cu/Al-HT (3 : 1)) promotes the cycloaddition of azides and alkynes, without any sacrificial reducing agents and ligands, because of its ability to stabilize Cu(I) [43].

Another way to immobilize the catalyst is through the covalent attachment of an accelerating chelating ligand to a resin material. In this way, undecyl alcohol mono-(benzimidazoylmethyl)-bis(pyridylmethyl) amine ligand (BimPy2) derivative may be attached to NovaPEG amine resin [44]. The incorporation of chelating ligands such as 2-pycolylamine, histamine, bispicolylamine, or bispyridylamine to silica or silica-coated magnetite nanoparticles allows the alkyne–azide cycloaddition under different conditions such as aqueous *t*-BuOH, DMF, water or in the absence of solvents at room temperature or under microwave [45].

An ionic liquid containing copper(I), especially in [bmim]$BF_4$ or and the basic ionic liquid [bmim]OH, catalyzes the reaction of α-azido ketones and terminal alkynes in various solvents. The presence of L-proline (10 mol%) may be needed [46]. This catalytic ionic liquid can be recovered and reused for at least five runs without any loss in its activity [47].

#### 3.2.1.5 Copper Nanocatalysis

Nanocatalysis has attracted more attention in recent years. Nanomaterials can combine the advantages of both heterogeneous and homogeneous catalyses and can offer unique activity with high selectivity. Catalytic activity of Cu species is remarkably improved by controlling the particle size.

Cheap, stable, and recyclable heterogeneous CuO nanoparticles and sphere-shaped CuO hollow nanoparticles show an excellent catalytic activity in the Huisgen cycloaddition reaction of azides and acetylenes, being compatible with a wide range of functional groups and reaction conditions involving air [48, 49].

Conventional copper nanoparticles seem to be able to only catalyze the click reaction in organic solvents. Polyvinylpyrrolidone-coated copper(I) oxide nanoparticles ($Cu_2O$-NPs) represent an alternative [50].

Cu nanoparticulates in Guar gum, Cu(0) nanoparticles stabilized on activated montmorillonite, copper supported on $SiO_2$ nanoparticles, PEI/Cu (polyethyleneimine/Cu) nanoparticles supported onto the $SiO_2$ surface, act as heterogeneous catalysts for the click reaction at room temperature in ethanol, water, $H_2O$:EtOH (v:v = 1 : 1), DMSO, respectively, in the presence of tertiary amines or carboxylic acids in some cases. Triazoles are synthesized in good to excellent yields in rather short times. The catalysts may be easily recovered and reused up to the third or fourth run [51–53].

Unsupported copper nanoparticles catalyze the 1,3-dipolar cycloaddition of azides to aliphatic or aromatic acetylenes under mild conditions, THF, and 65 °C, in short reaction times (0.5–24 h) and in the presence of $Et_3N$ (1 equiv.). The corresponding 1,2,3-triazoles are obtained in excellent yields [54].

Stable nanometric copper clusters are efficient ligand-free catalysts, probably due to their great surface area [55]. In the reaction between benzyl azide and prop-2-yn-1-ol, copper nanocluster catalysis affords 100% conversion after 18 h [9].

Nanoporous copper (CuNPore) catalysts with tunable nanoporosity catalyze the click reaction of phenylacetylene and benzyl azide with a 2 mol% loading, in toluene at 65 °C for 2 h without using any supports and bases. The use of CuNPore catalysts with approximately 25 and 30 nm ligament sizes gives moderate yields, but the use of a ligament size about approximately 40 nm affords the product in almost quantitative yield. The CuNPore recovered by simple filtration and acetone washing may be reused for multiple cycles without significant loss of catalytic activity [56].

Magnetically recoverable nanoparticles represent an easy and environmentally benign means for catalyst recovery. Zero-valent iron nanoparticles have been used as precursors to seed, reduce, and support another metal. Cu nanoparticles on the glutathione-functionalized nanoferrites with 10–25 nm particle size are an example of magnetic nanoparticles acting as a copper support. They catalyze the reaction of benzyl azide and phenyl acetylene in water, with MW heating at 120 °C, in 10 min, leading to triazoles in good yield [57].

Iron oxide nanoparticle–supported tris(triazolyl)–CuBr, in a 0.5 mol% loading, catalyzes the azide–alkyne cycloaddition in water at room temperature, leading to excellent yields. The catalyst can be used for another five catalytic cycles [58].

#### 3.2.1.6 **Other Metals as Catalysts**

Ruthenium complex $RuH_2(CO)(PPh_3)_3$ can catalyze the cycloaddition of a range of azides and alkynes to afford 1,4-disubstituted 1,2,3-triazoles with 100% selectivity. Aprotic solvents such as $CH_3NO_2$, $CH_3CN$, 1,2-dichloroethane (DCE), DMSO, dioxane, and THF were proved to be effective. The presence of electron-withdrawing groups such as ester groups in the arylacetylene has a negative effect. Aliphatic alkynes containing an OH or CN group showed good reactivities [59]. Moreover, $RuH(\eta^2\text{-}BH_4)(CO)(PCy_3)_2$ also selectively produces 1,4-disubstituted 1,2,3-triazoles with excellent yields [60].

The iridium-catalyzed azide–alkyne cycloaddition, [{Ir(cod)Cl}$_2$], reaction of electron-rich internal thioalkynes, in various polar and nonpolar solvents, including water, under mild reaction conditions (2 mol% of catalyst and room temperature), gives the corresponding 1,4-disubstituted 1,2,3-triazole system in 22–98% yields [61].

Au/TiO$_2$ can serve as an active catalyst in the Huisgen [3+2] cycloaddition in aqueous medium, yielding 1,4-disubstituted 1,2,3-triazoles regioselectively as the sole product, in short reaction times (45 min) with good to excellent yields. The catalyst may be recycled up to five times [62].

Ag(I) acetylide has been thought to catalyze the Huisgen cycloaddition; however, the addition of copper(I) salts is required to accomplish the cycloaddition and to obtain 1,4-disubstituted 1,2,3-triazoles [63]. Despite this, homogeneous complexes of type [$Ag^+(L_2)(X^-)$]catalyze the acetylene–azide cycloaddition. When *N,N*-diisopropyl(2-diisopropylphosphanyl)benzamide was used as ligand, the reaction conditions based on 2–2.5 mol% of catalyst, for 24 h, at 90 °C in toluene lead to conversions higher than 99% [64].

Charcoal impregnated with zinc is able to catalyze, in a 10 mol% loading, the cycloaddition of organic azides and alkynes in DMF at 50 °C, to provide the corresponding 1,4-disubstituted 1,2,3-triazoles and 1,4,5-trisubstituted 1,2,3-triazoles in good to excellent yields (64–95%). The catalyst can be recovered by filtration and reused at least, for three more cycles [65].

Bimetallic catalysts in which one metal can fine-tune or modify the catalytic properties of the other metal offer a considerable promise. In a Cu-Mn bimetallic catalyst, the reaction between CuO and $Mn_2O_3$ through electron transfer from $Mn^{3+}$ to $Cu^{2+}$, the formation of a $Mn^{4+}$ ion and a stabilized $Cu^+$ ion may take place [66].

#### 3.2.1.7 Nonconventional Energy Sources

Metallic copper efficiently catalyzes azide–alkyne cycloadditions under ultrasound or simultaneous ultrasound/microwave irradiation. Ultrasound favors mechanical depassivation in reactions involving metals and enhances both mass and electron transfer from the metal to the organic acceptor. It has been applied to accelerate the preparation of 1,2,3-triazole phthalimides via a 1,3-dipolar cycloaddition of phthalimidoethyl-azide [67].

Moreover, the combination of microwave and metals is possible if using well-dispersed fine metal particles in a polar high-boiling solvent [68].

### 3.2.2 Applications to Synthesis

#### 3.2.2.1 Regioselectivity of the Alkyne–Azide Cycloaddition

The regioselective formation of 1,4- and 1,5-disubstituted 1,2,3-triazoles can be accomplished by CuAAC reactions and ruthenium-catalyzed azide–alkyne [3+2] cycloaddition (RuAAC) reactions, respectively [69, 70]. Fokin and coworkers demonstrated that ruthenium complexes such as Cp*RuCl(PPh$_3$)$_2$

(Cp* = pentamethylcyclopentadienyl) catalyzed the Huisgen 1,3-dipolar cycloaddition; interestingly, the observed regioselectivity was reversed when compared with related copper(I) catalysis [71, 72].

Weinreb's group reported the preparation of 5-substituted 1,2,3-triazoles by the reaction of β-tosylethylazide with alkynes in the presence of ruthenium catalysts in refluxing benzene (Scheme 3.3) [73]. When the standard copper(I)-catalyzed protocol was used, it was found that the reaction of β-tosylethylazide with terminal alkynes led to the expected 4-substituted 1,2,3-triazoles.

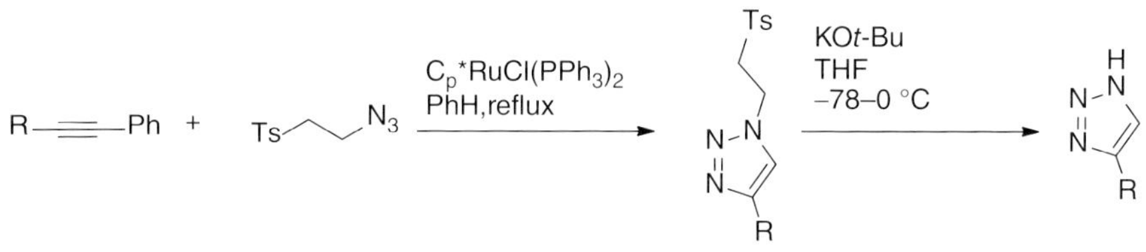


**Scheme 3.3** Synthesis of 5-substituted-1,2,3-triazoles.

#### 3.2.2.2 **Different Substitution Patterns on Triazole**

The synthesis of 1*H*-1,2,3-triazoles has been usually carried out by phenylacetylene and sodium azide. Other routes to 1*H*-1,2,3-triazoles include the reaction of sodium azide with nitroalkenes [74] and alkenyl bromides; the reaction of (*E*)-β-arylvinyl bromides and sodium azide [75], which is catalyzed by $Pd_2(dba)_3$ and Xantphos; the cycloaddition of propargyl cations derived from corresponding alcohols with azides [76]

Reports on the preparation of 1-monosubstituted 1,2,3-triazoles are limited: one protocol consists of the cycloaddition of azides to acetylene, acetylides, or vinyl compounds; aryl boronic acids and prop-2-ynoic acid or $CaC_2$; the decarboxylation of triazoles bearing a carboxylic acid substituent; the reaction of aliphatic halides (Cl and Br), sodium azide, and propiolic acid followed by a decarboxylation process [77].

The synthesis of 1,4-disubstituted 1,2,3-triazoles is often carried out by a conventional Huisgen cycloaddition between azides and terminal alkynes.

Regarding the synthesis of 1,4,5-trisubstituted triazoles, Wu reported that the Cu(I)–C bond in the cycloaddition intermediate 5-Cu(I)-1,2,3-triazole could be substituted *in situ* by an exogenous electrophile $E^+$ to produce 1,4,5-trisubstituted 1,2,3-triazoles [78]. The reaction between gold(I) acetylides and azides also furnishes 1,4-disubstitued-5-cuprotriazoles. The C–Cu species produced might react in a further Ullmann C–C coupling reaction and lead to the formation of 1,4,5-trisubstituted 1,2,3-triazoles directly [79, 80]. The reaction of dimethylphenylalkynylaluminum with benzylazide using copper iodide (10%) and $[Me_2N(CH_2)_2]2NMe$ (PMDTA, 10%) and 1 equiv. of THF in toluene at room temperature led to 1,4-disubstituted 5-alumino-1,2,3-triazoles in excellent yield [81]. The 1,4,5-trisubsituted 5-iodo-1,2,3-triazole has been prepared as a single product from 1-iodoalkyne in THF catalyzed by copper(I) iodide and a ligand [82].

#### 3.2.2.3 Strain-Promoted Cycloadditions

The strain-promoted [3+2] cycloaddition reaction of substituted cyclooctyne with aryl azides furnishes 1,4,5-trisubstituted 1,2,3-triazoles, which have become good bioorthogonal probes.

The azidophilic strained alkyne moiety of a cyclooctyne can be transiently masked via 1 : 1 complexation with a cationic copper(I) salt, $(MeCN)_4CuBF_4$. This method enabled the regioselective click reaction on terminal alkyne in a cyclooctyne derivative (Scheme 3.4) [83].

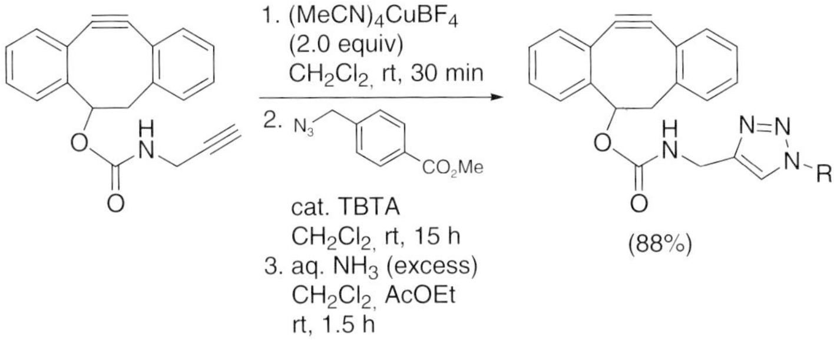


**Scheme 3.4** Regioselective 1,3-dipolar cycloaddition on a terminal alkyne.

Triazole-substituted titanocenes may be synthesized by a strain-driven 1,3-dipolar cycloadditions between azide-functionalized titanocenes and cyclooctyne. Titanocene derivative are interesting complexes since they are a promising class of cytotoxic compounds, efficient reagents and catalysts, and organometallic gelators. Azide-substituted cationic titanocenes reacted with cyclooctyne in $CH_2Cl_2$ at room temperature to give the expected product with yields between 75% and 92% [84].

#### 3.2.2.4 Sulfonyl Azides in Huisgen Cycloaddition

CuAAC could well be the most direct and practical route to 1-sulfonyl-1,2,3-triazoles. The cycloaddition between copper(I) acetylides and sulfonyl azides gives *N*-acylsulfonamides in aqueous medium [85] and amidines in the presence of secondary amines [86].

#### 3.2.2.5 Synthesis of Vinyl-1,2,3-Triazoles

The synthesis of 4-vinyl-1,2,3-triazoles, as monomers, have been designed by the efficient Cu(I)-catalyzed direct reaction of trimethylsilyl (TMS)-protected vinylacetylene with the corresponding organic azides. Although high yields of the monomer units were obtained, this strategy suffered from a number of drawbacks. Another strategy starts from but-3-yn-1-ol and but-3-yn-2-ol, which through the traditional Cu(I)-catalyzed click coupling resulted in high yields and regioselectivity for the formation of the triazole ring in both cases. A one-pot hydroxyl elimination using MsCl, $Et_3N$, and NaI was also found to give the desired alkene in good yield. The secondary alcohol underwent facile dehydration directly to the alkene, in the presence of PTSA or $POCl_3$/pyridine [87].

1-Vinyl-1,2,3-triazoles have been synthesized in moderate yields by the cycloaddition of vinyl azides and phenyl acetylene, using acetone or dioxane as solvent, in a flow microreactor where copper wire was inductively heated. The copper metal was directly and instantaneously heated inside the reactor, which results in a higher reactivity than with conventionally heated elemental copper [88].

#### 3.2.2.6 Triazole Derivative Ligands for Coordination Chemistry

In coordination chemistry, architectures containing the triazole heterocycle have been developed as monodentate and polydentate nitrogen ligands. Special attention has been given to (2-pyridyl)- and more flexible (2-picolyl)-1,2,3-triazoles. Thus, a library of selected 1,4-disubstituted 1,2,3-triazoles differently functionalized with pyridine, pyrimidine, and pyrazine have been synthesized by the reaction of aryl, heteroaryl, and heteroarylmethyl azides with alkynes. Three different copper(I) sources, that is, $CuSO_4$/sodium ascorbate, $CuBr(PPh_3)_3$, and $(CuOTf)_2{\cdot}C_6H_6$, were used to promote the reactions [89].

The copper-catalyzed azide–alkyne cycloaddition is particularly suitable for $^{18}F$-labeling of sensitive biomolecules. Three different radiolabeled alkyne precursors react with azido-functionalized peptides in radiochemical yields (RCYs) of 54–99% and an overall reaction time of 30 min. The reaction is catalyzed by copper sulfate in combination with ascorbic acid or sodium ascorbate. Addition of bathophenanthroline disulfonate (CuI stabilizing agent) accelerates the cycloaddıtıon [90].

#### 3.2.2.7 Tetrazole Synthesis

Since 1901, conventional synthesis of 5-substituted 1*H*-tetrazoles has been reported to proceed via [3+2] cycloaddition between azide and nitriles. Drawbacks from this procedure are the use of expensive and toxic azide, highly moisture-sensitive reaction conditions, strong Lewis acids, and hydrazoic acid.

Advances toward the click concept of this reaction have been tried using stoichiometric amounts of inorganic salts and metal complexes, Zn, $BF_3{\cdot}OEt_2$, $Pd(OAc)_2/ZnBr_2$, $Yb(OTf)_3$, $Zn(OTf)_3$, $AlCl_3$, and $Pd(PPh_3)_4$ as catalysts, $TMSN_3$ [91], or TBAF. However, the homogeneous process supposes the tedious separation and recovery of the catalysts. Recently, several heterogeneous catalytic systems have been proposed, but these methods require a large excess of sodium azide, longer reaction times, and expensive metals. Moreover, the cycloaddition is too slow to be synthetically useful except when strong electron-withdrawing groups activate the nitrile compounds [92].

#### 3.2.2.8 Synthesis of Chiral Triazoles

Chiral triazoles have been synthesized by the desymmetrization of prochiral molecules. Thus, a PyBox ligand and CuCl have been used to furnish a highly enantioselective CuAAC by means of desymmetrization of oxindole-based 1,6-heptadiynes, obtaining enantiomeric excesses as high as 98% [93].

#### 3.2.2.9 Synthesis of Triazoles with Luminescent Properties

Copper-catalyzed azide–alkyne cycloaddition has been used to couple the phenazine luminescent probe and an *N,N*-bis(2-picolyl)amine moiety in a 75% yield by using copper acetate (0.01 mol%) and sodium ascorbate in $CH_3CN/H_2O$ (10 : 1) [94]. The cycloaddition of 1,3-diethynyl-6-fluoroisoquinoline with 1-(2-azidoethyl)pyrrolidine has been applied to the synthesis of a soluble "click" fluorescent chemosensor (Scheme 3.5). The molecule displays good sensitivity toward heavy and transition metal ions such as $Zn^{2+}$, $Fe^{2+}$, or $Cu^{2+}$ in water at pH 7.0 [95].

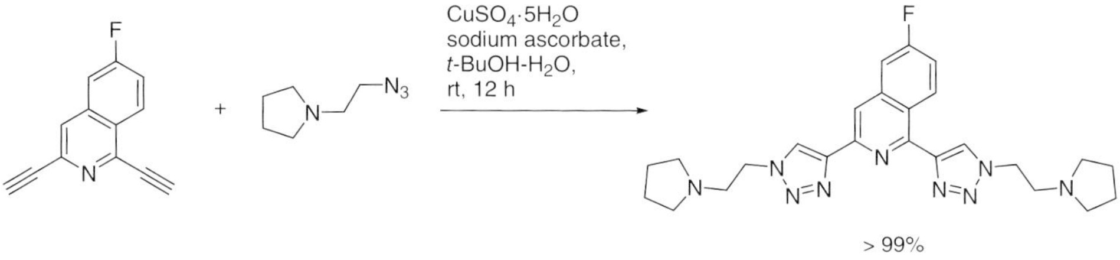


**Scheme 3.5** Sinthesis of a fluerescent chemosensor by a CuAAc.

New carbazole dyes have been synthesized in very high yields from functionalization of a nonfluorescent precursor, *N*-(4-azidophenyl)-carbazole, with several alkyne moieties [96].

#### 3.2.2.10 Synthesis of Triazole Libraries

The copper-catalyzed alkyne–azide cycloaddition is a very useful methodology for the preparation of triazole libraries. Thus, a library of triazole amphiphiles with a variety of functional polar "heads" and hydrophobic or superhydrophobic "tails" [97], a 96-product-sized library of 1,2,3-triazoles from organic azides and terminal alkynes [98], and a small library of triazole-conjugated coumarins have been synthesized [99].

#### 3.2.2.11 Synthesis of Phosphorylated Triazoles

Phosphorylated 1,2,3-triazoles have been obtained in excellent yields via the 1,3-cycloaddition of ω-phosphoryl azides to activated internal alkynes such as acetylenedicarboxylate and sodium azide to tetramethyl acetylenediphosphonate. These reactions proceed in water without any cosolvent or additive [100]. Moreover, silyl-protected borane complexes of phosphinoalkynes may give rise to chiral and achiral P-stereogenic 1,2,3-triazolyl-4-phosphines in moderate to good yields [101].

## 3.3 Other 1,3-Dipolar Cycloadditions

The 1,3-dipolar cycloaddition of pyrazolidinone-based dipoles (azomethine imides) to terminal alkynes with electron-withdrawing groups, promoted by the dinuclear copper complex $[Cu(\mu\text{-}OH)(tmen)]_2Cl_2$ (tmen = *N,N,N′,N′*-tetramethylethylenediamine) (Cu: 0.1 mol%), has been studied. *N,N*-bicyclic pyrazolidinone derivatives are obtained from this reaction in nonpolar and low-polar solvents (Scheme 3.6) [102].

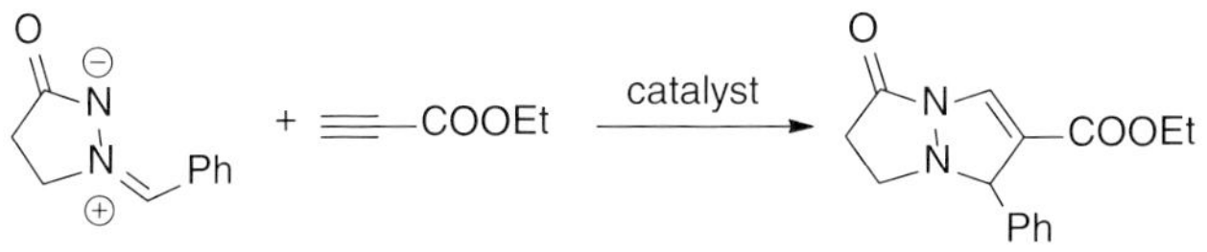


**Scheme 3.6** 1,3-Dipolar cycloaddition of an azomethine imine with a terminal alkyne.

Azomethine imines react with propiolates under copper catalysis to form *N,N*-bicyclic pyrazolidinone. Different catalysts have been used in this reaction: CuI in the presence of *N*-methyldicyclohexylamine [103]; copper(I)-exchanged zeolites [102, 104]; copper hydroxide catalyst, $Cu(OH)_x/Al_2O_3$ (Cu: 1.5 mol%) [105]; or the efficient copper(I) acetate [106].

The 1,3-dipolar cycloadditions of diversely substituted nitrilimines with dipolarophiles may form a variety of pyrazoles. In this way, the cycloadditions of *N*-[4-nitrophenyl]-*C*-[2-furyl] nitrilimine with styrene, 2-propyne-1-ol, or vinyl acetate in ethanol afforded the corresponding pyrazole products with complete regioselectivity and good yields [107].

The copper(I)-catalyzed 1,3-dipolar cycloaddition reaction between alkynes and nitrile oxides reported by Sharpless [108] and Hansen *et al.* [109] has made considerable progress in the synthesis of isoxazoles. This methodology enables the preparation of unsymmetrical 3,5-disubstituted isoxazoles with specific regioselectivity at satisfactory rates and yields. A large number of *ortho*-substituted arylboronic acids were evaluated for their ability to accelerate nitrile oxides [3+2] cycloaddition to alkynoic acids [110]. Copper-doped silica cuprous sulfate (CDSCS, 0.05 mol%) as a nanocatalyst and $NaHCO_3$ in a solution of *i*-PrOH/$H_2O$ (1 : 1, V/V) have been used in the reaction of diverse alkynes and *in situ* generated nitrile oxide [111].

## 3.4 Atom Economy and Simplicity of Procedures in Multicomponent Reactions

Multicomponent CuAAC methods can be used to rapidly generate a family of complex alkyl-, benzyl-, or aryl-substituted 1,2,3-triazole systems, avoiding handling of organic azides, generating them *in situ*, and making the click process more eco-friendly and safe.

### 3.4.1 Reaction Conditions

#### 3.4.1.1 Copper Compounds as Catalysts

1-Monosubstituted 1,2,3-triazoles are synthesized in moderate yields from a variety of aliphatic halides (Cl and Br), sodium azide, and propiolic acid by a click cycloaddition/decarboxylation process in DMF at 100 °C for 0.5 h, using CuI (0.2 equiv.) and sodium ascorbate (0.4 equiv.) as catalyst and in the presence of $Cs_2CO_3$ (0.5 equiv.) [77].

The reaction of aryl/alkyl halides, alkynes, and sodium azide in water may be catalyzed by a heterogeneous copper(I) catalyst, oyster shell powders (OSPs)-CuBr (2.5 mol%), and microwave irradiation (480 W and 70 °C). The catalyst could be easily recovered from the reaction mixture by a simple filtration and reused at least eight times without significant loss of its catalytic activity. Chitin and protein molecules on OSP particles surface seem to play important roles in the chelation of the CuBr species [112].

#### 3.4.1.2 Copper Complexes with Nitrogen- and Phosphorous-Donating Ligands

The three-component cycloaddition of benzyl bromide and sodium azide with phenylacetylene in water at room temperature is catalyzed by the water-soluble Cu(I)-NHC complex [(SIPr)CuCl] (0.5 mol%), affording a 98% yield of triazole within 48 h. The catalyst could be recovered by a simple extraction [113].

The three-component coupling of sulfonyl azides, alkynes, and amines, leading to the formation of *N*-sulfonyl amidines in moderate to good yields, is promoted by the robustly porous metal–organic framework MOF-$Cu_2I_2$ (BTTP4) (BTTP4 = benzene-1,3,5-triyl triisonicotinate) in a 2.4% loading. The catalytic system can be recycled by simple filtration and reused at least four times without any loss in yield [114]. Another metal–organic framework Cu (BTC)-L-[Pd] (BTC = benzene-1,3,5-tricarboxylate; L = 4-substituted pyridine) has been used to catalyze a tandem Sonogashira/click procedure. The reaction of 2-iodobenzylbromide, sodium azide, and a variety of alkynes produces 8*H*-[1,2,3]triazolo[5,1-*a*]isoindoles with good yields in the presence of 2 equiv. of potassium carbonate, 5 mol% of catalyst, and at 50 °C [115].

Catalysts systems based on Cu(I) and Cu(II) (1.0 mol% loading) species in combination with benzimidazole salts (1.2 mol%) catalyze the multicomponent reaction of phenylacetylene, benzyl bromide, and sodium azide at room temperature in water [116].

#### 3.4.1.3 Immobilized Copper Species

The synthesis of 1,2,3-triazole derivatives by a three-component coupling of alkyl (benzyl) halides or aryl boronic acids, sodium azide, and terminal alkynes over copper(II) sulfate supported on alumina (Cu/$Al_2O_3$, 10 mol%) has been developed [117].

The synthesis of β-hydroxy triazoles by the three-component reaction of a variety of epoxides, sodium azide, and diverse terminal alkyne reactions performed in water has been catalyzed by a copper(II) hydrotalcite catalyst, with 3 : 1 (Cu:Al) mole ratio [118].

Thiazolidinedione – triazole hybrids may be synthesized by the one-pot reaction based on a Knoevenagel condensation followed by an alkyne – azide cycloaddition, between thiazolidine-2,4-dione, substituted aryl aldehydes, propargyl bromide, and substituted aryl azides using piperidine. The catalytic system is composed of $CuSO_4 \cdot 5H_2O$ and sodium ascorbate in PEG-400 [119].

#### 3.4.1.4 Copper Nanocatalysis

Cu(I) nanoparticles supported on poly(4-vinylpyridine) (P4VPy) is as an efficient and recyclable catalyst for the regioselective multicomponent synthesis of 1,4-disubstituted 1*H*-1,2,3-triazoles from α-halo ketones, sodium azide, and terminal alkynes [120]. In a similar way, Cu(I) is generated *in situ* from copper sulfate, [P4VPy]$CuSO_4$ in the presence of sodium ascorbate and 1,4-disubstituted 1,2,3-triazoles are synthesized from benzyl halides, sodium azide, and a terminal alkyne. This polymeric and therefore heterogeneous catalyst can be recovered by simple filtration and recycled for six consecutive runs [121]. Furthermore, copper nanoparticles (mainly as $Cu_2O$) on activated carbon (0.5 mol% CuNPs/C) catalyze the reaction of organic halides, aryldiazonium salts, anilines, and epoxides in water to form triazoles [122]. 1,4-Dihydroxyanthraquinone-copper(II) nanoparticles immobilized on silica gel [$AQ_2$Cu(II)-APSi$O_2$] (5 mol%) catalyze the regioselective reaction of epoxides with wide range of terminal alkynes and sodium azide at 25 °C in water, to form triazoles in 2.5 h with good yields. Furthermore, the catalyst could be recovered by simple filtration, and it was recycled up to 10 times [123]. The synthesis of a variety of propargylamines from aldehydes, amines, and alkynes is catalyzed effectively by copper nanoparticles on $TiO_2$ (0.5 mol%) at 70 °C under solvent-free conditions in moderate to high yields. The catalyst can be handled in air and can be easily recovered by filtration and reused over four consecutive cycles [124].

Cu(0) nanoparticles in the nanopores of modified montmorillonite catalyze the three-component coupling reactions of aldehydes, amines, and alkynes toward propargylamines [125].

$Fe_3O_4$-nanoparticle-supported Cu(II) – β-cyclodextrin complex (15 mol%) catalyzes the synthesis of 1,2,3-triazoles via the cross-coupling reaction of aryl boronic acids with azides and subsequent click reaction with alkynes at room temperature in air and in water with excellent yields. Nanoparticles were collected by a magnet and, after washing, could be reused for at least four catalytic cycles [126].

## 3.5 Summary and Conclusions

Although the click concept can refer to many different reaction types, it is usually associated to the synthesis of 1,4-disubstituted triazoles, via the Cu-catalyzed cycloaddition reaction of azides and alkynes. In order to achieve high regioselectivities, good yields, and easy isolation procedures, numerous reaction conditions have been tested. Thus, different activation conditions have been reported: conventional methodologies, by using metal catalysts, as salts, complexes, nanoparticles, or anchored on polymeric structures, and also nonconventional procedures, by using ultrasound, microwave, or photoinduction. Nevertheless, the most widely used conditions are based on the use of Cu(II) salts and a reducing agents, such as sodium ascorbate, and research is still being conducted on this context in order to develop simple and efficient conditions for accessing triazole derivatives in a straightforward manner.

## References


1. Srihari, P., Dutta, P., Rao, R.S., Yadav, J.S., Chandrasekhar, S., Thombare, P., Mohapatra, J., Chatterjee, A., and Jain, M.R. (2009) *Bioorg. Med. Chem. Lett.*, **19**, 5569–5572.
2. Duan, M., Zhang, L.H., and Li, J. (2008) *Chem. J. Chin. Univ.*, **29**, 2118–2120.
3. Huisgen, R. (1984) in *1,3-Dipolar Cycloadditional Chemistry* (ed A. Padwa), John Wiley & Sons, Inc., New York, p. 1.
4. Rostovtsev, V.V., Green, L.G., Fokin, V.V., and Sharpless, K.B. (2002) *Angew. Chem. Int. Ed.*, **41**, 2596–2599.
5. Shao, C., Wang, X., Zhang, Q., Luo, S., Zhao, J., and Hu, Y. (2011) *J. Org. Chem.*, **76**, 6832–6836.
6. Shao, C., Cheng, G., Su, D., Xu, J., Wang, X., and Hu, Y. (2010) *Adv. Synth. Catal.*, **352**, 1587–1592.
7. Shao, C., Wang, X., Xu, J., Zhao, J., Zhang, Q., and Hu, Y. (2010) *J. Org. Chem.*, **75**, 7002–7005.
8. Wang, K., Bi, X., Xing, S., Liao, P., Fang, Z., Meng, X., Zhang, Q., Liu, Q., and Ji, Y. (2011) *Green Chem.*, **13**, 562–565.
9. Wang, D., Li, N., Zhao, M., Shi, S., Ma, C., and Chen, B. (2010) *Green Chem.*, **12**, 2120–2123.
10. Gonda, Z. and Novák, Z. (2010) *Dalton Trans.*, **39**, 726–729.
11. Wang, D., Zhao, M., Liu, X., Chen, Y., Li, N., and Chen, B. (2012) *Org. Biomol. Chem.*, **10**, 229–231.
12. Lal, S. and Díez-González, S. (2011) *J. Org. Chem.*, **76**, 2367–2373.
13. Lal, S., McNally, J., White, A.J.P., and Díez-González, S. (2011) *Organometallics*, **30**, 6225–6232.
14. García-Álvarez, J., Díez, J., and Gimeno, J. (2010) *Green Chem.*, **12**, 2127–2130.
15. Asano, K. and Matsubara, S. (2010) *Org. Lett.*, **12**, 4988–4991.
16. Liu, M. and Reiser, O. (2011) *Org. Lett.*, **13**, 1102–1105.
17. Bevilacqua, V., King, M., Chaumontet, M., Nothisen, M., Gabillet, S., Buisson, D., Puente, C., Wagner, A., and Taran, F. (2014) *Angew. Chem. Int. Ed.*, **53**, 5872–5876.
18. Michaels, H.A. and Zhu, L. (2011) *Chem. Asian J.*, **6**, 2825–2834.
19. Ozkal, E., Llanes, P., Bravo, F., Ferrali, A., and Pericás, M.A. (2014) *Adv. Synth. Catal.*, **356**, 857–869.
20. Jiang, L., Wang, Z., Bai, S.-Q., and Andy Hor, T.S. (2013) *Dalton Trans.*, **42**, 9437–9443.
21. Fabbrizzi, P., Cicchi, S., Brandi, A., Sperotto, E., and Koten, G. (2009) *Eur. J. Org. Chem.*, **2009**, 5423–5430.
22. Markalain Barta, J. and Díez-González, S. (2013) *Molecules*, **18**, 8919–8928.
23. Li, L., Lopes, P.S., Rosa, V., Figueira, C.A., Lemos, M.A.N.D.A., Duarte, M.T., Avilés, T., and Gomes, P.T. (2012) *Dalton Trans.*, **41**, 5144–5154.
24. Díez-González, S. and Nolan, P. (2008) *Angew. Chem.*, **120**, 9013–9016; *Angew. Chem. Int. Ed.* (2008) **47**, 8881–8884.
25. Díez-González, S., Correa, A., Cavallo, L., and Nolan, S.P. (2006) *Chem. Eur. J.*, **12**, 7558–7564.

26. Teyssot, M.L., Chevry, A., Traïkia, M., El-Ghozzi, M., Avignant, D., and Gautier, A. (2009) *Chem. Eur. J.*, **15**, 6322–6326.
27. Berg, R., Straub, J., Schreiner, E., Mader, S., Rominger, F., and Straub, B.F. (2012) *Adv. Synth. Catal.*, **354**, 3445–3450.
28. Buckley, B.R., Dann, S.E., Heaney, H., and Stubbs, E.C. (2011) *Eur. J. Org. Chem.*, **2011**, 770–776.
29. Han, X., Weng, Z., Young, D.J., Jin, G.-X., and Hor, T.S.A. (2014) *Dalton Trans.*, **43**, 1305–1312.
30. Sau, S.C., Roy, S.R., Sen, T.K., Mullangi, D., and Mandala, S.K. (2013) *Adv. Synth. Catal.*, **355**, 2982–2991.
31. Hohloch, S., Scheiffele, D., and Sarkar, B. (2013) *Eur. J. Inorg. Chem.*, **2013**, 3956–3965.
32. Gibard, C., Avignant, D., Cisnetti, F., and Gautier, A. (2012) *Organometallics*, **31**, 7902–7908.
33. Cruchter, T., Harms, H., and Meggers, E. (2013) *Chem. Eur. J.*, **19**, 16682–16689.
34. Worrell, B.T., Ellery, S.P., and Fokin, V.V. (2013) *Angew. Chem. Int. Ed.*, **52**, 13037–13041.
35. Casarrubios, L., de la Torre, M.C., and Sierra, M.A. (2013) *Chem. Eur. J.*, **19**, 3534–3541.
36. Gerken, J.B., Rigsby, M.L., Ruther, R.E., Pérez-Rodríguez, R.J., Guzei, I.A., Hamers, R.J., and Stahl, S.S. (2013) *Inorg. Chem.*, **52**, 2796–2798.
37. Rad, M.N.S., Behrouz, S., Movahedian, A., Doroodmand, M.M., Ghasemi, Y., Amini, S.R., Gandomania, A.-R.A., and Rezaiee, R. (2013) *Helv. Chim. Acta*, **96**, 688–701.
38. López-Ruiz, H., de la Cerda-Pedro, J.E., Rojas-Lima, S., Pérez-Pérez, I., Rodríguez-Sánchez, R., Santillan, B.V., and Coreño, O. (2013) *Arkivoc*, 139–164.
39. Chassaing, S., Sido, A.S.S., Alix, A., Kumarraja, M., Pale, P., and Sommer, J. (2008) *Chem. Eur. J.*, **14**, 6713–6721.
40. Ul Islam, R., Taher, A., Choudhary, M., Witcomb, M.J., and Mallick, K. (2015) *Dalton Trans.*, **44**, 1341–1349.
41. Roy, S., Chatterjee, T., and Islam, S.M. (2013) *Green Chem.*, **15**, 2532–2539.
42. Hosseinzadeh, R., Sepehrian, H., and Shahrokhi, F. (2012) *Heteroat. Chem*, **23**, 415–421.
43. Namitharan, K., Kumarraja, M., and Pitchumani, K. (2009) *Chem. Eur. J.*, **15**, 2755–2758.
44. Presolski, S.I., Mamidyala, S.K., Manzenrieder, F., and Finn, M.G. (2012) *ACS Comb. Sci.*, **14**, 527–530.
45. Megia-Fernandez, A., Ortega-Muñoz, M., Lopez-Jaramillo, J., Hernandez-Mateo, F., and Santoyo-Gonzalez, F. (2010) *Adv. Synth. Catal.*, **352**, 3306–3320.
46. Singh, H., Sindhu, J., and Khurana, J.M. (2013) *J. Iran. Chem. Soc.*, **10**, 883–888.
47. Ahmadya, A.Z., Heidarizadeha, F., and Keshavarza, M. (2013) *Synth. Commun.*, **43**, 2100–2109.
48. Ganesh Babu, S. and Karvembu, R. (2013) *Catal. Surv. Asia*, **17**, 156–176.
49. Woo, H., Kang, H., Kim, A., Jang, S., Park, J.-C., Park, S., Kim, B.S., Song, H., and Park, K.H. (2012) *Molecules*, **17**, 13235–13252.
50. Zhang, Z., Dong, C., Yang, C., Hu, D., Long, J., Wang, L., Li, H., Chen, Y., and Konga, D. (2010) *Adv. Synth. Catal.*, **352**, 1600–1604.
51. Borah, B.J., Dutta, D., Saikia, P.P., Barua, N.C., and Dutta, D.K. (2011) *Green Chem.*, **13**, 3453–3460.
52. Ciyabi Hashjin, M., Ciyabi, R., Baharloui, M., Hosseini, G., and Tavakoli, H. (2012) *Chin. J. Chem.*, **30**, 223–227.
53. Veerakumar, P., Velayudham, M., Lub, K.-L., and Rajagopal, S. (2011) *Catal. Sci. Technol.*, **1**, 1512–1525.
54. Alonso, F., Moglie, Y., Radivoy, G., and Yus, M. (2010) *Eur. J. Org. Chem.*, **2010**, 1875–1884.
55. Durán, L., van Maarseveen, J.H., and Rothenberg, G. (2005) *Adv. Synth. Catal.*, **347**, 811–815.
56. Jin, T., Yan, M., Menggengateer, Minato, T., Bao, M., and Yamamotoa, Y. (2011) *Adv. Synth. Catal.*, **353**, 3095–3100.
57. Nasir Baig, R.B. and Varma, R.S. (2012) *Green Chem.*, **14**, 625–632.
58. Wang, D., Etienne, L., Echeverria, M., Moya, S., and Astruc, D. (2014) *Chem. Eur. J.*, **20**, 4047–4054.

59. Nian Liu, P., Siyang, P.H., Zhang, L., Kai San Tse, S., and Jia, G. (2012) *J. Org. Chem.*, 77, 5844–5849.
60. Nian Liu, P., Li, J., Su, F.H., Ju, K.D., Zhang, L., Shi, C., Sung, H.H.Y., Williams, I.D., Fokin, V.V., Lin, Z., and Jia, G. (2012) *Organometallics*, **31**, 4904–4915.
61. Ding, S., Jia, G., and Sun, J. (2014) *Angew. Chem. Int. Ed.*, **53**, 1877–1880.
62. Boominathan, M., Pugazhenthiran, N., Nagaraj, M., Muthusubramanian, S., Murugesan, S., and Bhuvanesh, N. (2013) *ACS Sustainable Chem. Eng.*, **1**, 1405–1411.
63. Proietti Silvestri, I., Andemarian, F., Khairallah, G.N., Yap, S.W., Quach, T., Tsegay, S., Williams, C.M., O'Hair, R.A.J., Donnelly, P.S., and Williams, S. (2011) *J. Org. Biomol. Chem.*, **9**, 6082–6088.
64. McNulty, J. and Keskar, K. (2012) *Eur. J. Org. Chem.*, **2012**, 5462–5470.
65. Meng, X., Xu, X., Gao, T., and Chen, B. (2010) *Eur. J. Org. Chem.*, **2010**, 5409–5414.
66. Yousuf, S.K., Mukherjee, D., Singh, B., Maity, S., and Taneja, S.C. (2010) *Green Chem.*, **12**, 1568–1572.
67. da Silva, M.T., de Oliveira, R.N., Valença, W.O., Barbosa, F.C.G., da Silva, M.G., and Camara, C.A. (2012) *J. Braz. Chem. Soc.*, **23**, 1839–1843.
68. Sarmiento-Sánchez, J.I., Ochoa-Terán, A., and Rivero, I.A. (2011) *Arkivoç*, 177–188.
69. Kwak, J.M., Moon, J.S., Seo, S., Choi, J.I., Sampath, V., Lee, H.-Y., and Koh, H.Y. (2014) *Bull. Korean Chem. Soc.*, **35**, 3675–3678.
70. Zhang, L., Chen, X., Xue, P., Sun, H.H.Y., Williams, I.D., Sharpless, K.B., Fokin, V.V., and Jia, G. (2005) *J. Am. Chem. Soc.*, **127**, 15998–15999.
71. Boren, B.C., Narayan, S., Rasmussen, L.K., Zhang, L., Zhao, H., Lin, Z., Jia, G., and Fokin, V.V. (2008) *J. Am. Chem. Soc.*, **130**, 8923–8930.
72. Johansson, J.R., Lincoln, P., Nordén, B., and Kann, N. (2011) *J. Org. Chem.*, **76**, 2355–2359.
73. Yap, A.H. and Weinreb, S.M. (2006) *Tetrahedron Lett.*, **47**, 3035–3038.
74. Quiclet-Sire, B. and Zard, S.Z. (2005) *Synthesis*, **19**, 3319–3326.
75. Barluenga, J., Valdes, C., Beltrn, G., Escribano, M., and Aznar, F. (2006) *Angew. Chem.*, **118**, 7047–7050; *Angew. Chem. Int. Ed.* (2006) **45**, 6893–6896.
76. Zhang, H., Tanimoto, H., Morimoto, T., Nishiyama, Y., and Kakiuchi, K. (2013) *Org. Lett.*, **15**, 5222–5225.
77. Zhang, Z. and Kuang, C. (2013) *Chin. J. Chem.*, **31**, 1011–1014.
78. Wu, Y., Deng, J., Li, Y., and Chen, Q.-Y. (2005) *Synthesis*, 1314–1318.
79. Cai, Q., Yan, J., and Ding, K. (2012) *Org. Lett.*, **13**, 3332–3335.
80. Jeyachandran, R., Potukuchi, H.K., and Ackermann, L. (2012) *Beilstein J. Org. Chem.*, **8**, 1771–1777.
81. Zhou, Y., Lecourt, T., and Micouin, L. (2010) *Angew. Chem. Int. Ed.*, **49**, 2607–2610.
82. (a) Hein, J.E., Tripp, J.C., Krasnova, L.B., Sharpless, K.B., and Fokin, V.V. (2009) *Angew. Chem.*, **121**, 8162–8165; (b) Hein, J.E., Tripp, J.C., Krasnova, L.B., Sharpless, K.B., and Fokin, V.V. (2009) *Angew. Chem. Int. Ed.*, **48**, 8018–8021.
83. Yoshida, S., Hatakeyama, Y., Johmoto, K., Uekusa, H., and Hosoya, T. (2014) *J. Am. Chem. Soc.*, **136**, 13590–13593.
84. Gansäuer, A., Okkel, A., Schwach, L., Wagner, L., Selig, A., and Prokop, A. (2014) *Beilstein J. Org. Chem.*, **10**, 1630–1637.
85. Cassidy, M.P., Raushel, J., and Fokin, V.V. (2006) *Angew. Chem. Int. Ed.*, **45**, 3154–3157.
86. Bae, I., Han, H., and Chang, S. (2005) *J. Am. Chem. Soc.*, **127**, 2038–2039.
87. Takizawa, K., Nulwala, H., Thibault, R.J., Lowenhielm, P., Yoshinaga, K., Wooley, K.L., and Hawker, C.J. (2008) *J. Polym. Sci., Part A: Polym. Chem.*, **46**, 2897–2912.
88. Kupracz, L., Hartwig, J., Wegner, J., Ceylan, S., and Kirschning, A. (2011) *Beilstein J. Org. Chem.*, **7**, 1441–1448.
89. Bolje, A., Urankar, D., and Kosmrlj, J. (2014) *Eur. J. Org. Chem.*, **2014**, 8167–8181.
90. Kettenbach, K., Schieferstein, H., and Ross, T.L. (2014) *Biomed Res. Int.*, **2014**, 1–16.

91. Gaydou, M. and Echavarren, A.M. (2013) *Angew. Chem. Int. Ed.*, **52**, 13468–13471.
92. García-Fernández, A., Díez, J., Gamasa, M.P., and Lastra, E. (2014) *Eur. J. Inorg. Chem.*, **2014**, 917–924.
93. Song, T., Li, L., Zhou, W., Zheng, Z.-J., Deng, Y., Xu, Z., and Xu, L.W. (2015) *Chem. Eur. J.*, **21**, 554–558.
94. Jardim, G.A.M., Calado, H.D.R., Cury, L.A., and da Silva Júnior, E.N. (2015) *Eur. J. Org. Chem.*, **2015**, 703–709.
95. Midya, G.C., Paladhi, S., Bhowmik, S., Sahaa, S., and Dash, J. (2013) *Org. Biomol. Chem.*, **11**, 3057–3063.
96. Hörner, A., Volz, D., Hagendorn, T., Fürniss, D., Greb, L., Rönicke, F., Nieger, M., Schepersc, U., and Bräse, S. (2014) *RSC Adv.*, **4**, 11528–11534.
97. Solovyeva, V.A., Vu, K.B., Merican, Z., Sougrat, R., and Rodionov, V.O. (2014) *ACS Comb. Sci.*, **16**, 513–517.
98. Jlalia, I., Beauvineau, C., Beauvière, S., Önen, E., Aufort, M., Beauvineau, A., Khaba, E., Herscovici, J., Meganem, F., and Girard, C. (2010) *Molecules*, **15**, 3087–3120.
99. Arya, A., Kumar, V., Mathur, D., Singh, S., Brahma, R., Singh, R., Singh, S., Sharma, G.L., Parmar, V.S., and Prasada, A.K. (2015) *J. Heterocycl. Chem.*, **52**, 1–14.
100. Artyushin, O.I., Matveeva, E.V., Bushmarinov, I.S., and Odinets, I.L. (2012) *Arkivoc*, 252–263.
101. Veillard, R., Bernoud, E., Abdellah, I., Lohier, J.-F., Alayrac, C., and Gaumont, A.-C. (2014) *Org. Biomol. Chem.*, **12**, 3635–3640.
102. Oishi, T., Yoshimura, K., Yamaguchi, K., and Mizuno, N. (2010) *Chem. Lett.*, **39**, 1086–1087.
103. (a) Suarez, A., Downey, C.W., and Fu, G.C. (2005) *J. Am. Chem. Soc.*, **127**, 11244–11245; (b) Shintani, R. and Fu, G.C. (2003) *J. Am. Chem. Soc.*, **125**, 10778–10779.
104. Keller, M., Sido, A.S.S., Pale, P., and Sommer, J. (2009) *Chem. Eur. J.*, **15**, 2810–2817.
105. Yoshimura, K., Oishi, T., Yamaguchi, K., and Mizuno, N. (2011) *Chem. Eur. J.*, **17**, 3827–3831.
106. Shao, C., Zhang, Q., Cheng, G., Cheng, C., Wang, X., and Hu, Y. (2013) *Eur. J. Org. Chem.*, 6443–6448.
107. Bakavoli, M., Moeinpour, F., Davoodnia, A., and Morsali, A. (2010) *J. Mol. Struct.*, **969**, 139–144.
108. Himo, F., Lovell, T., Hilgraf, R., Rostovtsev, V.V., Noodleman, L., Sharpless, K.B., and Fokin, V.V. (2005) *J. Am. Chem. Soc.*, **127**, 210–216.
109. Hansen, T.V., Wu, P., and Fokin, V.V. (2005) *J. Org. Chem.*, **70**, 7761–7764.
110. Zheng, H., McDonald, R., and Hall, D.G. (2010) *Chem. Eur. J.*, **16**, 5454–5460.
111. Soltani Rad, M.N., Behrouz, S., and Faghihi, M.A. (2014) *J. Iran. Chem. Soc.*, **11**, 361–367.
112. Xiong, X., Cai, L., Jiang, Y., and Han, Q. (2014) *ACS Sustainable Chem. Eng.*, **2**, 765–771.
113. Wang, W., Wu, J., Xia, C., and Li, F. (2011) *Green Chem.*, **13**, 3440–3445.
114. Yang, T., Cui, H., Zhang, C., Zhang, L., and Su, C.-Y. (2013) *Inorg. Chem.*, **52**, 9053–9059.
115. Arnanz, A., Pintado-Sierra, M., Corma, A., Iglesias, M., and Sánchez, F. (2012) *Adv. Synth. Catal.*, **354**, 1347–1355.
116. Jianming, L., Muwen, L., Yuanyuan, Y., Meihuan, Y., and Kelei, Z. (2012) *Chin. J. Chem.*, **30**, 644–650.
117. Mukherjee, N., Ahammed, S., Bhadra, S., and Ranu, B.C. (2013) *Green Chem.*, **15**, 389–397.
118. Prasad, A.N., Thirupathi, B., Raju, G., Srinivas, R., and Reddy, B.M. (2012) *Catal. Sci. Technol.*, **2**, 1264–1268.
119. Sindhu, J., Singh, H., Khurana, J.M., Sharma, C., and Aneja, K.R. (2015) *Chin. Chem. Lett.*, **26**, 50–54.
120. Albadi, J., Keshavarz, M., Shirini, F., and Vafaie-nezhad, M. (2012) *Catal. Commun.*, **27**, 17–20.
121. Karimi Zarchi, M.A. and Nazem, F. (2014) *J. Iran. Chem. Soc.*, **11**, 1731–1742.
122. (a) Alonso, F., Moglie, Y., Radivoy, G., and Yus, M. (2011) *J. Org. Chem.*, **76**, 8394–8405; (b) Alonso, F., Moglie, Y., Radivoy, G., and Yus, M. (2012) *Heterocycles*, **84**, 1033–1044.

123. Sharghi, H., Khoshnood, A., Mahdi Doroodmand, M., and Khalifeh, R. (2012) *J. Iran. Chem. Soc.*, **9**, 231–250.
124. Albaladejo, M.J., Alonso, F., Moglie, Y., and Yus, M. (2012) *Eur. J. Org. Chem.*, 3093–3104.
125. Borah, B.J., Borah, S.J., Saikia, L., and Dutta, D.K. (2014) *Catal. Sci. Technol.*, **4**, 1047–1054.
126. Kaboudin, B., Mostafalua, R., and Yokomatsub, T. (2013) *Green Chem.*, **15**, 2266–2274.

# 4
# Synthesis of Substituted 1,2,3-Triazoles through Organocatalysis

*Kengadarane Anebouselvy and Dhevalapally B. Ramachary*

## 4.1
## Introduction

1,2,3-Triazoles possess importance in various research areas, ranging from medicinal chemistry or crop protection to material sciences. Their features make them pharmaceutically important molecules [1]. They are very stable to both reduction and oxidation and also to hydrolysis in acidic as well as basic conditions due to their high aromatic stabilization. They have a high dipole moment (about 5 D) and participate actively in hydrogen bond formation as well as in dipole–dipole and π stacking interactions, due to which they bind easily with the biological targets and possess improved solubility. They exhibit a wide range of biological properties, such as anticancer, antituberculosis, antibacterial, and antivirus. Furthermore, these substances have been shown to be valuable amide surrogates in bioactive molecules, because of their physicochemical properties and remarkable metabolic stability [1]. Although 1,2,3-triazoles are not naturally occurring compounds, the “amide triazole bioequivalence” was exploited for the development of HIV-1 protease inhibitors, potential anticancer agents, and non-nucleoside reverse-transcriptase inhibitors. The application of fused and nonfused 1,2,3-triazole frameworks is not only limited to the biological field but also being extended to other divisions of the fine chemical industries, for example, corrosion inhibitors, dyes, agrochemicals, and photostabilizers [1].

As a result, the most popular reaction that has been adapted to produce the 1,2,3-triazole moiety is the 1,3-dipolar cycloaddition also known as *Huisgen cycloaddition*, between a terminal alkyne **1** and an azide **2** under thermal conditions (Scheme 4.1a). Despite the fact that this reaction was first discovered at the start of the twentieth century, the detailed mechanistic study was described by Huisgen in the 1960s. The classical thermal Huisgen [3+2]-cycloaddition between azides **2** and alkynes **1** was of not much value initially in synthetic applications due to poor regioselectivity (forming **3** and **4**), low chemical yield, and elevated temperatures, but resurfaced after the findings of Sharpless and Meldal, as a prominent example for a “click-”concept reaction in 2001 [2]. This

*Click Reactions in Organic Synthesis*, First Edition. Edited by Srinivasan Chandrasekaran.

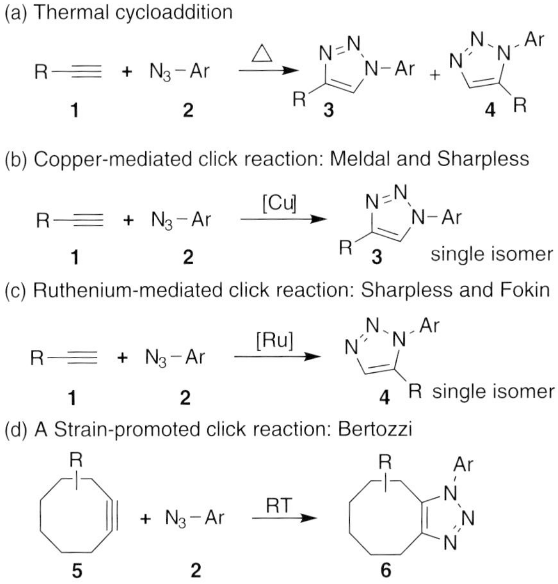


**Scheme 4.1** Thermal-, metal-, and strain-promoted 1,2,3-triazole synthesis.

copper(I)-catalyzed version of azide–alkyne cycloaddition reaction (CuAAC) yields regiospecifically 1,4-disubstituted 1,2,3-triazoles **3** under mild conditions in high yields (Scheme 4.1b) [2]. This approach has been widely used in different areas of science such as oligonucleotide synthesis, bioconjugation, DNA labeling, drug discovery, and construction of bolaamphiphilic structures.

Later, Ru(II)-catalyzed azide–alkyne cycloaddition reaction (RuAAC) was developed for the regiospecific synthesis of isomeric 1,5-disubstituted 1,2,3-triazoles **4** (Scheme 4.1c) [3]. Unlike CuAAC reaction, which worked only on terminal alkynes, RuAAC was successfully applied to internal alkynes, thus giving access to fully substituted 1,2,3-triazole derivatives also.

Due to the toxicity of the heavy metals to living cells and biomolecules such as DNA, the application of these methods are limited in biomaterial chemistry and chemical biology. As a metal-free approach, Bertozzi and coworkers have developed a strain-promoted 1,3-dipolar cycloaddition reaction (SPAAC) between azide **2** and cyclooctyne **5** to form 1,2,3-triazoles **6** for bioconjugate chemistry, which provided only very limited regioselectivity (Scheme 4.1d) [4]. Moreover, due to the complexity of the required substrates, this method was not suitable for straightforward synthetic applications.

Although numerous methodologies have been designed to construct these moieties by using metal and nonmetal approaches, there is still a huge demand to design a straightforward efficient methodology to obtain selectively the triazole core by metal-free chemistry. More and more researchers are being attracted into this field of 1,2,3-triazole synthesis via organocatalytic approaches, obviously due to their wide array of importance. Recently, emerging organocatalytic methods have certain advantages over the respective metal-mediated versions as they are potentially greener, eco-friendly, insensitive to oxygen and water, more sustainable, and considered to be nontoxic toward biological systems. The mammoth growth in the field of organocatalytic reactions has paved the way for an alternate synthetic approach that can eventually reach large-scale applications for the synthesis of various pharmaceuticals. Further, due to limited availability of structurally diverse alkynes, attention was focused onto the simple carbonyl compounds, considering carbonyl group as the nucleophilic source for generating an active species for addition to the azide. A handful of articles have been reported wherein activated carbonyls were converted to preformed enolates, which then underwent addition to azide to generate 1,2,3-triazoles. We have put our maximum efforts in gathering the overall progress of research carried out in this field, from the beginning until now and have presented herein under six major divisions, categorized on the basis of the nature and the quantity of the catalyst used and the mechanistic aspects, in chronological order.

## 4.2 Preformed-Enolate-Based Synthesis of Substituted 1,2,3-Triazoles

Dimroth first described, in 1902, the synthesis of 1,4-disubstituted 5-amino-1,2,3-triazoles **9**, by the reaction of an organic azide, **2**/**7**, with an acetonitrile **8**, in the presence of stoichiometric amounts of sodium ethoxide in boiling ethanol, which gave excellent yields for aryl azides **2** (Scheme 4.2) [5]. Long hours of heating were required to carry out the cycloaddition for benzyl azides and even greater difficulties were faced for alkyl azides.

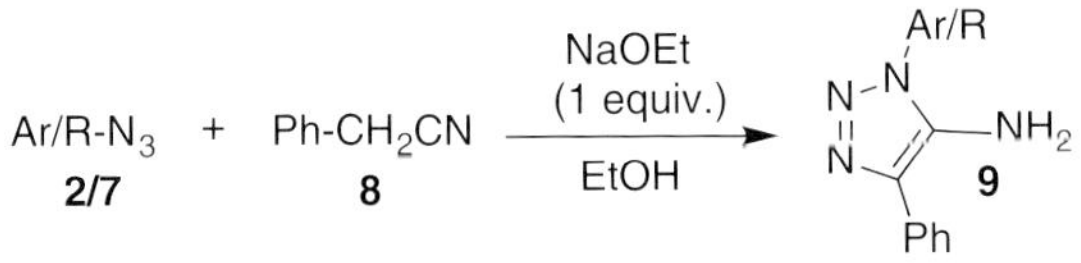


**Scheme 4.2** Synthesis of 5-amino-1,2,3-triazoles.

Nonetheless, the usefulness of the method was somewhat limited due to the lability of triazole or active methylene compound under the vigorous reaction conditions employed. Thus, 1-substituted 5-amino-1*H*-1,2,3-triazoles are liable to undergo a Dimroth rearrangement to yield 4-substituted amino derivatives and also alkyl or benzyl azides with certain active methylene compounds give only moderate or poor yields of the desired products. Some success in improving the yields was achieved by the use of potassium *t*-butoxide as base at room temperature, which gave good yields of 5-amino-1*H*-1,2,3-triazoles **9** from benzyl and *n*-hexyl azides with phenylacetonitrile **8** (Scheme 4.3) [5], but with methyl ketones, yields of the product were variable due to dimerization of the ketones and other side reactions.

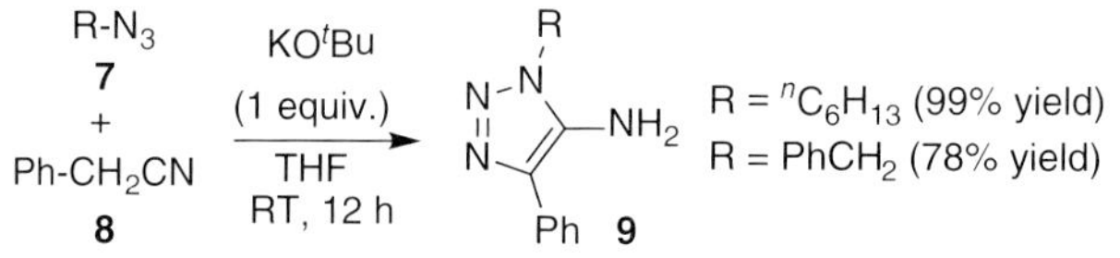

Scheme 4.3 Potassium *tert*-butoxide mediated synthesis of 5-amino-1,2,3-triazoles.

Just like the aryl azides **2**, the vinyl azides **11** are more reactive in their cycloadditions with active methylene compounds than are alkyl azides. In 1970, Alfred Hassner *et al.* developed a general synthetic approach to l-vinyl-1,2,3-triazoles **12** and **14**, (Scheme 4.4) [6] by the cycloaddition of active methylene compounds **10** or **13** with vinyl azides **11** (or their precursors, the β-haloalkyl azides), in the presence of 1 equiv. of an alkoxide (NaOMe). Decarboxylation of the 1-vinyl-5-substituted 1,2,3-triazole-4-carboxylic acids **14** synthesized from the reaction of ethyl acetoacetate or ethyl benzoylacetate with vinyl and β-haloalkyl azides led to 1-vinyl-5-substituted l,2,3-triazoles in almost quantitative yield [6].

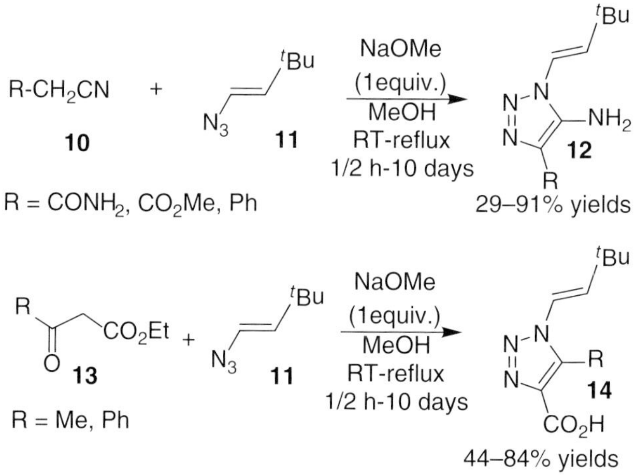

Scheme 4.4 Synthesis of 1-vinyl-1,2,3-triazoles.

In 1981, Smith *et al.* prepared 5-methyl-1,4-diphenyl-1,2,3-triazole **17** (Scheme 4.5) as an intermediate in the synthesis of 5-(diazomethyl)-1,4-diphenyl-1,2,3-triazole, which was used for the study on fragmentation of a diazocompound to carbene [7].

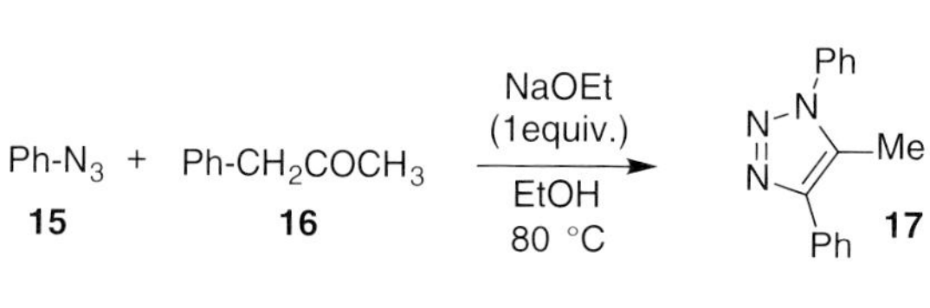

Scheme 4.5 Synthesis of 5-methyl-1,4-diphenyl-1,2,3-triazole.

Later in 1986, Townsend *et al.* prepared 5-amino-1-(2,3,5-tri-*O*-acetyl-β-D-ribofuranosyl)-1,2,3-triazole-4-carboxamide **19** [8], which is a crucial intermediate for the synthesis of 2-azacoformycin, by the 1,3-dipolar cycloaddition of cyanoacetamide **10a** and 2,3,5-tri-*O*-benzoyl-β-D-ribofuranosyl azide **18** under strongly basic conditions (Scheme 4.6). A complete removal of the benzoyl protecting groups from the crude nucleoside product (partial removal of these groups had occurred during the course of the reaction) was effected and then followed by an *in situ* acetylation afforded the triazole **19** in 37% overall yield [8].

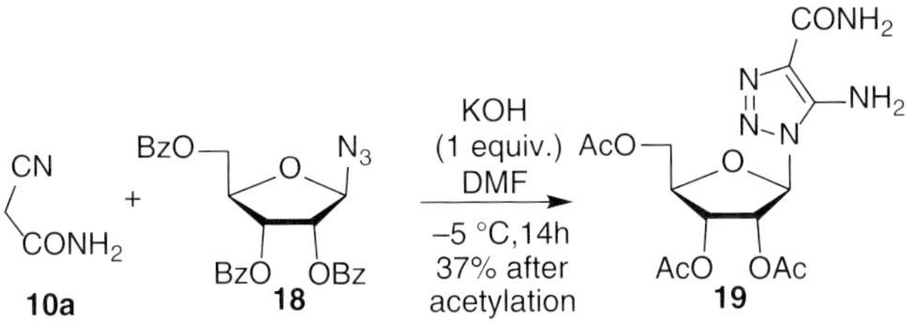


**Scheme 4.6** Base-promoted 1,3-dipolar cycloaddition of cyanoacetamide and 2,3,5-tri-*O*-benzoyl-β-D-ribofuranosyl azide.

In 1991, Wright *et al.* reported a procedure for the preparation of substituted 1-benzyl-1*H*-1,2,3-triazoles **21** and **23** from benzyl azides **20** under very mild conditions (Scheme 4.7) [9]. Benzyl azides **20** reacted with active methylene compounds in DMSO induced by potassium carbonate at 35–40 °C to give 1-benzyl-1*H*-1,2,3-triazoles **21** and **23** usually in good yield. Acetonitrile derivatives **10** gave 5-amino-1-benzyl-1*H*-1,2,3-triazoles **21,** whereas diethyl malonate gave 5-hydroxy-1-benzyl-1*H*-1,2,3-triazoles. 1*H*-1,2,3-Triazole-4-carboxylate esters and 1*H*-1,2,3-triazole-4-ketones were obtained from ethyl acetoacetate and β-diketones, respectively. Benzyl methyl ketone reacted to give a 5-methyl-4-phenyl-1*H*-1,2,3-triazole, but acetone and acetophenone failed to react. Other active methylene compounds that did not react under these reaction conditions included ethyl cyanoacetate, ethyl fluoroacetate, and ethyl nitroacetate.

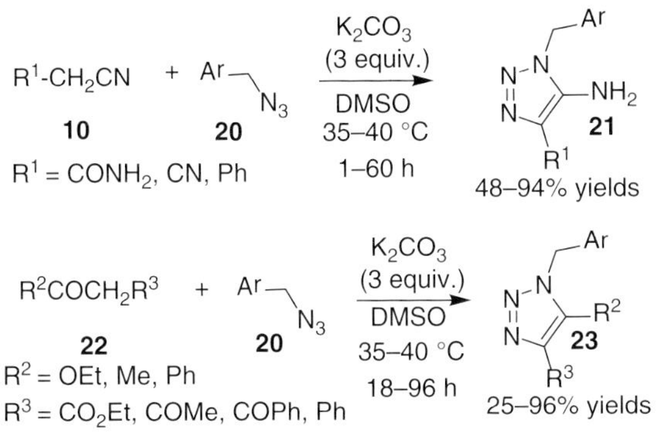


**Scheme 4.7** Synthesis of 5-amino-1-benzyl-1*H*-1,2,3-triazoles.

In 2002, Philip Jones and Mark Chambers treated an equimolar mixture of the aryl azide (2-azidobenzoic acid) with the corresponding isoxazole acetonitrile **24** (heteroaryl = isoxazole) in the presence of sodium ethoxide at 80 °C to prepare the desired core skeleton, triazoloquinazolinone **26** [10] (Scheme 4.8), which was used in a rapid analog synthetic strategy for C-5-substituted 1,2,3-triazolo-[1,5-*a*]quinazolines. In continuation, after a few years, Obushak *et al.* reported the cyclization of aryl azides with (4-aryl-1,3-thiazol-2-yl)acetonitriles, 1,3-benzothiazol-2-ylacetonitrile, 1*H*-benzimidazol-2-ylacetonitrile, and 2-benzothiazolylacetone in methanol in the presence of sodium methoxide to give high yields of triazole derivatives **27**–**29**, and **31**, respectively, with a heterocyclic substituent at C-4 (Schemes 4.8 and 4.9) [11, 12].

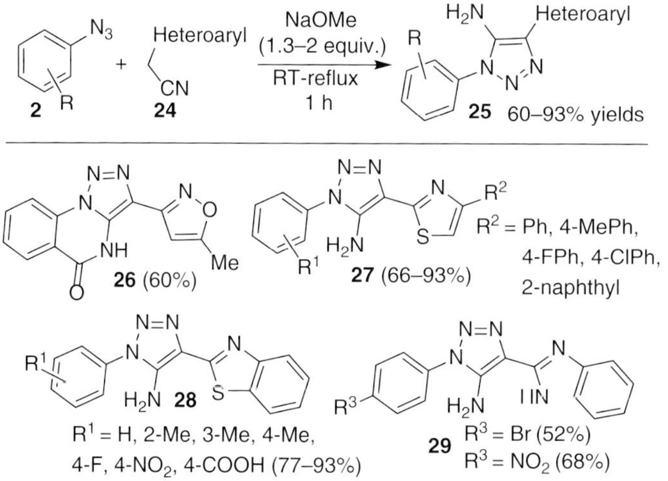


**Scheme 4.8** 1*H*-1,2,3-Triazole synthesis from heterocyclic CH-active compounds.

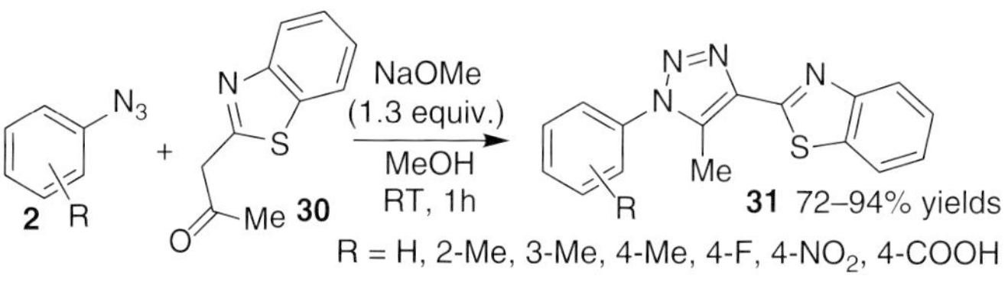


**Scheme 4.9** 1*H*-1,2,3-Triazole synthesis from 2-benzothiazolylacetone.

The same authors, in 2009, used β-keto sulfones and β-nitrile sulfones **32** as the building blocks for the 1,2,3-triazole synthesis in the cycloaddition with aryl azides **2** [13]. It was shown that sulfone reagents undergo base-mediated cyclization under mild conditions (at room temperature) to give 1,2,3-triazoles **33** in moderate to excellent yields (Scheme 4.10).

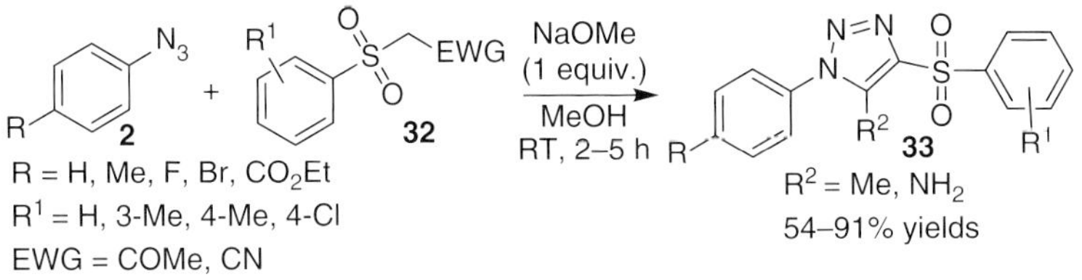


**Scheme 4.10** Synthesis of aryl-1*H*-1,2,3-triazol-4-yl sulfones.

A year later, Sergio Bacchi *et al.* reported a safe and reliable synthesis of aryl 1,2,3-triazoles **35** from the corresponding anilines **34** via intermediate aryl azides **2**, using a continuous-flow process (Scheme 4.11) [14]. The method was applied to a variety of substrates with good to excellent yields, without the need to isolate the reactive and possibly unstable intermediates, which were constantly kept at low concentration in the matrix environment. The tight control of the reaction variables combined with the advantages offered by the continuous-flow application for the reaction scale-up made this process an undoubtedly synthetic utility.

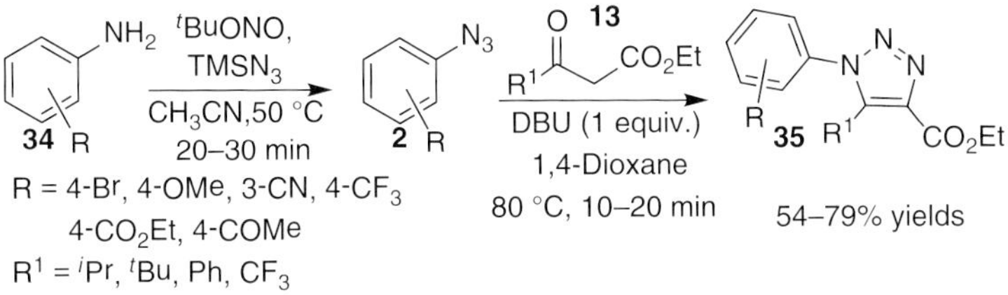


**Scheme 4.11** Synthesis of aryl 1,2,3-triazoles in a continuous-flow reactor.

In 2012, Song Cao *et al.* devised a novel, one-pot, three-component approach for the synthesis of 1,4,5-trisubstituted 1,2,3-triazoles **37** through the cycloaddition of a wide range of primary alcohols **36** with sodium azide and active methylene ketones **22** at 80 °C in the presence of *N*-(*p*-toluenesulfonyl)imidazole (TsIm), tetrabutylammonium iodide (TBAI), and triethylamine (TEA) in DMF/DMSO (Scheme 4.12) [15]. Potassium hydroxide was used as the base for inducing the cycloaddition. The mild reaction conditions, high yields, and one-pot reaction without the necessity to isolate the unstable and hazardous

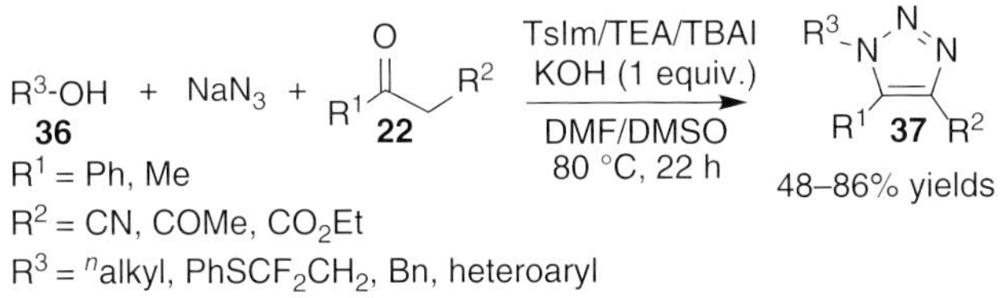


**Scheme 4.12** One-pot, three-component synthesis of 1,4,5-trisubstituted 1,2,3-triazoles starting from primary alcohols.

azides (especially aliphatic azide) are the main merits of this method, the most remarkable feature being easy access to fully substituted 1,2,3-triazoles from commercially available primary alcohols **36**.

Around the same time, Bakulev *et al.* showed that 1-trifluoromethyl-substituted 1,3-dicarbonyl compounds **38** undergo 100% regioselective cyclization with alkyl and aryl azides **2**/**7** in the presence of TEA at 70–80 °C for 5 h to form 4-acyl-5-trifluoromethyl-1,2,3-triazoles **39** (Scheme 4.13) [16]. Notably, the solvent-free reaction conditions proceeded smoothly and faster. The use of any solvents and other bases lowered the yield of the target products. The yields of the triazoles **39** depend mainly on the nature of substituents on the azide component. The use of azides bearing electron-withdrawing groups, such as 4-nitrophenyl and 4-chlorophenyl, afforded the triazoles in high yields. On the contrary, triazoles bearing *p*-methoxyphenyl or especially benzyl groups in position 1 were obtained in lower yields 12% and 9%, respectively.

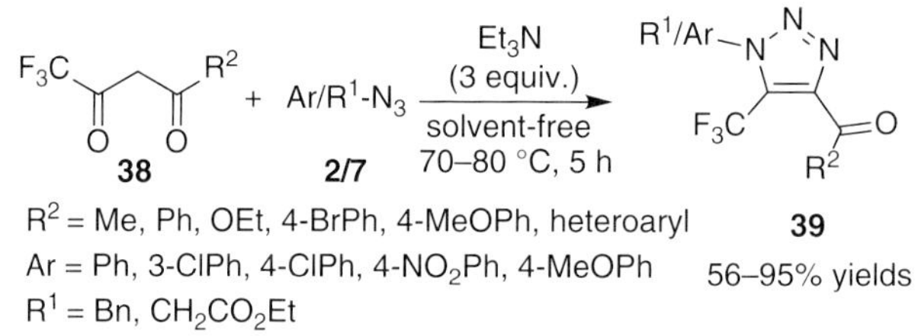


**Scheme 4.13** Regioselective synthesis of 5-trifluoromethyl-1,2,3-triazoles via $CF_3$-directed cyclization.

## 4.3 Preformed-Enamine-Based Synthesis of Substituted 1,2,3-Triazoles

In 1965, Pocar *et al.* prepared 1,3-diene-2-amines **42** by heating morpholine **41** with ketal derivatives of methyl $^n$alkyl ketones **40** and studied their cycloaddition reaction with aryl azides, yielding the corresponding morpholino triazoline, which on acidic treatment gave the triazole **43** (Scheme 4.14) [17].

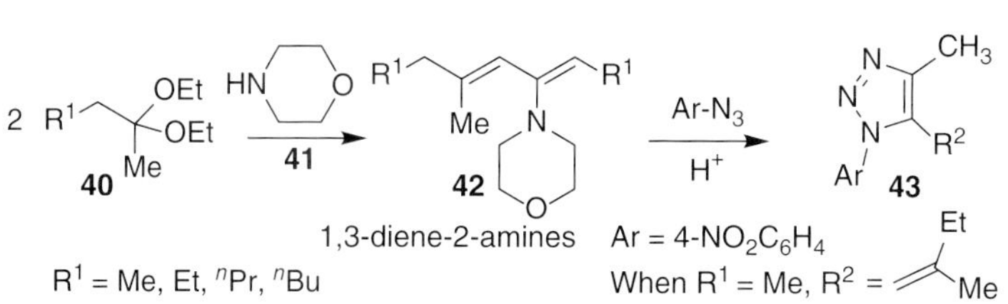


**Scheme 4.14** Reaction of dienamines with aryl azides.

Later, in the next year, Vigevani *et al.* prepared the dienamine **44**, which on reaction with 4-nitrophenylazide furnished the corresponding triazolinic adduct, which after treatment with 25% acetic acid produced the corresponding triazole **45** (Scheme 4.15) [18].

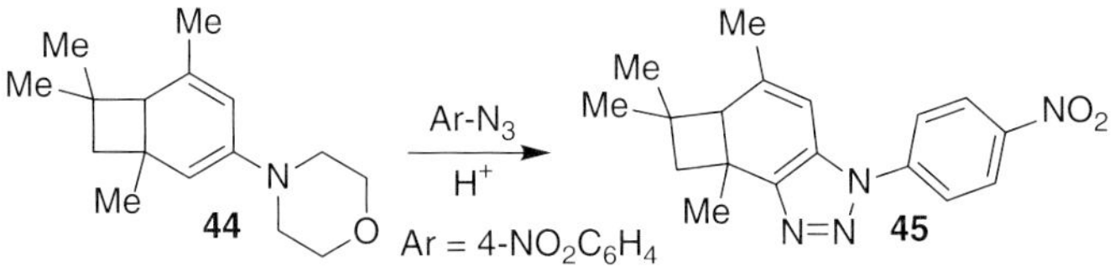


**Scheme 4.15** Reaction of cyclic dienamine with aryl azides.

In 1986, Gelmi *et al.* converted a mixture of stereoisomers of the aldehyde **46** exo–endo in the ratio 95 : 5 into the corresponding enamines by treatment with secondary amines, which underwent quick cycloaddition with 4-nitrophenylazide to furnish the triazole **48** via the triazoline intermediate **47** by a rearrangement at room temperature (Scheme 4.16) [19].

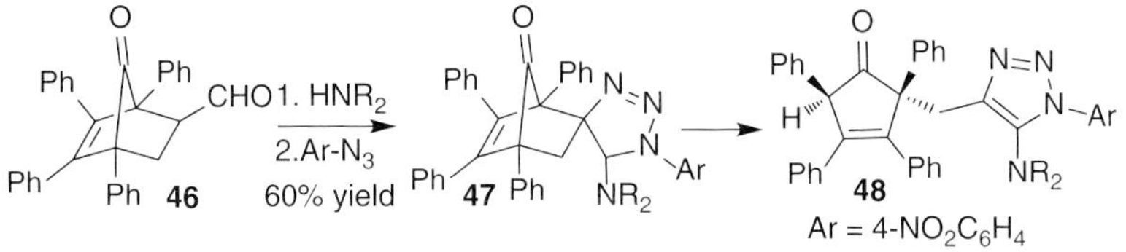


**Scheme 4.16** Reaction of enamines with aryl azides.

In the following decade, Emanuela Erba *et al.* prepared 1-aryl-4,5-dihydro-5-morpholino-1,2,3-triazoles **51** in a one-pot procedure, by reaction of an aldehyde **49** with morpholine **41** and the appropriate aryl azide in benzene solution at room temperature (Scheme 4.17) [20, 21]. The formation of the compounds **51** occurred through cycloaddition of the azide to the enamine **50** produced *in situ*.

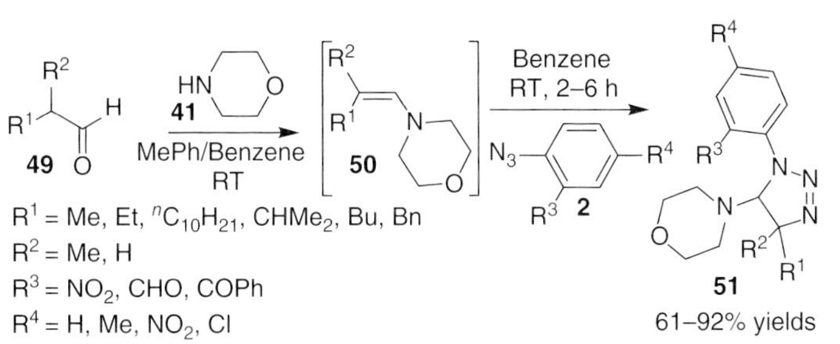


**Scheme 4.17** Synthesis of 1-aryl-4,5-dihydro-5-morpholinotriazoles.

In 2000, Helmut Quast *et al.* deprotonated 2-alkyl-1-methylquinolinium hexafluorophosphates **52** by sodium or potassium hydride to afford solutions of 2-alkylidenedihydroquinolines **53**, which on 1,3-dipolar cycloaddition with phenyl azide **15** yielded the spirocyclic products **54** (Scheme 4.18) [22].

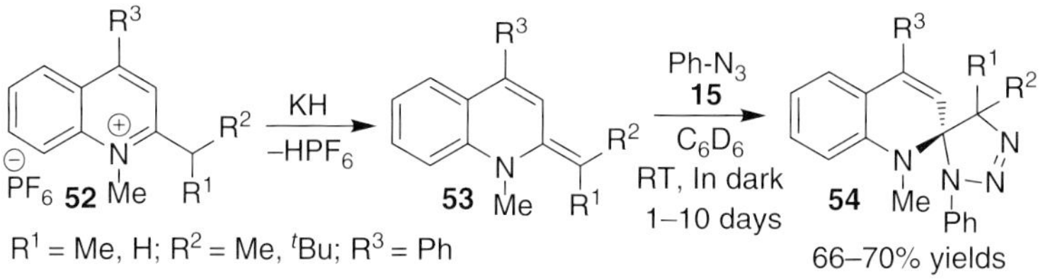


**Scheme 4.18** 1,3-Dipolar cycloaddition reactions of 2-alkylidenedihydroquinolines and phenyl azide.

In 2005, Gerhard Maas *et al.* investigated the cycloaddition of organic azides with some conjugated enamines of the 2-amino-1,3-diene, 1-amino-1,3-diene, and 2-aminobut-1-en-3-yne type. 2-Morpholinobuta-1,3-diene **55** underwent regioselective [3+2]-cycloaddition, with several electrophilic azides Ar-$N_3$ (Ar = 4-nitrophenyl, ethoxycarbonyl, tosyl, phenyl) to form 5-alkenyl-4,5-dihydro-5-morpholino-1*H*-1,2,3-triazoles, which were further transformed into 1,5-disubstituted 1*H*-triazoles **56** (Scheme 4.19) [23].

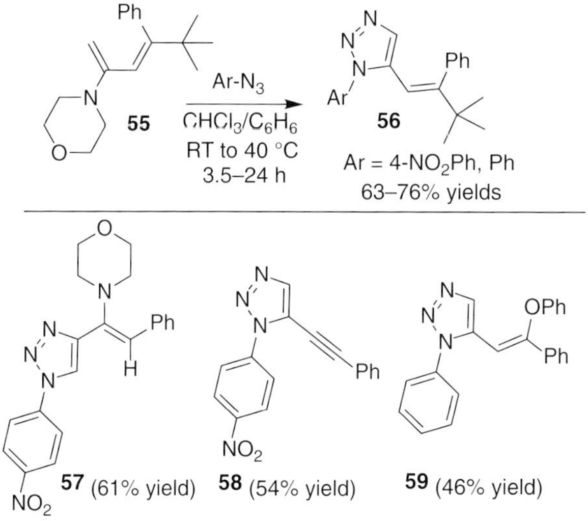


**Scheme 4.19** 1,3-Dipolar cycloaddition reactions of organic azides with morpholinobuta-1,3-dienes.

Very recently, in 2014, Bakulev *et al.* showed that the reactions of β-azolylenamines **60** with sulfonyl azides **61** in acetonitrile furnished the 1*H*-4-(azol-5-yl)-1,2,3-triazoles **62** in 50–93% yields (Scheme 4.20) [24]. β-Benzoylenaminones and β-nitroenamine of type **60** also reacted with tosyl

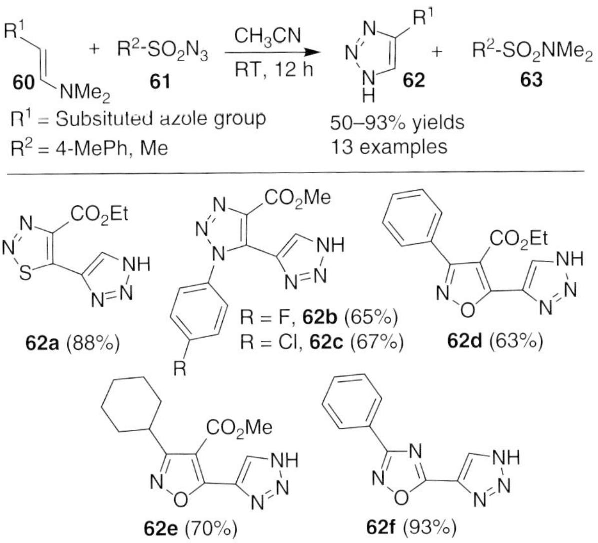


**Scheme 4.20** Reactions of β-azolylenamines with sulfonyl azides for the synthesis of N-unsubstituted 1,2,3-triazoles.

azide or mesyl azide to form the same type of products **62**, providing a general and efficient method for the synthesis of N-unsubstituted 1,2,3-triazoles **62**, via the 5-amino-1,2,3-triazoline intermediates, which generated the 1*H*-1,2,3-triazoles **62** by elimination of the corresponding *N,N*-dimethylamine followed by hydrolysis of Ts or Ms group.

## 4.4 Synthesis of Substituted 1,2,3-Triazoles via Catalytic Enolate Intermediates

In 2013, Ayoob Bazgir *et al.* reported 1,1,3,3-tetramethylguanidine (TMG)-catalyzed synthesis of 1,2,3-triazoles via the cycloaddition reaction of azides and CH acids **22** in ethanol at 30 °C, employing simple experimental procedure, short reaction times, mild reaction conditions to provide good yields (Scheme 4.21) [25]. In the actual procedure, a mixture of CH acid **22** (1 mmol), azide **2** (1 mmol), and TMG (15 mol%) in EtOH (5 ml) was stirred for an appropriate time at 30 °C. After completion, the solvent was removed under reduced pressure and the residue was washed with ether (5 ml) and recrystallized from $CHCl_3$/*n*-hexane (1 : 3) to afford the pure products **64**. A wide diversity of CH acids **22** and aryl azides **2** were acceptable for this reaction. The reaction of cyclic 1,3-diketones **65** with aryl azides **2** generated the bicyclic or tricyclic triazoles **66** in good isolated yields.

In the same year, Khurana *et al.* published [26] a novel environmentally friendly protocol for the synthesis of 1,4,5-trisubstituted 1,2,3-triazoles **68** by the reaction

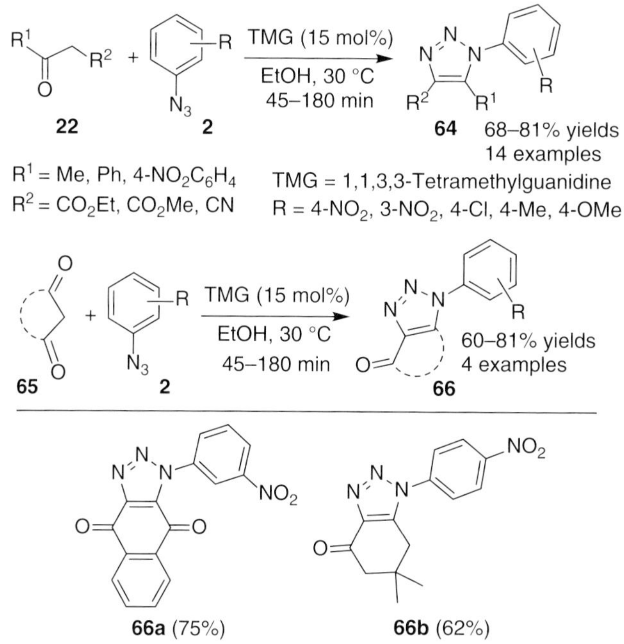


**Scheme 4.21** TMG (1,1,3,3-tetramethylguanidine)-catalyzed synthesis of 1,2,3-triazoles.

of various aryl, alkyl, and benzyl azides **2**/**7** with active methylene compounds **67** in a 1,8-diazabicyclo[5.4.0]undec-7-ene (DBU)–water system under conventional heating, ultrasonic, and microwave irradiation conditions (Scheme 4.22). They tested the recyclability of the catalyst and the solvent for the reactions under ultrasonic irradiation. The product was separated from the reaction mixture by

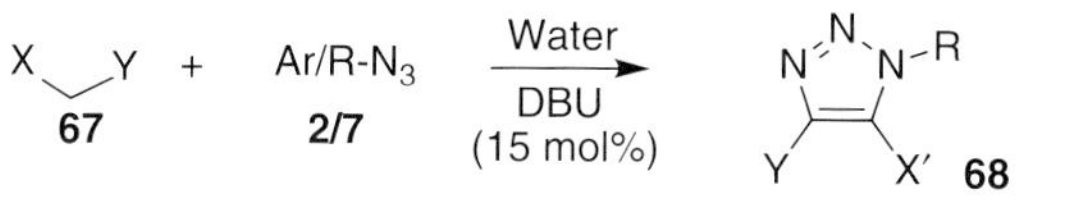


**67a** X = Y = COMe
**67b** X = COMe, Y = $CO_2Et$
**67c** X = COMe, Y = $CO_2Me$
**67d** X = Y = CN
**67e** X = Y = COPh

**68a** X′ = Me, Y = COMe
**68b** X′ = Me, Y = $CO_2Et$
**68c** X′ = Me, Y = $CO_2Me$
**68d** X′ = $NH_2$, Y = CN
**68e** X′ = Ph, Y = COPh

Ar = Ph, various substituted phenyl
R = alkyl, Bn

20 examples

Conditions
A: 60 °C, 5–80 min, 78–95% yields
B: )))), RT, 5–70 min, 75–93% yields
C: Microwave 250 W, 40 °C, 60–360 s, 72–94% yields

**Scheme 4.22** Recyclable DBU-$H_2O$ catalytic system for synthesis of 1,4,5-trisubstituted 1,2,3-triazoles.

simple filtration, which allowed quick recovery of the catalyst and the solvent for reuse in the next run. For few runs of the reaction, the catalyst showed no substantial reduction in activity; therefore, this system can act as an excellent recyclable reaction medium for the synthesis of trisubstituted triazoles in good yield. These protocols offered several advantages in terms of operational simplicity, tolerance of variety of alkyl/aryl azides and active methylene compounds, recyclability of both catalyst and solvent, easy workup, short reaction time, and good yields of product.

Very recently, Ramachary *et al.* reported an elegant synthesis of 1,4-disubstituted 1,2,3-triazoles **70** from enolizable aldehydes **69** and aryl azides **2** with 10 mol% of DBU as catalyst through a [3+2]-cycloaddition reaction under metal-free conditions (Scheme 4.23) [27]. The reactions were carried out in DMSO at room temperature and required only 30 min for completion to furnish the product with high rate and selectivity. The same reaction was also catalyzed by 10 mol% of nonamine bases $K_2CO_3$ and $KO^tBu$ to furnish the 1,2,3-triazoles **70** but only in moderate to good yields. The reaction was found to be general for aldehydes **69** with electron-rich and -poor aromatic substituents. Aryl azides containing functional groups, such as $NO_2$, $CO_2Et$, CN, $CF_3$, CHO, halogen, alkyl, and OMe, at different positions of the aromatic ring, furnished the expected 1,2,3-triazoles **70** in excellent to good yields, which were dependent on the nature of the substituent at the para position of **2**, high yields for electron-withdrawing groups, and slightly decreasing yields for alkyl and electron-donating groups. Aliphatic azides **7** showed poor reactivity. The reaction proceeds via the formation of an enolate, by the interaction of the catalyst with the aldehyde **69**

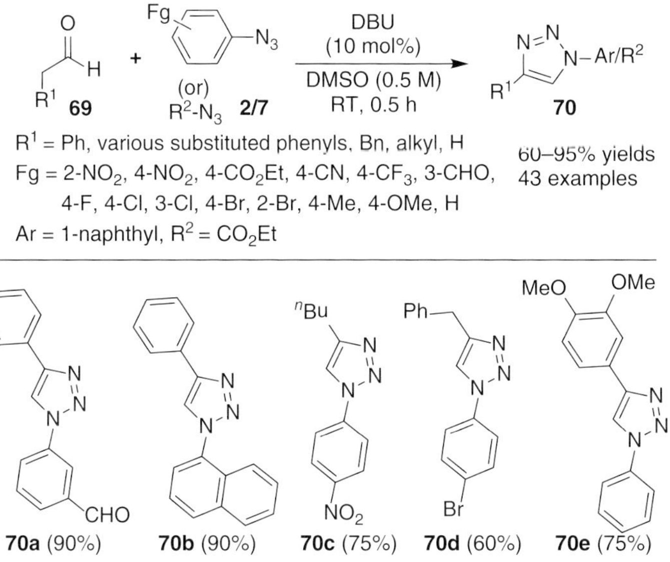


**Scheme 4.23** Organocatalytic azide–aldehyde [3+2]-cycloaddition for regioselective synthesis of 1,4-disubstituted 1,2,3-triazoles.

(refer Scheme 4.26). The triazoline intermediate formed from the enolate and azide undergoes a base-induced $H_2O$ elimination to furnish the 1,4-disubstituted 1,2,3-triazole **70.**

After reporting on the synthesis of 1,4-disubstituted 1,2,3-triazoles **70** from enolizable aldehydes **69**, Ramachary and coworkers extended their methodology toward trisubstituted triazoles **72** starting from enolizable ketones **71** [28]. They developed an enolate-mediated organocatalytic azide–ketone [3+2]-cycloaddition (OrgAKC) reaction of a variety of enolizable arylacetones and deoxybenzoins with aryl azides for the synthesis of fully decorated 1,4-diaryl-5-methyl(alkyl)-1,2,3-triazoles in excellent yields with high regioselectivity at 25 °C under 10 mol% of DBU catalysis (Scheme 4.24). The reaction was found to be general with a variety of aryl azides containing both electron-donating and -withdrawing groups with a slight decrease in reaction rate for *ortho*-substitution and also for electron-donating group substitution. It was found that the reaction was not affected by the substituents present on the aryl group of the ketone **71**. DBU-catalyzed reaction of **71** with benzyl/acyl/tosyl/mesyl azides did not furnish the triazoles, but the same reaction under $KO^tBu$ catalysis furnished the corresponding triazoles from benzyl and acyl azides. In the case of β-tetralone, the yield was found to decrease with the presence of the electron-donating substituent. By utilizing this OrgAKC strategy, the mGluR1 antagonist **72d** was successfully synthesized in 90% yield [28]. This reaction also follows the same enolate pathway as in the case of enolizable aldehydes **69**.

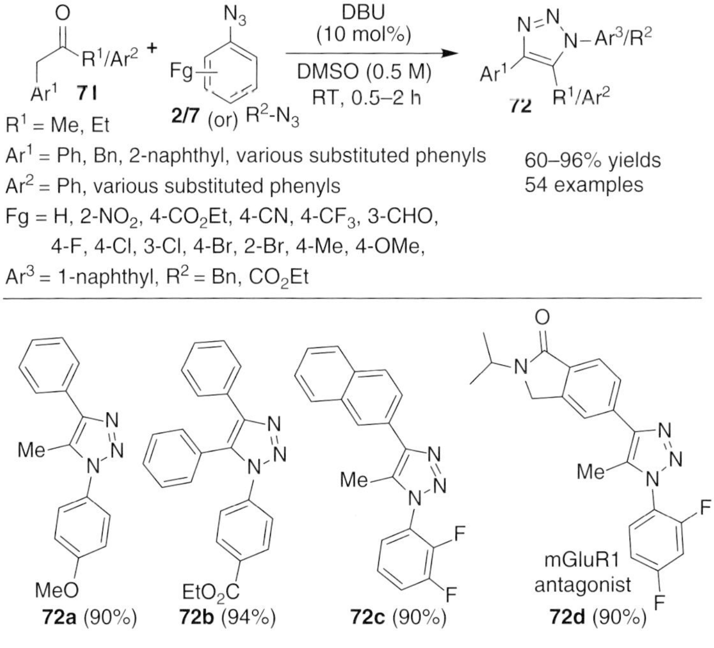


**Scheme 4.24** Organocatalytic azide–ketone [3+2]-cycloaddition for regioselective synthesis of fully decorated 1,2,3-triazoles.

The same group in their subsequent paper discussed a carbonate-catalyzed synthesis of 5-amino-1,2,3-triazoles **9** and **74–75** from substituted acetonitriles **73** and azides **2**/**7** in the presence of 10 mol% of $Cs_2CO_3$ in DMSO + $H_2O$ (7 : 3) at room temperature for 0.5–2 h (Scheme 4.25) [29]. The 5-amino-1,2,3-triazoles **9** and **74–75** were obtained with very high regioselectivity. The reaction was found to tolerate various substituents on the phenyl ring of the aryl acetonitriles **73**. Various substituted aryl azides generated the products in good to excellent yields. Even though benzyl and alkyl azides did not furnish the products under these conditions, the corresponding 5-amino-1,2,3-triazoles **74b**, **75** were obtained via KO$^t$Bu catalysis at room temperature.

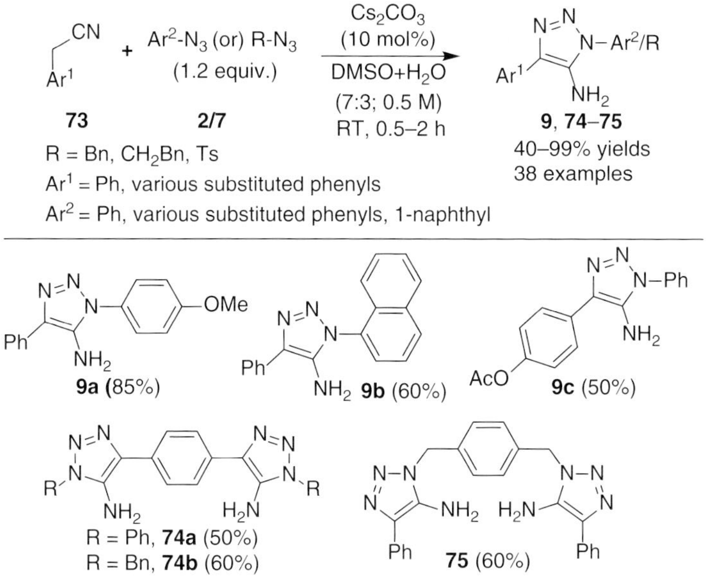


**Scheme 4.25** Carbonate-catalyzed synthesis of 5-amino-1,2,3-triazoles.

## 4.5 General Mechanistic Aspects of Enolate Route

The possible general mechanism for the enolate-mediated synthesis of 1,2,3-triazoles from active methylene compounds, enolizable aldehydes **69**, enolizable ketones **71**, and substituted acetonitriles **73** is illustrated in Scheme 4.26. First, the reaction of the catalyst (base) with the CH acid **76** generated the enolate **A**, which on *in situ* treatment with Ar-$N_3$ **2** selectively furnished the adduct 1,2,3-triazolines **B** through a concerted [3+2]-cycloaddition or stepwise amination–cyclization reaction. Adduct **B** further transformed into the 1,2,3-triazole **77** through the rapid elimination of water under ambient conditions induced by the basic nature of the catalyst.

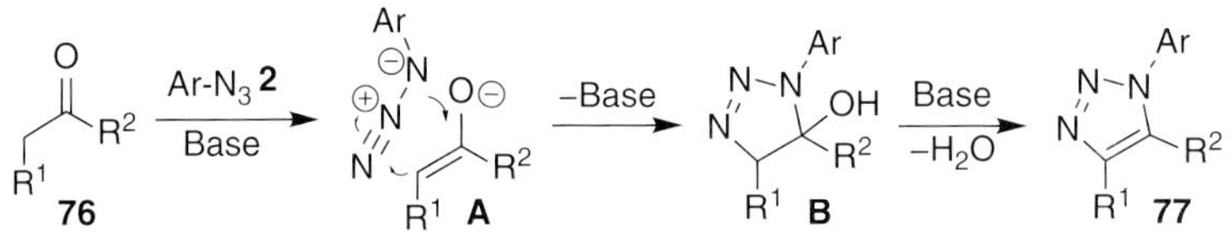


**Scheme 4.26** General mechanism for base-catalyzed 1,2,3-triazole synthesis.

## 4.6
## Synthesis of Substituted 1,2,3-Triazoles via Enamine Intermediates

For the first time, in 2008, Ramachary *et al.* reported a synthesis of fused triazoles **80** and **81** starting from Hagemann's ester **78**, catalyzed by proline **79** (Scheme 4.27) [30]. The cyclic enone **78** on treatment with tosyl azide **61a** in the presence of proline **79** furnished the *NH*-1,2,3-triazoles **80** in excellent yields. The reaction proceeded via a dienamine intermediate, which underwent cycloaddition with the azide. The triazole compound was formed by elimination of the catalyst. For substituted Hagemann's ester, a solvent-induced tosyl group hydrolysis took place to afford the *NH*-triazole **80**. In the case of simple enones, the tosyl group was intact under the reaction conditions as can be seen in triazole **81a**. Aliphatic as well as aromatic substituted Hagemann's esters furnished the corresponding fused *NH*-triazoles in good yields. Surprisingly, when the catalyst was changed from proline to a benzyl amine, the sole product obtained was α-diazo ketone. This reaction has given inspiration to many other chemists to work in this area for further development.

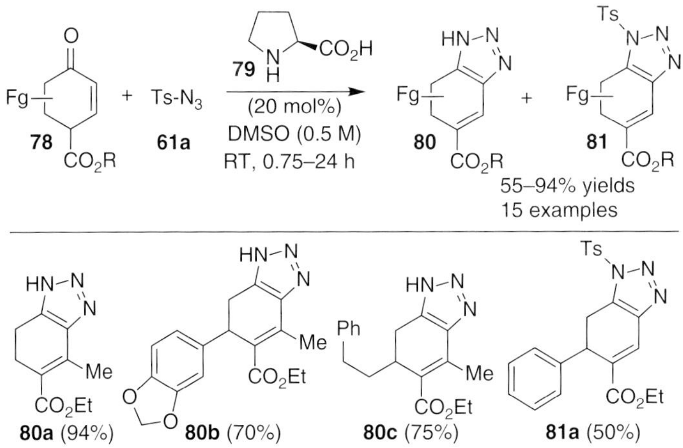


**Scheme 4.27** Synthesis of *NH*-1,2,3-triazoles via push–pull dienamine intermediate.

In 2011, Wang and coworkers developed the regioselective organocatalytic enamide–azide cycloaddition reaction for the synthesis of 1,4,5-trisubstituted

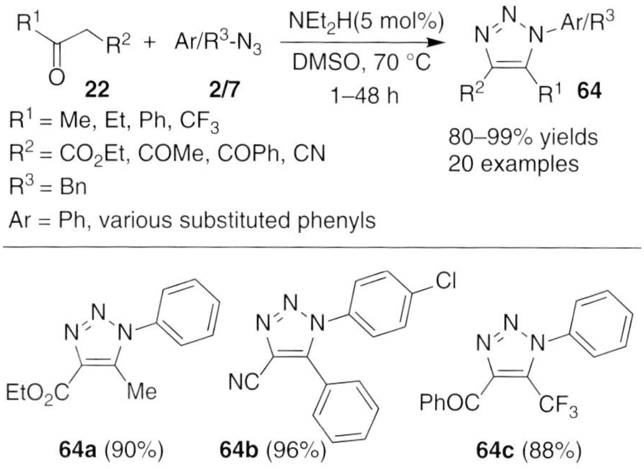


**Scheme 4.28** Organocatalytic enamide–azide cycloaddition for regiospecific synthesis of 1,4,5-trisubstituted 1,2,3-triazoles.

triazoles **64** from active methylene compounds **22** and organic azides **2**/**7** in the presence of 5 mol% of diethylamine in DMSO at 70 °C (Scheme 4.28) [31]. The reaction was found to work with a variety of active methylene compounds **22** and is applicable to a variety of aryl azides containing different substituents. The only aliphatic azide, benzyl azide was reported to afford the expected product in good yield. The triazole synthesis needed considerably less time when electron-poor azides were used in comparison with electron-rich analogs. A number of aryl ketones and ketoesters have been employed. As a result, multisubstituted 1,2,3-triazoles **64** were obtained in high to excellent yields (80–99%). Significantly, none of the regioisomers were formed. Moreover, this catalytic system also tolerated many synthetically useful functional groups, such as nitrile, ketone, and ester functional groups, which might be manipulated for accessing more sophisticated heterocyclic compounds. The authors have proposed an enamine mechanism as well as an enolate mechanism for the reaction.

Simultaneously, Pons-Bressy *et al.* reported an efficient organocatalytic method to synthesize highly substituted 1,2,3-triazoles **83**–**85** starting from unactivated ketones **82** and aryl azides **2** (Scheme 4.29) [32]. Proline **79** was chosen as the organocatalyst for this reaction, with the acid moiety being crucial for the elimination step as mentioned in the Ramachary's cycloaddition (see Scheme 4.27). The reaction required 20 mol% of proline **79** and a 2 : 1 ratio between ketone **82** and azide **2** in dichloromethane as solvent. The reaction needed to be run for a prolonged time (5–6 days) under thermal conditions (80 °C in a sealed tube) for completion. The long reaction time was reduced to 1 h by microwave irradiation at 80 °C, which proved to be equally high yielding. A wide range of aryl azides and ketones were used, with cyclic ketones giving the best results. It was found that the reaction worked only with aryl azides possessing electron-donating substituents. The reaction proceeded with high regio- and chemoselectivity with both cyclic and acyclic ketones. Notably, dissymmetrical cyclic ketones afforded the product

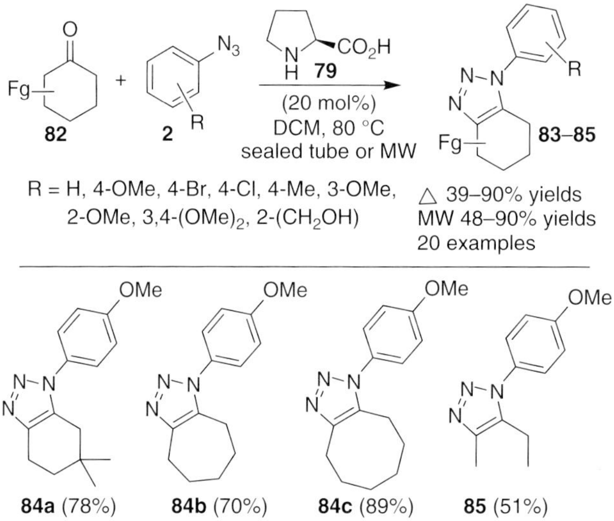


**Scheme 4.29** Organocatalytic synthesis of 1,2,3-triazoles from unactivated ketones and aryl azides.

**84a** with high levels of regioselectivity. Acyclic ketones gave lower yields of product **85**, under the same conditions, but acetophenone was unreactive.

After a while, Wang and coworkers developed an enamine-catalyzed strategy using pyrrolidine **86** (10–20 mol%) as the catalyst and DMSO as the solvent for an efficient synthesis of highly substituted 1,2,3-triazoles **83–85** and **87–88** (Scheme 4.30) [33]. The reaction was carried out at 80 °C and took 10–30 h for completion. This strategy has been applied to a broad range of carbonyl compounds and

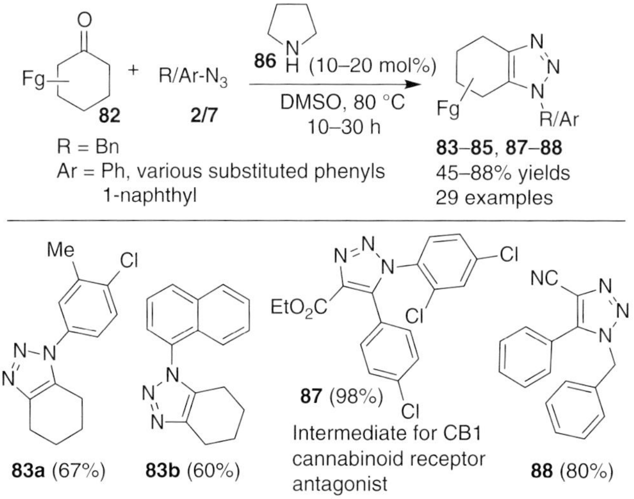


**Scheme 4.30** Pyrrolidine-catalyzed synthesis of highly substituted 1,2,3-triazoles.

phenyl azide, thereby providing efficient access to a vast pool of highly substituted 1,2,3-triazoles. In particular, the employment of commonly used and commercially available carbonyl compounds has allowed the introduction of a diverse set of functional groups, such as alkyl, aryl, nitrile, ester, and ketone groups, at the 1-, 4-, or 5-positions of the 1,2,3-triazole scaffold. Most significantly, the reaction process exhibits complete regioselectivity, with no other regioisomer formation. Drawback of this method is the less/no reactivity of aliphatic azides. Nonetheless, the triazole **88** was obtained in 80% yield by reaction of benzyl azide with active β-ketonitrile. By utilizing this method, Wang *et al.* prepared the CB1 cannabinoid receptor antagonist intermediate **87** in 98% yield.

The group of Alves–Paixao used β-enaminone–azide cycloaddition [34] for the synthesis of arylselanyl-1*H*-1,2,3-triazole-4-carboxylates **90** by reaction of azidophenyl arylselenides **89** with β-keto-esters **22** (Scheme 4.31). The reactions were performed under mild conditions using catalytic amount of $Et_2NH$ (1 mol%) in DMSO as solvent at room temperature for 3–24 h, and the corresponding products were obtained in good to excellent yields. The reaction was reported to work for a range of substituted β-keto-esters **22**, affording high yields of the desired products **90**. The position of the arylselanyl group does not seem to influence the yield of the reaction. Besides, azidophenyl arylselenides containing electron-donating group and electron-withdrawing group at the aromatic ring gave excellent yields of selenium-triazole-carboxylates **90**. Microwave experiments reduced considerably the time of these reactions to few minutes. Remarkably, similarly to β-keto-esters, ketone with a cyano group in the α-position (benzoylacetonitrile) also efficiently reacted with azidophenyl arylselenide **89** to generate selenium-triazole-carbonitrile **90d** in 91% yield, however, using 10 mol% of catalyst. The authors have postulated that the reaction

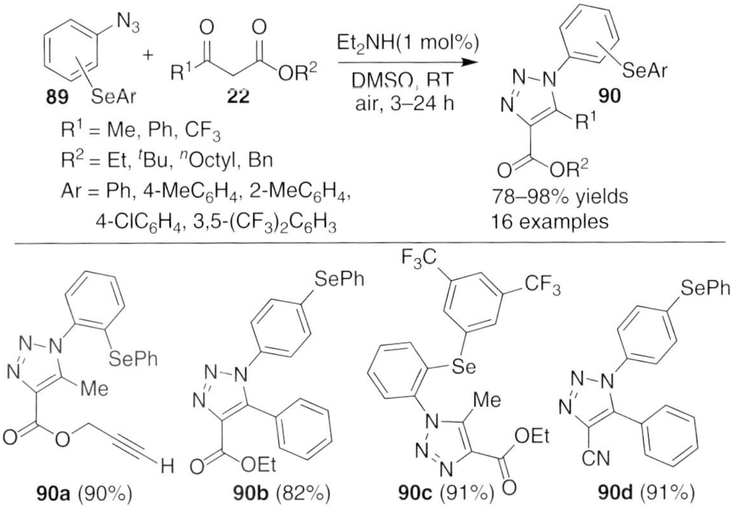


**Scheme 4.31** Synthesis of arylselanyl-1*H*-1,2,3-triazole-4-carboxylates by organocatalytic cycloaddition.

proceeded via an enamine intermediate but, at the same time, the possibility for an enolate mechanism cannot be ruled out.

In 2013, Jian Wang reported an organocatalytic dienamine-mediated 1,3-dipolar cycloaddition reaction between various azides with unsaturated aldehydes **91** for a straightforward synthesis of 4-alkenyl-1,2,3-triazoles **92** (Scheme 4.32) [35]. The reaction required 10 mol% of DBU in addition to the secondary amine catalyst, diethylamine in 10 mol%. The authors found that the yield diminished considerably in the absence of either secondary amine catalyst or DBU. The alkenyl triazole synthesis was found to be general regardless of the electron-donating or -withdrawing substituents on the α,β-unsaturated aldehydes **91** and aryl azides **2**. Aliphatic azides such as benzyl azide showed poor reactivity. This triazole synthesis was reported to work well with a long aliphatic chain containing aldehyde and π-extended unsaturated aldehyde. Initially, condensation of diethylamine and unsaturated aldehyde **91** formed the iminium species, which was converted to dienamine intermediate by base (DBU)-induced deprotonation. Regioselective cycloaddition took place between the *in situ* generated dienamine and the aryl azide to furnish the triazoline species, which eliminated diethylamine to yield the 4-alkenyl-1,2,3-triazole **92**.

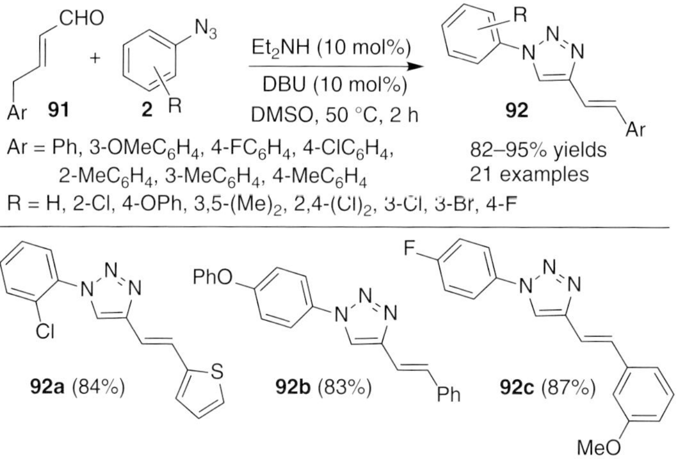


**Scheme 4.32** Diethylamine-catalyzed cycloaddition of azides to unsaturated aldehydes for the synthesis of 1,4-disubstituted 1,2,3-triazoles.

In 2013, Ramachary and Shashank reported the synthesis of trisubstituted 1,2,3-triazoles **93** via dienamine intermediate by treating the cyclic enones **78** with aryl azides **2** in the presence of 10 mol% of pyrrolidine **86** as the catalyst (Scheme 4.33) [36]. The reaction was performed in DMSO as solvent at room temperature. Both substituted and simple Hagemann's esters reacted in a similar manner. Various substituents on the aryl ring of the azide component were tolerated. Both electron-deficient ($NO_2$, $CF_3$, $CO_2Et$, CN, and Br) and neutral ($CH_3$ and H) aryl azides (Ar-$N_3$) **2** furnished the bicyclic *N*-aryl-1,2,3-triazoles **93** in excellent to good yields. But electron-donating 4-$MeOPhN_3$ did not furnish the triazole even at higher temperatures. The triazoles **93** generated were subjected to aromatization with 2,3-dichloro-5,6-dicyano-1,4-benzoquinone (DDQ) to furnish a variety of benzotriazoles **94** as shown in Scheme 4.33.

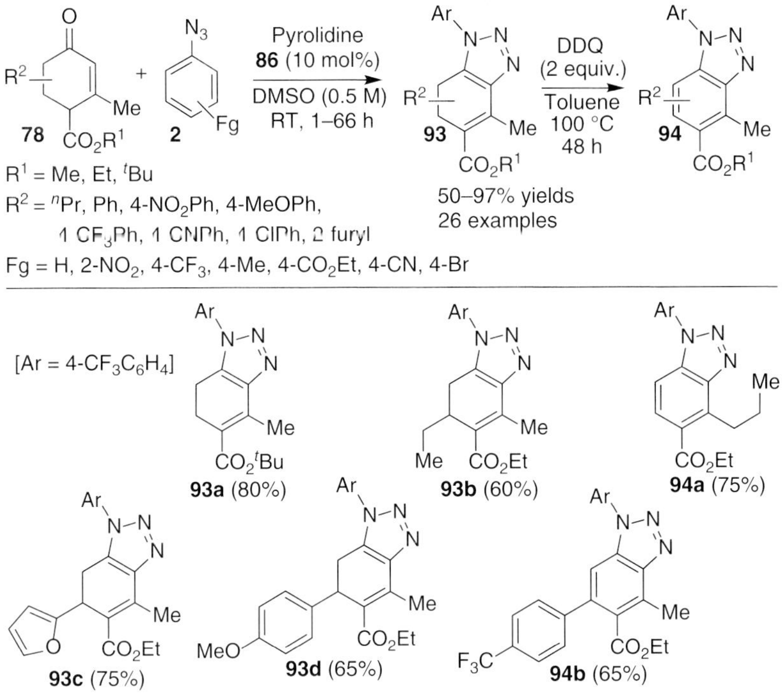


**Scheme 4.33** Synthesis of 1,4,5-trisubstituted 1,2,3-triazoles via dienamine intermediate.

The organocatalytic synthesis of substituted 1,2,3-triazoles **83** and **96** in water was reported by the group of Wang in the same year (Scheme 4.34) [37]. They designed a catalyst **95** with suitable hydrophobic groups that could bring reactants together and thus force the reaction to occur in hydrophobic pockets. These micelles could shield the enamine intermediate from water, thereby resulting in a reaction that is comparable to those carried out in organic solvents. The prolinamide catalyst **95** bearing two ⁿoctyl chains gave the best yield for the synthesis of substituted triazoles in water. The reactions of both activated (1,3-diphenyl-1,3-propanedione, ethyl acetoacetate, and ethyl benzoylacetate) and unactivated methylene compounds (various substituted cyclohexanones, cycloheptanone, and cyclooctanone) with aryl azides in water at 80 °C furnished the corresponding trisubstituted triazoles **83** and **96** in excellent yields. The regioselective outcome remained unchanged for this green methodology when compared to the reactions carried out in organic solvent. The reactions were rarely affected by the electronic nature of the substituents on the aromatic rings. Electron-withdrawing, electron-donating, or electron-neutral groups on the phenyl ring of azides did not affect

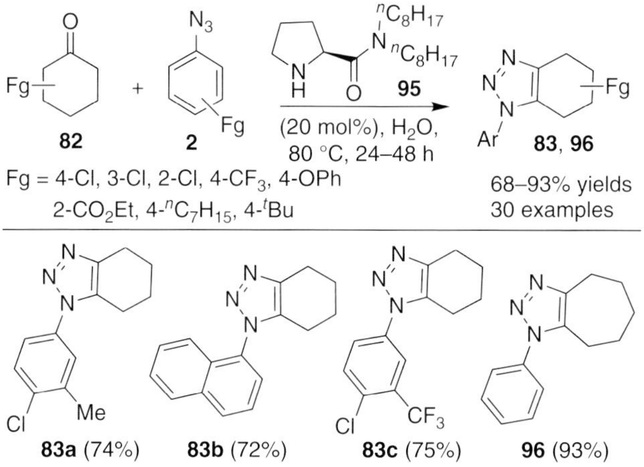


**Scheme 4.34** Organocatalytic 1,3-dipolar cycloaddition reactions of ketones and azides with water as a solvent.

the reaction. It is worth mentioning that a naphthalene ring was also suitable for this system to afford the desired product **83b** (72%).

As an extension, the group of Alves–Paixao used β-oxo-amides **97** in organocatalytic cycloaddition with aryl azidophenyl selenides **89** (Scheme 4.35) [38]. The cycloaddition reactions were performed under mild conditions, with β-oxo-amides **97** and aryl azidophenyl selenides **89** in the presence of a catalytic amount of $Et_2NH$ (5 mol%), in DMSO as solvent at room temperature for 2–4 h

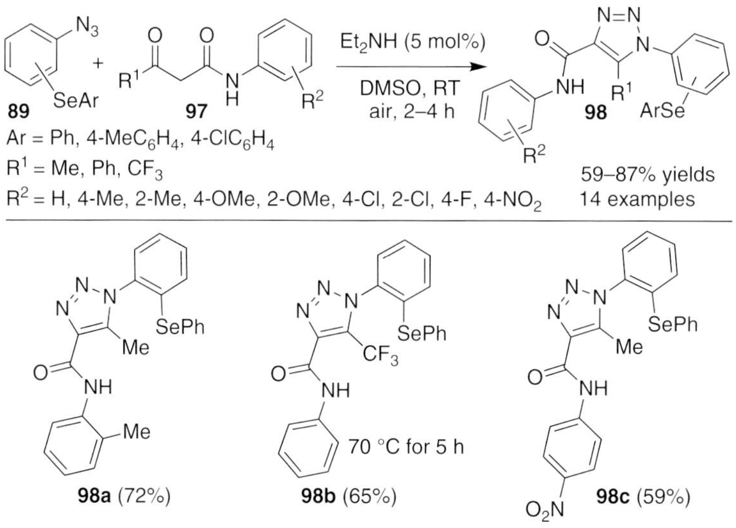


**Scheme 4.35** Organocatalytic synthesis of (arylselanyl)phenyl-1*H*-1,2,3-triazole-4-carboxamides by cycloaddition of azidophenyl arylselenides and β-oxo-amides.

and the corresponding products **98** were obtained in good to excellent yields. This method tolerated a range of substituents either in the β-oxo-amides or in the aryl azidophenyl selenides. The presence of electron-withdrawing substituents on the aryl group of the amide moiety seems to decrease the yield (**98c**). As discussed in the earlier case, here too, the position of the arylselanyl group does not have any effect on the yield of the reaction.

Recently, in 2014, Wang *et al.* reported a regioselective synthesis of trisubstituted triazoles **100** starting from allyl ketones **99** using a 2 : 1 mixture of aryl azide **2** and allyl ketone **99** in the presence of 10 mol% of diethylamine as the catalyst in DMSO at 80 °C (Scheme 4.36) [39]. The reaction worked well with aliphatic, aromatic, and heteroaromatic allyl ketones. Further, both aliphatic and

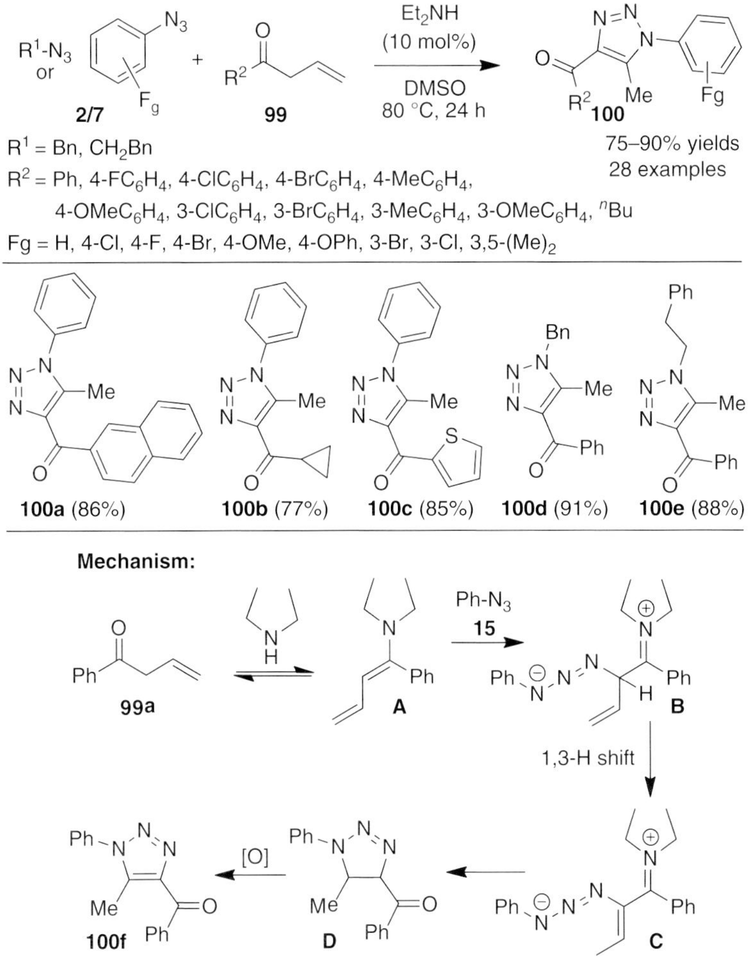


**Scheme 4.36** Organocatalytic 1,3-dipolar cycloaddition reaction of allyl ketones with azides for the synthesis of 1,4,5-trisubstituted 1,2,3-triazoles.

aromatic azides were compatible and gave excellent yields for substituted triazoles **100**. The substitution pattern of allyl aryl ketones **99** could be varied successfully: electron-withdrawing, neutral, and electron-donating substituents were tolerated and resulted in high yields. Remarkably, allyl heteroaryl ketones exhibited good reactivities and generated the triazoles **100c** in high yields. The authors proposed a plausible mechanism as shown in Scheme 4.36 to explain the reaction. Initially, the allyl ketone **99a** reacted with the catalyst diethylamine to form the dienamine intermediate **A**, which being an electron-rich olefinic partner reacted with phenyl azide **15** to furnish the intermediate **B**. Subsequent 1,3-H shift allowed conversion of intermediate **B** to intermediate **C**, which then underwent an intramolecular addition to afford the triazoline intermediate **D**. Eventually, aerobic oxidation of the intermediate **D** generated the final product **100f**.

In 2015, the same group reported a 1,3-dipolar cycloaddition reaction of commercially available aldehydes **69** with azides and secondary amines through a one-pot strategy (Scheme 4.37) [40]. The reaction was done by taking a mixture of azide **2**/**7** (1 equiv.), aldehyde **69** (2 equiv.), and pyrrolidine **86** (1.0 equiv.) in THF (0.5 ml) and stirring at 50 °C for 12 h. After the reaction was complete, *m*-CPBA (1.5 equiv.) was added at 0 °C and the mixture warmed up to room temperature for 1 h to furnish 1,4-disubstituted 1,2,3-triazoles **70** and **101** in good to excellent yields and high levels of regioselectivity. Using this methodology, they accomplished the formal synthesis of Trypanosoma cruzi *trans*-sialidase (TcTS) inhibitor **70i**. A mixture of the corresponding sugar azide, 3-phenylpropanal and pyrrolidine **86** was allowed to react under the standard conditions to lead to an intermediate triazole, which on deacetylation in the presence of sodium methoxide furnished the target TcTS inhibitor **70i** (Scheme 4.37, 90% yield).

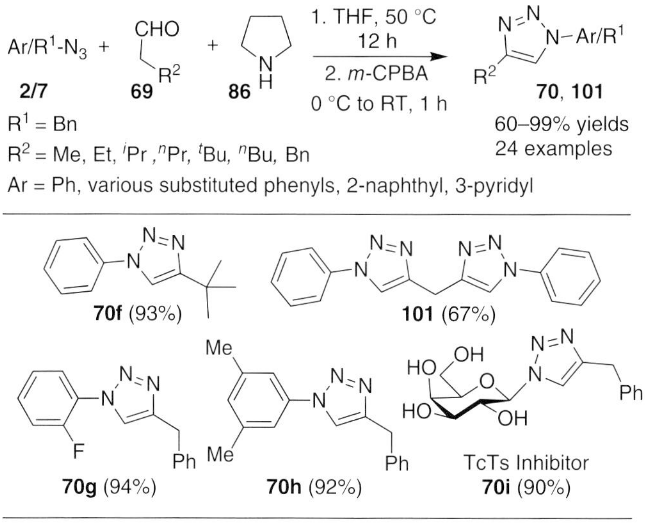


**Scheme 4.37** One-pot synthesis of 1,4-disubstituted 1,2,3-triazoles.

## 4.7
## General Mechanistic Aspects of Enamine Route

The general plausible mechanism proposed for this enamine-based triazole synthesis is shown in Scheme 4.38. The first step is most likely the condensation of the organocatalyst with the carbonyl compound **22** to generate the enamine intermediate **A**, which acts as the electron-rich olefinic partner and reacts with the aryl azide **2** to form the intermediate **B**. The cycloaddition adduct, intermediate **B** undergoes a 1,3-hydride shift to generate the triazoline intermediate **C**. The zwitterion **D** formed from the triazoline then eliminates the organocatalyst to furnish the 1,2,3-triazole **64**.

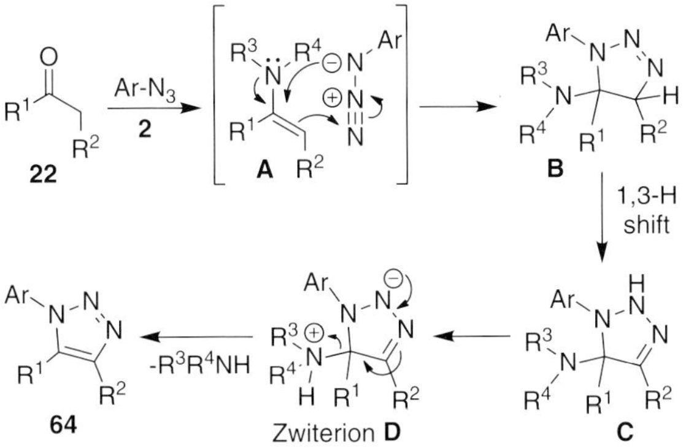


**Scheme 4.38** General mechanism for enamine-mediated 1,2,3-triazole synthesis.

## 4.8
## Synthesis of Substituted 1,2,3-Triazoles via Iminium Intermediate

Wang *et al.* very recently reported the synthesis of 1,4,5-trisubstituted 1,2,3-triazoles **104** in good yields with high levels of regioselectivity from α,β-unsaturated ketones **102** (Scheme 4.39) [41] through iminium activation. In their procedure, they used the α,β-unsaturated ketones **102** and the azides **2**/**7** in the ratio of 1 : 2 for best yields and the reactions were performed in DMSO solvent at 80 °C for 72 h. The reaction was general for aromatic α,β-unsaturated ketones **102**, for both electron-donating and electron-withdrawing groups, the corresponding products were obtained in good to high yields (77–89%) irrespective of the substitution pattern. Moreover, the heterocyclic rings such as furan and thiophene rings were also tolerated in the substrates, affording the desired products in high yields (84–87%). Even for less reactive α,β-unsaturated alkyl ketones, good yields were obtained (75–72%). The aryl azides **2**, containing electron-donating groups and electron-withdrawing groups, were all tolerated for this reaction, giving the 1,2,3-triazoles in high yields. This reaction worked well for heteroaromatic azides also. Even for alkyl azides, the reaction proceeded smoothly to give the triazole **104c** in 89% yield.

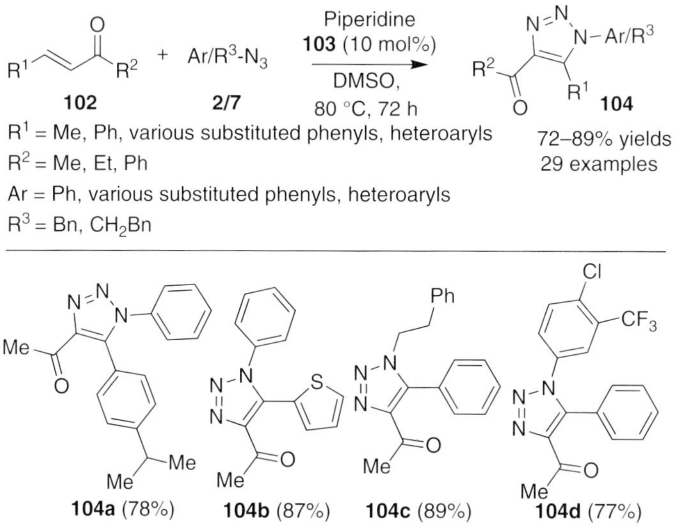


**Scheme 4.39** Organocatalytic 1,3-dipolar cycloaddition of α,β-unsaturated ketones with azides for the synthesis of 1,4,5-trisubstituted 1,2,3-triazoles.

The authors supported the reaction with a possible mechanism via an iminium intermediate (Scheme 4.40). In the first step, the α,β-unsaturated ketone **102a** reacts with the catalyst piperidine **103** to generate the iminium intermediate **A**. Cycloaddition between the iminium species **A** and the azide **15** generated the triazoline intermediate **B**, which on hydrolysis of the iminium center and subsequent air oxidation of the triazoline moiety in **C** resulted in the formation of the fully substituted triazole **104e**.

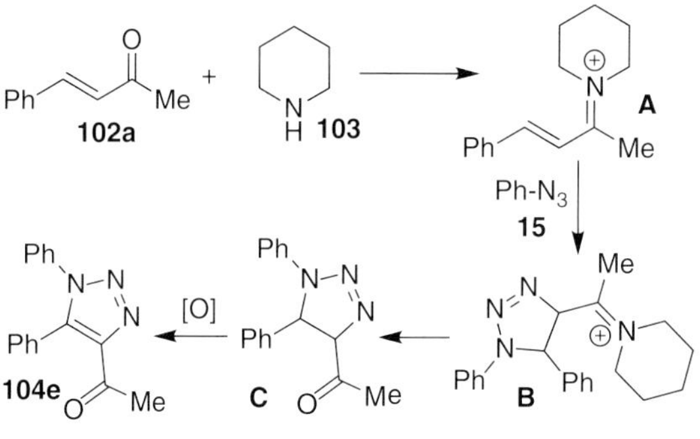


**Scheme 4.40** Mechanism of 1,2,3-triazole formation via iminium intermediate.

## 4.9 Miscellaneous Routes for the Synthesis of 1,2,3-Triazoles

Although various research efforts have resulted in alternative cyclization strategies, even now alkynes are frequently employed for the formation of

1,2,3-triazoles. In 2010, Fokin *et al.* reported the use of terminal alkynes **105** to generate reactive acetylide intermediates *in situ*, which reacted with aryl azides **2** in DMSO in the presence of catalytic tetraalkylammonium hydroxide under mild, transition-metal-free conditions, resulting in exclusive formation of 1,5-diarylsubstituted 1,2,3-triazoles **106** and **107** (Scheme 4.41) [42]. The reaction is experimentally simple and is not sensitive to atmospheric oxygen and moisture. Most aryl and heteroaryl azides and terminal alkynes readily participated in this 1,5-diaryl-substituted 1,2,3-triazole synthesis. $NMe_4OH$ was found to be the preferred base, due to the favorable yield and the shorter reaction time. Electronic properties of both the reactants seem to significantly influence the outcome of the reaction. Notably, the reaction showed that electron-deficient aryl azides, as well as aryl azides with sterically demanding *ortho*-substituents, afforded lower yields. In the case of acetylenes, aryl groups containing strong electron-withdrawing groups afforded low yields, possibly due to the reduced nucleophilicity of the acetylide anion. It is worth mentioning that base-labile functionalities were tolerated under the reaction conditions. Disappointingly, alkyl acetylenes failed to react under the current conditions, probably due to their lower acidity.

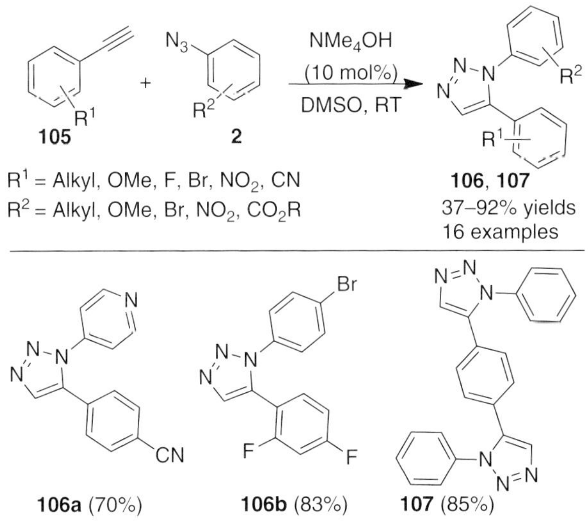


**Scheme 4.41** Tetramethylammonium-hydroxide-catalyzed synthesis of 1,5-disubstituted 1,2,3-triazole.

A while later, Schubert *et al.* showed that [3+2]-cycloaddition reactions of aryl azides **2** and silylated alkynes **108** in aqueous media yielded 1,5-disubstituted 4-(trimethyl-silyl)-1*H*-1,2,3-triazoles **109** (Scheme 4.42) [43]. Strikingly, the usage of water as the reaction medium entails economic advantages. The formation of 1,5-isomer is highly favored in this metal-free cycloaddition. The authors illustrated the high regio-preference induced by the trimethylsilyl (TMS) moiety. All

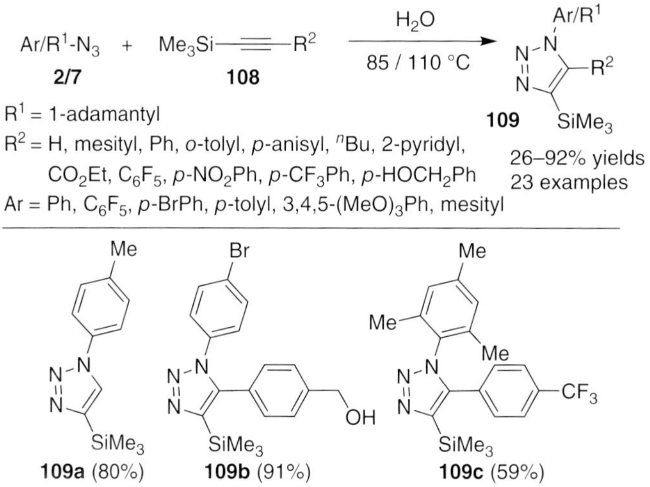


**Scheme 4.42** Synthesis of 1,5-disubstituted 4-(trimethylsilyl)-1*H*-1,2,3-triazoles.

three substituents have a convergent influence on the regioselectivity, whereas the TMS group has the most powerful effect, substituting the 4-position in the triazole ring exclusively.

In 2012, Mark J. Kurth *et al.* developed an operationally simple, one-pot, two-step cascade method to afford quinazolino[1,2,3]triazolo[1,4]benzodiazepines **112** (Scheme 4.43) [44]. Fascinatingly, this unique, atom-economical transformation engaged five reactive centers (amide, aniline, carbonyl, azide, and alkyne) and employed environmentally benign iodine as a catalyst. The method seemed to proceed via sequential quinazolinone-forming condensation and intramolecular azide–alkyne 1,3-dipolar cycloaddition reactions. The substrate

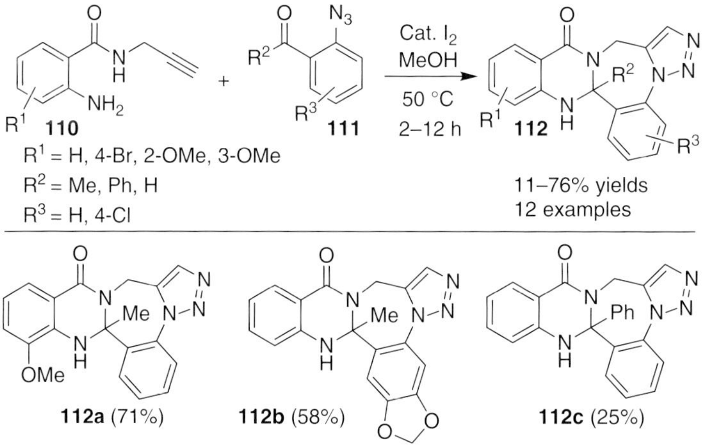


**Scheme 4.43** One-pot, two-step synthesis of 1,5-fused-1,2,3-triazoles.

scope showed that $R^1$ group substitution has significant effect on the product yield. The electron-donating substituents in **110** were well tolerated by this protocol, whereas electron-withdrawing substituent (Br) resulted in lower yield. Not surprisingly, when $R^2$ is a phenyl group in the azide, the lowest yields (**112c**, 25%) were obtained, as a consequence of steric congestion.

In the subsequent year, Xiuling Cui *et al.* published an efficient synthetic methodology for preparing 1,5-disubstituted 1,2,3-triazoles based on an unprecedented cascade Michael addition/deacylative diazo transfer/cyclization reaction of primary amines **113**, propynones **114**, and sulfonyl azides (Scheme 4.44) [45]. Of the various bases tested for the reaction, $LiO^tBu$ afforded the best results. The authors employed a straightforward one-pot strategy to synthesize 1,5-disubstituted 1,2,3-triazoles **115** directly from the aliphatic amines **113** and propynones **114**, without isolating the intermediate enaminone. A mixture of **113** and **114** in toluene was stirred at 80 °C under an air atmosphere until the starting materials were consumed completely, after which the reaction mixture was cooled to ambient temperature. $TsN_3$ and $LiO^tBu$ were added and then stirred at room temperature for 30 min to obtain the products **115** in similar yields compared to the two-step reaction. Various functional groups, such as terminal alkynes and azides, were showed to be well tolerated in this transformation compared with the 1,3-dipolar azide–alkyne cycloaddition. Moreover, chiral triazoles could be obtained in high yield from readily available chiral α-amines. Remarkably, both aryl- and aliphatic-substituted propynones **114** gave the corresponding triazoles **115** in good yields.

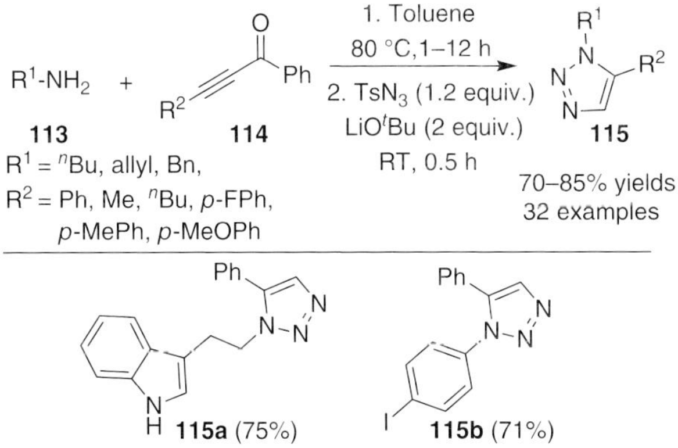


**Scheme 4.44** Multicomponent cascade reaction for regiospecific synthesis of 1,5-disubstituted 1,2,3-triazoles.

Pathak *et al.* developed a metal-free route to 1,5-disubstituted 1,2,3-triazolylation at C1, C2, C3, C4, and C6 of pyranosides generating 1,5-triazolylated monosaccharides and 1,5-triazole-linked disaccharides for the first time. Exocyclic vinyl sulfones **117** and **118** derived from styrene epoxide and monotosylated glycerol were reacted with six different azidopyranosides **116** to generate 1,5-disubstituted 1,2,3-triazoles **119** and **120** (Scheme 4.45) [46]. The reaction was

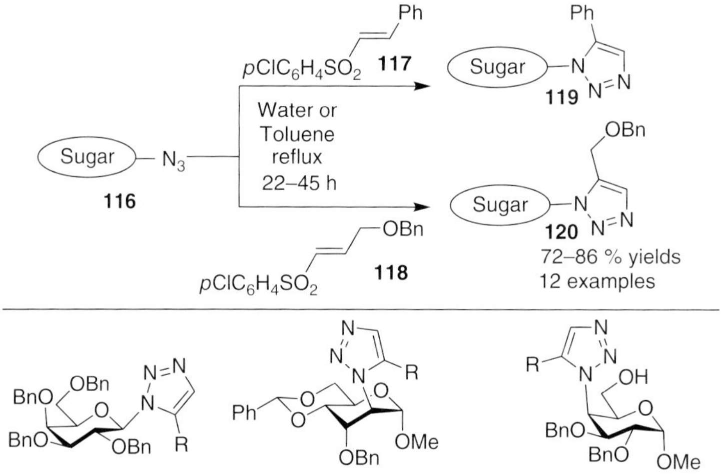


**119a** (R = Ph, no product) **119b** (R = Ph, 72%) **119c** (R = Ph, 72%)
**120a** (R = $CH_2OBn$, 78%) **120b** (R = $CH_2OBn$, 85%) **120c** (R = $CH_2OBn$, 83%)

**Scheme 4.45** Synthesis of 1,5-disubstituted 1,2,3-triazolylated carbohydrates.

performed mostly in water at elevated temperature without any metal catalyst to afford regioselectively 1,5-disubstituted triazolylated pyranosides **119** and **120** in high yields 72–86%. Steric bulk around the azido and vinyl sulfone groups seemed to play a significant role in deciding the outcome of the reactions.

Tanimoto *et al.* described a regioselective rapid triazole synthesis at low temperature. Organic azides and propargyl cations generated by acids from alcohols **121** furnished fully substituted 1*H*-1,2,3-triazoles **122** (Scheme 4.46) [47]. Most reactions were reported to be performed within 5 min not only at room temperature but also at −90 °C. Both terminal and internal alkynes were acceptable, and

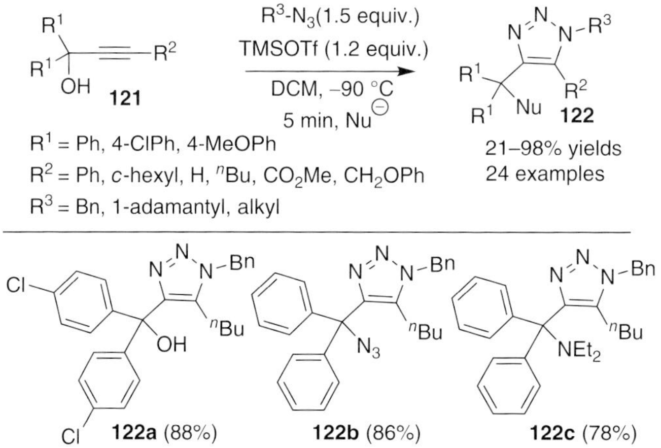


**122a** (88%) **122b** (86%) **122c** (78%)

**Scheme 4.46** Propargyl-cation-mediated rapid synthesis of fully substituted 1,2,3-triazoles.

even the sterically bulky substituents afforded the products smoothly. As for the substituents on alcohols **121**, R[1] being electron-deficient aryl was found to be effective as a phenyl group **122a**, whereas electron-donating aryl furnished low yield. The authors have demonstrated various types of three-component coupling reactions and indicated the presence of allenylaminodiazonium intermediates. As a result of the strong reactivities of both propargyl cations and diazonium intermediates, trisubstituted triazoles functionalized with additional nucleophiles **122b** and **122c** could also be obtained.

Very recently, Dehaen group reported a metal-free three-component reaction to synthesize 1,4,5-trisubstituted 1,2,3-triazoles **125** from readily available building blocks, such as nitroalkanes **123**, aldehydes **124**, and organic azides **2**/**7**, which involved sequential nitroalkene formation and addition of organic azides (Scheme 4.47) [48]. The reaction featured an excellent substrate scope, and the products were obtained with high yield and regioselectivity. Many functional groups, such as aldehydes, ketones, esters, sulfones, halides, amides, and phosphonates, were well tolerated under these reaction circumstances. Aliphatic aldehydes afforded somewhat lower yields, possibly due to the undesired aldol condensation under the employed reaction conditions. This three-component reaction was also applicable to acetal protected aldehydes, which gave access to triazole heterocycles **125a** that cannot be synthesized by other methods. Diethyl chloroacetal yielded **125b** in 52% yield with good regioselectivity. The authors exemplified the versatility of this method by synthesizing a metal-free tetraarylporphyrin functionalized with four fully substituted 1,2,3-triazole groups.

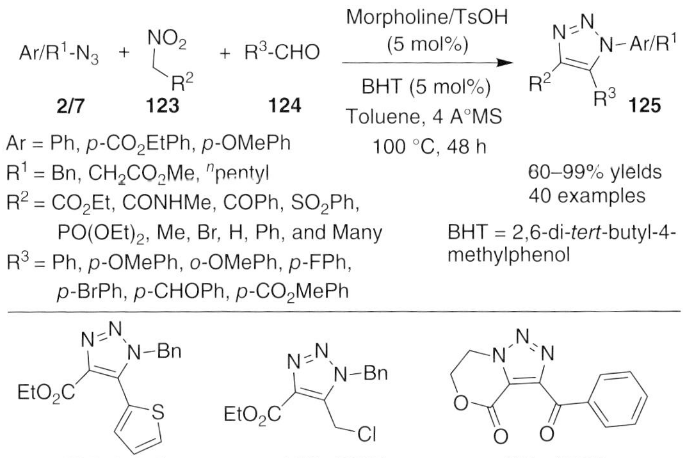


**Scheme 4.47** Synthesis of 1,4,5-trisubstituted 1,2,3-triazoles via three-component reaction.

Wang group presented their interesting discovery of a novel aerobic oxidative intermolecular azide–zwitterion reaction catalyzed by DBU (Scheme 4.48) [49]. The authors have demonstrated that the merger of the Lewis base, DBU and

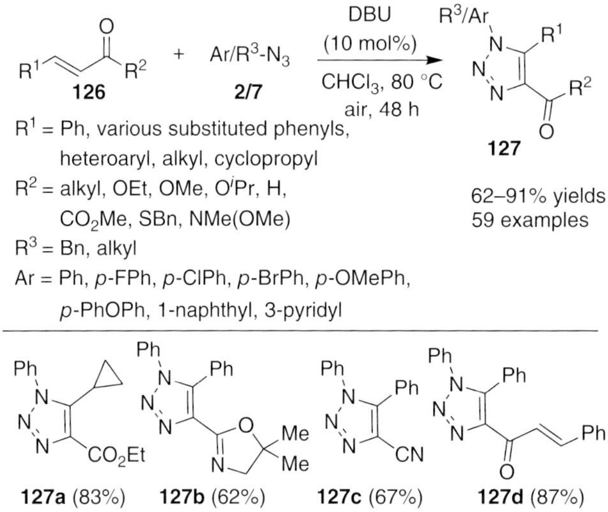


**Scheme 4.48** Lewis-base-catalyzed synthesis of 1,4,5-trisubstituted 1,2,3-triazoles via azide–zwitterion cycloaddition.

electron-deficient olefins **126** generated reactive zwitterion intermediates, which readily participated in cycloaddition reactions with an array of azides **2**/**7**, thus providing facile entry to fully or highly substituted 1,2,3-triazole frameworks **127**. The reaction showed an excellent substrate scope, and the products **127** were obtained in high yields with excellent regioselectivities. Utilizing this methodology, the authors accomplished the synthesis of CB1 cannabinoid receptor antagonist, by preparing the intermediate **87** in 78% yield. A variety of electron-deficient alkenes participated in this efficient cycloaddition to furnish a wide range of fully or highly substituted 1,2,3-triazoles **127**.

The mechanism proposed by the authors is depicted in (Scheme 4.49), where DBU first reacts to some extent with the starting α,β-unsaturated ester **126a** to form the zwitterion intermediate **A**. Subsequent addition of **A** to **15** resulted in

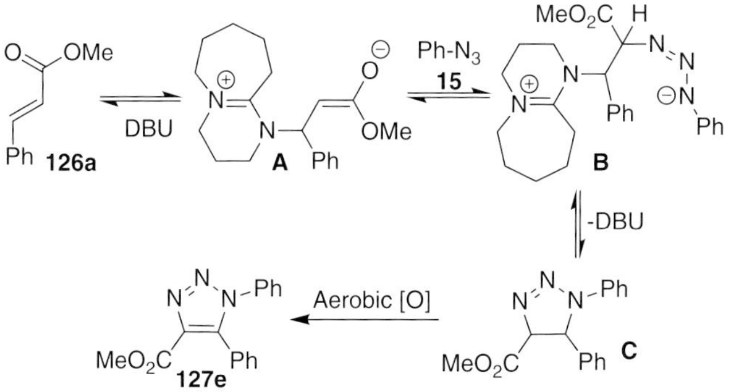


**Scheme 4.49** Mechanism for 1,2,3-triazole synthesis via azide–zwitterion cycloaddition.

the formation of the intermediate **B**, which underwent an $S_N2$ reaction together with elimination of DBU to generate the cyclic triazoline intermediate **C**, which on subsequent aerobic oxidation produced the desired 1,2,3-triazole **127e**.

Recently, Paixao *et al.* developed a metal-free methodology for the regioselective synthesis of 1,4-disubstituted 1,2,3-triazoles **70**, by applying a novel inverse electron-demand-1,3-dipolar cycloaddition approach (Scheme 4.50) [50]. This one-pot reaction can be accomplished with various alkylidene malononitriles and aromatic azides (electron-neutral and different electron-deficient) in the presence of a base. Knoevenagel condensation of malononitrile **128** and an aldehyde **69** generated the alkylidene malononitrile, which on deprotonation with DBU furnished a vinylogous carbanion, which operated as the electron-rich olefinic partner in the cycloaddition to azides. Finally, an elimination step delivered the final product, recycling malononitrile. Substitution patterns of the electron-withdrawing groups on the aryl azides have little effect on the reactivity. Unfortunately, aryl azides containing electron-donating groups as well as benzyl azides were not compatible with this transformation. This new strategy showed an excellent tolerance to a set of aliphatic aldehydes. Notably, the one-pot protocol enhanced the yield drastically. Gratifyingly, this concept was extended to include the formation of optically active triazoles.

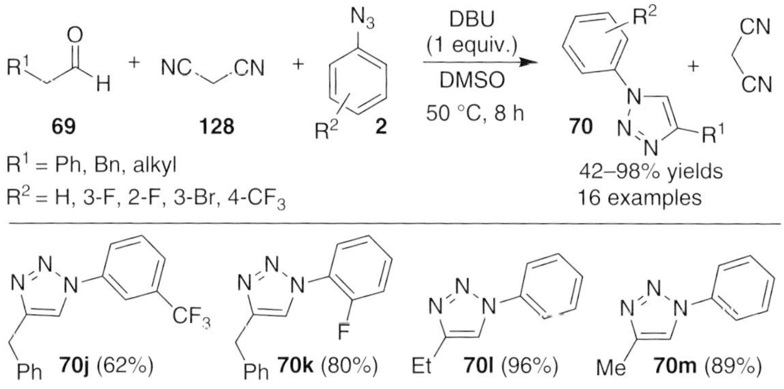


**Scheme 4.50** One-pot strategy for synthesis of 1,4-disubstituted 1,2,3-triazoles.

Yuhang Zhang *et al.* demonstrated a novel synthetic approach toward 1,4-disubstituted 1,2,3-triazoles **130** by C–N and N–N-bond formation, under transition-metal-free and mild conditions (Scheme 4.51) [51]. Significantly, the operationally simple transformation was executed in one-pot from commercially available anilines **113**, ketones **129**, and *N*-tosylhydrazine in the presence of molecular iodine to regioselectively afford 1,4-disubstituted 1,2,3-triazoles **130** in high yields, without the use of azides. The substrate scope is broad and the reaction could be readily scaled up to gram scale, offering a practical approach for the production of diverse 1,2,3-triazoles. Even though, the method displayed good tolerance toward aryl ketones **129** containing a range of electron-withdrawing groups, including $CF_3$, $NO_2$, and $CO_2H$, the aryl ketones

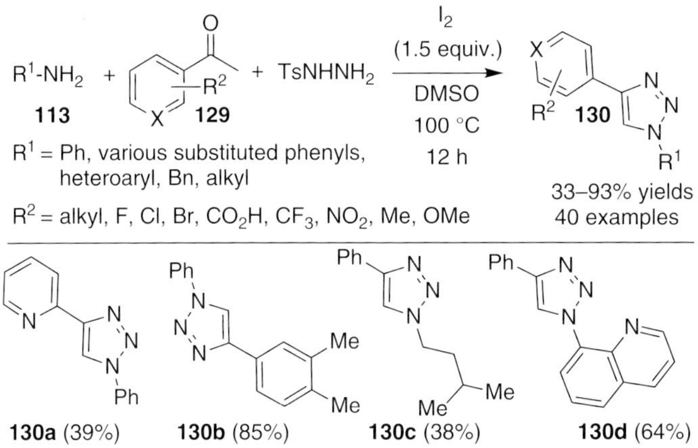


**Scheme 4.51** Synthesis of 1,4-disubstituted 1,2,3-triazoles in one pot from ketones, *N*-tosylhydrazines, and amines.

that bore electron-donating groups still showed better reactivity and gave higher yields. Steric hindrance played little role in this transformation, and both *ortho*- and *meta*-substituted aryl ketones presented excellent reactivity. In the case of anilines **113**, compared to electron-rich anilines, electron-deficient anilines showed better reactivity.

The possible mechanism for the reaction is shown in Scheme 4.52 [51], the first step being the formation of *N*-tosylhydrazone **A**, by the reaction of an arylketone (acetophenone **129a**) with *N*-tosylhydrazine. Subsequent α-iodination of the *N*-tosylhydrazone **A** formed the intermediate **B**. Elimination of HI from **B** gave the key intermediate 1-tosyl-2-vinyldiazene **C**, which underwent aza-Michael addition with aniline **113a** to generate the intermediate **D**. Cyclization of the intermediate **D** generated the triazoline intermediate **E**, which on iodination resulted in the formation of the intermediate **F**. Finally, aromatization by dehydroiodination of the intermediate **F** afforded the triazole product **130e**.

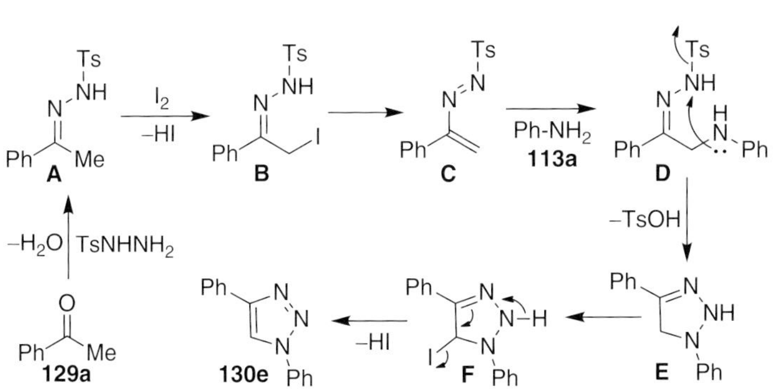


**Scheme 4.52** Mechanism for $I_2$ mediated 1,2,3-triazole synthesis from ketones, *N*-tosylhydrazines, and amines.

Applying similar strategy, Shun Yi Wang *et al.* developed an efficient $I_2$ (20 mol%)/TBPB-mediated oxidative formal [4+1]-cycloaddition of *N*-tosylhydrazones **131** with anilines **113** via C–N/N–N bond formation and S–N bond cleavage, representing a simple, general, and efficient approach for the construction of 1,2,3-triazoles **132** under metal-free and azide-free conditions (Scheme 4.53) [52]. The methodology was successfully applied to a wide range of anilines **113** and *N*-tosylhydrazones **131**, showing broad functional group tolerance. The mechanistic pathway followed by this reaction more or less resembles the same, which is explained in Scheme 4.52.

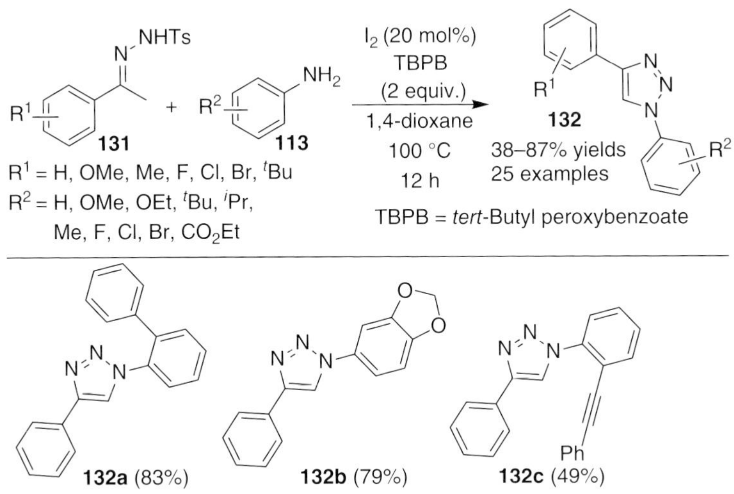


**Scheme 4.53** $I_2$/TBPB-mediated synthesis of 1,4-disubstituted 1,2,3-triazoles.

Guan *et al.* utilized nitroolefins **133** in a *p*-TsOH-mediated 1,3-dipolar cycloaddition with sodium azide for the synthesis of 4-aryl-*NH*-1,2,3-triazoles **134** (Scheme 4.54) [53]. *p*-TsOH was discovered to be a vital additive in this type of 1,3-dipolar cycloaddition. Utilizing this novel cycloaddition reaction, the authors were able to synthesize, in good yields, 4-aryl-*NH*-1,2,3-triazoles ($R^2$ = H) **134** as well as 4,5-disubstituted triazoles **134c**, employing disubstituted nitroolefins in the latter case. The protocol tolerated a wide range of aromatic nitroolefins bearing electron-withdrawing and electron-donating groups and could be applied on gram scale. At present, this methodology, however, is only applicable to aromatic nitroolefins **133**.

Most recently, Alcaide *et al.* described the generation of a highly active 1,2-dipole $Tf_2C{=}CH_2$ *in situ*, which immediately reacted at room temperature with an azide to afford previously unknown 4-trifluoromethanesulfonyl-1,2,3-triazoles **136** and **137** through a stepwise [3+2]-cycloaddition reaction from simple starting materials (Scheme 4.55) [54]. Remarkably, this mild and powerful uncatalyzed protocol is highly regio- and chemoselective. This metal-free method was applicable not only to aliphatic azides but also to aromatic ones. While the reactions of

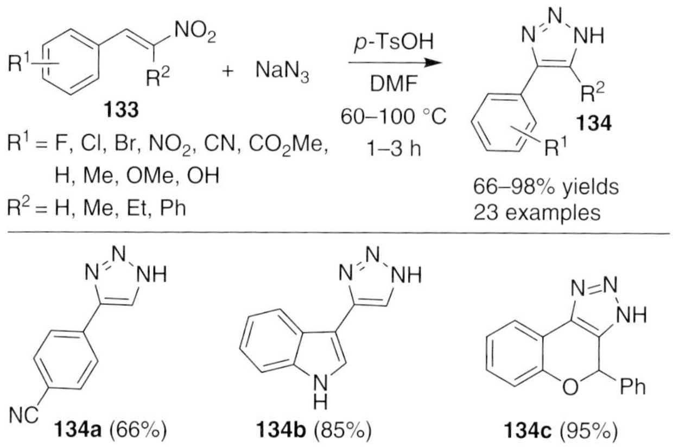


**Scheme 4.54** *p*-TsOH-mediated synthesis of 4-aryl-*NH*-1,2,3-triazoles.

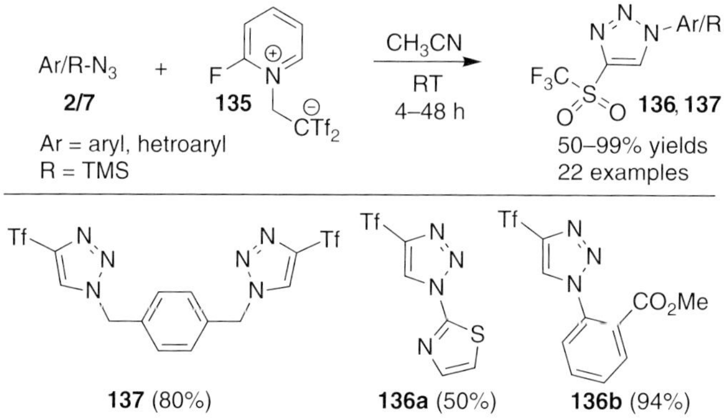


**Scheme 4.55** Synthesis of 4-trifluoromethanesulfonyl-1,2,3-triazoles.

alkyl azides were instantaneous, the reactions of their aromatic counterparts took several hours at room temperature. The authors observed that TMS azide was also a suitable starting material, resulting in the formation of 4-triflyl triazole lacking the TMS group.

Xihe Bi *et al.* reported a general cascade hydroazidation and alkyne–azide 1,3-dipolar cycloaddition of diynes **138** using silver catalysis, which constitutes a general method for the convenient synthesis of diverse pharmaceutically relevant 1,5-fused-1,2,3-triazole frameworks, including the fused heterocyclic units of piperidine, piperazine, morpholine, diazepine, and isoquinoline (Scheme 4.56) [55]. A wide variety of diynes **138** participated in the reaction with trimethylsilyl azide (TMS-$N_3$) in the presence of $H_2O$, affording the corresponding 1,5-fused-1,2,3-triazoles **139** in good to excellent yields. This protocol is operationally simple with a broad substrate scope, good functional group tolerance, and high reaction efficiency, thus providing easy access to various fused 1,2,3-triazoles.

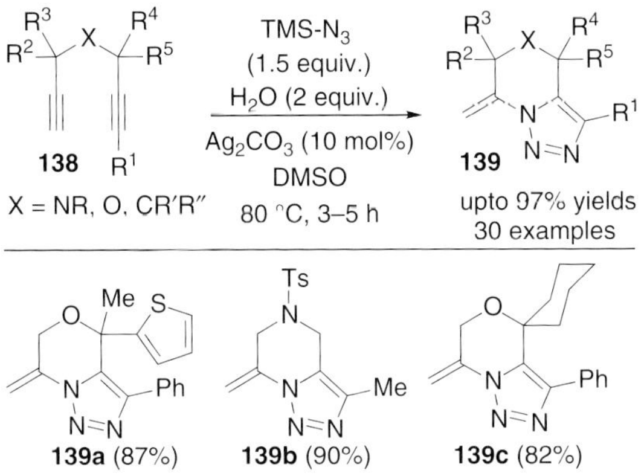


**Scheme 4.56** Silver-catalyzed access to synthesis of 1,5-fused-1,2,3-triazoles.

Shun-Yi Wang *et al.* disclosed a novel synthetic approach toward 1,5-disubstituted 1,2,3-triazoles **141** and 1,4-disubstituted 1,2,3-triazoles **142** by aerobic oxidative cycloaddition of α-chlorotosylhydrazone **140** with primary aryl amine **113** (Scheme 4.57) [56]. Significantly, the reaction proceeded smoothly to afford 1,4-disubstituted 1,2,3-triazoles **142** and 1,5-disubstituted 1,2,3-triazoles **141** under catalyst-free, metal-free, azide-free, and peroxide-free conditions. Strikingly, this transformation provided selective synthesis of both regioisomers of 1,2,3-triazoles.

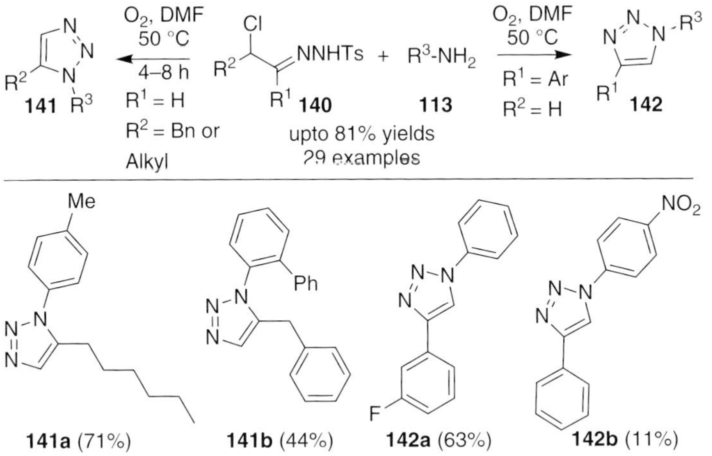


**Scheme 4.57** Synthesis of 1,4-disubstituted and 1,5-disubstituted 1,2,3-triazoles via cycloaddition of α-chlorotosylhydrazones with arylamines.

The authors have explained the proposed reaction pathway as shown in Scheme 4.58 [56]. First, chlorotosylhydrazone **140** gets easily converted to the azoalkene **A**. The 1,4-addition of aniline **113a** to the azoalkene **A** gave intermediate **B**, which on

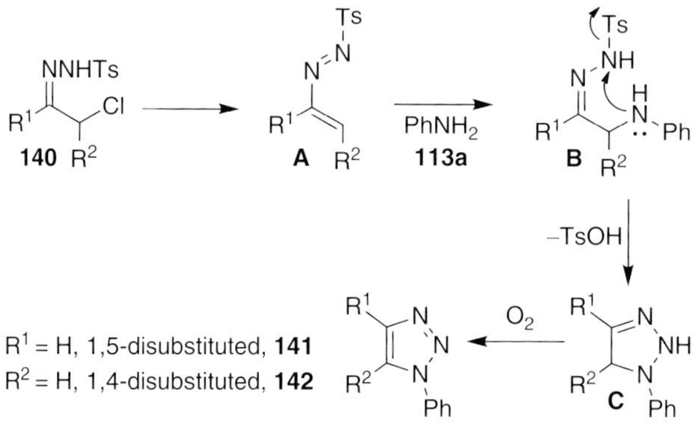


**Scheme 4.58** Mechanism for the synthesis of 1,4-disubstituted and 1,5-disubstituted 1,2,3-triazoles from α-chlorotosylhydrazones.

intramolecular cyclization generated the triazoline intermediate **C**. Subsequent oxidation by $O_2$ converted the triazoline intermediate **C** into 1,2,3-triazole.

## 4.10 Conclusions

While the recent developments in the metal-free catalytic synthesis of substituted 1,2,3-triazoles is impressive at large, the development of more sustainable variants is highly warranted since the click reaction already found significant applications in pharmaceuticals to materials. Discovery of room-temperature reactions under the friendly metal-free catalysts with employing reduced azide loading could be of immense interest in industrial applications. In this scenario, the aforementioned methods provide rapid entry to access functionalized 1,2,3-triazoles in both academically and industrially. As shown in this chapter, many of these metal-free amine-catalyzed [3+2]-cycloaddition reactions have a broad substrate scope, good functional group tolerance, simple operation, and high reaction rate and efficiency, thus providing an easy access to various functionalized 1,2,3-triazoles. These organocatalytic reactions open up new prospects in synthetic, material, and pharmaceutical chemistry for further exploiting the synthetic power of amine-catalyzed enolate/enamine-mediated click reactions."

## Acknowledgments

We thank all our past and present lab members and collaborators for their significant contributions, which are reflected in the references. We thank Ms. K. S. Shruthi for proofreading of this chapter. All the schemes presented in this chapter were "reprinted (adapted) with permission from corresponding journals" and full citation is given in the reference section.

## References


1. (a) Tome, A.C. (2004) *Sci. Synth.*, **13**, 415–601; (b) Kallander, L.S., Lu, Q., Chen, W., Tomaszek, T., Yang, G., Tew, D., Meek, T.D., Hofmann, G.A., Schulz Pritchard, C.K., Smith, W.W., Janson, C.A., Ryan, M.D., Zhang, G.F., Johanson, K.O., Kirkpatrick, R.B., Ho, T.F., Fisher, P.W., Mattern, M.R., Johnson, R.K., Hansbury, M.J., Winkler, J.D., Ward, K.W., Veber, D.F., and Thompson, S.K. (2005) *J. Med. Chem.*, **48**, 5644–5647; (c) Angell, Y.L. and Burgess, K. (2007) *Chem. Soc. Rev.*, **36**, 1674; (d) Tam, A., Arnold, U., Soellner, M.B., and Raines, R.T. (2007) *J. Am. Chem. Soc.*, **129**, 12670; (e) Ito, S., Satoh, A., Nagatomi, Y., Hirata, Y., Suzuki, G., Kimura, T., Satow, A., Maehara, S., Hikichi, H., Hata, M., Kawamoto, H., and Ohta, H. (2008) *Bioorg. Med. Chem.*, **16**, 9817; (f) Ito, S., Hirata, Y., Nagatomi, Y., Satoh, A., Suzuki, G., Kimura, T., Satow, A., Maehara, S., Hikichi, H., Hata, M., Ohta, H., and Kawamoto, H. (2009) *Bioorg. Med. Chem. Lett.*, **19**, 5310; (g) Tsuritani, T., Mizuno, H., Nonoyama, N., Kii, S., Akao, A., Sato, K., Yasuda, N., and Mase, T. (2009) *Org. Process Res. Dev.*, **13**, 1407; (h) Holub, J.M. and Kirshenbaum, K. (2010) *Chem. Soc. Rev.*, **39**, 1325; (i) Agalave, S.G., Maujan, S.R., and Pore, V.S. (2011) *Chem. Asian J.*, **6**, 2696–2718; See also: (j) (2011) *Chem. Asian J.*, **6** (10). Special edition devoted to click chemistry; (k) Fujinaga, M., Yamasaki, T., Kawamura, K., Kumata, K., Hatori, A., Yui, J., Yanamoto, K., Yoshida, Y., Ogawa, M., Nengaki, N., Maeda, J., Fukumura, T., and Zhang, M.-R. (2011) *Bioorg. Med. Chem.*, **19**, 102; (l) Valverde, I.E., Bauman, A., Kluba, C.A., Vomstein, S., Walter, M.A., and Mindt, T.L. (2013) *Angew. Chem. Int. Ed.*, **52**, 8957; (m) Lauria, A., Delisi, R., Mingoia, F., Terenzi, A., Martorana, A., Barone, G., and Almerico, A.M. (2014) *Eur. J. Org. Chem.*, **2014**, 3289–3306; (n) Li, S. and Huang, Y. (2014) *Curr. Med. Chem.*, **21**, 113.
2. (a) Tornoe, C.W., Christensen, C., and Meldal, M. (2002) *J. Org. Chem.*, **67**, 3057–3064; (b) Rostovtsev, V.V., Green, L.G., Fokin, V.V., and Sharpless, K.B. (2002) *Angew. Chem.*, **114**, 2708–2711; *Angew. Chem. Int. Ed.* (2002) **41**, 2596–2599; (c) Lee, L.V., Mitchell, M.L., Huang, S.J., Fokin, V.V., Sharpless, K.B., and Wong, C.H. (2003) *J. Am. Chem. Soc.*, **125**, 9588–9589; (d) Speers, A.E., Adam, G.C., and Cravatt, B.F. (2003) *J. Am. Chem. Soc.*, **125**, 4686–4687; (e) Binder, W.H. and Kluger, C. (2006) *Curr. Org. Chem.*, **10**, 1791–1815; (f) Whiting, M., Muldoon, J., Lin, Y.C., Silverman, S.M., Lindstrom, W., Olson, A.J., Kolb, H.C., Finn, M.G., Sharpless, K.B., Elder, J.H., and Fokin, V.V. (2006) *Angew. Chem.*, **118**, 1463–1467; *Angew. Chem. Int. Ed.* (2006) **45**, 1435–1439; (g) Nandivada, H., Jiang, X., and Lahann, J. (2007) *Adv. Mater.*, **19**, 2197–2208; (h) Lutz, J.F. and Zarafshani, Z. (2008) *Adv. Drug Delivery Rev.*, **60**, 958–970; (i) Hein, J.E., Tripp, J.C., Krasnova, L.B., Sharpless, K.B., and Fokin, V.V. (2009) *Angew. Chem.*, **121**, 8162–8165; *Angew. Chem. Int. Ed.* (2009) **48**, 8018–8021.
3. (a) Zhang, L., Chen, X., Xue, P., Sun, H.H.Y., Williams, I.D., Sharpless, K.B., Fokin, V.V., and Jia, G. (2005) *J. Am. Chem. Soc.*, **127**, 15998–15999; (b) Rasmussen, L.K., Boren, B.C., and Fokin, V.V. (2007) *Org. Lett.*, **9**, 5337–5339; (c) Boren, B.C., Narayan, S., Rasmussen, L.K., Zhang, L., Zhao, H., Lin, Z., Jia, G., and Fokin, V.V. (2008) *J. Am. Chem. Soc.*, **130**, 8923–8930; (d) Rasolofonjatovo, E., Theeramunkong, S., Bouriaud, A., Kolodych, S., Chaumontet, M., and Taran, F. (2013) *Org. Lett.*, **15**, 4698; (e) Ding, S., Jia, G., and Sun, J. (2014) *Angew. Chem. Int. Ed.*, **53**, 1877–1880.
4. (a) Agard, N.J., Preschner, J.A., and Bertozzi, C.R. (2004) *J. Am. Chem. Soc.*, **126**, 15046–15047; (b) Agard, N.J., Baskin, J.M., Prescher, J.A., Lo, A., and Bertozzi, C.R. (2006) *ACS Chem. Biol.*, **1**, 644–648; (c) Baskin, J.M. and Bertozzi, C.R. (2007) *QSAR Comb. Sci.*, **26**, 1211–1219; (d) Johnson, J.A., Baskin, J.M., Bertozzi, C.R., Koberstein, J.T., and Turro, N.J. (2008) *Chem. Commun.*, 3064–3066; (e) Sletten, E.M. and

Bertozzi, C.R. (2008) *Org. Lett.*, **10**, 3097–3099; (f) Laughlin, S.T., Baskin, J.M., Amacher, S.L., and Bertozzi, C.R. (2008) *Science*, **320**, 664–667; (g) Jewett, J.C., Sletten, E.M., and Bertozzi, C.R. (2010) *J. Am. Chem. Soc.*, **132**, 3688–3690; (h) Jewett, J.C. and Bertozzi, C.R. (2010) *Chem. Soc. Rev.*, **39**, 1272–1279; (i) Gold, B., Dudley, G.B., and Alabugin, I.V. (2013) *J. Am. Chem. Soc.*, **135**, 1558–1569.

5. Lieber, E., Rao, C.N.R., and Rajkumar, T.V. (1959) *J. Org. Chem.*, **24**, 134–135.
6. Labbe, G. and Hassner, A. (1970) *J. Heterocycl. Chem.*, **7**, 361–366.
7. Bedford, C.D., Bruckmann, E.M., and Smith, P.A.S. (1981) *J. Org. Chem.*, **46**, 679–686.
8. Acevedo, O.L., Krawczyk, S.H., and Townsend, L.B. (1986) *J. Org. Chem.*, **51**, 1050–1058.
9. Cottrell, I.F., Hands, D., Houghton, P.G., Humphrey, G.R., and Wright, S.H.B. (1991) *J. Heterocycl. Chem.*, **28**, 301–304.
10. Jones, P. and Chambers, M. (2002) *Tetrahedron*, **58**, 9973–9981.
11. Pokhodylo, N.T., Matiychuk, V.S., and Obushak, M.D. (2009) *Chem. Heterocycl. Compd.*, **45**, 245–247.
12. Pokhodylo, N.T., Matiychuk, V.S., and Obushak, N.B. (2009) *Chem. Heterocycl. Compd.*, **45**, 483–488.
13. Pokhodylo, N.T., Matiychuk, V.S., and Obushak, M.D. (2009) *Synthesis*, **14**, 2321–2323.
14. Stazi, F., Cancogni, D., Turco, L., Westerduin, P., and Bacchi, S. (2010) *Tetrahedron Lett.*, **51**, 5385–5387.
15. Jin, G., Zhang, J., Fu, D., Wu, J., and Cao, S. (2012) *Eur. J. Org. Chem.*, **2012**, 5446–5449.
16. Rozin, Y.A., Leban, J., Dehaen, W., Nenajdenko, V.G., Muzalevskiy, V.M., Eltsov, O.S., and Bakulev, V.A. (2012) *Tetrahedron*, **68**, 614–618.
17. Bianchetti, G., Croce, P.D., and Pocar, D. (1965) *Tetrahedron Lett.*, **25**, 2039–2041.
18. Bianchetti, G., Pocar, D., Croce, P.D., Gallo, G.G., and Vigevani, A. (1966) *Tetrahedron Lett.*, **15**, 1637–1642.
19. Almirante, N., Gelmi, M.L., Marelli, P., and Pocar, D. (1986) *Tetrahedron*, **42**, 57–62.
20. Battistini, M., Erba, E., and Pocar, D. (1993) *J. Chem. Soc., Perkin Trans. 1*, 339–342.
21. Erba, E., Pocar, D., and Valle, M. (1999) *J. Chem. Soc., Perkin Trans. 1*, 421–426.
22. Quast, H., Ivanova, S., Peters, E.M., and Peters, K. (2000) *Eur. J. Org. Chem.*, **2000**, 507–520.
23. Brunner, M., Maas, G., and Klarner, F.G. (2005) *Helv. Chim. Acta*, **88**, 1813–1825.
24. Efimov, I., Bakulev, V., Beliaev, N., Beryozkina, T., Knippschild, U., Leban, J., Jin, F.Z., Eltsov, O., Slepukhin, P., Ezhikova, M., and Dehaen, W. (2014) *Eur. J. Org. Chem.*, **2014**, 3684–3689.
25. Ahmadi, F., Tisseh, Z.N., Dabiri, M., and Bazgir, A. (2013) *C.R. Chim.*, **16**, 1086–1090.
26. Singh, H., Sindhu, J., and Khurana, J.M. (2013) *RSC Adv.*, **3**, 22360–22366.
27. Ramachary, D.B., Shashank, A.B., and Karthik, S. (2014) *Angew. Chem. Int. Ed.*, **53**, 10420–10424.
28. Shashank, A.B., Karthik, S., Madhavachary, R., and Ramachary, D.B. (2014) *Chem. Eur. J.*, **20**, 16877–16881.
29. Krishna, P.M., Ramachary, D.B., and Peesapati, S. (2015) *RSC Adv.*, **5**, 62062–62066.
30. Ramachary, D.B., Ramakumar, K., and Narayana, V.V. (2008) *Chem. Eur. J.*, **14**, 9143–9147.
31. Danence, L.J.T., Gao, Y., Li, M., Huang, Y., and Wang, J. (2011) *Chem. Eur. J.*, **17**, 3584–3587.
32. Belkheira, M., Abed, D.E., Pons, J.M., and Bressy, C. (2011) *Chem. Eur. J.*, **17**, 12917–12921.
33. Wang, L., Peng, S., Danence, L.J.T., Gao, Y., and Wang, J. (2012) *Chem. Eur. J.*, **18**, 6088–6093.
34. Seus, N., Goncalves, L.C., Deobald, A.M., Savegnago, L., Alves, D., and Paixao, M.W. (2012) *Tetrahedron*, **68**, 10456–10463.
35. Li, W., Jia, Q., Du, Z., and Wang, J. (2013) *Chem. Commun.*, **49**, 10187–10189.
36. Ramachary, D.B. and Shashank, A.B. (2013) *Chem. Eur. J.*, **19**, 13175–13181.
37. Yeung, D.K.J., Gao, T., Huang, J., Sun, S., Guo, H., and Wang, J. (2013) *Green Chem.*, **15**, 2384–2388.

38. Seus, N., Goldani, B., Lenardão, E.J., Savegnago, L., Paixão, M.W., and Alves, D. (2014) *Eur. J. Org. Chem.*, **2014**, 1059–1065.
39. Li, W., Du, Z., Huang, J., Jia, Q., Zhanga, K., and Wang, J. (2014) *Green Chem.*, **16**, 3003–3006.
40. Jia, Q., Yang, G., Chen, L., Du, Z., Wei, J., Zhong, Y., and Wang, J. (2015) *Eur. J. Org. Chem.*, **2015**, 3435–3440.
41. Li, W., Du, Z., Zhanga, K., and Wang, J. (2015) *Green Chem.*, **17**, 781–784.
42. Kwok, S.W., Fotsing, J.R., Fraser, R.J., Rodionov, V.O., and Fokin, V.V. (2010) *Org. Lett.*, **12**, 4217–4219.
43. Kloss, F., Kohn, U., Jahn, B.O., Hager, M.D., Gorls, H., and Schubert, U.S. (2011) *Chem. Asian J.*, **6**, 2816–2824.
44. Guggenheim, K.G., Toru, H., and Kurth, M.J. (2012) *Org. Lett.*, **14**, 3732–3735.
45. Cheng, G., Zeng, X., Shen, J., Wang, X., and Cui, X. (2013) *Angew. Chem. Int. Ed.*, **52**, 13265–13268.
46. Kayet, A. and Pathak, T. (2013) *J. Org. Chem.*, **78**, 9865–9875.
47. Zhang, H., Tanimoto, H., Morimoto, T., Nishiyama, Y., and Kakiuchi, K. (2013) *Org. Lett.*, **15**, 5222–5225.
48. Thomas, J., John, J., Parekh, N., and Dehaen, W. (2014) *Angew. Chem. Int. Ed.*, **53**, 10155–10159.
49. Li, W. and Wang, J. (2014) *Angew. Chem. Int. Ed.*, **53**, 14186–14190.
50. Ali, A., Correa, A.G., Alves, D., Schpector, J.Z., Westermann, B., Ferreira, M.A.B., and Paixao, M.W. (2014) *Chem. Commun.*, **50**, 11926–11929.
51. Chen, Z., Yan, Q., Liu, Z., and Zhang, Y. (2014) *Chem. Eur. J.*, **20**, 17635–17639.
52. Cai, Z.J., Lu, X.M., Zi, Y., Yang, C., Shen, L.J., Li, J., Wang, S.Y., and Ji, S.J. (2014) *Org. Lett.*, **16**, 5108–5111.
53. Quan, X.J., Ren, Z.H., Wang, Y.Y., and Guan, Z.H. (2014) *Org. Lett.*, **16**, 5728–5731.
54. Alcaide, B., Almendros, P., and Milla, C.L. (2015) *Chem. Commun.*, **51**, 6992–6995.
55. Ning, Y., Wu, N., Yu, H., Liao, P., Li, X., and Bi, X. (2015) *Org. Lett.*, **17**, 2198–2201.
56. Bai, H.W., Cai, Z.J., Wang, S.Y., and Ji, S.J. (2015) *Org. Lett.*, **17**, 2898–2901.

# 5
# Applications of the Cu-Catalyzed Azide–Alkyne Cycloaddition (CuAAC) in Peptides

*Freek A. B. M. Hoogstede and Floris P. J. T. Rutjes*

## 5.1
## Introduction

After the discovery of the cycloaddition between azides and alkynes [1], neither its inventor Rolf Huisgen nor any other chemist at that time could have foreseen that this transformation would become one of the most widely applied chemical reactions in the first decades of the twenty-first century. It required, however, that more than 30 years after the invention, two groups independently discovered that the presence of a catalytic amount of a Cu(I) species has a highly beneficial effect on the reaction [2]. The Cu catalysis causes that the reaction proceeds very rapidly at room temperature (instead of elevated temperatures), in a completely regioselective manner (the 1,4-isomer is selectively formed, while previously a 1 : 1 mixture of regioisomers was obtained) and in virtually any solvent including water. These factors also made that this reaction is currently known as the *click reaction* [3], albeit a more official name is Cu-catalyzed azide–alkyne cycloaddition or CuAAC reaction (Figure 5.1). Both names have been used throughout this chapter.

The robustness of the reaction, in combination with the fact that the cycloaddition between azides and alkynes under these mild conditions gives virtually no cross-reactivity with any other functional group, has turned this reaction into the widely applied reaction that it currently is. Also, due to the fact that the triazole functional group is considered an isostere of the amide functional group, many of the conjugations that previously would be carried out by amide bond formation now proceed by using the CuAAC reaction. This reaction has been reviewed multiple times during the past decade [4], albeit none of the reviews so far has exclusively focused on applications in peptides. In this chapter, we focus on applications of the CuAAC reaction in combination with peptides and proteins. Since the number of applications in the peptide and protein field is overwhelming, we have decided to make a choice of the different strategies that have been pursued over the years and will not provide a comprehensive overview of all the examples that have been published. We start with highlighting approaches in which the

*Click Reactions in Organic Synthesis*, First Edition. Edited by Srinivasan Chandrasekaran.

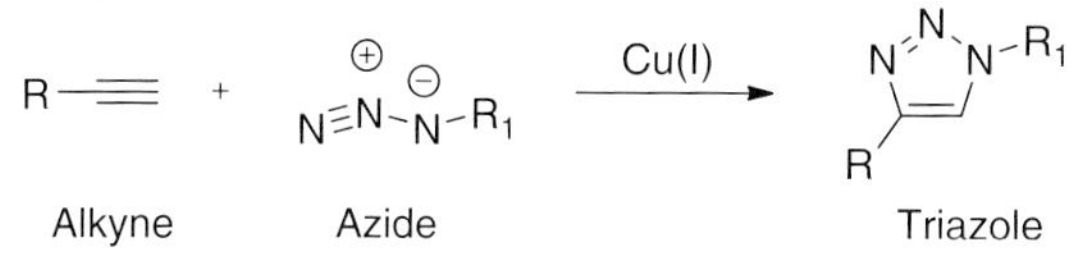


**Figure 5.1** The CuAAC or click reaction.

click reaction has been used as a tool to conjugate peptides, containing either an azide or an alkyne functional handle on one of the side chains or at one of the termini, with molecules in the broadest sense containing the complementary functional group. The second part of the chapter is dedicated to describe examples of peptides in which the backbone has been modified by incorporating a triazole instead of an amide functional group, where the properties of the resulting peptidomimetics have been evaluated. Furthermore, since the conditions under which the CuAAC reaction is carried out are essentially identical, we have not focused on the cycloaddition itself, but more on the approach and the concept behind the strategy.

## 5.2 CuAAC-Mediated Peptide Conjugation Strategies

The most common application of the Huisgen click reaction in peptides is the conjugation with tags or labels, for example, detection or activity-based protein profiling. This is generally achieved by integration of an unnatural amino acid containing an alkyne or azide functional group or by modification of existing reactive residues to incorporate those functional groups. The tag or label is functionalized with its reactive counterpart, and due to the bioorthogonality of both the alkyne and azide functional groups, the Huisgen click reaction can be performed *in vitro* or even *in vivo* to give the peptide that is then conjugated with the tag or label of interest.

One approach to achieve site-specific labeling has been explored by Chin *et al.* [5] by introducing unnatural amino acids containing an alkyne or azide functionality into recombinant myoglobin using an orthogonal tRNA/tRNA synthetase pair. Instead of using phenylalanine derivatives, which cannot be used to replace aliphatic amino acids since the difference in hydrophobicity may cause misfolding or loss of protein function, derivatized lysine derivatives were used (Figure 5.2). Thus, both lysine derivatives **1** and **2** were successfully incorporated at position 4 in recombinant myoglobin. With alkyne **1** at this position, biotin-azide **3** and fluorophore By3 (**4**) were coupled to myoglobin with an efficiency of 90–100% as was determined by ESI mass spectroscopy.

Another example of clickable labels includes fluorescent coumarins synthesized by Mertens and Gütschow [6]. By incorporating both alkyne and acid functionalities, coumarins **5** can be clicked to azides and also coupled to amine functional groups (Figure 5.3). More specifically, this allows for conjugation to azido amino

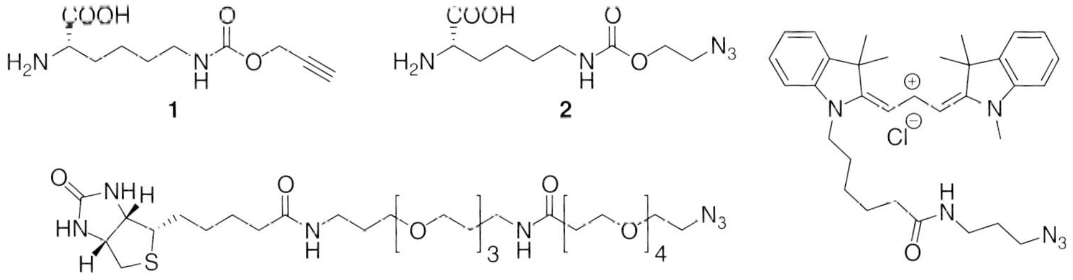

1 2 3 4

**Figure 5.2** Examples of clickable amino acids and tags.

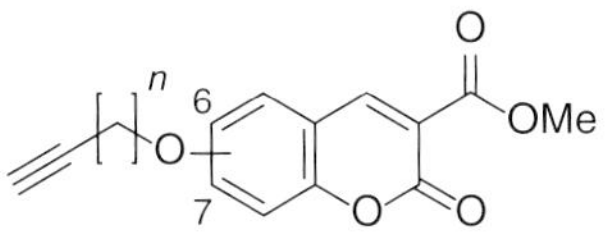

5

**Figure 5.3** Clickable fluorescent coumarins.

acids or another label (e.g., biotin-azide) and to either the N-terminus or amine-containing side chains resulting in labeled amino acids to be used in peptide synthesis. Coumarins **5** remained fluorescent regardless of the length of the alkyne side chain ($n = 1, 2,$ or 3) or azide counterpart (Boc-Phe-$N_3$, Boc-Ala-$N_3$, or biotin azide), position on the aromatic ring (6 or 7), or coupled amino acid (Fmoc-Lys-OH side chain, Fmoc-Dap-OH side chain, Ala-OMe, Phe-Ome, or Leu-OMe).

In addition to the introduction of fluorescent labels, peptides may also be conjugated to polymers. Conjugation of peptides with poly(ethylene glycol) (PEG) will give rise to the so-called PEGylated peptides, which often display higher *in vivo* stability and longer circulation times than their non-PEGylated counterparts. Many different conjugation strategies have been used for PEGylation of proteins and peptides. The synthesis of PEGylated lipopeptides has been particularly difficult due to the diverse properties of the peptide side chains, high molecular mass of the resulting polymer, and the amphiphilic character of the conjugate resulting in difficulties during purification. Jølck, Berg, and Andresen explored the synthesis of PEGylated lipoprotein **6** via the Huisgen click reaction (Figure 5.4) [7]. The lipoprotein was synthesized on a solid support to contain (*S*)-2-amino-4-pentynoic acid and at the N-terminus an L-2,3-diaminopropionic

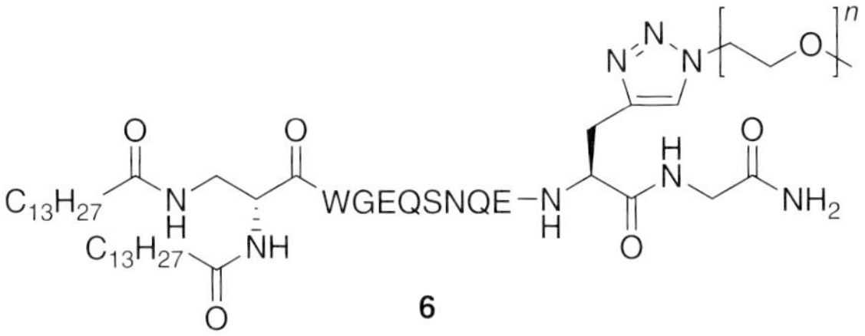

6

**Figure 5.4** A PEGylated peptide.

acid, which was then diacylated with myristic acid. While still on the resin, the alkyne was reacted with MeO-PEG-azide of different lengths ($n = 350$, 1000, or 2000). The first one resulted in complete PEGylation within 1 h, whereas the latter two gave 80% and 50% PEGylation after 48 h, respectively. The use of solid-phase synthesis resulted in most contaminants being rinsed away after each step and allowed for verification of possible amino acid deletions before PEGylation. Altogether, the method provided a straightforward synthesis of PEGylated lipopeptides requiring only a single purification step at the end.

Lu, Shi, and Shoichet functionalized a copolymer with a peptide sequence [8]. They synthesized an amphiphilic copolymer consisting of a hydrophobic poly(2-methyl-2-carboxytrimethylene carbonate-*co*-D,L-lactide) and a hydrophilic PEG azide–terminated pendant chain, which self-assembled in aqueous solution to form nanoparticles presenting the azide groups on the surface. These azides were conjugated to an N-terminally 4-pentynoic acid–modified KGRGDS peptide to give the polymer–peptide hybrid **7** (Figure 5.5). The peptide contains the Arg-Gly-Asp sequence (RGD) through which integrins interact with the extracellular matrix [9]. The KGRGDS nanoparticles were produced to target $\alpha_v\beta_1$ integrin, which interacts with fibronectin (an RGD-containing protein) in the cases of corneal epithelium injuries. As such the KGRGDS nanoparticles were tested in a competitive assay of rabbit corneal epithelial cell surface receptors, and while it showed some nonspecific binding, it exhibited stronger inhibition than soluble KGRGDS. More generally, conjugation of peptides with various kinds of polymers has been successfully achieved using the CuAAC reaction [10].

A particularly relevant class of biomolecules comprises the so-called glycoproteins or glycopeptides. As with protein–protein (such as RGD–integrin) interactions, the interactions between carbohydrates and proteins facilitate many important biological processes [11, 12]. Monovalent carbohydrate–protein interactions show very weak binding affinities ($K_d \approx 10^{-3}$ to $10^{-6}$ M) [7], while multivalent interactions frequently enhance the affinity due to the so-called *glycoside cluster effects* [13, 14]. In order to study multivalent carbohydrate–protein interactions, multivalent glycoclusters are required. Polymers, dendrimers, benzene scaffolds, monosaccharides, and others can be used as the base to affix multiple carbohydrate units. Chen *et al.* [15] chose the cyclic decapeptide cyclo-[GPKAKGPKKK] where the Lys 3, 5, 8, and 10 residues were used for functionalization with propiolic acid to yield the tetravalent alkyne **8a**, while

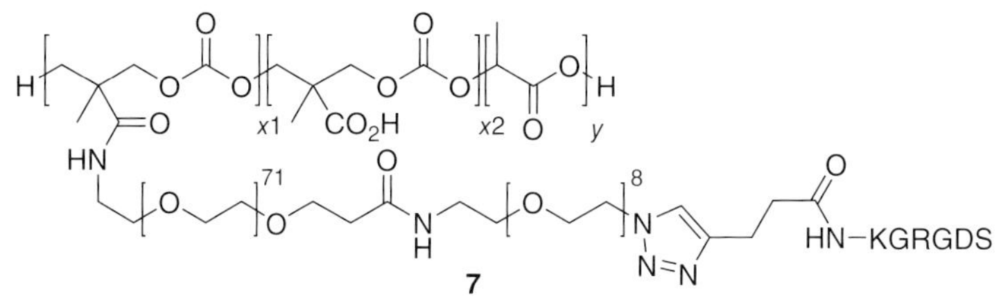


**7**

**Figure 5.5** A polymer–peptide hybrid.

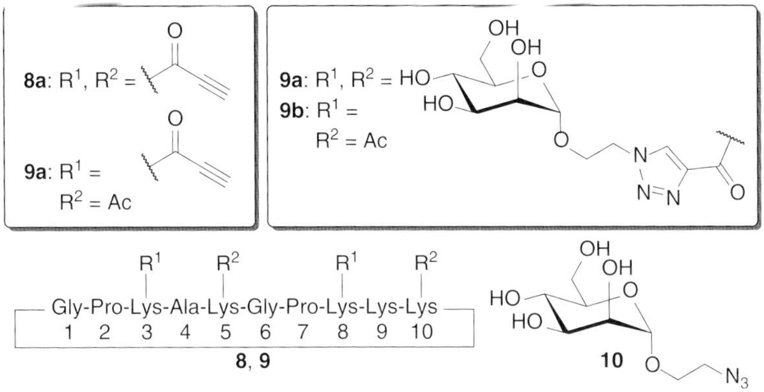


**Figure 5.6** Triazole-linked glycopeptide structures.

functionalization of Lys 3 and 8 with propiolic acid with acetyl capped Lys 5 and 10 yielded divalent alkyne **9a** (Figure 5.6). Both **8a** and **9a** were coupled to 2-azidoethyl α-D-mannopyranoside (**10**) via the CuAAC reaction to give glycopeptides **8b** and **9b**. Using surface plasmon resonance (SPR), with the target protein Concanavalin A immobilized on the sensor chip, the binding constants (corrected for valency) of the di- and tetravalent glycopeptides were determined, indicating a threefold increase in binding affinity for the tetravalent to the divalent glycoprotein.

Peptide C34, a 34-mer peptide derived from the C ectodomain of HIV-1 inner envelope glycoprotein gp41, is an HIV-1 entry inhibitor active by blocking viral membrane fusion at nanomolar concentrations ($IC_{50}$ = 7.0 nM) [16]. The peptide, however, is poorly soluble, and glycosylation at the natural glycosylation site (N637 in gp41, N25 in the 34-mer) improved its solubility with only moderate loss of activity ($IC_{50}$ = 14.3 nM). Because it was envisaged that the carbohydrate portion of the molecule could protect against protease digestion, Wang *et al.* [16] explored analogs of this inhibitor. Their tests showed that glycosylation reduced the digestion from 85% to approximately 60%. The amide bond, however, linking the sugar to the peptide is vulnerable to glycoamidase-catalyzed deglycosylation leaving the peptide once again fully vulnerable to peptidase digestion. To negate the vulnerability to deglycosylation, a triazole linker was used to replace the amide bond. While the amide-linked glycoprotein was fully deglycosylated after 1 h, the triazole-linked glycoprotein did not show any digestion, thereby demonstrating the improved *in vitro* stability of the triazole against glycoamidase-catalyzed deglycosylation. Peptidase digestion of the triazole-linked glycopeptide was similar to that of the amide-linked peptide. Although inhibition activity was measured at a third of that of the unglycosylated peptide ($IC_{50}$ = 21.0 nM), the improved solubility and stability of the triazole-containing drug candidate promise an overall improvement of efficacy.

The fact that the triazole is a bioisostere of the amide bond was also shown by the chemoenzymatic process that was used to couple the oligosaccharide to the

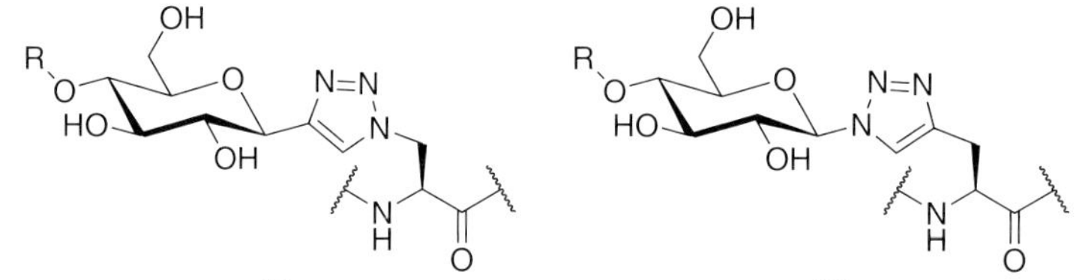


11 12

**Figure 5.7** Triazole-linked glycopeptides derived from C34.

monosaccharide peptide to form the glycopeptides used (Figure 5.7). Two dipeptides were synthesized in which the monosaccharide was either C-linked (**11**) or N-linked (**12**) to the saccharide. Both were found to be substrates for *endo*-β-*N*-acetylglucosaminidase from *Arthrobacter* (Endo-A) with 80% conversion for **12** and 52% conversion for compound **11**. Albeit only being a single example, it shows that the N-linked triazole is a better analog of the amide bond than its regioisomeric counterpart. It also shows the importance of the directionality of the mimic.

Inspired by work of the Kessler group, showing that glycosylated cyclic RGD peptides possessed improved pharmacokinetic behavior than the nonglycosylated native peptides [17], Kuijpers *et al.* studied the effect of replacing the amide linkage by the bioisosteric triazole linkage in the glycosylated RGD derivatives[18] (Figure 5.8). First, the stability of compounds **13** were tested under various conditions (Table 5.1). From the results, it was concluded that triazole-linked **13b** is more stable in a range of basic and acidic conditions, making it a more versatile building block for preparation of stable analogs of the common asparagine-linked *N*-glycoproteins. Secondly, solid phase synthesis of cyclic peptides **16** was successfully performed from both amino acids **13a** and **13b**, although incorporation of the amide-linked glycoamino acid proceeded

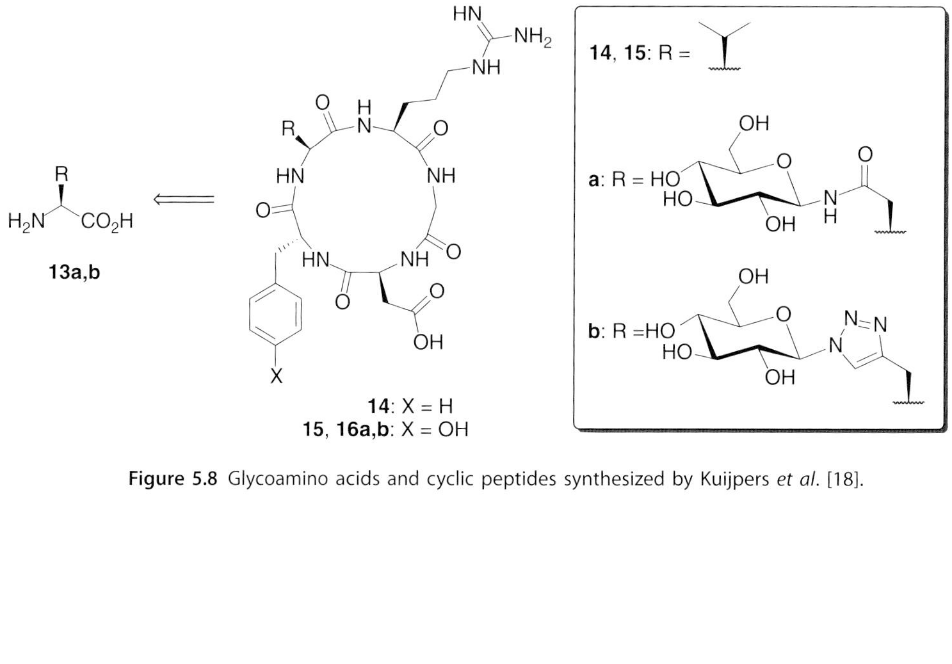


**Figure 5.8** Glycoamino acids and cyclic peptides synthesized by Kuijpers *et al.* [18].

**Table 5.1** Stability tests for **13a** and **13b**, respectively.

| Conditions | Temperature | Time | Stability | |
|---|---|---|---|---|
| | | | 13a | 13b |
| 2 M HCl (aq) | Room temperature | 3 d | ++ | ++ |
| 2 M HCl (aq) | Reflux | 2 h | ++ | + |
| 4 M HCl (aq) | Reflux | 3 h | ++ | −− |
| 2 M NaOH (aq) | Room temperature | 3 d | ++ | − |
| 2 M NaOH (aq) | Reflux | 3 h | ++ | −− |
| 2.6 M HCl/EtOAc | Room temperature | 3 d | ++ | + |
| $Et_3N$ | Reflux | 3 h | ++ | + |

at a much slower rate. As mentioned before, integrins interact with the extracellular matrix through the RGD sequence. One specific integrin, $\alpha_v\beta_3$, plays an important role in tumor-induced angiogenesis and tumor metastasis and is therefore of interest in tumor targeting [19]. To test the affinity for the integrin, a solid-phase $\alpha_v\beta_3$-binding assay of cyclopeptides **14–16** showed only relatively small differences in $IC_{50}$ values (65, 144, 238 , and 144 nM, respectively) indicating that side-chain modification had only limited effect on the $\alpha_v\beta_3$-binding affinity. Therefore, as such, this work provides additional evidence for the assumption that 1,2,3-triazoles acts as bioisosteres of their amide counterparts.

Besides the conjugation with stable tags, polymers, or saccharides, peptides can be coupled to radioactive labels. $^{18}$F-Labeled sugars have been used in positron emission tomography (PET) scan due to $^{18}$F decay by positron emission. Since glucose accumulates much stronger in cancer cells than in regular cells due to the upregulated metabolism, tumors can be visualized by PET using $^{18}$F-labeled sugar derivatives. In order to improve the targeting, $^{18}$F-glycosides can be coupled to peptides, such as RGD for targeting tumors overexpressing integrin. The half-life of $^{18}$F, however, is less than 110 min, calling for rapid late-stage fluorination. The Huisgen CuAAC reaction can fulfill this requirement, as was shown by Vala *et al.* [20]. An azide functional group was introduced onto the sugar at the anomeric center, resulting in the corresponding 2-azido ethyl glycoside **18a** (Figure 5.9). In order to introduce the required alkyne functionality in the peptide, selective propargylation was carried out at the most nucleophilic residue (cysteine). With a reaction time of 1 h and a yield of 97%, oligopeptide **17** and fluoroglycoside

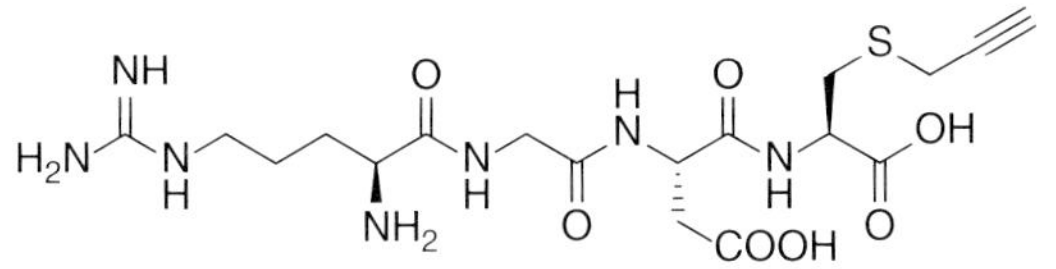


**17**

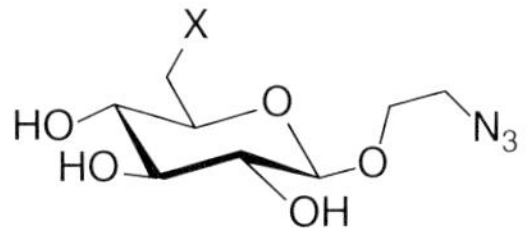

**18a**: X = OH
**18b**: X = F

**Figure 5.9** Late-stage fluorination through the click reaction.

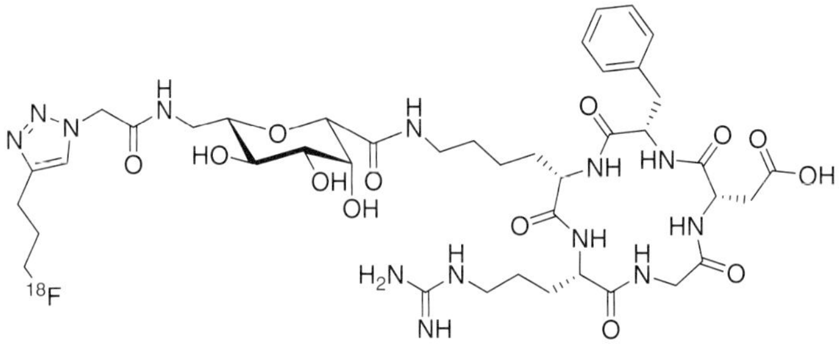

19

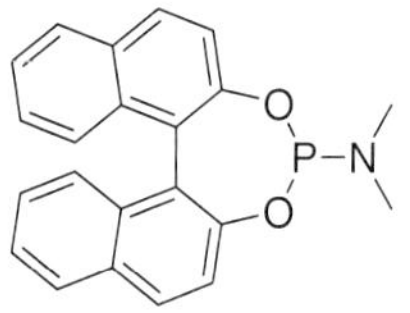

20

**Figure 5.10** Late-stage fluorination click reaction of an RGD derivative.

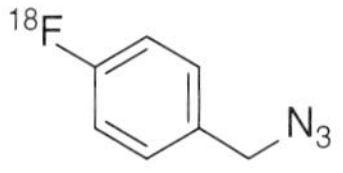

21

**Figure 5.11** Azide-functionalized $^{18}F$-label for conjugation with alkyne-containing peptides.

**18b** were then coupled successfully, fulfilling the time constraint in excellent yield.

A well-studied $^{18}F$-fluoroglycoside synthesized originally by Siemens uses the Huisgen click reaction to couple 5-[$^{18}F$]fluoro-1-pentyne in the late-stage fluorination of RGD-K5 azide resulting in [$^{18}F$]-RGD-K5 (**19**, Figure 5.10) [21–24]. Mirfeizi *et al.* [25] further improved the Siemens method raising the radiochemical yield from 20% to 35% by optimization of all steps involved and in particular of the click reaction itself. By use of MonoPhos ligand **20** instead of TBTA, the amount of RGD-K5 azide, Na ascorbate, $CuSO_4$, and ligand could be reduced by 40-, 160-, 300-, and 50-fold, respectively, for the same amount of pentynyl tosylate used in the synthesis of [$^{18}F$]-fluoropentyne.

Another example of an $^{18}F$-label that has been incorporated in alkyne-containing peptides through the Huisgen click reaction involves 1-(azidomethyl)-4-[$^{18}F$]-fluorobenzene (**21**, Figure 5.11), which was published by Thonon *et al.* [26]. One advantage of such an aromatic fluoride over the previously shown aliphatic ones is the improved the resistance toward *in vivo* defluorination [27].

## 5.3 CuAAC-Mediated Peptide Backbone Modification Strategies

The metabolic stability of peptides is generally relatively low, which can be a drawback for therapeutic use. Approaches to increase the metabolic stability include backbone modifications, such as the introduction of amide bond isosteres (e.g., sulfonamides), N-alkylation, incorporation of nonproteinogenic or D-amino acids, and the use of dialkylated amino acids. More recently, the replacement of amide bonds with triazole isosteres has also become a viable strategy to modify

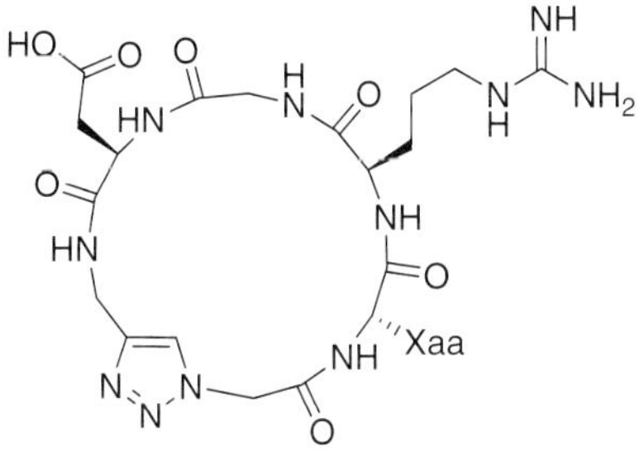

**22**

**Figure 5.12** Backbone-modified analog of a cyclic RGD peptide.

the properties of peptides. Not only does it lead to metabolically more stable peptides but it generally also has an influence on the conformation of the peptide and hence its biological activity.

As mentioned before, the RGD–integrin interactions can be exploited to the benefit of targeting integrin-rich cells, such as $\alpha_v\beta_3$ in tumor cells. Kessler *et al.* designed and synthesized a series of cyclic RGD-containing peptides [17] of which cyclo[-Arg-Gly-Asp-(D)-Phe-Lys-] (c(RGDfK)) was used as comparison to cyclic triazole-containing backbone-modified peptides synthesized by Liu *et al.* [28]. These cyclo[-Arg-Gly-Asp-Ψ(triazole)-Gly-Xaa-] peptides (**22**, Figure 5.12) were synthesized using SPPS, cyclized, and tested on toxicity versus human umbilical vein endothelial cell line, human lung carcinoma cell line SPC-A-1, and hepatocellular carcinoma cell line HepG2. The results clearly showed that the compounds with Lys, Gly-Gly, Gly, Pro, Asn, and Glu as Xaa were the most cytotoxic ones for all three cell lines and with comparable $IC_{50}$ values between 0.21 and 0.93 μM as those of the positive control c(RGDfK) ($IC_{50}$ between 0.23 and 0.28). With Ala, no Xaa, Phe, Arg, Met, or Cys, the cyclic peptidomimetic was much less cytotoxic than c(RGDfK), while Leu, Ile, and His variants exhibited no obvious cytotoxicity. In addition, Arg and Trp variants showed significantly different $IC_{50}$ values for different cell lines. It was concluded that the replacement of the amide bond by a triazole allowed for straightforward and mild synthesis of the target peptidomimetics seemingly without interfering with their native function.

Inhibitors of apoptosis proteins (IAPs) are a class of key regulators of apoptosis characterized by containing one to three baculoviral IAP repeat (BIRs) domains. Some of these IAPs inhibit tumor necrosis factor–mediated apoptosis increasing tumor viability. An endogenous inhibitor of IAPs is second mitochondria-derived activator of caspase (Smac). The interaction between IAPs and Smac is mediated by the BIR domain(s) in IAPs and the four N-terminus amino acid residues (AVPI) of Smac. Sun *et al.* [29] previously designed PSmac-21 (**23**), made by cyclization of two (N-Me)AKPF tetrapeptides through two amide bonds formed between the Lys side-chain amino groups and the C-terminus. This (N-Me)AKPF binds more effectively to a specific IAP BIR3 domain than the Smac tetrapeptide AVPI. Based on PSmac-21, two new cyclopeptidic Smac mimetics **24** and **25** were synthesized (Figure 5.13). The two amide bonds formed in the cyclization were replaced with

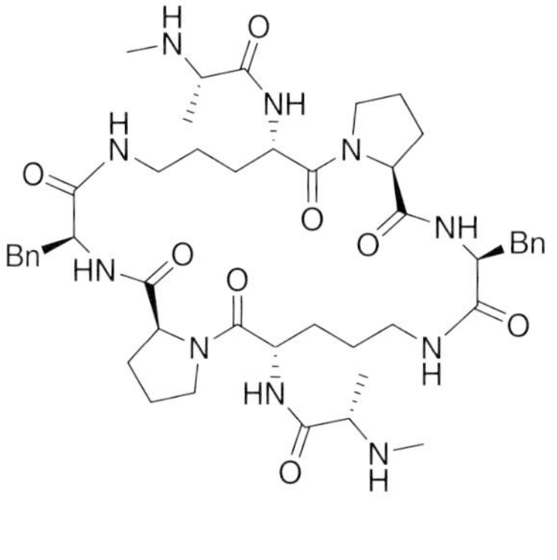

**23**

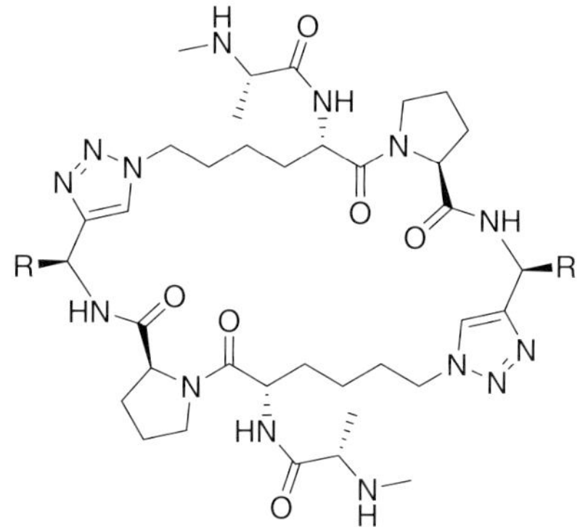

**24**: R = Ph
**25**: R = Bn

**Figure 5.13** Triazole-containing backbone-modified analogs of PSmac-21 (**23**).

a 1,4-substituted 1,2,3-triazole, and at the same time, compound **24** was synthesized to contain phenylglycine instead of the phenylalanine residues in **23** and **25**. In binding affinity studies of **23–25** to multiple different IAP BIR domains, peptidomimetic **25** in all but one case exhibited higher $IC_{50}$ values than **23**, while the $IC_{50}$ values of **24** were all significantly lower (2–17 times) than those of **23**. When tested against two cancer cell lines (MDA-MB-231 and SK-OV-3), compound **24** was five to eight times more potent than compound **25**. Considering that peptidomimetic **25** is the direct triazole analog of **23**, the triazole linkage is seemingly the cause for its worse performance. Compound **24**, however, contains the same triazole moieties but performs significantly better.

An example of a 1,4-triazole-containing peptidomimetic that is not geared toward biological activity is the heterocyclic peptide nanotube studied by Horne *et al.* [30]. Earlier related work from the Ghadiri group showed that cyclic oligopeptides with an even number of alternating D- and L-α-amino acids form a flat ring with the amide bonds in the backbone being oriented perpendicular to the plane of the ring [31, 32]. From D-leucine ε-amino acid **26** was synthesized and coupled with L-phenylalanine to form the cyclic tetramer **27** (Figure 5.14). Mass spectroscopy showed the existence of a noncovalent dimer, and from $^1$H NMR, the apparent association constants for the formation of the dimeric ($K_2 = 8.6 \times 10^4\ M^{-1}$) and higher-order aggregates ($K_n = 3.8 \times 10^4\ M^{-1}$) were determined. X-ray diffraction shows that three out of four amide bonds are involved in direct hydrogen bonding, while the fourth one forms an indirect hydrogen bond via an ethanol molecule. While the triazole ring was positioned perpendicular to the peptide ring, it was not found to participate in any hydrogen bonding.

Sansalvamide A (San A) has been used extensively as a template for the synthesis of cytotoxic compounds. Davis *et al.* [33] based their research on San A amide **28**, differing from San A by the amide bond connecting the two consecutive leucines instead of an ester bond (Figure 5.15). They synthesized 13 analogs of cyclopeptide

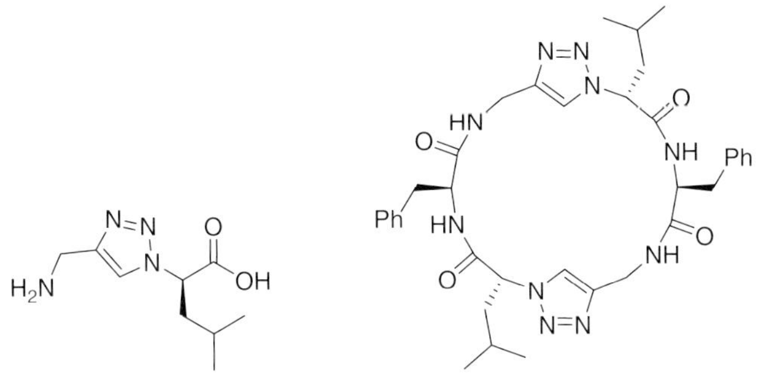

26 27

Figure 5.14 Triazole-containing peptides self-assembling to nanotubes.

**28**, containing a triazole, an oxazole, a thiazole, or a pseudoproline moiety. Four of these analogs (**39** to **32**) were tested on their cytotoxicity relative to **28**. Triazole **29** exhibited cytotoxicity at the same level as **28** with the thiazole derivative as a close second one. Pseudoproline **32** showed a significant decrease in cytotoxicity, which was explained by the *cis*-locked pseudoproline moiety disturbing the shape of *trans*-bonded **28**. While thiazole **31** exhibited a cytotoxicity close to that of **28**, oxazole **30** scored nearly as low as the pseudoproline, explained by Davis *et al.*, with the difference in flexibility between the two compounds.

The hydrogen bonding capacity of 1,4-triazoles as amidomimetics has been explored by Proteau-Gagné *et al.* [34] by systematically replacing the amide bonds in the pentapeptide Leu-enkephalin (**33**), a ligand for the delta opioid receptor (Figure 5.16). In earlier research, the group replaced the last amide bond in the oligopeptide by an *E*-alkene inducing a large decrease in affinity ($K_i = 116$ nM) [35], while replacement by the hydrogen bond accepting *N*-methyl amide (**34**) or ester (**35**) resulted in peptidomimetics with similar affinity ($K_{i,CONMe} = 12.6$ nM, $K_{i,COO} = 11.9$ nM) [36] as compared to Leu-enkephalin itself ($K_i = 6.3$ nM) [30]. When a triazole moiety was used instead, an affinity was observed ($K_i = 89$ nM) that ranked between the values for the *E*-alkene and ester, *N*-methyl amide and

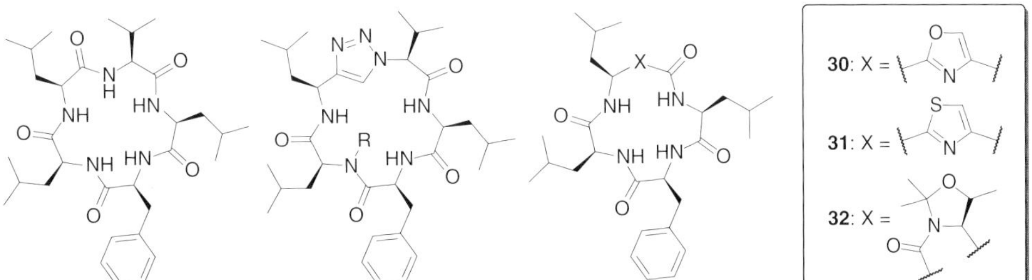


28 29 30-32

Figure 5.15 Cyclopeptides and triazole-backbone modifications.

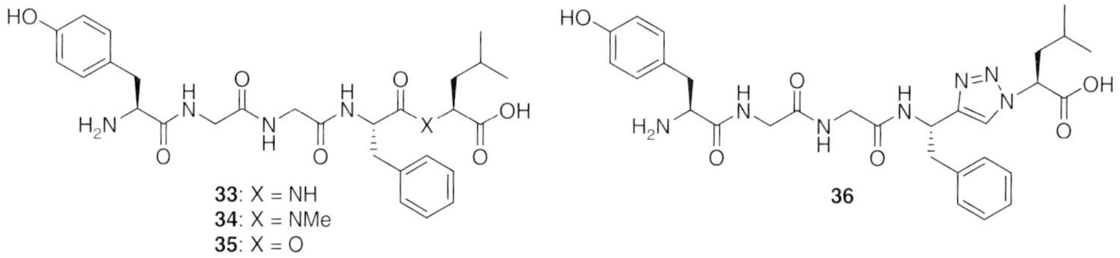


**Figure 5.16** Leu-enkephalin (**33**) and a triazole-modified mimic.

amide bond, which was taken as evidence for the hydrogen-bond-accepting properties of triazole **36**.

In an attempt to improve the half-life of an inhibitor of the gastrin-releasing peptide receptor (GRPr), which is overexpressed in a variety of clinically relevant tumors including prostate and breast cancer, Valverde *et al.* [37] used what they termed a *triazole scan*. The inhibitor that was subject of those studies was [$^{117}$Lu]DOTA-PEG$_4$-[Nle$^{14}$]BBN(7–14) (**37**, Figure 5.17, which is the radiolabeled variant of the minimal binding sequence (H-QWAVGHLM-NH$_2$) of the peptide bombesin. By systematically replacing the peptide amide bonds of the radiolabeled peptide and testing the half-life and uptake of the original and nine triazole variants of the compound, it was determined which of these compounds were the best candidates for a binding assay. From the assay, half-life, and uptake experiments, it was clear that the introduction of a single 1,2,3,-triazole heterocycle between Gly$^{11}$ and His$^{12}$, namely **38**, improved the half-life more than threefold (from 5 to 17 h) while retaining the uptake (after 4 h from 27.7% to 28.3%) and GRPr affinity ($K_D$ from $2.0 \pm 0.6$ to $3.1 \pm 1.0$ nM). This compound was also tested *in vivo* where it showed improved stability and uptake indicating that a triazole as amide mimic can improve stability while retaining or even improving *in vivo* efficacy.

Angelo and Arora explored the synthesis of the so-called triazolamers, peptide mimics in which all amide bonds in the backbone are replaced with 1,4-disubstituted 1,2,3-triazoles (Figure 5.18). They previously developed a multistep sequence, involving (i) diazotransfer onto a free amine followed by (ii) a CuAAC reaction with a protected aminoalkyne and (iii) subsequent deprotection to yield

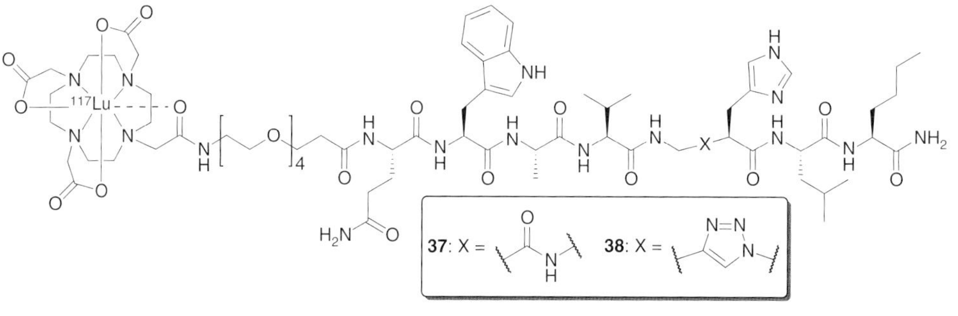


**Figure 5.17** A radiolabeled triazole-containing bombesin peptidomimetic.

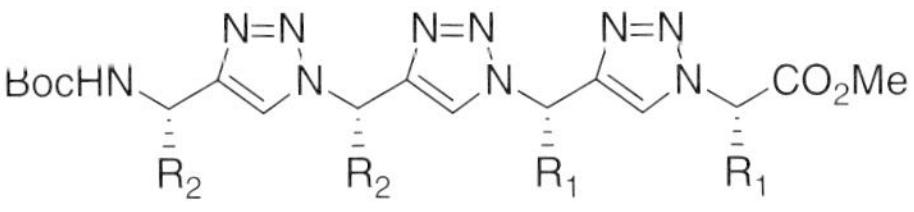

**39**: $R_1$= Bn, $R_2$= $(CH_2)_4NH(Z)$

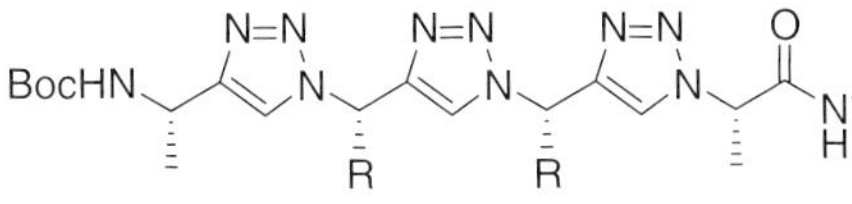

**40**: R= Bn

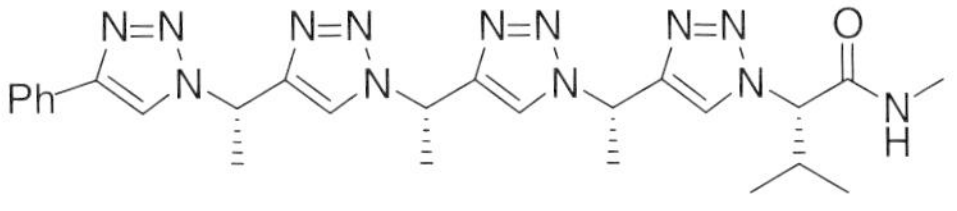

**41**

**Figure 5.18** Examples of triazolamers.

the free amine, which can then be repeated from step one. By choosing the proper functional end groups, application of this sequence yielded triazolamer trimers in overall yields of 6–12% [38]. In a second paper, these yields were improved to 62% for trimer **39** by sequential single-pot synthesis in solution [39]. In addition to the initial solution-phase approach, solid-phase synthesis was explored as this would allow for the synthesis of libraries of triazolamers and hopefully acceptable yields for the corresponding tetramers. The catalyst used for the diazotransfer and click reaction in the solution-phase synthesis was $CuSO_4 \cdot 5H_2O$, which, when used in the solid-phase synthesis, led to an overall yield of 37% for trimer **40** and 33% for tetramer **41**. The use of $ZnCl_2$ as the diazotransfer catalyst improved the overall yields to 57% for **40** and 78% for **41** showing the ability of both the solution- and solid-phase synthesis to yield triazole oligomers.

As an important secondary structure in peptides, the α-helix has also been the subject of 1,2,3-triazole-containing peptidomimetics. Horne *et al.* [40] used ε-amino acid **42** containing a 1,4-disubstituted 1,2,3-triazole as a dipeptide surrogate in α-helical coiled coils (Figure 5.19). As a basis, oligopeptide pLI-GCN4 (**43**) was modified at three different locations by replacing dipeptides $K_8L_9$, $K_{15}L_{16}$, or $E_{22}L_{23}$ with ε-amino acid **42** to give proteins **44–46**. By the minima at 208

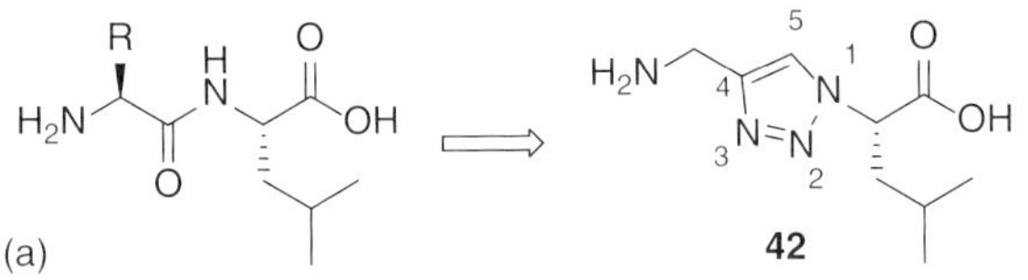

(a) **42**

Ac-RMKQIEDKLEEILSKLYHIENELARIKKLLGER-OH (**43**)
Ac-RMKQIED-XEEILSKLYHIENELARIKKLLGER-OH (**44**)
Ac-RMKQIEDKLEEILS-XYHIENELARIKKLLGER-OH (**45**)
Ac-RMKQIEDKLEEILSKLYHIEN-XARIKKLLGER-OH (**46**)

(b)

**Figure 5.19** (a) Native dipeptide and the triazole derivative. (b) Sequences of pLI-GCN4 (**43**) and modified peptides **44–46**.

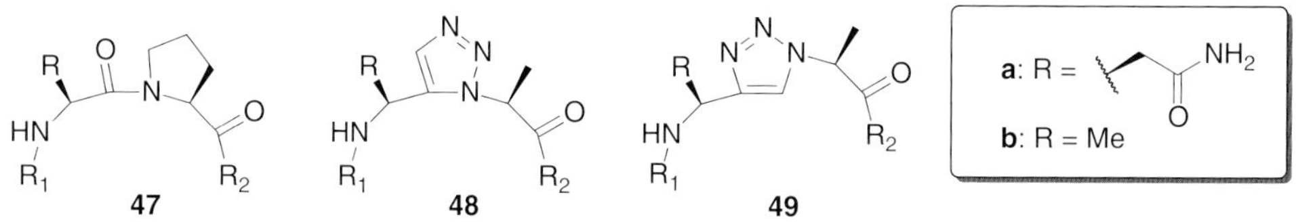


Figure 5.20 Triazole-based β-turn mimetics.

and 222 nm in the CD spectra, α-helical secondary structure was confirmed and denaturation was tested by monitoring the molar ellipticity at 222 nm. Peptides **45** and **46** showed denaturation at 36 and 61 °C, respectively, while peptide **44** showed no full denaturation up to 96 °C, similarly to the parent peptide. By crystallography, it was determined that the triazole fully participates in α-helical backbone hydrogen bonding in peptide **45** with the $N^2$ as the hydrogen acceptor and the $C^5$-H as the hydrogen donor.

Tam *et al.* [41] explored the use of the 1,5-triazole as a mimic for the *cis*-prolyl peptide bond encountered in the Asn-Pro (**47a**) and Ala-Pro turn (**47b**) (Figure 5.20). While regular amino acids can be used as surrogates for *trans*-prolyl peptide bonds, *cis*-prolyl peptide bond surrogates have to be made synthetically. They reasoned that 1,5-disubstituted 1,2,3-triazoles **48** would make good substitutes due to the similar size and ease of synthesis with Ru(II) catalysts allowing for the regioselective synthesis of the 1,5-isomer without the use of high temperatures [42, 43]. The isostericity was tested by incorporating the mimic into both bovine pancreatic ribonuclease (RNase A; 124 residues) and its N113A variant replacing the *cis*-prolyl turn located at AA113-Pro114. In both cases, the $T_m$ of the mimic was comparable to the wild-type enzyme and its N113A variant, respectively. In order to show the importance of the 1,5-isomer over the 1,4-isomer in this mimic, 1,4-disubstituted 1,2,3-triazoles **49** were also synthesized and incorporated into both enzymes. The 1,4-isomer mimics showed a lower $T_m$ compared to both the wild-type, N113A variant and their 1,5-isomer mimics, showing a greater dissimilarity and therefore to be an inferior mimic to the 1,5-isomer. All mimics, however, retained full catalytic activity, which is a strong indication of retention of tertiary protein structure [44].

Another approach to mimicking β-turn formation using a triazole was presented by Castellucci and Tomasini, who explored the synthesis of foldamers based on oligomers containing a 4-carboxy-5-methyloxazolidin-2-one (Oxd) and a 1,4-triazole (Tri) moiety (Figure 5.21) [45]. The new dipeptide mimic favors a bent conformation proposed to be similar to a β-turn by combination of a 3,4-disubstituted oxazolidin-2-one ring and a 1,4-disubstituted triazole ring, connected by a flexible methylene group. The bend was demonstrated to exist by ROESY analysis of Boc-D-Phe-L-Oxd-Tri-C=O-D-Val-Ome (**50**) and Boc-D-Phe-L-Oxd-Tri-$CH_2$-D-Val-OMe (**51**) showing a cross-peak between the triazole C–H and the Boc *tert*-butyl group.

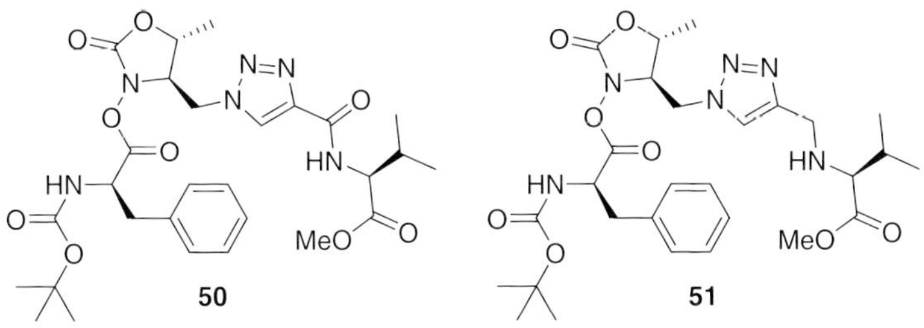


**Figure 5.21** β-Turn mimetics containing a 1,4-triazole.

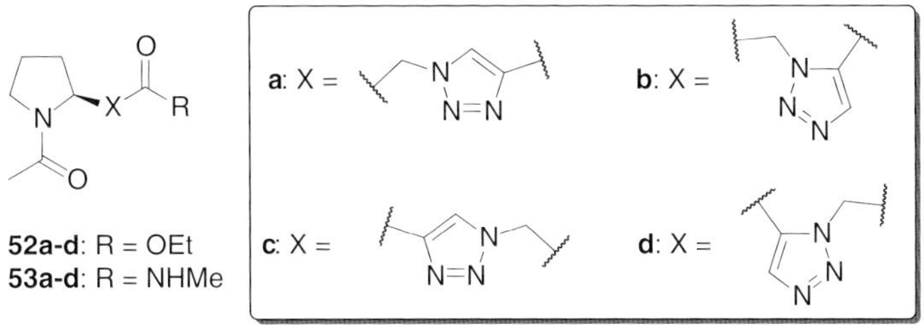


**Figure 5.22** Triazole-mimetics to study β-turn *cis/trans*-isomerization.

*Cis/trans*-isomerization plays a crucial role in various biological processes. One example is the opening of an ion channel by *cis/trans*-isomerization of a proline in a hinge position of the 5-HT3 receptor [46]. Another example is the common Gly-Pro-Gly-Arg motif in the HIV-1 gp120 V3 loop switching from a type II β-turn to a type VI β-turn, requiring a *cis*-proline peptide bond, as the key step before HIV-1 becomes infective [47]. Paul *et al.* [48] explored the *cis/trans*-isomerization of triazole analogs of Pro-Gly by synthesizing compounds **52–53** (Figure 5.22). The *cis*-fractions of the compounds were measured with NMR and compared to Ac-Pro-Gly-OMe for compounds **52** and to Ac-Pro-Gly-NHMe for compounds **53**. Compounds **52d** and **53d** showed similar *cis*-fractions to their reference compounds, whereas compounds **52b,c** showed substantially higher *cis*-fractions of 30%, which **53c** retained and **53b** did not lose completely. On the other hand, compound **52a** displayed a lower tendency for the *cis*-isomer than the reference dipeptide. Potentially, these triazole–proline dipeptides could be used for further exploration of analogs of peptides that show *cis/trans*-isomerization in their function.

Besides *cis/trans*-isomerization, some peptides possess a locked *cis*-amide bond as part of their functionality. One such peptide is sunflower trypsin inhibitor 1 (SFTI-1) from *Helianthus annuus*. Its 14-amino-acid backbone GRCTKSIPPICFPD is cysteine-bridged and head-to-tail cyclized, and the bond of interest is the Ile-*cis*-Pro amide bond. Tischler *et al.* [49] modified the open-chain variant (SFTI-1[1,14]) at residues 7–9 (Ile-cis-Pro-Pro) to Ile-Ala-Pro (**54**) and

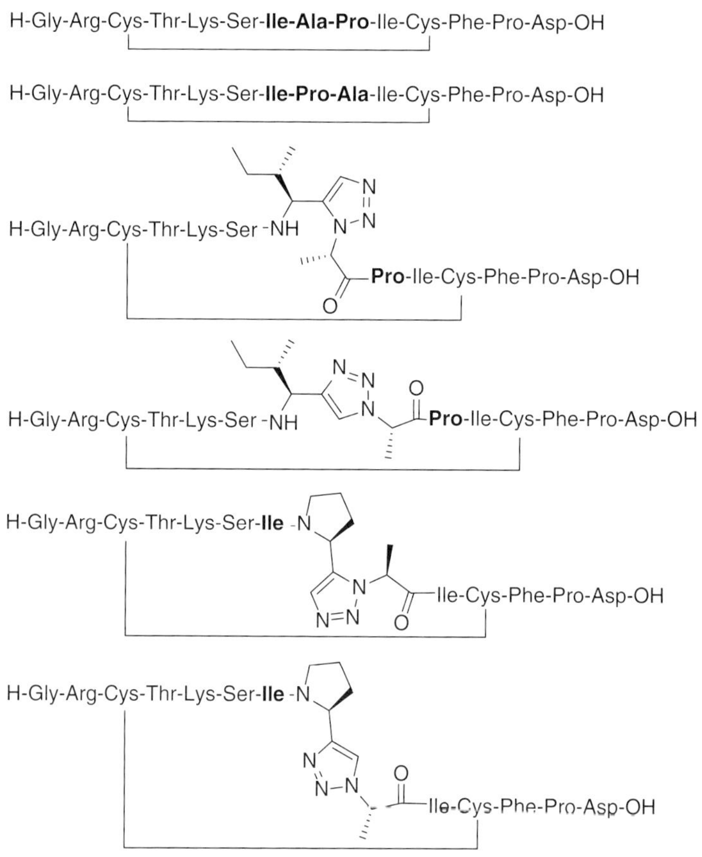


**Figure 5.23** Triazole-containing locked conformations of SFTI-1.

Ile-Pro-Ala (**55**) as reference peptides to four triazole analogs of SFTI-1[1,14] **56–59** (Figure 5.23). *Wild-type* SFTI-1[1,14] has an inhibitor constant of 0.2 nM, and any mismatch of the Ile-*cis*-AA-*trans*-AA conformation such as in **54**, **57**, and **58** shows a drastic reduction of binding affinity with inhibitor constants of 178, 302, and 255 nM, respectively. Peptidomimetics **55**, **56**, and **59** with the correct *cis*- and *trans*-bonds show closer inhibitor constants of 3.2, 34, and 6.3 nM, respectively. The value for **56** can be explained by the importance of the hydrogen bond between Tyr39 of the trypsin and amino acid 8 of the inhibitor. While nitrogens 2 and 3 of the triazole possess hydrogen-bond-accepting abilities, there is a mismatch in position of the acceptor when the inhibitor is bonded as modeled *in silico*. This research has shown how triazoles as amide isosteres can be used for the generation of locked *cis*- and *trans*-conformations with additional benefit of allowing the use of any amino acid residue after proline without loss of the *cis*-conformation.

## 5.4 Conclusions

The mild conditions required for the CuAAC reaction and its compatibility with functional groups and reaction media such as water have turned this into a widely applicable reaction. The reaction has also been successfully applied for the covalent and side-specific conjugation of peptides and proteins with fluorescent tags, polymers, saccharides, and radiolabels. Depending on the solubility of the substrates, the reaction can be carried out in organic solvents, in water, or even under physiological conditions. Owing to these advantages, the CuAAC reaction has developed into an important conjugation reaction in the field of chemical biology. One drawback, however, is that a catalytic amount of Cu(I) is required, which is sometimes difficult to remove from the product or causes adverse effects in *in vitro* applications due to the toxicity of the metal ions. This may be circumvented by applying strain-promoted variations [50], but these in turn require much more complex and bulky acetylenes. Besides conjugation of peptides, the CuAAC reaction has been successfully applied to alter the backbone of peptides taking advantage of the resemblance between triazole and amide functionalities. Amide replacement by a triazole may lead to metabolically more stable peptides, but also to modified properties due to the slightly altered conformation.

In this chapter, we have highlighted the versatility of the click reaction and the success so far in modifying peptides and proteins. Considering the enormous number of applications in this field and the ease with which this reaction can be applied, we anticipate that in the years to come, the success will only increase and also conceptually new applications remain to be discovered.

## References


1. Huisgen, R. (1989) *Pure Appl. Chem.*, **61**, 613–628.
2. (a) Rostovtsev, V.V., Green, L.G., Fokin, V.V., and Sharpless, B.K. (2002) *Angew. Chem. Int. Ed.*, **41**, 2596–2599; (b) Tornøe, C.W., Christensen, C., and Meldal, M. (2002) *J. Org. Chem.*, **67**, 3057–3064.
3. Kolb, H.C., Finn, M.G., and Sharpless, K.B. (2001) *Angew. Chem. Int. Ed.*, **40**, 2004–2021.
4. (a) Meldal, M. and Tornøe, C.W. (2008) *Chem. Rev.*, **108**, 2952–3015 For some excellent reviews, see:; (b) Bock, V.D., Hiemstra, H., and van Maarseveen, J.H. (2006) *Eur. J. Org. Chem.*, **2006**, 51–68; (c) 2010) *Chem. Soc. Rev.*, **39** (4), 1221–1408, special issue on click-chemistry.
5. Nguyen, D.P., Lusic, H., Neumann, H., Kapadnis, P.B., Deiters, A., and Chin, J.W. (2009) *J. Am. Chem. Soc.*, **131**, 8720–8721.
6. Mertens, M. and Gütschow, M. (2014) *Synthesis*, **46**, 2191–2200.
7. Jølck, R.I., Berg, R.H., and Andresen, T.L. (2010) *Bioconjugate Chem.*, **21**, 807–810.
8. Lu, J., Shi, M., and Shoichet, M.S. (2009) *Bioconjugate Chem.*, **20**, 87–94.
9. Ruoslahti, E. (1996) *Annu. Rev. Cell Dev. Biol.*, **12**, 697–715.
10. For an additional example, see e.g.: Lutz, J.-F., Börner, H.G., and Weichenhan, K. (2006) *Macromolecules*, **39**, 6376–6383.
11. Bertozzi, C.R. (2001) *Science*, **291**, 2357–2364.

12. Lis, H. and Sharon, N. (1998) *Chem. Rev.*, **98**, 637–674.
13. Mammen, M., Choi, S.-K., and Whitesides, G.M. (1998) *Angew. Chem. Int. Ed.*, **37**, 2754–2794.
14. Ohta, T., Miura, N., Fujitani, N., Nakajima, F., Niikura, K., Sadamoto, R., Guo, C.-T., Suzuki, T., Suzuki, Y., Monde, K., and Nishimura, S.-I. (2003) *Angew. Chem. Int. Ed.*, **42**, 5186–5189.
15. Chen, Y.-X., Zhao, L., Huang, Z.-P., Zhao, Y.-F., and Li, Y.-M. (2009) *Bioorg. Med. Chem. Lett.*, **19**, 3775–3778.
16. Huang, W., Groothuys, S., Heredia, A., Kuijpers, B.H.M., Rutjes, F.P.J.T., van Delft, F.L., and Wang, L.-X. (2009) *ChemBioChem*, **10**, 1234–1242.
17. Haubner, R., Finsinger, D., and Kessler, H. (1997) *Angew. Chem. Int. Ed. Engl.*, **36**, 1374–1389.
18. Kuijpers, B.H.M., Groothuys, S., Soede, A.C., Laverman, P., Boerman, O.C., van Delft, F.L., and Rutjes, F.P.J.T. (2007) *Bioconjugate Chem.*, **18**, 1847–1854.
19. Eliceiri, B.P. and Cheresh, D.A. (1999) *J. Clin. Invest.*, **103**, 1227–1230.
20. Vala, C., Chrétien, F., Balentova, E., Lamandé-Langle, S., and Chapleur, Y. (2011) *Tetrahedron Lett.*, **52**, 17–20.
21. Kolb, H., Walsh, J., Chen, G., Gangadharmath, U., Kasi, D., Scott, P., Haka, M., Collier, T., and Padgett, H. (2009) *J. Nucl. Med.*, **50** (Suppl. 2), 1939.
22. Doss, M., Kolb, H.C., Zhang, J.J., Bélanger, M.-J., Stubbs, J.B., Stabin, M.G., Hostetler, E.D., Alpaugh, R.K., von Mehren, M., Walsh, J.C., Haka, M., Mocharla, V.P., and Yu, J.Q. (2012) *J. Nucl. Med.*, **53**, 787–795.
23. McParland, B.J., Miller, M.P., Spinks, T.J., Kenny, L.M., Osman, S., Khela, M.K., Aboagye, E., Coombes, R.C., Hui, A.-M., and Cohen, P.S. (2008) *J. Nucl. Med.*, **49**, 1664–1667.
24. Kenny, L.M., Coombes, R.C., Oulie, I., Contractor, K.B., Miller, M., Spinks, T.J., McParland, B., Cohen, P.S., Hui, A.-M., Palmieri, C., Osman, S., Glaser, M., Turton, D., Al-Nahhas, A., and Aboagye, E.O. (2008) *J. Nucl. Med.*, **49**, 879–886.
25. Mirfeizi, L., Walsh, J., Kolb, H., Campbell-Verduyn, L., Dierckx, R.A., Feringa, B.L., Elsinga, P.H., de Groot, T., Sannen, I., Bormans, G., and Celen, S. (2013) *Nucl. Med. Biol.*, **40**, 710–716.
26. Thonon, D., Kech, C., Paris, J., Lemaire, C., and Luxen, A. (2009) *Bioconjugate Chem.*, **20**, 817–823.
27. Magata, Y., Lang, L., Kiesewetter, D.O., Jagoda, E.M., Channing, M.A., and Eckelman, W.C. (2000) *Nucl. Med. Biol.*, **27**, 163–168.
28. Liu, Y., Zhang, L., Wan, J., Li, Y., Xu, Y., and Pan, Y. (2008) *Tetrahedron*, **64**, 10728–10734.
29. Sun, H., Liu, L., Lu, J., Qiu, S., Yang, C.-Y., Yi, H., and Wang, S. (2010) *Bioorg. Med. Chem. Lett.*, **20**, 3043–3046.
30. Horne, W.S., Stout, C.D., and Ghadiri, M.R. (2003) *J. Am. Chem. Soc.*, **125**, 9372–9376.
31. Ghadiri, M.R., Granja, J.R., Milligan, R.A., McRee, D.E., and Khazanovich, N. (1993) *Nature*, **366**, 324–327.
32. Hartgerink, J.D., Granja, J.R., Milligan, R.A., and Ghadiri, M.R. (1996) *J. Am. Chem. Soc.*, **118**, 43–50.
33. Davis, M.R., Singh, E.K., Wahyudi, H., Alexander, L.D., Kunicki, J.B., Nazarova, L.A., Fairweather, K.A., Giltrap, A.M., Jolliffe, K.A., and McAlpine, S.R. (2012) *Tetrahedron*, **68**, 1029–1051.
34. Proteau-Gagné, A., Rochon, K., Roy, M., Albert, P.-J., Guérin, B., Gendron, L., and Dory, Y.L. (2013) *Bioorg. Med. Chem. Lett.*, **23**, 5267–5269.
35. Proteau-Gagné, A., Bournival, V., Rochon, K., Dory, Y.L., and Gendron, L. (2010) *ACS Chem. Neurosci.*, **1**, 757–769.
36. Rochon, K., Proteau-Gagné, A., Bourassa, P., Nadon, J.F., Coîté, J., Bournival, V., Gobeil, F., Guérin, B., Dory, Y.L., and Gendron, L. (2013) *ACS Chem. Neurosci.*, **4**, 1204–1216.
37. Valverde, I.E., Bauman, A., Kluba, C.A., Vomstein, S., Walter, M.A., and Mindt, T.L. (2013) *Angew. Chem. Int. Ed.*, **52**, 8957–8960.
38. Angelo, N.G. and Arora, P.S. (2005) *J. Am. Chem. Soc.*, **127**, 17134–17135.
39. Angelo, N.G. and Arora, P.S. (2007) *J. Org. Chem.*, **72**, 7963–7967.
40. Horne, W.S., Yadav, M.K., Stout, C.D., and Ghadiri, M.R. (2004) *J. Am. Chem. Soc.*, **126**, 15366–15367.

41. Tam, A., Arnold, U., Soellner, M.B., and Raines, R.T. (2007) *J. Am. Chem. Soc.*, **129**, 12670–12671.
42. Zhang, L., Chen, X., Xue, P., Sun, H.H.Y., Williams, I.D., Sharpless, K.B., Fokin, V.V., and Jia, G. (2005) *J. Am. Chem. Soc.*, **127**, 15998–15999.
43. Majireck, M.M. and Weinreb, S.M. (2006) *J. Org. Chem.*, **71**, 8680–8683.
44. Knowles, J.R. (1987) *Science*, **236**, 1252–1258.
45. Castellucci, N. and Tomasini, C. (2013) *Eur. J. Org. Chem.*, **2013**, 3567–3573.
46. Lummis, S.C.R., Beene, D.L., Lee, L.W., Lester, H.A., Broadhurst, R.W., and Dougherty, D.A. (2005) *Nature*, **438**, 248–252.
47. Wittelsberger, A., Keller, M., Scarpellino, L., Patiny, L., Acha-Orbea, H., and Mutter, M. (2000) *Angew. Chem. Int. Ed.*, **39**, 1111–1115.
48. Paul, A., Bittermann, H., and Gmeiner, P. (2006) *Tetrahedron*, **62**, 8919–8927.
49. Tischler, M., Nasu, D., Empting, M., Schmelz, S., Heinz, D.W., Rottmann, P., Kolmar, H., Buntkowsky, G., Tietze, D., and Avrutina, O. (2012) *Angew. Chem. Int. Ed.*, **51**, 3708–3712.
50. See e.g.: Jewett, J.C. and Bertozzi, C.R. (2010) *Chem. Soc. Rev.*, **39**, 1272–1279.

# 6
# Synthesis of Diverse Carbohydrate-Based Molecules using Click Chemistry

*Anoop S. Singh, Kunj B. Mishra, Amrita Mishra, and Vinod K. Tiwari*

## 6.1
## Introduction

Carbohydrate moiety, due to its potential polyfunctional nature, rigidity, and chirality, possesses many unique stereochemical and functional features, essential for induction of selectivity and chiral discrimination in various chemical, metabolic, and recognition processes [1–3]. Hence, functionality and structural variations are the key features due to which sugar-based molecules prove to be the interesting scaffolds for various purposes ranging from chemical biology to catalysis. In recent years, Cu(I)-catalyzed version of the Huisgen [4]–Meldal [5]–Sharpless [6] 1,3-dipolar cycloaddition reaction between terminal alkynes (**1**) and azides (**2**) (known as *CuAAC* or *Click Chemistry*) for the regioselective construction of 1,4-disubstituted triazoles (**3**) has emerged as a powerful strategy in organic synthesis for adjoining two entirely different building blocks, enabling an easy access of simple to complex molecular-level architectures, which find exponentially expanding applications in diverse disciplines of science ranging from organic synthesis to material sciences and chemical biology to catalysis. Alternatively, changing the catalysis from Cu(I) to Ruthenium catalysis [7], similar cycloaddition of **1** and **2** affords regioselectively opposite regioisomer **4** (Scheme 6.1).

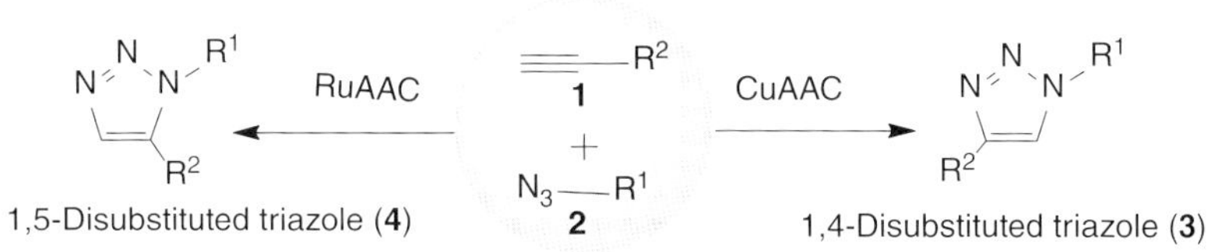


**Scheme 6.1** Regioselectivity in azide–alkyne [3+2] cycloaddition reaction.

This highly efficient, regiospecific, simple to perform, modular, and wide-scope reaction creates no or inoffensive by-products, involving readily available starting materials and reagents that can be conducted in easily removable or benign

*Click Reactions in Organic Synthesis*, First Edition. Edited by Srinivasan Chandrasekaran.

solvents with high chemoselectivity under mild and compatible reaction conditions, which are completely complementary for the development of novel pharmaceutical leads and thus satisfy all the essential conditions of "Sharpless Click-Chemistry Concept" [8]. The impact of click chemistry is increasing exponentially and particularly more useful in glycoscience [9–15]. Compatibility of the carbohydrate with "Click Chemistry" enables the rapid construction of simple to complex glycoconjugates of widespread applications, which is presented in this chapter.

## 6.2 Cu-Catalyzed Click Chemistry in the Synthesis of Diverse Glycoconjugates

The Cu(I)-catalyzed azide–alkyne 1,3-dipolar cycloaddition is a proven synthetic tool through combining the properties of two distinct molecular entities with distinct properties by coupling them to produce novel glycoconjugates of simple to complex architecture under mild reaction conditions [12–16]. $CuSO_4$/ascorbic acid–catalyzed coupling of 1-azido-1-deoxy-2,3,4,6-tetra-*O*-acetyl-β-D-glucopyranose **5** with 6-propargyl-1,2-3,4-di-*O*-ispropylidene-α-D-galactopyranose **6** in water affords regioselective triazole-linked disaccharide analog 7 (Scheme 6.2) [17].

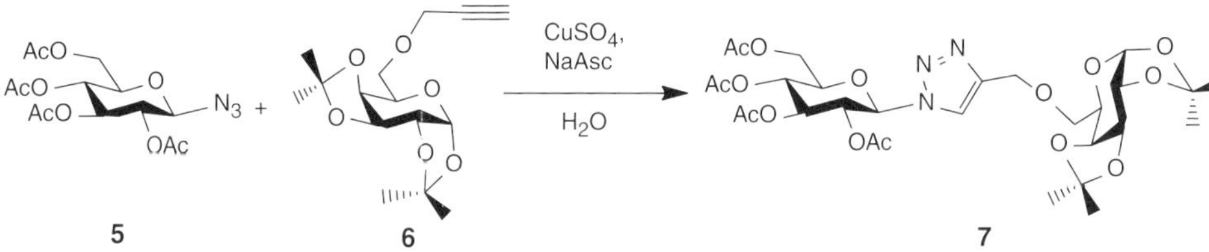


5 6 7

Scheme 6.2 Synthesis of triazole-linked disaccharide analog 7.

Click-supported MCRs are nowadays recognized as a vital tool for the synthesis of rare and multistep products in one-pot, which are utilized as precursor of potential fused heterocycle systems. We developed a two-step synthetic protocol of morpholine-fused [5,1-*c*]-triazoles from sugar alkynes. Triazolyl *azido* alcohol was obtained from sugar alkynes **8** via oxirane ring opening of epichlorohydrin in the presence of $NaN_3$ and subsequent azidation of chlorohydroxy triazoles under one-pot condition along with bis-triazolyl glycoconjugates as side product. Azido alcohol, **9** on propargylation followed by subsequent metal-free intramolecular cyclization, was converted into morpholine-fused [5,1-*c*]-triazoles **10** in good to excellent yield (Scheme 6.3) [18].

Cu click is widely utilized for creation of nature scaffold mimics. In this regard, several triazole-linked ethisterone glycoconjugates **13** were synthesized via CuAAC of azido sugars **12** with ethisterone **11**. Recently, multivalent and divalent triazole-bridging ethisterone–peptide conjugates were created to compete for androgen receptor binding and found useful against prostate cancer. Thus,

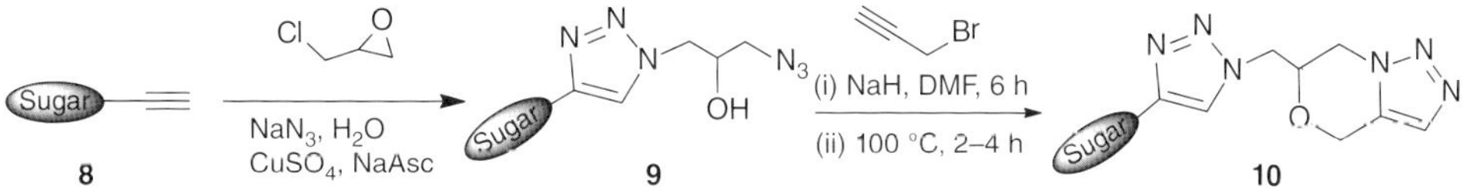


**Scheme 6.3** Click-inspired synthesis of morpholine-fused triazolyl glycoconjugates.

numerous sugar azides were clicked with ethisterone, which generated triazolyl ethisterone glycoconjugates in excellent yields (Scheme 6.4) [19]. Furthermore, glycosyl azidohydroxy triazole **14** on clicking with ethisterone gave bis-triazolyl ethisterone glycoconjugate **15** of potential therapeutic values [20].

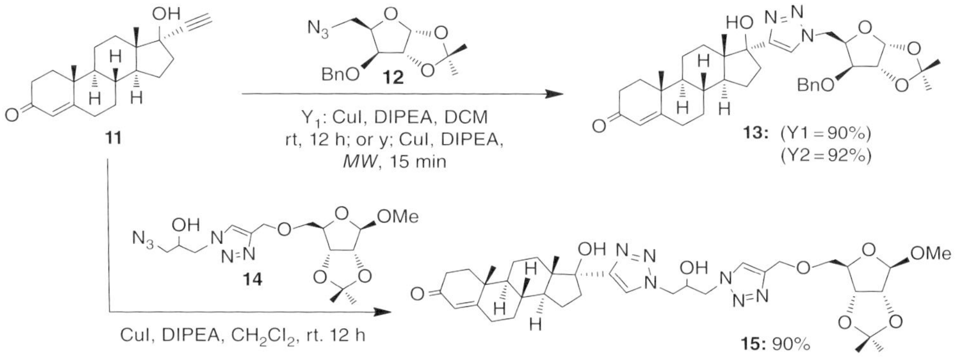


**Scheme 6.4** Synthesis of ethisterone glycoconjugates via Cu-catalyzed click chemistry.

Similarly, CuAAC was utilized for the successful conjugation of different carbohydrates with noscapine, “a phthaisoquinoline alkaloid” obtained from opium latex, to afford glycoconjugates **18**. Due to increased potency of 7-*O*-alkylated noscapine derivatives, sugars were attached at seventh position of noscapine via generating hydroxy functionality by selective demethylation using $NaN_3$/NaI and subsequent introduction of propargyl group followed by copper click with azido sugars **17** (Scheme 6.5) [21].

In addition of these, our group has recently introduced triazole skeleton at the β-position (C-5) of glycosyl ester. The conjugate addition of ammonia to glycosyl olefinic ester followed by diazo transfer reaction in one-pot fashion afforded glycosyl-β-azido ester **19**. A diverse range of triazolyl glycoconjugates was achieved in high yields via clicking glycosyl-β-azido ester with various alkynes (Scheme 6.6) [22].

Similar click protocol was employed for synthesis of ferrocene-based triazolyl glycoconjugates **23** by reacting ferrocenyl azide **22** and terminal sugar-based acetylenic compound **21** under click condition. Resultant ferrocenyl glycoconjugate **23** is supposed to be applicable for the study of protein–carbohydrate interaction using electrochemical biosensing techniques (Scheme 6.7) [23].

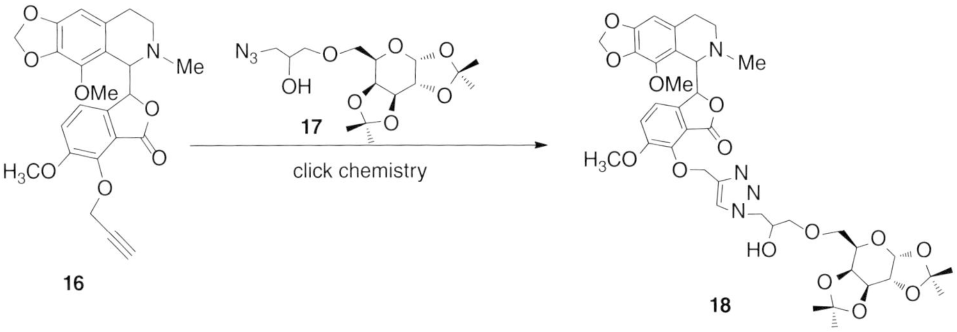


**Scheme 6.5** Click-inspired synthesis of noscapine glycoconjugates.

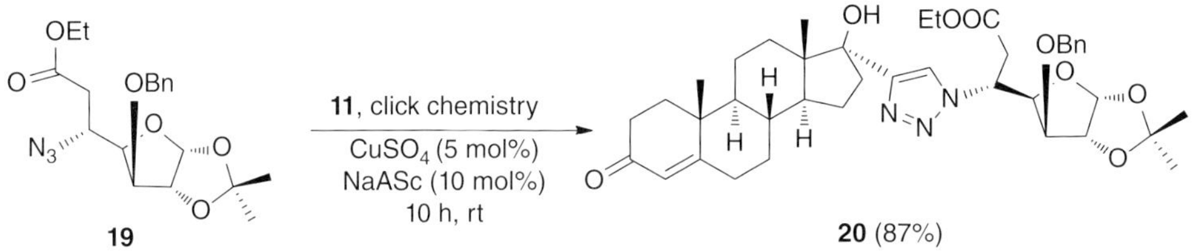


**Scheme 6.6** Synthesis of triazolyl glycoconjugates.

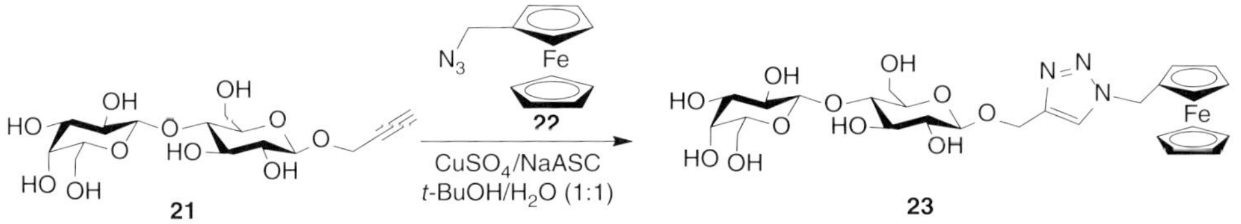


**Scheme 6.7** Synthesis of ferrocene-containing triazolyl glycoconjugates.

The coupling of a sugar azide **5** to a non-phloem-mobile insecticide possessing acetylene functionality **24** converted it into phloem-mobile insecticide **26** (GTF) using CuAAC reaction (Scheme 6.8). GTF was found mobile in the sieve tubes, confirmed via phloem-mobility test in *Ricinus communis,* although compound **26** displayed lower insecticidal activity against the third-instar larvae of *Plutella xylostella* [24].

Three different series of diverse triazolyl *O*-benzylquercetin glycoconjugates were synthesized *using* CuAAC for investigating them against leishmaniasis. After incorporation of propargyl group at three different positions (C-3 or C-5 or both C-3 and C-5) of benzylated quercetin, resultant *O*-benzylquercetin alkynes **27** were treated with a variety of sugar azides by copper click (Scheme 6.9). All synthesized glycoconjugates **28** were evaluated for antileishmanial activity against the promastigotes and amastigotes of *Leishmania donovani. Triazolylated* monosaccharides linked to 3,7,3′,4′-*O*-tetrabenzylquercetin skeleton **28a**

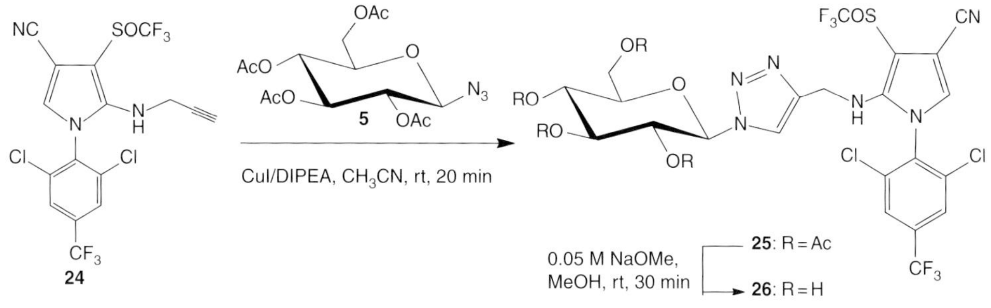

Scheme 6.8 Synthesis of glycosyl triazoles as potential insecticidal activity.

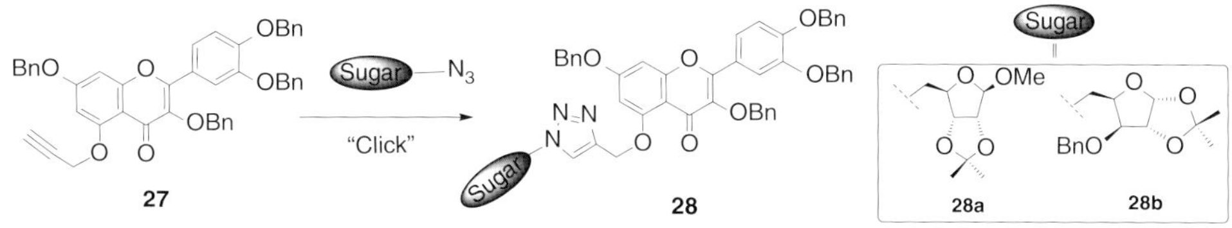

Scheme 6.9 Synthesis of antileishmanial triazolyl *O*-benzylquercetin glycoconjugates.

and **28b** exhibited efficient activity in comparison to other developed triazolyl glycoconjugates [25].

Click coupling of a coumarin-based alkyne **29,** "a well-known HIV-1 PR inhibitor" with the HIV-1 RT inhibitor azidothymidine (AZT) **30,** afforded a series of dual HIV-1 PR/HIV-1 RT inhibitors **31** (Scheme 6.10). The developed triazoles displayed inhibition levels of 52–60% and 94–99% at 50 μM against HIV-1 PR and HIV-1 RT, respectively. These analogs unveil an attractive way for the development of such glycoconjugates having dual inhibition properties [26].

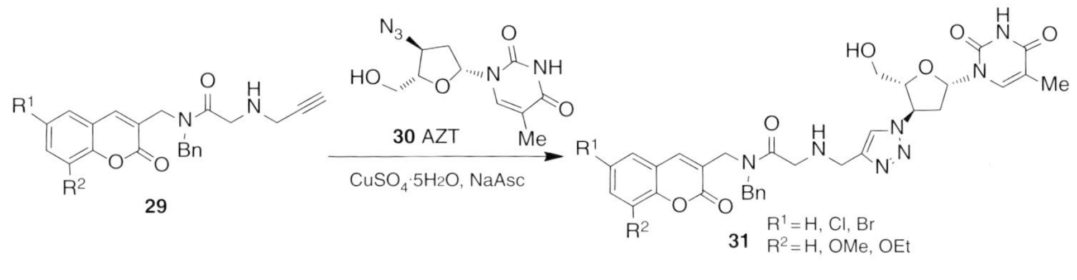

Scheme 6.10 Click-inspired synthesis of dual HIV-1 PR/HIV-1 RT inhibitor.

Glycohybrid of peptido lactam **34** was developed using CuAAC reaction of sugar azide **33** with α-acetamido-α-propargyl-β-lactam **32** (Scheme 6.11). Glycohybrid **35**, obtained from β-1-azido-L-fucose, was bound to *Ulex Europaeus Lectin-1* (UEL-I) after conformational change around a rotatable triazolylmethylene moiety [27].

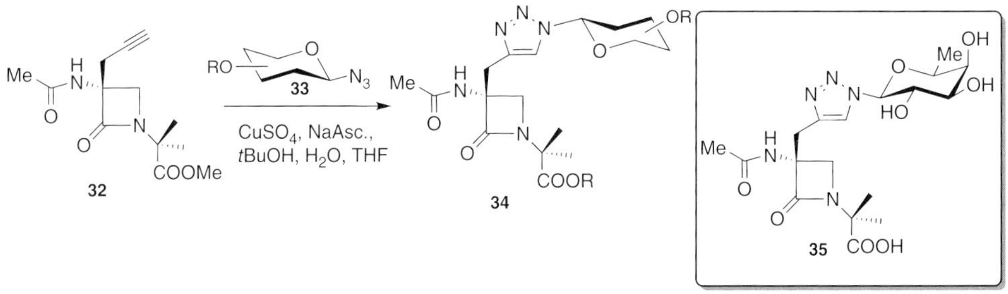


**Scheme 6.11** Synthesis of glycopeptide hybrid lactam mimetics using CuAAC.

A library of α-D- and β-D-glucopyranosyl triazoles **38–41** have been developed utilizing copper click (Scheme 6.12). An interesting fact observed in this synthesis is that the β-anomeric azide **5** takes 10–45 min to complete the reaction, whereas α-anomeric azide **37** requires 45–120 min under similar conditions. Difference in reaction rates was correlated with the anomeric effect on the dipole character of the anomeric azides [28].

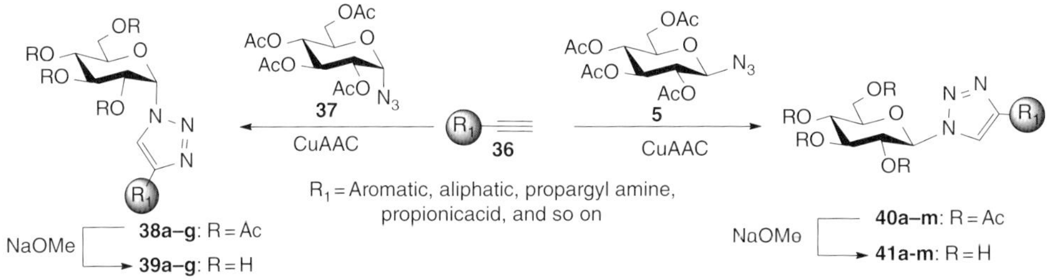


**Scheme 6.12** Click-inspired synthesis of α- and β-D-glucopyranosyl triazoles.

Galactosyl azides (**43** or **44**) were separately employed under *MW*-induced click coupling with different alkynes (**42**) and afforded high yield of triazoles **45** and **46**. Compound **46b** showed maximum *Trypanosoma cruzi trans*-sialidase (TcTs) inhibitory activity with 37% inhibition, while compound **45b** was found as the most powerful inhibitor for *in vitro* trypanocidal activity against trypomastigote forms of *T. cruzi* Y strain in hundreds of micrometer range (Scheme 6.13) [29].

An assembly of 1,2,3 triazole containing sialyl-mimetic neoglycoconjugates **47–50** were created by clicking azide containing sugars and amino acid with alkyne-functionalized sialic acid, where glycoconjugate **50** was found to be the most powerful TcTS inhibitor with $IC_{50}$ in the 260 μM range [30]. The chemistry was utilized for the oligomerization of azido–alkyne-functionalized galactose, which afforded a series of triazole-linked linear oligomers and macrocycles as promising TcTs inhibitory activities (Figure 6.1) [31].

A significant effect of multivalency was observed on selectivity and affinity of glycosidases. The monovalent iminosugar-based glycocluster **52**, developed via

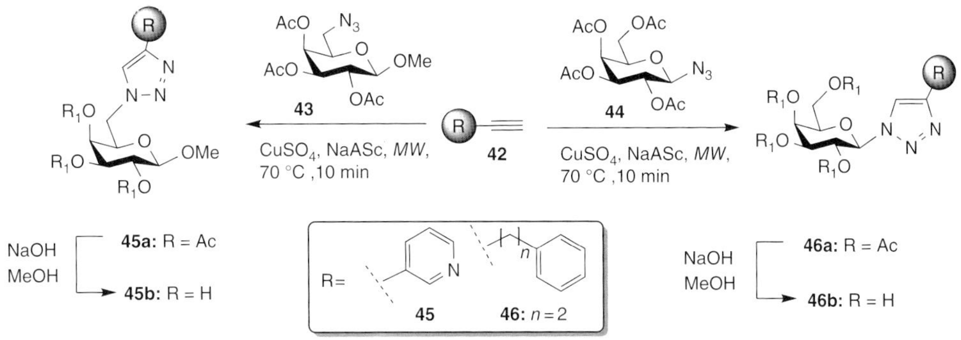


**Scheme 6.13** Synthesis of galactose-derived TcTs inhibitors using CuAAC reaction.

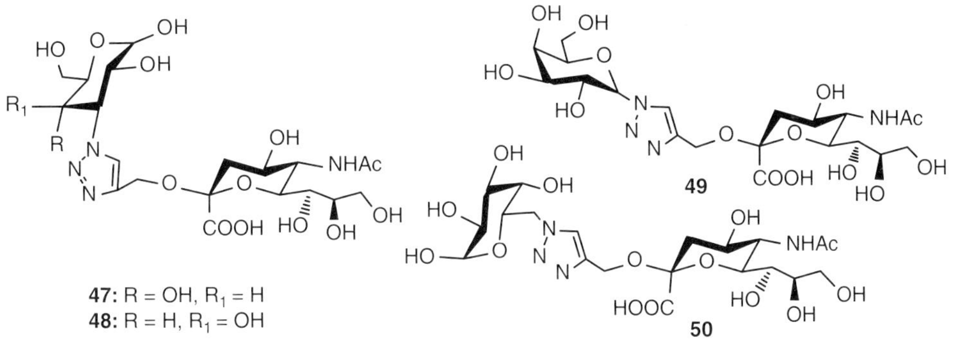


**Figure 6.1** Structure of triazolyl glycoconjugates as TcTS inhibitory activity.

azide–alkyne click reaction (Scheme 6.14), displayed promising inhibition activity of β-glucosidase and α-mannosidase [32]. Chemistry was extended for an easy access of higher-order iminosugar-based glycocluster.

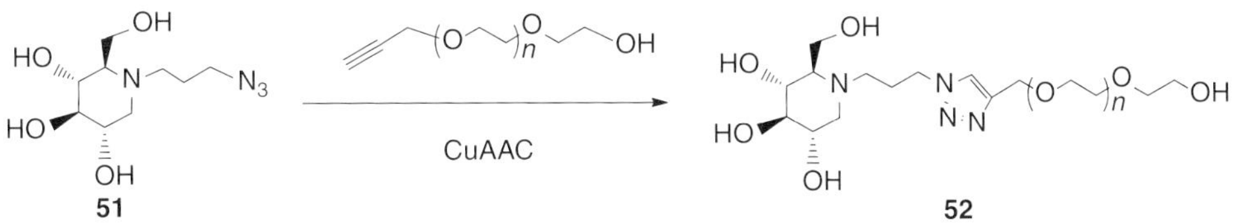


**Scheme 6.14** Synthetic multivalent iminosugars prepared by click reaction.

The triazolyl glycoconjugates displayed an impressive effect for inhibiting PTPs enzymes and are considered as potential therapeutic agents. A *MW*-accelerated CuAAC of sugar alkyne **53** with azide **54** followed by demethylation afforded triazolyl glycoconjugate **56** (Scheme 6.15). The molecule **56** ($IC_{50}$ = 5.1 μM) was identified as a selective and potent inhibitor of PTP1B [33].

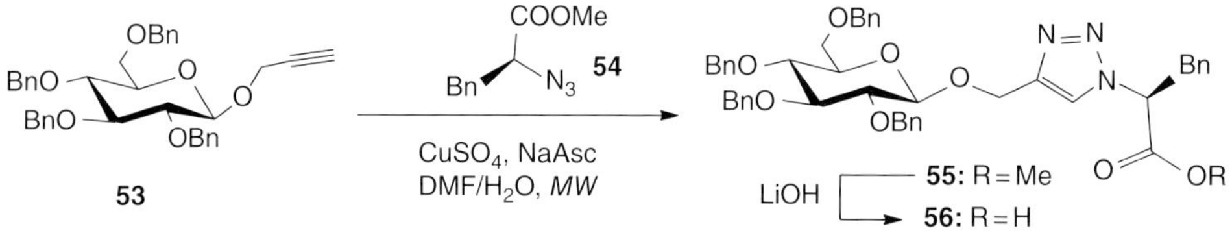


Scheme 6.15 Triazolyl glycoconjugates as selective PTP1B inhibitors.

Oleanolic-acid-based triazolyl glycoconjugates were further examined for their glycogen phosphorylase inhibitory activity, where compound **57** ($IC_{50}$ = 1.14 μM) showed good inhibition activity [34]. Similarly, pentacyclic triterpene heteroconjugate **58**, obtained through clicking of sugar azide with propargylated fatty acid, displayed moderate inhibition against RMGPa ($IC_{50}$ = 26 μM) [35]. Some simple glycosyl triazoles (**59** and **60**) were found as a potential inhibitor of glycogen phosphorylase (Figure 6.2) [36, 37].

A sugar-dependent *in vitro* antiproliferative activity was observed along with inhibition of S6 ribosomal protein phosphorylation. The 6-alkylated perillyl glucoside **61** was converted into the corresponding 6′-triazole perillyl glucoside **62** via treating with 3-azido propanol using CuAAC in 82% yield (Scheme 6.16) [38].

D-Mannose-based alkyne, on similar clicking with *o*-, *m*-, and *p*-substituted *azido* benzoic acid esters, afforded triazolyl glycoconjugates **65**–**67** (Scheme 6.17). These molecules were evaluated for their *in vitro* cytotoxic activity against seven cancer cell lines, where compound **65** showed increased activity ($IC_{50}$ = 23.8 μM) than precursor **63** ($IC_{50}$ = 93.2 μM) for CEM lymphoblasts. Also, deacetylation of compound **65** results in dramatic loss of activity [39].

Glycosyl triazole-linked 18β-glycyrrhetinic acids were developed by clicking per-*O*-acetylated glycosyl azides with propargyl ester of 18β-glycyrrhetinic acid (Figure 6.3). Subsequent de-*O*-acetylation of these molecules afforded potential

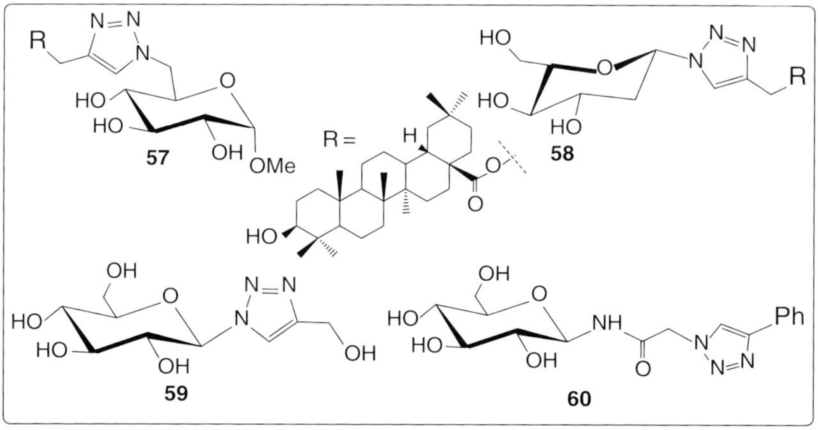


Figure 6.2 Diverse triazolyl glycoconjugates as glycogen phosphorylase inhibitors.

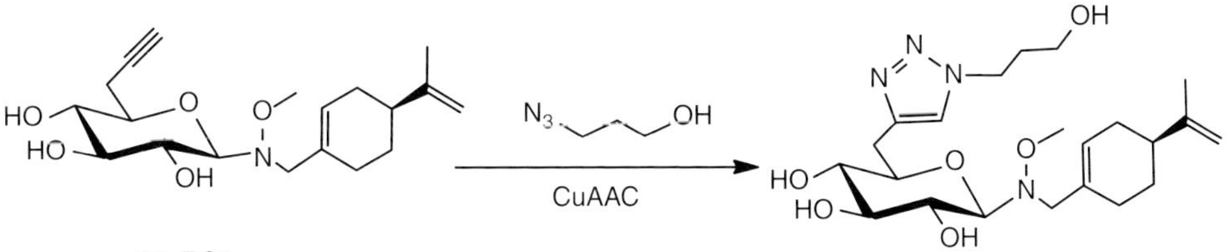


**61**, PG2

**62**, PG15

**Scheme 6.16** Synthesis of triazolyl glycoconjugate as antiproliferative activity.

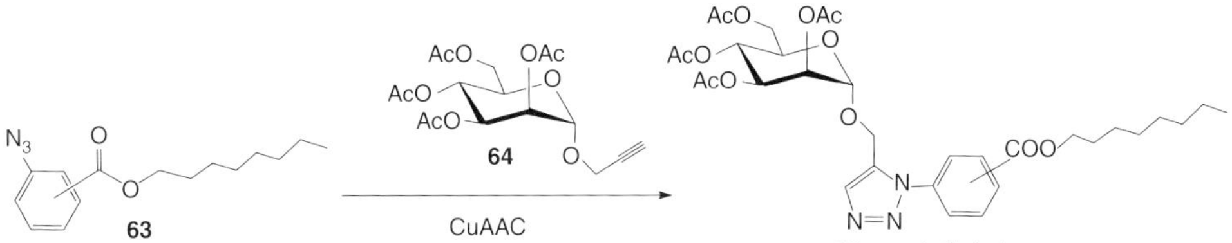


**65**: *o*-substituted
**66**: *m*-substituted
**67**: *p*-substituted

**Scheme 6.17** Synthesis of glycosyl triazoles as anticancer agents.

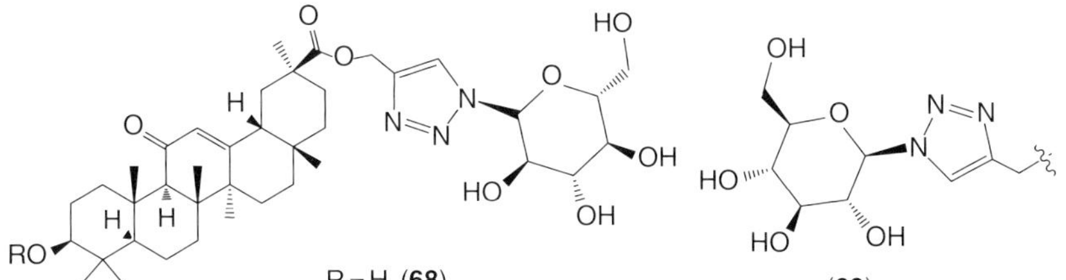


R = H (**68**)

(**69**)

**Figure 6.3** Chemical structures of anticancer agent created via triazole linkage.

anticancer agents when examined against human cervical cancer cells (HeLa) and normal kidney epithelial (NKE) cells. Compounds **68** and **69** displayed promising anticancer activities for HeLa cells with ($IC_{50}$: 13.76 ± 5.30) and ($IC_{50}$: 21.50 ± 2.24), respectively [40].

Sahoo *et al.* synthesized triazole-linked divalent glycoconjugates via CuAAC of alkyne-functionalized divalent building blocks with per-*O*-acetylated glycosyl azides (Scheme 6.18). Resultant glycoconjugates (**71**) on selective deprotection and transformation of the nitro- and cyano-groups to α- or β-amines can be used in peptide synthesis to produce unique divalent glycopeptide mimics having potential biological applications [41].

Sen *et al.* used basic alumina as solid support for a novel one-pot synthesis of triazolyl *N*-carboxamides **74** from glycosyl amines **72** without any solvent, ligand, or base (Scheme 6.19). Also, a number of biologically important unsymmetrical bis-1,2,3-triazoles were obtained by consecutive domino-click reaction [42].

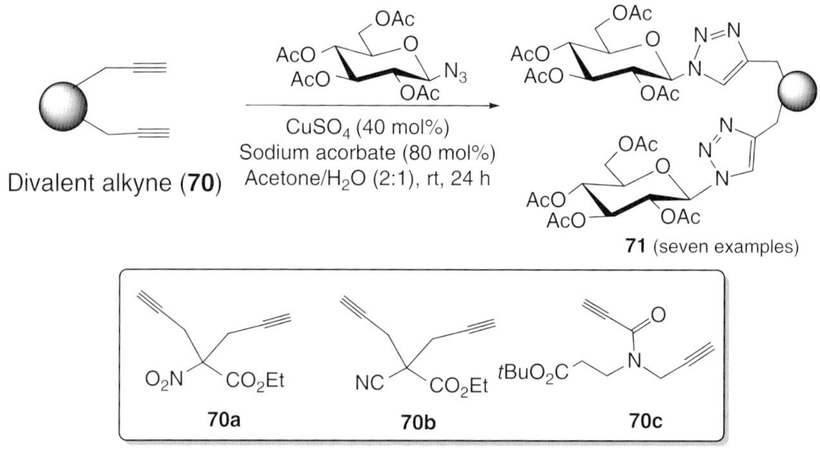


Scheme 6.18 Synthesis of triazole-linked divalent glycoamino acid mimics.

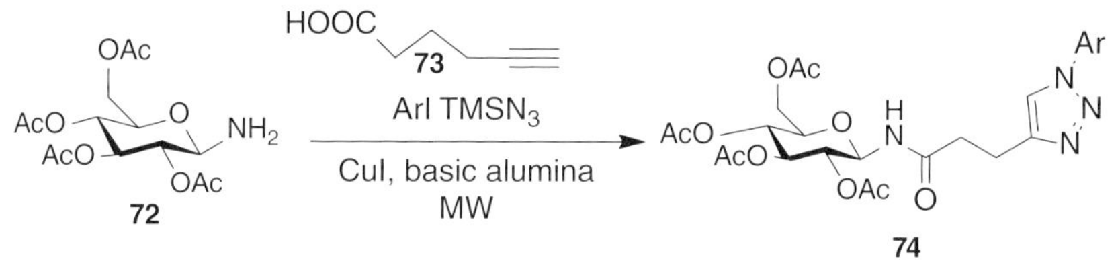


Scheme 6.19 Multicomponent synthesis of Triazolyl *N*-carboxamides.

Stereochemical diversity in carbohydrate part of the glycoconjugates interrogated the carbonic anhydrases active-site topology and generated various potent inhibitors, which were found selective toward *h*CA-IX. In this context, compound **75**, obtained by coupling of sugar azide with *p*-substituted sulfonamide-based alkyne, exhibited potential activity and selectivities against the CA-IX [43]. A library of carbohydrate-based small molecules was developed and evaluated for CA inhibitory activity in cell-based models. Two of them, **76a** and **76b**, inhibited CA IX and XII within narrow range $K_i$ values (5.3–11.2 nM) (Figure 6.4) [44].

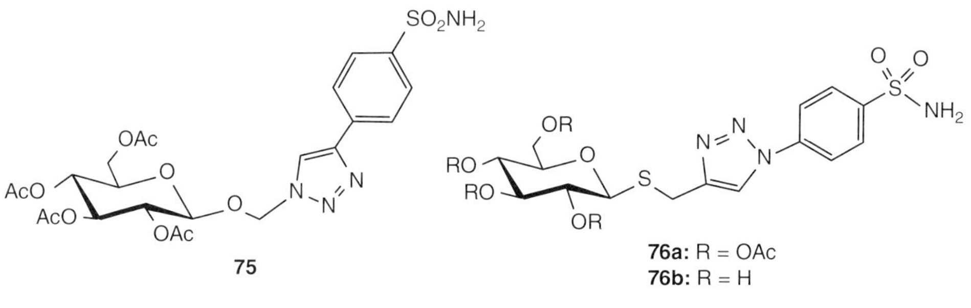


Figure 6.4 Chemical structures of carbonic anhydrase inhibitors explored via click chemistry.

The glycosyl neuraminidase inhibitors treated influenza directly by blocking the neuraminidase enzyme, which stimulates breaking of sialic acid from glycoproteins and glycolipids of the host cells. Thus, various triazole-linked sialic acid analogs were developed to treat resistant influenza viruses. A library of C-4 triazole-substituted zanamivir derivatives was prepared using click chemistry (Scheme 6.20), where compounds **79** ($IC_{50} = 6.4\,\mu M$) substantially showed maximum anti-AIV activity comparable to zanamivir ($IC_{50} = 2.8\,\mu M$) [45].

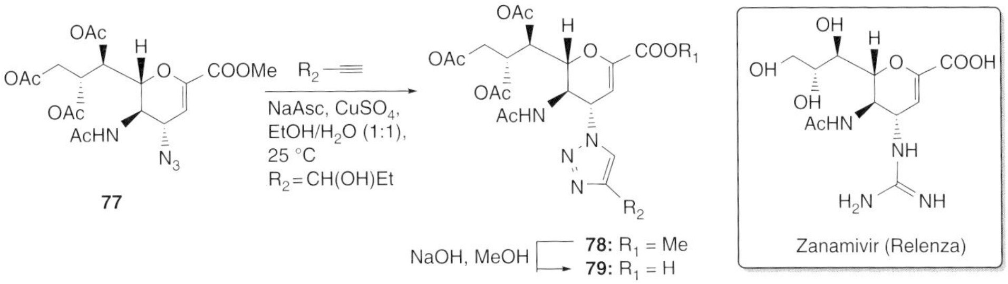


**Scheme 6.20** Synthesis of zanamivir-based anti-AIV agent.

Similarly, standard drug DANA was docked for NEU3 enzyme, which revealed about the presence of big hydrophobic pocket near the C9 position of the sialic acid. Therefore, a library of mammalian NEU3 enzyme inhibitors was synthesized by modifying DANA at C9 and N5Ac positions using click reaction. *N*-5-Azidoacetyl- and 9-azido-9-deoxy DANA derivatives were synthesized from Neu5Ac, which was clicked with different alkyne-functionalized molecules to afford diverse triazolo-sialosides **80** ($20 \pm 10\,\mu M$), hexyl **81** ($23 \pm 4\,\mu M$), and phenoxymethyl **82** ($45 \pm 3\,\mu M$) (Figure 6.5).

An attractive library of nonhydrolyzable 1,2,3-triazole derivatives of sialic acid was developed using CuAAC reaction of glycosyl azide **83** with different terminal alkynes, and the developed series was examined for neuraminidase inhibitory activity. Compound **84**, obtained by clicking of azide **83** with dec-1-yne, was recognized as potential lead compound ($IC_{50} = 28\,\mu M$) than standard Neu5Ac2en. Similarly, other multivalent derivative **85** ($IC_{50} = 17\,\mu M$) and dendrimer **86**

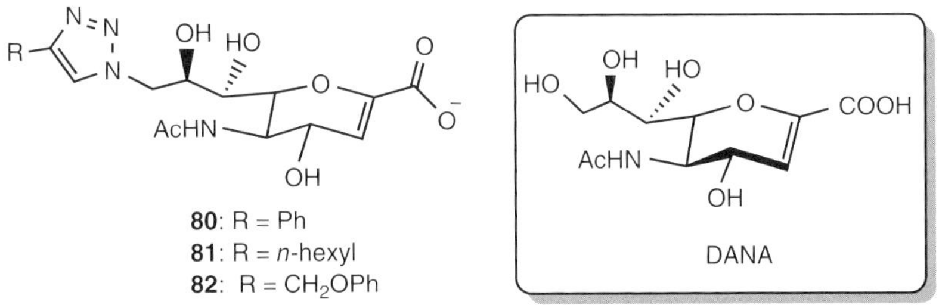


**Figure 6.5** Neuraminidase inhibition activity of developed glycosyl triazoles.

($IC_{50}$ = 20 μM) were also screened (Scheme 6.21) [46]. Similar azide–alkyne click coupling of azide-functionalized neuraminic acid with different nucleoside-based alkynes was extended for the creation of bacterial sialyltransferase inhibitor [47].

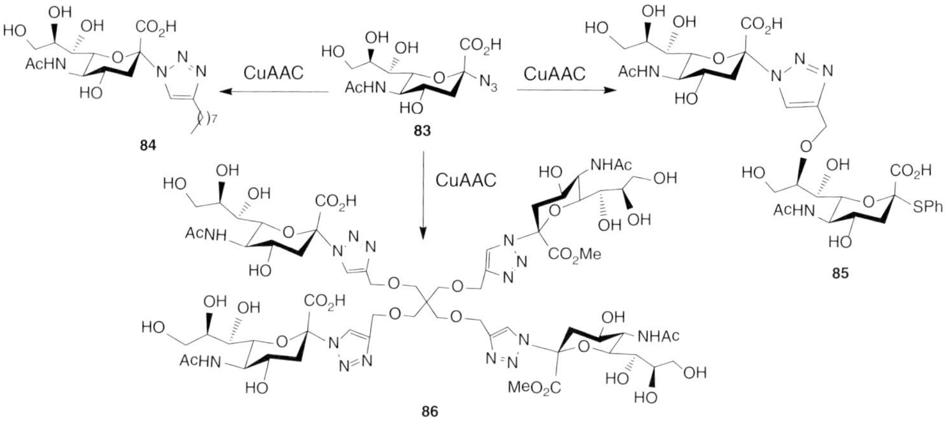


**Scheme 6.21** Triazole-linked sialic-acid-based neuraminidase inhibitors.

6-Triazolo-6-deoxy eugenol glucoside **87** displayed promising activity against the Gram-negative bacteria Salmonella typhimurium, which revealed by *in vitro* cytotoxicity evaluation on mouse spleen cells (Figure 6.6) [48]. Triazole-linked admantyl glycoconjugate **88**, obtained on clicking adamantane-based alkyne and sugar azide, displayed potential broad-spectrum activity against both bacterial and fungal strains [49]. Similarly, triazole-bearing cellulose nanocrystals were developed for photodynamic inactivation of *Mycobacterium smegmatis* [50].

A new class of (1,6)-oligomannose analogs with triazole as bridges displaying promising ManT inhibitory activity was achieved with the aid of click chemistry [51]. An interesting conjugated system **90**, developed by clicking azide **89** with

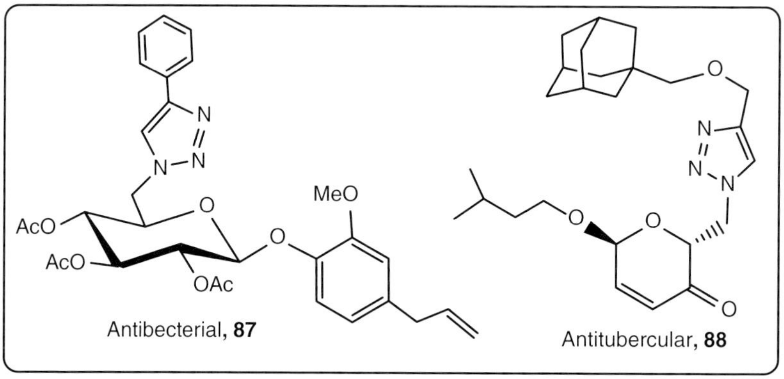


**Figure 6.6** Triazolyl glycoconjugates as antifungal and antibacterial agents.

alkyne **64** followed by deactylation (Scheme 6.22) displayed ManT activity (95% inhibition at 1 mM concentration) of the mycobacterial membranes by TLC analysis of the full range of radioactive lipids incorporated [$^{14}$C]-Man from GDP-[$^{14}$C]-Man [52]. Similarly, triazole-linked 1,6-oligomannoside **91** displayed promising inhibitory activity against ManT. Hexa- and octa-mannosides **91** with $IC_{50}$ values ranging between 0.14 and 0.22 mM were found to be good ManT inhibitors [53].

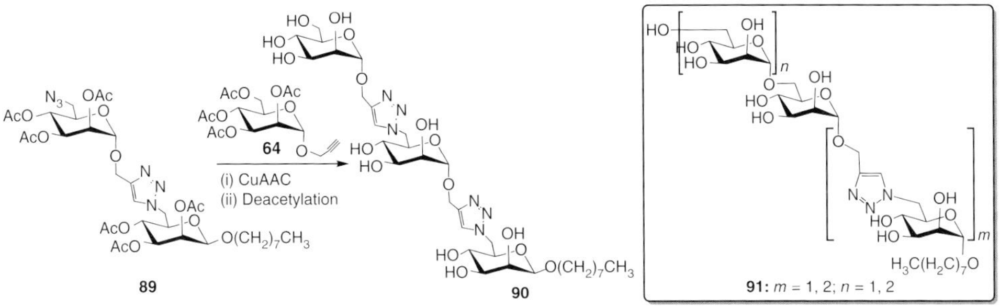


**Scheme 6.22** Triazole-linked glycoconjugate against ManT activity.

The click-inspired coupling of α-tocopherol-azide **93** with glyco-alkynes **92** afforded high yield of α-tocopherol-based triazolyl glycoconjugate **94,** which was recognized to be a radical-scavenging activity comparable to α-tocopherol (Scheme 6.23) [54]. However, the glycoconjugate, obtained by similar clicking of α-tocopherol alkyne with sugar azide, was recognized as unstable and air-sensitive under normal condition.

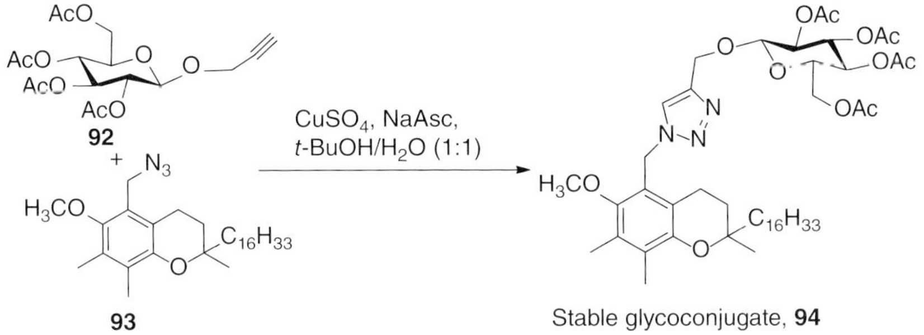


**Scheme 6.23** Synthesis of α-tocopherol-based triazolyl glycoconjugate.

Dondoni group introduced a successful conversion of C-40 secondary hydroxyl group of rapamycin into the azido group **95**, which on conjugation with various unprotected propargyl-*O*-glycosides and other glycosyl alkynes (e.g., alkyne **96**) afforded high yield of triazole-bridged rapamycin glycoconjugates (**97**) (Scheme 6.24) [55].

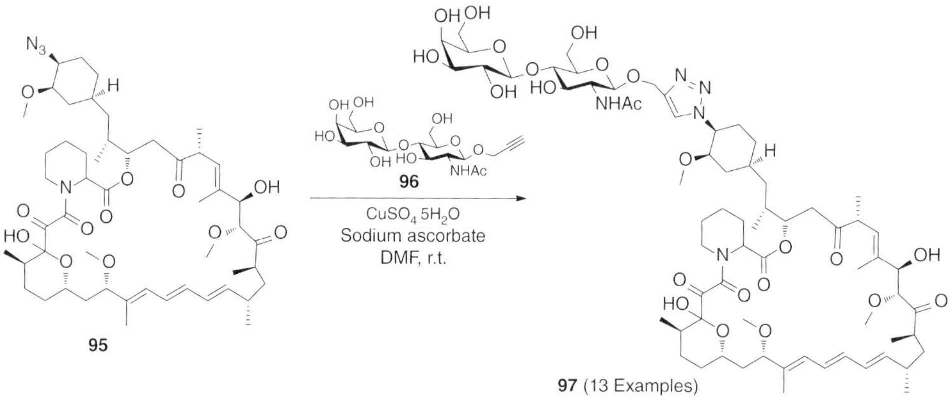


**Scheme 6.24** Click-inspired synthesis of rapamycin triazolyl glycoconjugate.

Carbohydrates, being important metabolic molecules, are taken by tumor cells more actively. Thus, for several decades, radiolabeled carbohydrate derivates have been successfully employed as imaging probes with single-photon emission computed tomography (SPECT) and positron emission tomography (PET). Due to the inertness of azide and alkyne groups, higher decay-corrected radiochemical yield, higher specific activity, shorter synthesis time, and stability of triazolyl ring under physiological conditions, CuAAC was considered as the most favorite choice for the development of carbohydrate-based imaging probes. Short-lived positron emitter ($^{18}$F, $t_{1/2} = 109.8$ min) $^{18}$F-labeled triazole-linked glycoconjugate **105** was developed by Kim *et al.* via click reaction, and the protocol was patented for radiolabeling [56]. Prante designed the $^{18}$F-glycosyl azide **98,** which was clicked with peptide moieties such as RGD or neurotensin derivatized with L-propargylglycine for radiolabeling purpose [57]. Resultant [$^{18}$F]FGlc-RGD **103** showed better blood clearance and stability in $\alpha_v\beta_3$-integrin expressing tumors during *in vivo* investigations with U87MG-bearing mice. Following the same method, Fischer *et al.* synthesized [$^{18}$F]fluorodeoxyglucosyl folate **100** to image the folate receptor, which is overexpressed in various tumor entities (Scheme 6.25) [58].

A click-based approach for direct derivatization of peptides $^{18}$F labeled [2-azidoethyl]-6-fluoroglucoside has recently been reported [59]. A diarylpyrazole glycoconjugate **107** was developed by Prante *et al.* using ligation of a terminal alkyne **106** with 2-deoxy-2-[$^{18}$F]fluoroglucosylazide. The glycoconjugate **107** displayed excellent affinity and selectivity toward NTS1 overexpressing tumors (Scheme 6.26) [60].

The methodology was further extended by Prante *et al.* for the development of $^{18}$F-labeled RGD glycopeptide with favorable biokinetics. A series of RGD glycopeptides was developed with various 6-fluoroglycosyl residues from monosaccharide to disaccharide units. RGD glycopeptides **108** showed high affinity to $\alpha_v\beta_3$

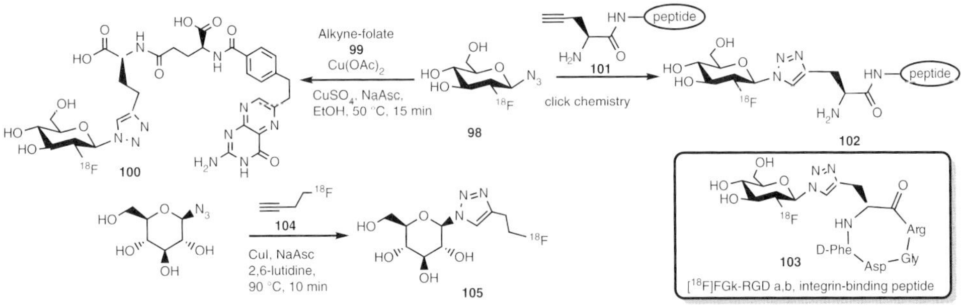


**Scheme 6.25** $^{18}F$-glycosylation of sugar azide and subsequent clicking with peptide moieties.

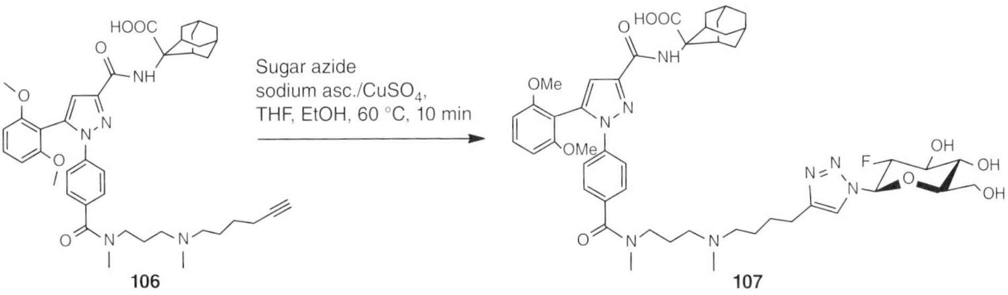


**Scheme 6.26** Synthesis of diarylpyrazole glycoconjugate.

(11–55 nM), $\alpha_v\beta_5$ (6–14 nM) and to $\alpha_v\beta_3$-positive U87MG cells (90–395 nM). [$^{18}$F]2Glc-RGD showed specific tumor uptake *in vivo*, whereas [$^{18}$F]6Glc-RGD **109** showed significantly decreased liver and kidney uptake by PET relative to the 2-[$^{18}$F]fluoroglucosyl analog (Scheme 6.27) [61].

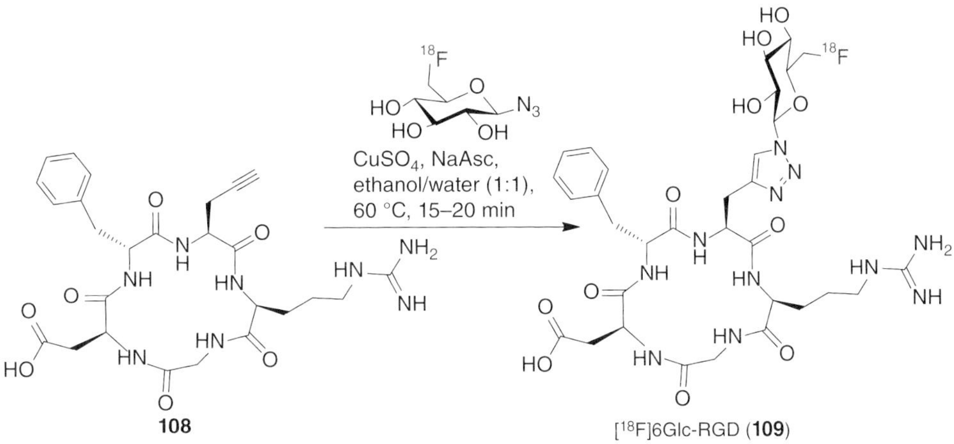


**Scheme 6.27** Synthesis of $^{18}F$-fluoroglycosylation of alkyne-bearing RGD peptides.

A fascinating application of click chemistry is well demonstrated in the development of carbohydrate-based chelates of medicinal and radiopharmaceutical interest. A one-pot Click-to-Chelate protocol for regular triazole chelators **114** was developed by Mindt *et al.*, which includes click reaction of L-propargyl glycine **110** with azide-functionalized D-galactose followed by $[^{99m}TcCO_3]^+$ addition to give $^{99m}$Tc-tricarbonyl complexes **114** (Scheme 6.28) [62]. Using Mindt Click-to-Chelate protocol, Fernández *et al.* developed a stable $^{99m}Tc(CO)_3$–glucose–glycine complex **112**, functionalized at the C-2 position, under one-pot condition with more than 90% radiochemical purity (Scheme 6.28). Biodistribution studies of the complex **115** showed low blood and liver uptake, moderate tumor uptake, rapid urinary excretion, and good retention in C57BL/6 mice bearing induced Lewis murine lung carcinoma [63].

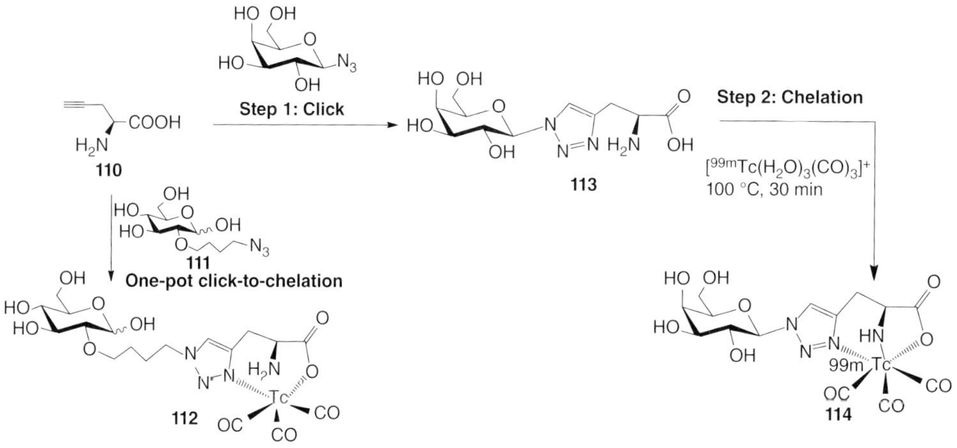


**Scheme 6.28** Click-to-chelate procedure for a regular click ligand.

The strategy was employed to develop hypoxia imaging agents for the assessment of tumor-oxygenation status [64] and bi- and tridentate click chelators for the $^{99m}$Tc-tricarbonyl labeling of 5-nitro-imidazoles [65]. A different approach of complexation of metal including radionuclei without involvement of triazole-N is also known. A series of functionalized pyridine–tetraacetic ligands attached to glucoside and maltoside scaffolds was achieved by clicking an alkyne **115** containing potent lanthanide chelate **117** with glycosyl azide **116** (Scheme 6.29) [66].

Similar chemistry was implemented to obtain the mannose and galactose-containing triazolyl glycoconjugates for different carbohydrate–lectin binding affinities with functionalized oligothiophene and Con A interaction [67]. Also, Tethering of Tc-chelating agents and Rh complexes onto a glucoside scaffold was achieved under click condition [68].

In addition to the wide application of click protocol in chemical biology, this highly facile method has also been well established for the development of chiral catalysis to be useful in asymmetric synthesis. A carbohydrate-based ligand **119**,

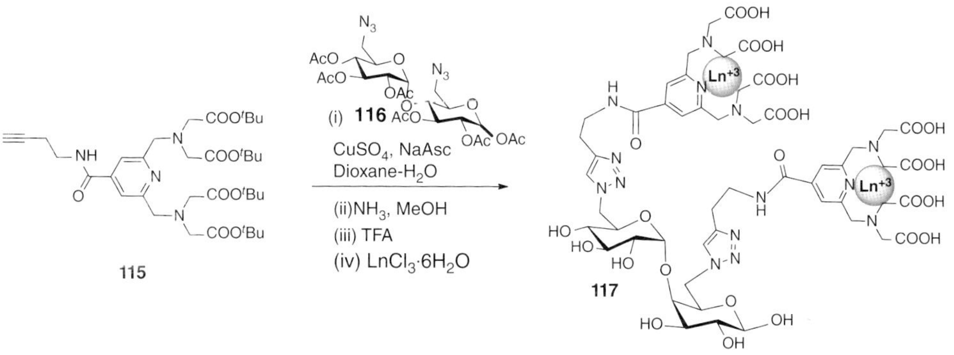


**Scheme 6.29** Synthesis of pyridine–tetraacetic acid glycoconjugates.

developed by clicking the alkyne with glycosyl azide **118** (Scheme 6.30), in the form of Pd complex, is useful for the allylic alkylation of 1,3-diphenyl propenyl acetate with dimethyl malonate [69].

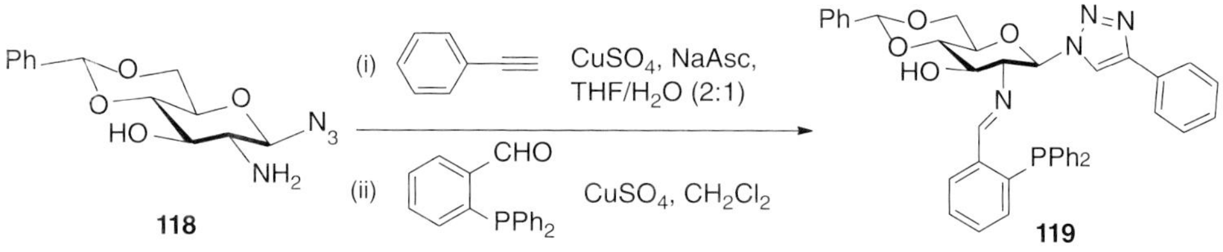


**Scheme 6.30** Synthesis of triazolyl glycoconjugate **119** as suitable ligand.

Wang *et al.* linked chiral pyrrolidine ring to sugar scaffold by clicking L-proline derived azide **123** with α-D-glucose-derived alkyne in the presence of CuI/DIPEA (Scheme 6.31) [70]. The resultant compound **124** successfully catalyzed Michael addition of ketones to nitrostyrenes with up to 98% yields, high diastereoselectivities (*syn*/*anti* 94 : 6 to 99 : 1), and high enantioselectivities (91% to >99%) at room temperature [70].

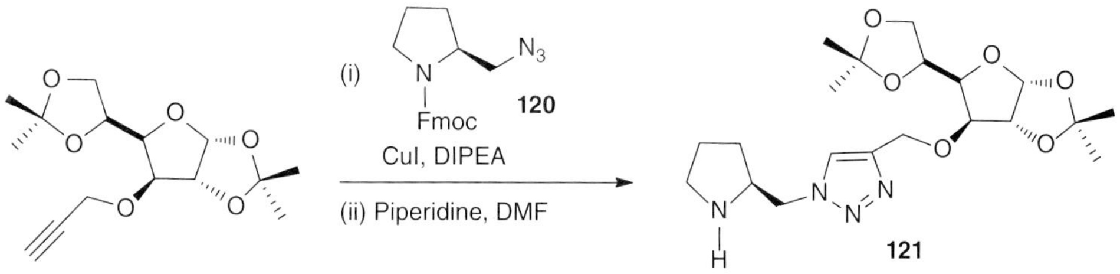


**Scheme 6.31** Synthesis of Michael addition catalyst **121** via click reaction.

Glucose-linked 1,2,3-triazolium ionic liquid **122** was prepared by copper(I)-catalyzed regioselective cycloaddition of a glucose azide with a glucose alkyne followed by quaternization with methyl iodide (Scheme 6.32). This IL was used as reusable ligand and solvent in copper(I)-catalyzed amination of aryl halides with aqueous ammonia [71]. Similar chemistry, that is, clicking of β-propargyl xyloside with phenyl azide, followed by acetate removal and methylation was employed for the syntheses of D-xylose-based ionic liquids [72].

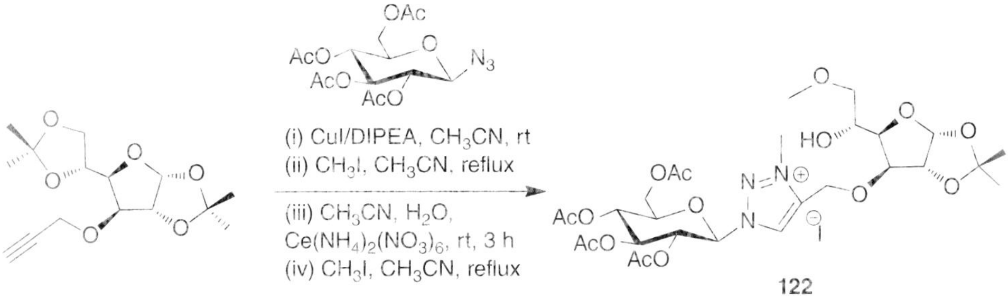


**Scheme 6.32** Preparation of glucose-linked 1,2,3-triazolium ionic liquids.

Furthermore, incorporation of sugar molecules using click chemistry would be a good strategy for capturing cations [73]. Utilizing fluorogenic dual-click chemistry, our group has recently developed sugar-based fluorescent chemosensors **125** and **126**, from D-glucose by using a number of high-yielding steps [74]. Thus, clicking of glycosyl-bis-alkyne **123** and **124** separately with fluorescent-deactivated 3-azidocoumarin delivered high yield of bis-triazolyl probes **125** and **126**, respectively (Scheme 6.33) [74]. Both sensors are highly selective for detection of $Cu^{2+}$ ions by the way of fluorescence quenching (detection limits were estimated as 6.99 lM for **125** and 7.30 lM for **126**), and DFT calculations reveal their similar optical behavior with akin mode of binding for **125**-$Cu^{2+}$ and **126**-$Cu^{2+}$ complexes.

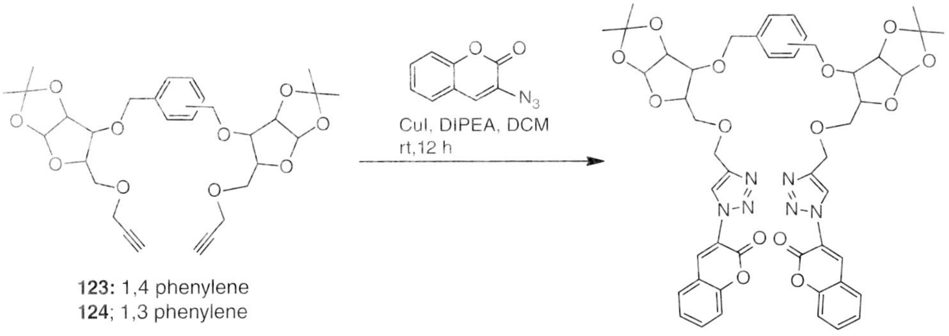


**123:** 1,4 phenylene
**124**; 1,3 phenylene

**125:** 1,4 phenylene (76%)
**126:** 1,3 phenylene (69%)

**Scheme 6.33** Click-inspired synthesis of D-glucose-derived chemosensor for $Cu^{2+}$ ions.

A fluorescent glycosyl azide **127** was clicked with alkyne **128** and afforded fluorescent triazolyl glycoconjugate **129** (Scheme 6.34). The developed fluorescent-labeled biomolecules **129** are likely to be an important tool for investigating various biological events. The fluorescent glycosyl conjugates were used to trace the translocation route of glycosyl pesticides within plants [75].

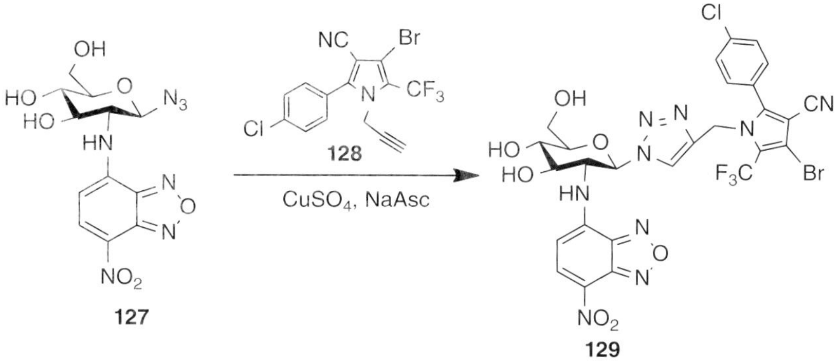


Scheme 6.34 Synthesis of sugar-based fluorescent-labeled biomolecules.

Click chemistry has proven its mettle as a valuable tool for understanding and mimicking biologically relevant lipid functions as well. Recently, a mechanochemical solvent-free strategy was applied to develop carbohydrate-anchored triazole-linked lipids, which interestingly self-assemble in different patterns (i.e., size and shape) [76]. In a recent report, Ugi four-component reaction was utilized by Labrada *et al.* to achieve double lipidic scaffolds with clickable azide or alkyne functionalities [77]. Trisaccharide alkyne **130** was conjugated with the lipid azide **131** by clicking in the presence of $Cu(Ac)_2$/NaAsc in $THF/H_2O$ to afford triazole-based neoglycolipid **132** (Scheme 6.35) [77].

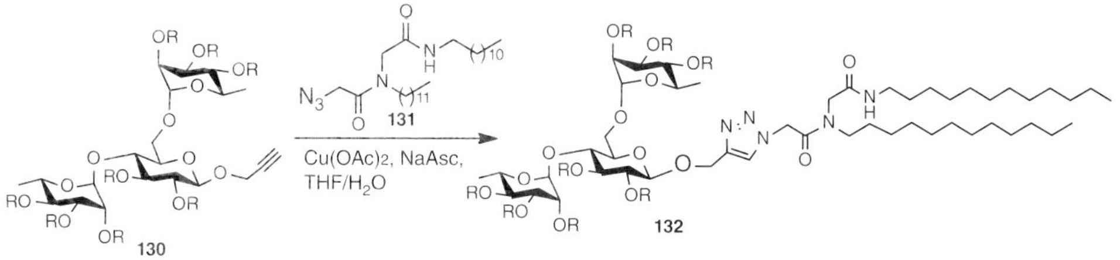


Scheme 6.35 Click-inspired synthesis of triazole-containing glycolipids.

α-GalCer triazole derivatives **135** were created in good yields using click coupling of lipid amide of α-GalCer containing an azido group **133** with various lipidic alkynes **134** (Scheme 6.36). Utilized variation in the length of chain in alkyne part displayed comparable stimulatory effect on cytokine production as α-GalCer, which exhibited a stronger Th2 cytokine response. Because IL-4 is an important cytokine for controlling autoimmune diseases such as type-1 diabetes, the developed glycosyl triazoles may be of significant use for treating these diseases [78].

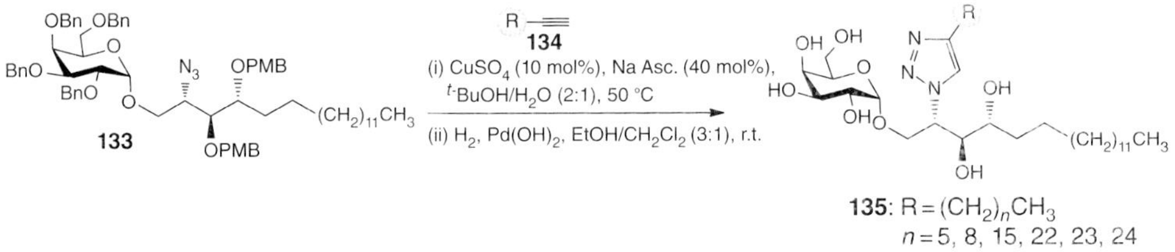


Scheme 6.36 Synthesis of 1,2,3-triazole-linked α-GalCer analogs.

Carbohydrate-based labeling reagents can be easily tagged with lipids by a simple click for functionalization of membrane surface. CuAAC successfully tagged mannose moieties on the surface of preformulated liposomes [79]. In addition to the multiple dimer glycolipids [80], recently, this chemistry was used to deliver a valuable complex glycolipid analog **138** by clicking protected alkyne-grafted analog of monophosphoryl lipid A, **136** with azide **137** (Scheme 6.37) [81].

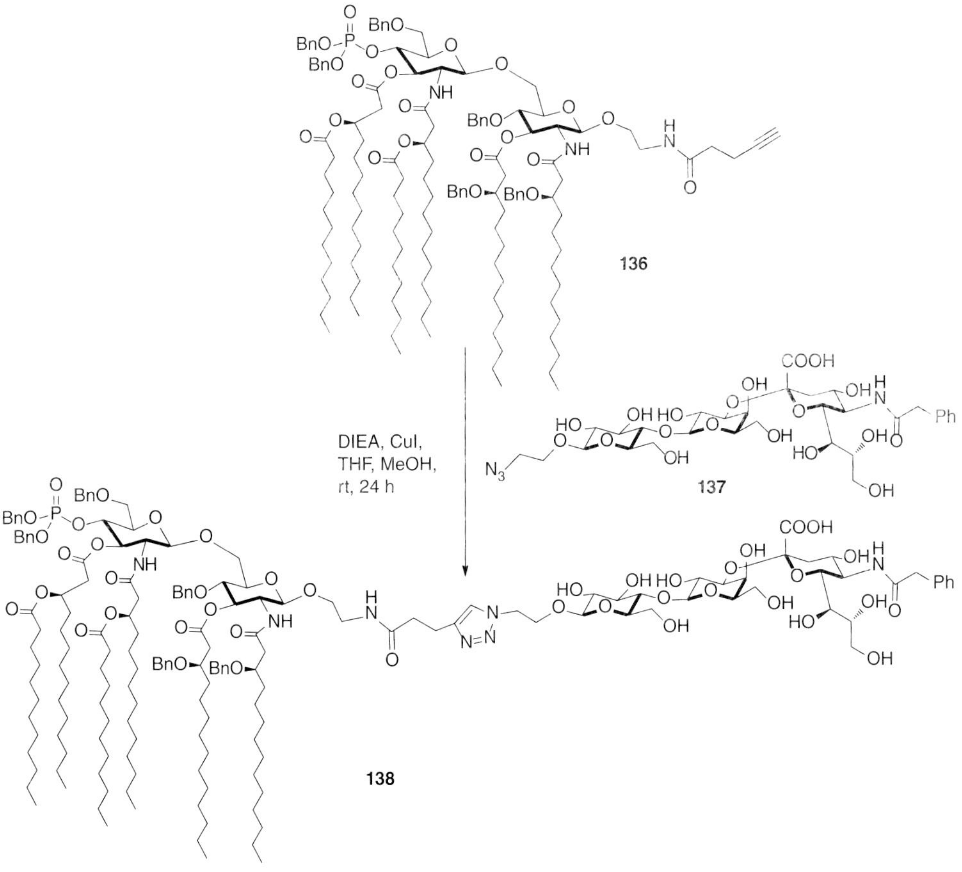


Scheme 6.37 Functionalization of alkyne-grafted analog of monophosphoryl lipid A.

## 6.3
## Synthesis of Carbohydrate-Based Simple to Complex Macrocycles

Multifunctionality, structural diversity, geometric structural rigidity, presence of several stereogenic centers, and other chemical and physicochemical properties make carbohydrate scaffold a fascinating choice to generate macrocycles of great importance [82]. Several *C*-2 and *C*-3 symmetric glyco-macrocycles were developed by cyclo-oligomerization of mannosyl-based mono-, di-, and tri-saccharides with suitably substituted anomeric azide and 4-propargyl ether at the two terminals using multiple-click strategy employing CuI/DBU (Scheme 6.38) [83]. The macrocycle **141** can bind with 2,4-hexadiyne-1,6-diol with a small association constant, but not to the aromatic systems, since openings of the cavities are too small for most of the molecules to enter, which was supported by computational studies [83].

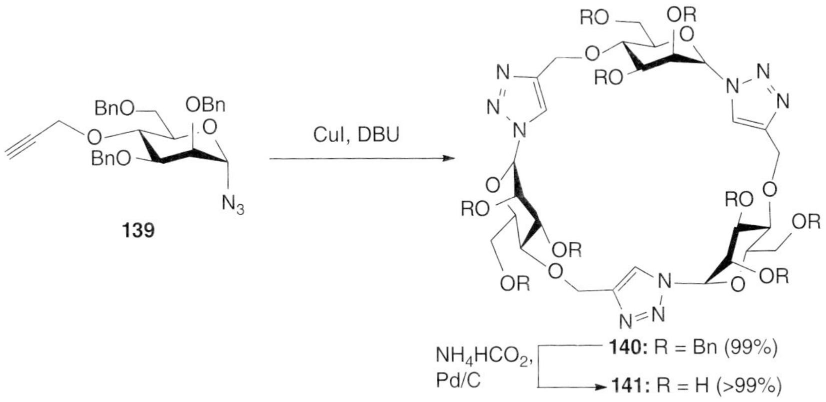


**Scheme 6.38** Click-inspired synthesis of *C*-2 and *C*-3 symmetric glyco-macrocyles.

Applying a similar protocol to various azido–alkyne-functionalized furanosides possessing ester (**142–143**), triazole-containing macrocycles (**144–145**) of various ring sizes were prepared [84]. This strategy has also been utilized for the development of cyclic peptides containing macrocycles used as building blocks for the assembly of oriented nanotubes (Scheme 6.39) [85].

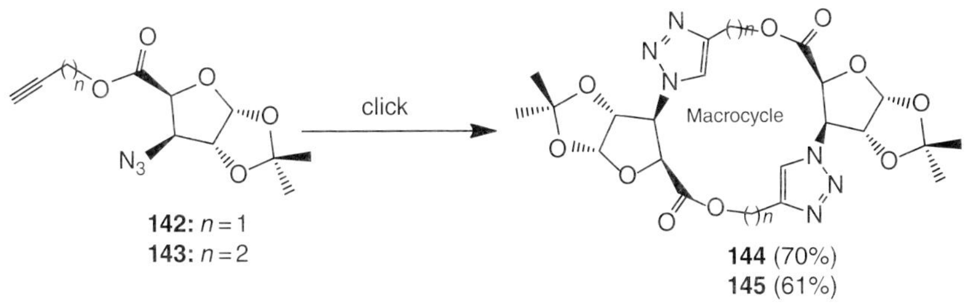


**Scheme 6.39** Cu(I)-catalyzed cyclooligomerization of azidoalkyne-functionalized furanosides.

Nilsson *et al.* synthesized sterically constrained bifunctional macromolecules **149** and **150** containing $C_2$-symmetry with blended features of two functionalities, amino acids and sugars together by coupling methyl 2-amino-6-azido-3,4-di-*O*-benzoyl-2,6-dideoxy-β-D-glucopyranoside **148** with two different *N*-propiolyl dipeptides (propiolyl-Tyr-Tyr-OH **146** and propiolyl-Arg(Mtr)-Tyr-OH) **147** followed by cyclodimerization using CuI/DIPEA in $CH_3CN$ (Scheme 6.40) [86]. The authors further developed a modular approach to fluorescent macrocycles as an extension to their work [87]. A click dimerization followed by ring-closing metathesis approach has been recently utilized for the synthesis of well-defined sugar-based macrocycle [88].

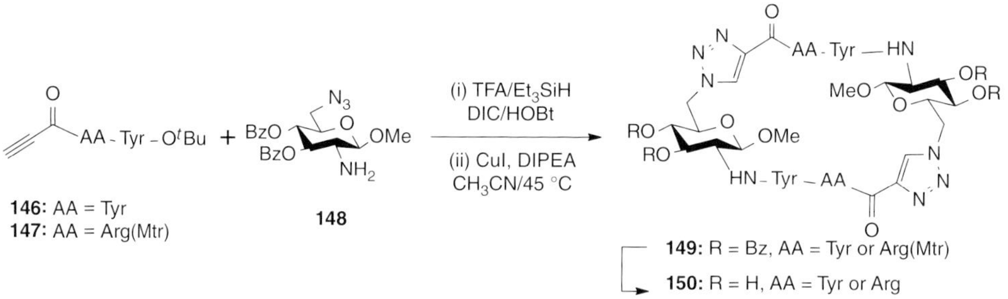


Scheme 6.40 Synthesis of macrocyclic carbohydrate/amino acid hybrids via CuAAC reaction.

Triazole-fused bicycles can be readily accessed via intramolecular cycloaddition through Cu(I)-catalyzed click. Cyclodextrins such as $C_n$- symmetric macromolecules containing two, four, and six C-glucopyranoside residues (**153a–c**) were successfully developed from bifunctionalized linear oligomers by clicking azide- and alkyne-functionalized sugars followed by Cu(I)-catalyzed intramolecular cyclization (Scheme 6.41) [89].

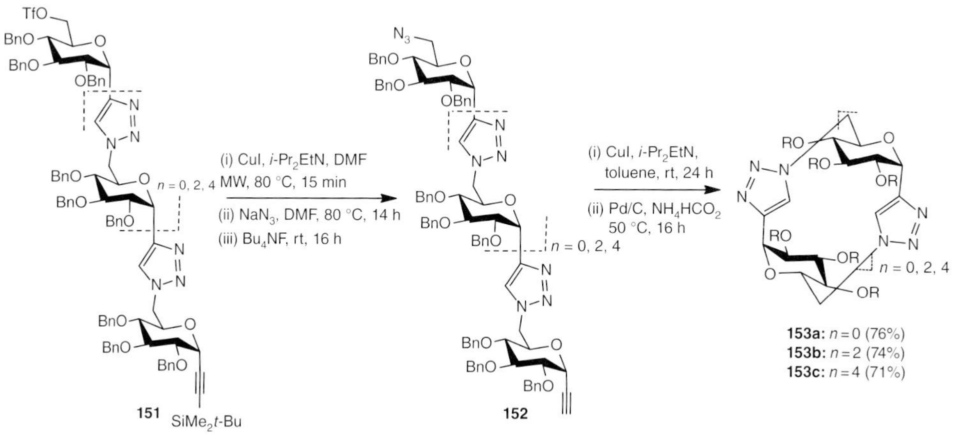


Scheme 6.41 Synthesis of $C_n$-symmetric triazole-linked cycloglucopyranosides.

A sucrose-derived $C_2$-symmetric macrocycle **158** was developed by Lewandowski and Jarosz via clicking lutidine-derived dialkyne **155** linkers with sucrose-derived azide **154** followed by activation of terminal-free hydroxyl groups and macrocyclization with ethylene diamine [90a]. Resultant macrocycle was obtained using template **159** in 20% yield while no product was found in the absence of **159** (Scheme 6.42). A similar protocol was further explored for an easy access of carbohydrate-based macrocycles [90b].

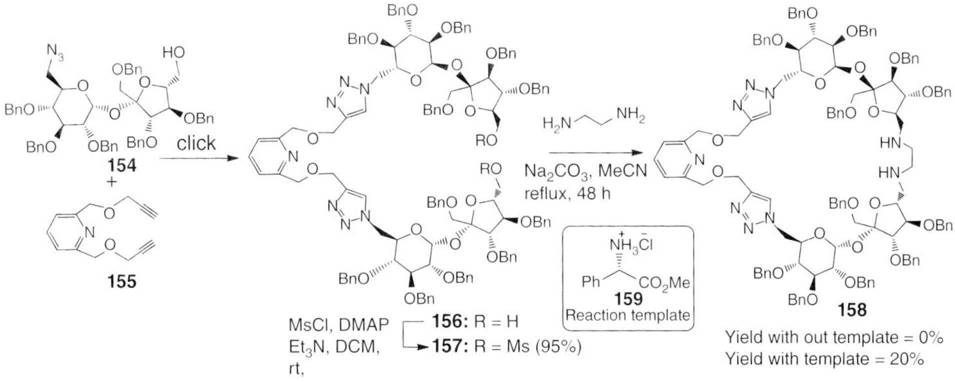


**Scheme 6.42** Amino-acid-templated macrocyclization to access sucrose-derived macrocycles.

The molecule **160** containing both azide furanoside and aromatic alkyne group on intramolecular Cu(I)-catalyzed cycloaddition constructed a strained monomeric 12-membered triazolophane **161**. Furthermore, several monomeric 12- to 17-membered triazolophanes fused to furanoside rings are prepared by furanoside-ring and peptide-appended azido–alkynes by click reaction (Scheme 6.43) [91].

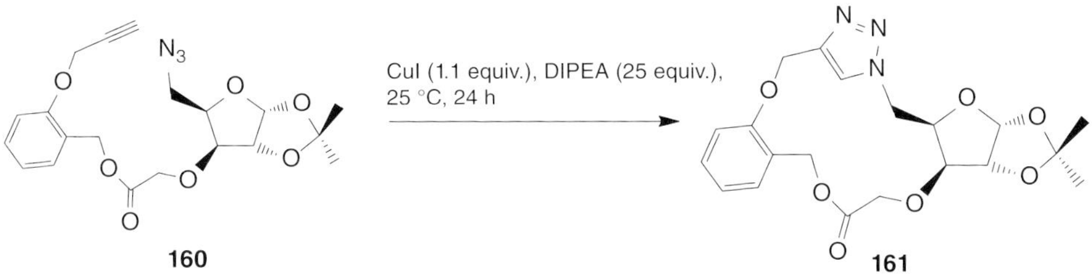


**Scheme 6.43** Synthesis of monomeric triazolophane from furanoside-tethered azido–alkyne.

The azide–alkyne cycloaddition of bisazide **162** with *p*-bis-propargyloxybenzene **163** using $CuSO_4$/NaAsc affords an interesting class of cyclophane-based macrocycle **164** in 56% yield (Scheme 6.44) [92]. The poor solubility of macrocycle

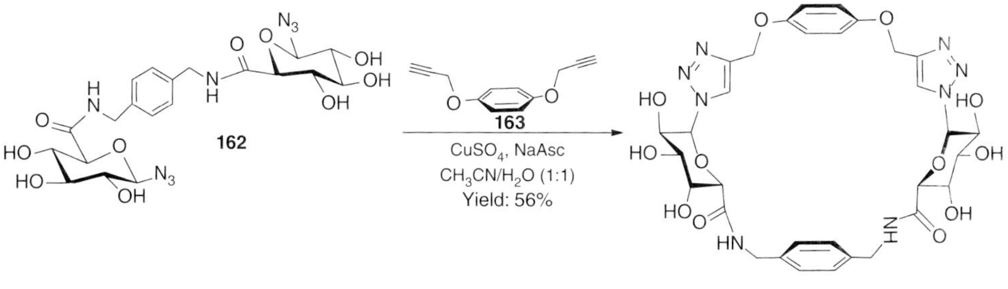


**Scheme 6.44** Synthesis of carbohydrate-based macrocycle.

**164** precluded an investigation in molecular recognition phenomena in aqueous system.

Cyclooligomerization leading to cyclic pseudo-oligosaccharides serves as a template for enzymatic silylation with TcTS in the presence of MUNANA [93]. Interestingly, sialic-acid-containing oligosaccharides **165** equipped with azide and alkyne functionalities at the two ends on consecutive cyclization reaction using CuI/DIPEA afforded good yield of sialic-acid-containing macrocycle **166** (Scheme 6.45) [94].

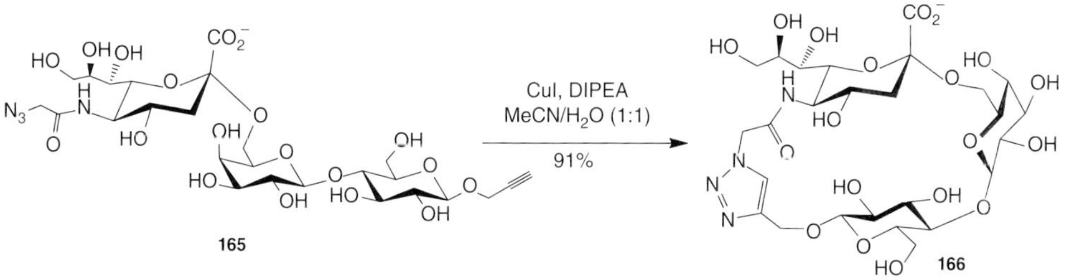


**Scheme 6.45** Chemoenzymatic synthesis of sialic-acid-containing macrocycle.

Schmidt *et al.* recently investigated IAD protocol utilizing the triazolylmethyl moiety in combination with *o*- and *m*-xylylene moiety as relatively rigid spacer system in intramolecular glycosylations. Thus, glycosyl alkyne **168** was first clicked with *O*-(2-azidomethylbenzyl)-protected acceptors **167** using CuI/DIPEA followed by PMB removal with DDQ affording the required donor–acceptor-linked triazole-containing disaccharide **170**. NIS/TMSOTf-mediated IAD of disaccharide **170** afforded the 15-membered macrocycle **171β** in 58% yield with greater than 98% β-selectivity. The anomeric selectivity depends on several factors; the deciding factor is presumed to be the ring size. Furthermore, clicking the alkyne **168** with acceptor **172** and subsequent IAD afforded the disaccharide macrocycle **174α** in 55% yield with α-selectivity (Scheme 6.46) [95].

Similarly, utilizing the *o*-azidobenzyl group, the author extended the "Click-to-IAD protocol" to create the triazole as rigid spacer [96]. Triazole-$N_3$ plays a part

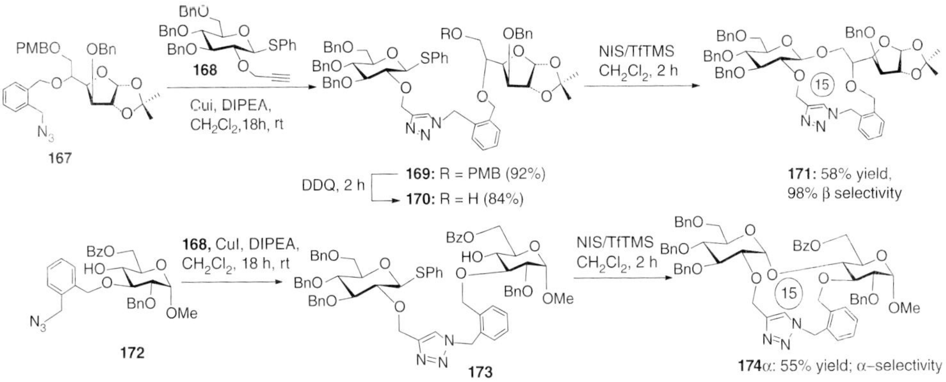


**Scheme 6.46** Intramolecular glycosidation through click-generated triazole as rigid spacer.

to stabilize the anomeric carbenium ion involved as part of six-membered ring formation that occurred from α- and/or the β-side; thus, the possible anchimeric assistance could be rationalized. Interestingly, the triazole ring was successfully deprotected from macrocycle disaccharide under Birch conditions (Na/Liq$NH_3$ at −75 °C) [96].

## 6.4 Click-Inspired Synthesis of Diverse Neoglycoconjugates

Triazole as linker between carbohydrate and peptide enables a straightforward route of accessing biologically privileged glycopeptides, representing approximately one-third of the approved biopharmaceutics [97]. Rutjes and coworkers developed a series of stable glycopeptides **177** via α- and β-triazole linker using $Cu(OAc)_2$/Na ascorbate–catalyzed clicking of mono/dipeptide alkynes/azides with mono/disaccharide azides/alkynes in water–*tert*-butanol system (Scheme 6.47) [98]. The work was extended to the synthesis of glycosylated cyclic arginine–glycine–aspartate derivatives **178** with a triazole linker that selectively targets $\alpha_v\beta_3$ integrin [99].

Considering the importance of sugar in cyclic peptides, a library of their mimics was reported via combined RCM-CuAAC protocol. Chemoselective ligation was subjected to flexible glycosyl transferases on nucleotide diphosphosugar libraries of glyco-randomized vancomycin analog, where carboxylic acid derivative was found to be twice effective as vancomycin against antibiotic-resistant strains [100]. A swift synthesis of MUC1 glycopeptide dendrimer was established using click chemistry [101]. Recently, Renaudet *et al.* reported the synthesis of cyclopeptide-based fucosylated glycodendrimers **180** with the aid of well-known oxime ligation protocol. The molecules are useful for a variety of chemical biology purposes such as the enzyme-linked lectin assay. The chemistry is established as both the oxime

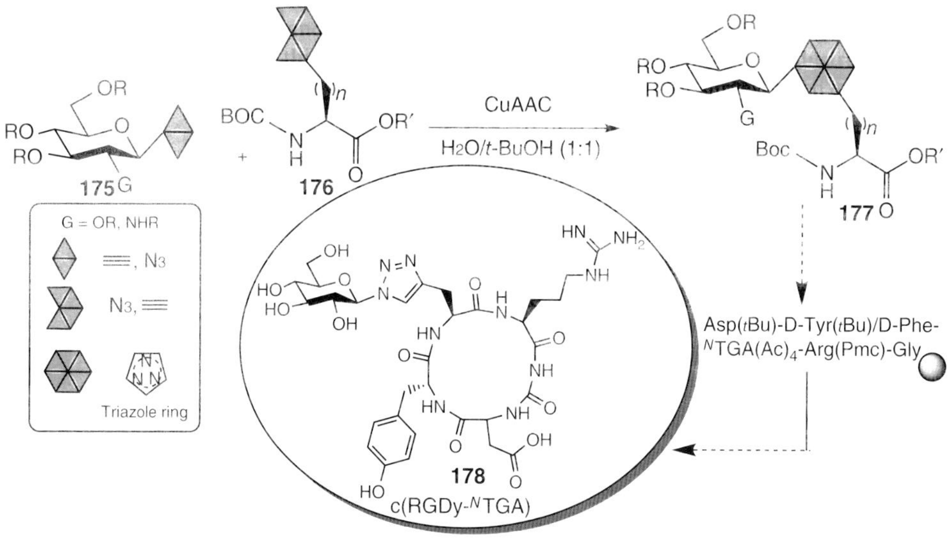


**Scheme 6.47** Click-inspired synthesis of cyclic arginine–glycine–aspartate-containing macrocycle.

ligation and the click chemistry were successfully used for an easy access of such glycodendrimers (Scheme 6.48) [102].

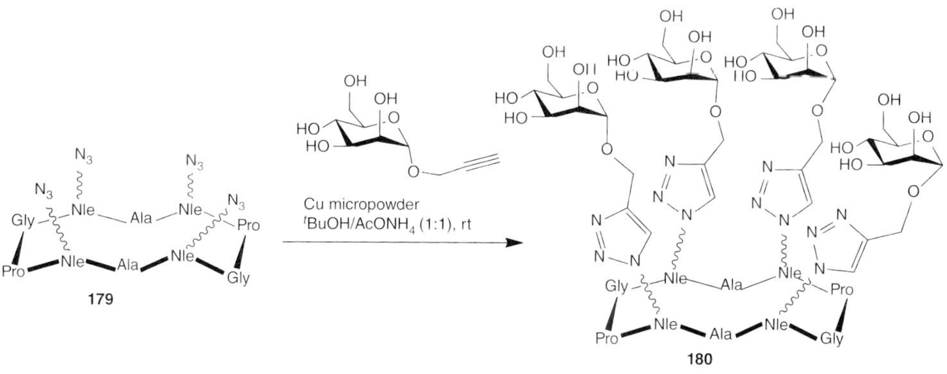


**Scheme 6.48** Click-inspired synthesis of cyclopeptide-based fucosylated glycodendrimers **180**.

Glycosylation of cysteine peptides was attempted using CuAAC methodology and a series of glycopeptides was developed [103]. Alanine-based peptides (**181a** and **181b**) having azido functionality clicked with *O*-propargylated sugar in the presence of Cu(I) catalyst under MW irradiation at 80 °C yielded almost quantitative amount of glycopeptides **182a** and **182b**. Furthermore, deacetylation of resultant peptidoglycans using NaOMe in methanol at pH 11.3 gave AFGP analogs

**183a** and **183b** in high yields (Scheme 6.49). Such types of glycopeptides appear to be a possible platform for the synthesis of biologically relevant neoglycopeptides [104].

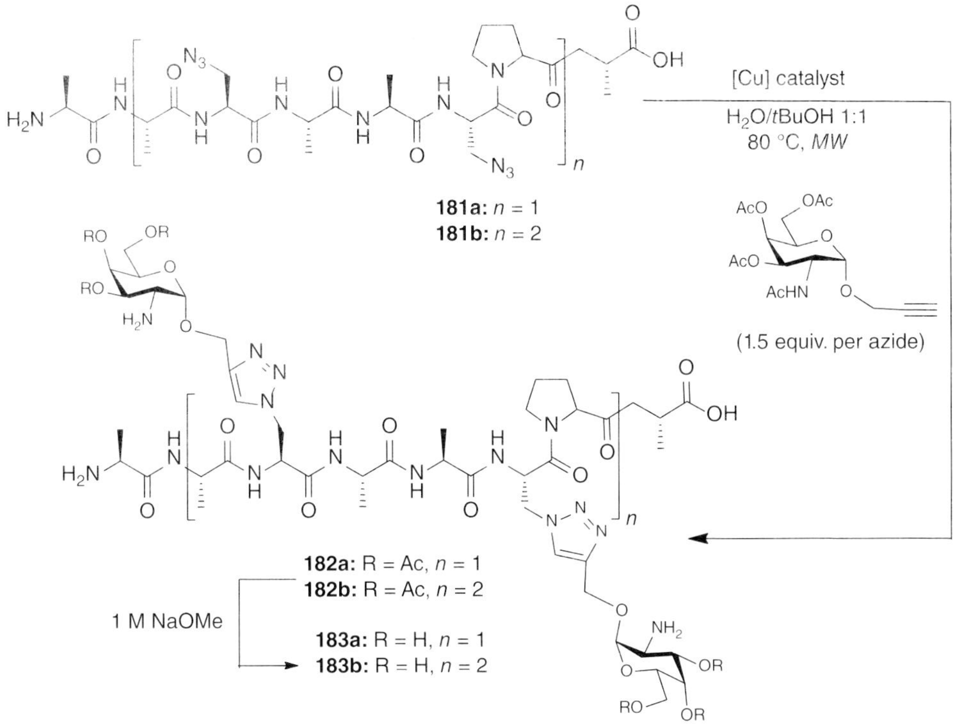


**Scheme 6.49** Click-inspired synthesis of neoglycopeptides.

Similarly, multivalent neoglycoconjugates of MUC1 were prepared in good yields via Cu(I) clicking of azido sugar with propargylated glycine peptides [105], while glycopeptide-containing azido group was successfully coupled with lipopeptides having alkyne spacer using parallel protocol [106].

Furthermore, the highly expeditious CuAAC reaction has also become a complementary tool for polymerization, providing an opportunity for introducing functionalized polymer chains with several glycosides [107, 108]. The cobalt-catalyzed chain transfer polymerization controls high chain-end fidelity and generates ω-end functional polymers with controlled molecular weight. TMS-protected alkyne monomer **184** was subjected to homopolymerization using CoBF/AIBN catalysts, which after the deprotection of alkyne group resulted in polymer **185** having a ω-terminal vinyl group, which was well functionalized by base-catalyzed Michael–thiol addition (Scheme 6.50) [109]. Consequently, alkyne-functionalized clickable backbone was reacted with various sugar azides using CuAAC condition and afforded glycopolymer **186a–c**.

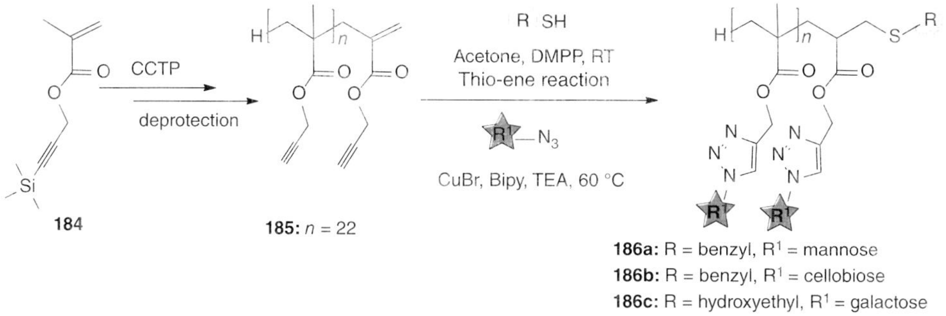


**Scheme 6.50** Synthesis of glycopolymer by combining click reaction with CCCTP technique.

A number of glycopolymers were designed using the CuAAC reaction with RAFT technique. For example, unique 4-vinyl-1,2,3-traizole monomer **188** was developed by Stenzel *et al.* utilizing the CuAAC and allowed to react under RAFT polymerization technique producing polymer **189** (Scheme 6.51) [110]. A similar work was explored for the synthesis of poly(diethylene glycol methyl ether methacrylate) block polymers where the terminal alkyne functionality was introduced on the hydroxyl-end groups, which on further reaction with galactosyl azide under click condition yielded a thermoresponsive polymer, an inhibitor of toxin ricin [111].

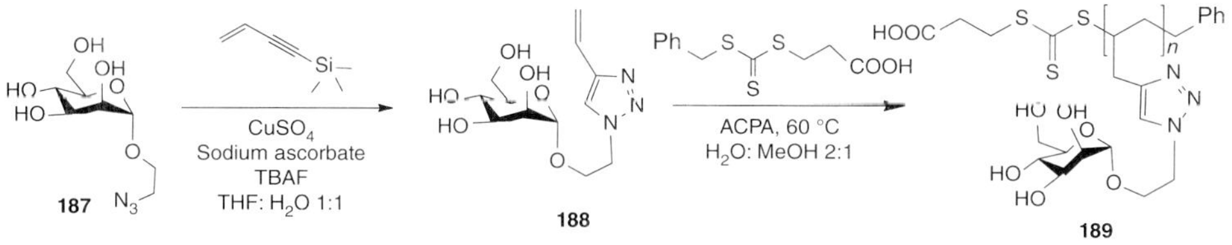


**Scheme 6.51** 4-Vinyl-traizole monomer by combining CuAAC and RAFT technique.

Glycopolymers with sugar-appended fluorescent backbone were studied for biosensing and carbohydrate–lectin interaction. Recently, lactosyl-bearing polyfluorene-based highly fluorescent glycopolymer showed prominent fluorescence quenching in the case of $Ca^{2+}$ ion probably due to the $Ca^{2+}$-mediated aggregation of polymer [112]. Further, a combined Click–Suzuki coupling was used to generate the carbazole-containing glycoconjugates [113]. Recently, a new divalent, ROMP-capable monomer was developed by Nguyen *et al.*, which has the ability to be polymerized to its corresponding neo-glycopolymer mimetic of the surface glycans on gp120 envelope spike of the HIV virus [114].

The click approach also served as an efficacious way to synthesize glycodendrimers, highly branched nanostructured, monodisperse polymers, which displayed enormous applicability in many areas, ranging from material science to drug delivery systems and also from biomedicines to catalysis purpose. CuAAC reaction was utilized to develop glycodendrimers having different core like β-CD based dendrimers which displaying itself as synthetic activators in cell adhesion and stimulation of monocyte/macrophage cell lines [115], multivalent water-soluble organic nanoparticles via "surface clicking" of alkynylated surfactant micelles with sugar azides [116], and multivalent presentation of mannose on hyperbranched polyglycerol including their interaction with lectin [117]. In addition, similar protocol was extended for the synthesis of giant glycodendrimer embodying 243 peripheral xylopyranoside moieties [118] and star-shaped G(0) dendrimers, which displayed lactosyl glycosides at the periphery of the rigid hexaphenyl benzene core [119].

Wide biological significance of glycoconjugated porphyrins including application in photodynamic therapy (PDT) encouraged us to develop porphyrin-cored glycodendrimers [120]. Thus, a series of glycodendrimers containing 8, 12, 16, and 24 β-D-glucopyranose units at the periphery by convergent methodology using click reaction was achieved. Various first- and second-generation azide-functionalized dendrimeric wedges **192** and porphyrin-cored glycodendrimer **193** containing 24 β-D-glucopyranose units at the periphery were obtained from the starting materials through a series of facile reaction sequences and then coupled with the tetraalkyne-functionalized porphyrin core **190** (Scheme 6.52) [121].

Furthermore, gold-nanoparticle-decorated glycodendrimer, known to catalyze the reduction of 4-nitrophenol to 4-aminophenol with $NaBH_4$ under aqueous conditions, has been reported recently [122].

The uttermost important classes such as *Glycoclusters* created a landmark for the study of carbohydrate–lectin processes involved in biologically relevant aspects. Simple yet competent CuAAC reaction miraculously made it viable to design high-affinity, spatially well-defined, and site-specific glycoclusters. In order to develop and embellish structurally manifold class of clusters with different properties, various propargyl-bearing *O*-glycosides and *C*-glycosides were affixed onto polyazido-functionalized carbohydrate and noncarbohydrate cores such as natural and unnatural amino acids, adamantine, calix[4]arene, cyclodextrin, resorcin[4]arene, and fullerene, with the aid of CuAAC reaction [123]. Following the same motif, glycosyl azides clicked with polyalkyne-functionalized core lead to evolution of easily hydrolyzable *N*-glycosyl triazoles, for example, $C_3$-symmetric (1-6)-*N*-acetyl-β-D-glucosamine octadecasaccharide [124], second-generation iminosugar clusters using trivalent dendrons [125], fullerene glycoconjugates [126]. *C*-Sialoside clusters were developed via clicking the diverse acetylene cores containing scaffolds coupled with sialic-acid-based azides with an expectation for interaction of these clusters with diverse biological targets. One of the oligovalent *C*-sialosides **196** is presented in Scheme 6.53 [127].

Recently, Ligeour *et al.* developed a series of mannose-centered galactoclusters (**199**–**201**) using CuAAC reaction. Alkyne functionality was introduced on free hydroxyl groups of pyranose mannose sugar **197**, which clicked with galactosyl azide **198** and afforded three different galactoclusters, known to exhibit high affinity for LecA (Scheme 6.54) [128].

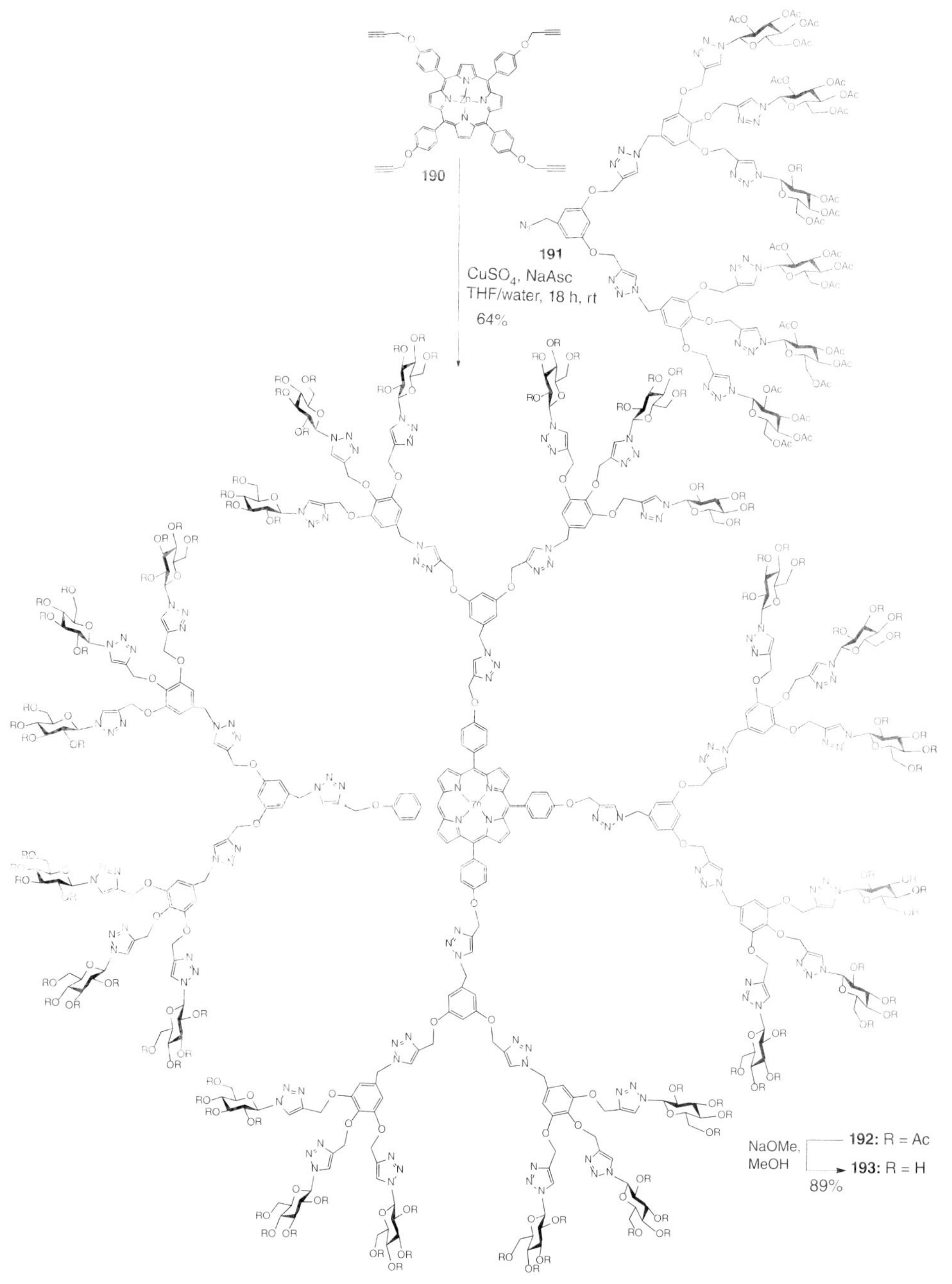


Scheme 6.52 Click-inspired synthesis of porphyrin core glycodendrimer.

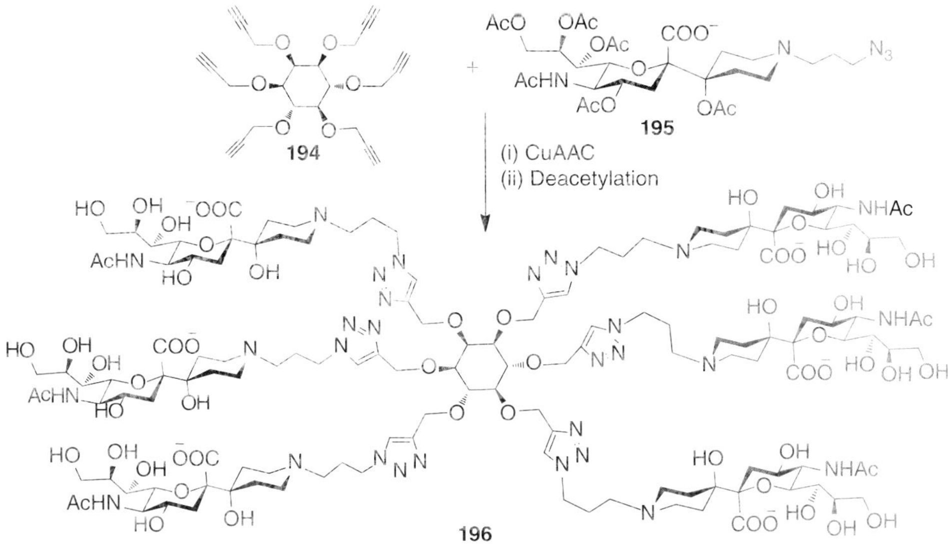

Scheme 6.53 Multivalent C-sialoside monomers via Click reaction.

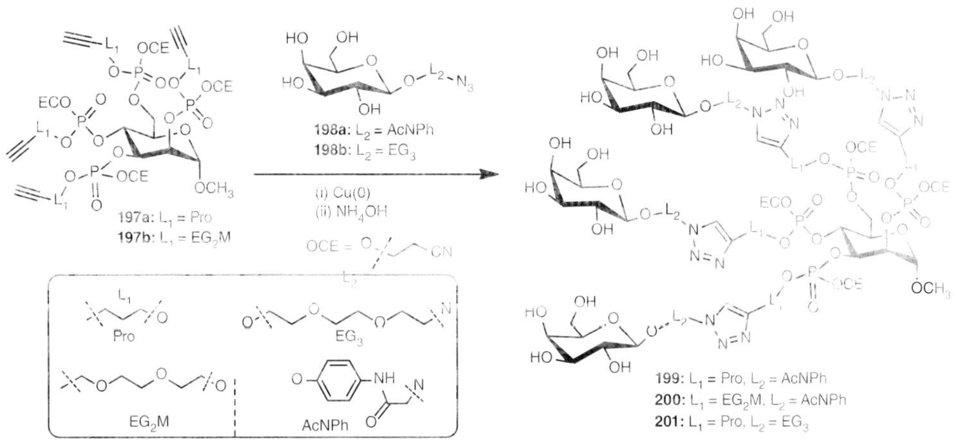

Scheme 6.54 Synthesis of biologically active mannose-centered tetragalactose clusters.

The triazole ring constructed by click chemistry is recognized as an ideal linker in bioconjugation, which is a widely useful protocol in pharmaceutical sciences for therapy and diagnostic tool. Important applications includes a click-based bioconjugation for the profiling of fucosylated glycoproteins [129], labeling and detecting from cultured bacteroidales [130], chemoselective attachment of carbohydrate to cowpea mosaic virus, fluorogenic 4-azidonaphthylfluorescein to label alkyne-functionalized proteins and glycoproteins on cells, and so on, [131, 132]. Also, the fluorescent labeling of cellular-glycan-bearing azide groups [133] and Cu(I)-click protocol is crucial for rapid labeling of cell-surface glycans on mammalian cells in culture [134]. In addition to these applications, a series of virus glycoconjugates were created to target HIV-1 using the oligomannose analogs **203a–d** as potential building blocks. First of all, the surface amino groups were acylated using excess of alkynyl *N*-hydroxysuccinimide ester, with the latter on the scaffold **202** being conjugated with selected oligomannose azides **203a–d** under click condition. In this synthetic strategy, the use of Cu complex **204** results in significant enhancement in the rate of cycloaddition reaction (Scheme 6.55) [135].

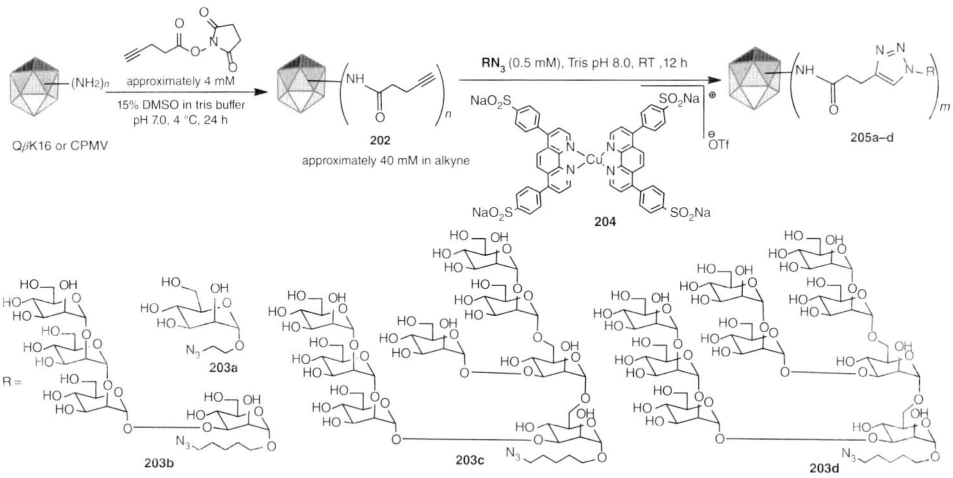


**Scheme 6.55** Click-inspired synthesis of virus glycoconjugates.

A series of fucosylated glycoclusters was reported via a combination of DNA solid-phase synthesis and click chemistry. The solid-supported mannose platform **207**, generated by *MW*-induced click reaction of propargylated solid support **206** and sugar azide, on linking with alkyne **208** followed by labeling and ammonia treatment afforded Cy3-oligonucleotide **209**. Alkyne **209** on clicking with fucosylated azide **210** followed by treatment with aqueous ammonia afforded triazole-linked fucosylated glycocluster **211** (Scheme 6.56) [136]. Evaluation of binding assays against PA-IIL for these fucomimetics explained the influence of the linker arm on biological activity, and mannose-based core was found to be better than the glucose- and galactose-based cores.

The glycoclusters firmly anchored to a conformationally rigidified calixarene have been argued as suitable systems for the study of carbohydrate–protein interactions [137]. Thus, considering the wide application of calix[4]arene glycocluster, the CuAAC protocol was utilized to constitute a general way for the easy attachment of various carbohydrate units to this polyfunctionalized substrates. For example, CuAAC coupling of calix[4]arene scaffold **212** with sugar azides **213** and **214** afforded triazole-linked calix[4]arene *O*-glycosides **215–217** (Scheme 6.57) [138]. Each cycloaddition proceeds with high yield and exclusively gives the 1,4-disubstituted triazole. Similarly, a similar chemistry was implemented for the

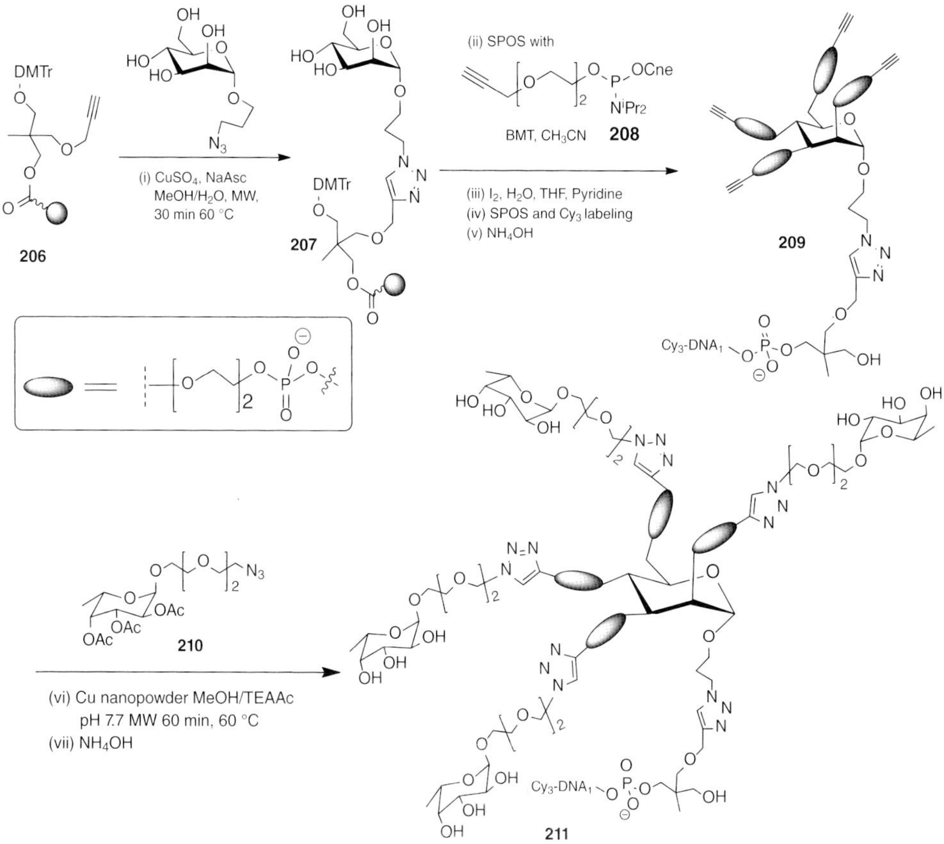


**Scheme 6.56** Crown-like tetra-fucosylated glycocluster-based on a mannose core.

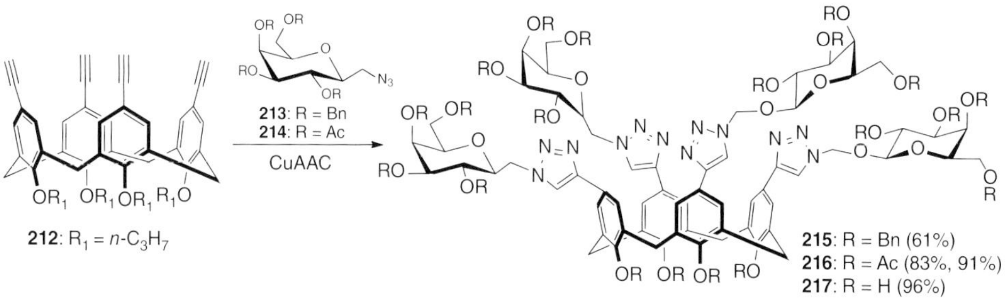


**Scheme 6.57** Synthesis of calix[4]arene glycocluster.

synthesis of series of partially and fully *O*-galactosylated calix[4]arene clusters in cone, partial cone, and 1,3-alternate conformation [139].

The authors exploit a suitable functionalized calix[4]arene platform and conceded the immobilization of triazole-tethered calix[4]arene *O*-glycosides on $TiO_2$ nanoparticles via utilizing the ability of carboxylate groups to form strong bonds with $TiO_2$ surfaces [140]. Thus, calix[4]arene **218** functionalized with *azido* and carboxyl groups at the upper and lower rims, respectively, was first grafted on $TiO_2$ nanoparticles and then conjugated with sialoside **219** under click condition to get the glyconanoparticle **220** (Scheme 6.58) [140].

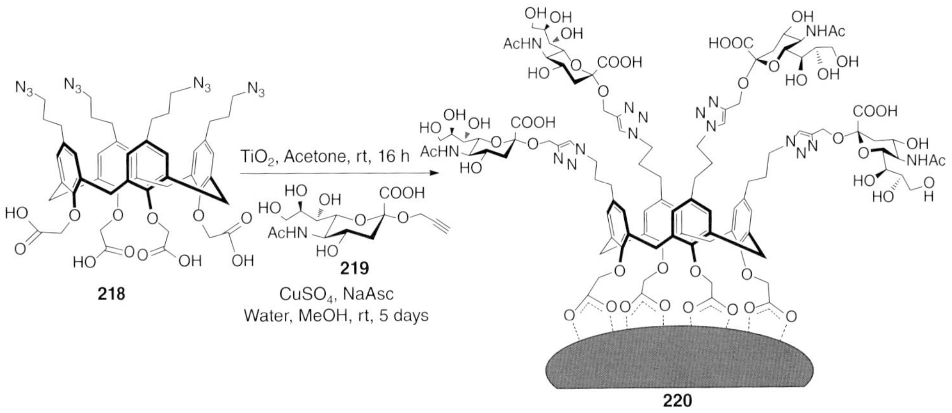


**Scheme 6.58** Click-inspired synthesis of calix[4]arene glycocluster using calix[4]arene platform.

Furthermore, the CuAAC combined with "thiol–ene Click" was successfully utilized for the synthesis of calix[4]arene *S*-glycosides [141]. Thus, tetraene- and tetrayne-functionalized calix[4]arene platform **221** with two different carbohydrate fragments at the lower and upper rims first underwent CuAAC coupling with galactosylmethyl azide **222** and then subsequent "thiol–ene coupling" with glucosyl thiol **223** successfully resulted in differentially glycosylated calixarene **225** (Scheme 6.59).

Recently, a link-spacer-controlled supramolecular chirality based on the self-assembly of the perylene bisimide glycoconjugates was reported. Two mannose-modified perylene bisimides with triazole as well as the amide bond as the link spacer in the bay position (**226** and **227**) were attained. Interestingly, the conjugates having triazole as the linker exhibited right-handed chirality; alternatively, the amide bond as the linker showed left-handed chirality in aqueous solution (Figure 6.7) [142].

Interestingly, the triazolyl azobenzene glycoconjugate could be switched between two geometrical isomeric states to change the spatial orientation of the

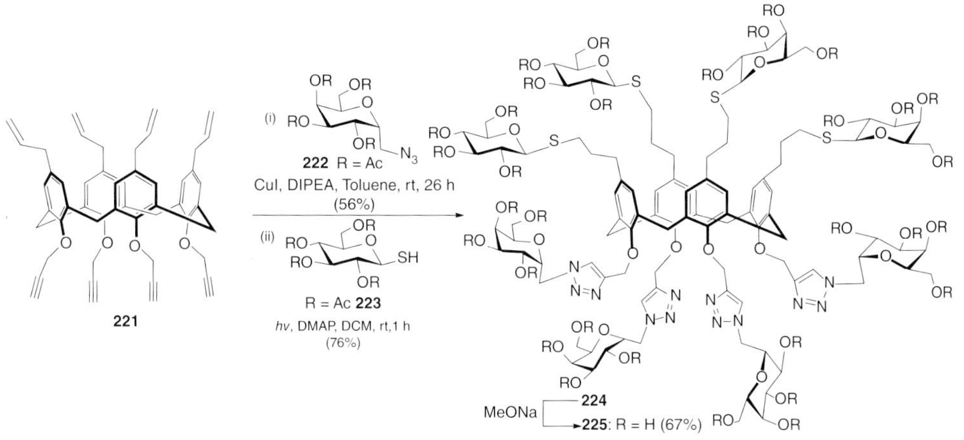


**Scheme 6.59** Synthesis of glycosylated calixarene using CuAAC.

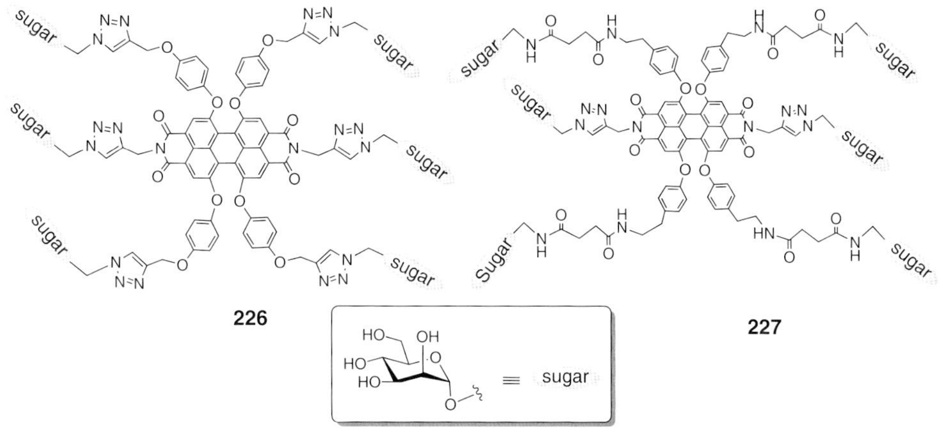


**Figure 6.7** A link-spacer-controlled supramolecular chirality based on self-assembly of the perylene bisimide glycoconjugates.

glycoconjugate ligands. The multivalency effect was detected in photoisomerization that would be useful for the fabrication of photo-switchable multivalent glyco-assembly for bio-evaluation [143].

## 6.5 Conclusion and Future Perspective

Cu(I)-catalyzed 1,3 dipolar cycloaddition of terminal alkynes and azides has emerged as a promising tool for an efficient and easy access of complex molecular-level structures regioselectively in good yields. This coupling technique deals with a wide range of applications in drug discovery, surface chemistry, polymer chemistry, material chemistry, cell labeling and radio labeling, and so on. Inertness of triazole ring toward several protection/deprotection reactions and other chemical transformations and high reactivity and selectivity of CuAAC makes it a potent method for coupling of carbohydrate-based azides and/or alkynes to access diverse carbohydrate mimetics. A broad range of polysaccharides, glycopeptides, glyco-oligonucleotides, glyco-arrays, glyco-macrocyles, glycoclusters, glycodendrimers, and glycopolymers with great biological and pharmaceutical relevance have been synthesized using CuAAC. Some successful carbohydrate-based organocatalysts, ligands, and ionic liquids, synthesized via click chemistry, have opened new paths for the development of sugar-derived asymmetric catalytic tools by click approach. Although the world of carbo-click has been enriched with a large number of synthetically and medically valuable glycol–triazolyl architectures and this number has been increasing day by day, still there are some unbeaten challenges in this field, which require development of new click catalysts. Moreover, development of solid-supported organocatalysts is still untouched by carbo-click, which needs to be explored.

## Acknowledgment

The authors thank Council of Scientific and Industrial Research (CSIR), New Delhi, and Banaras Hindu University for the funding and support.

## References

1. Varki, A. (1993) *Glycobiology*, **3**, 97–130.
2. Tiwari, V.K., Mishra, R.C., Sharma, A., and Tripathi, R.P. (2012) *Mini Rev. Med. Chem.*, **12**, 1497–1519.
3. Mishra, A., Mishra, N., and Tiwari, V.K. (2015) *Curr. Org. Synth.* doi: 10.2174/1570179412666150706185829
4. Huisgen, R. (1963) *Angew. Chem. Int. Ed. Engl.*, **2**, 565–598.
5. Tornoe, C.W., Christensen, C., and Meldal, M. (2002) *J. Org. Chem.*, **67**, 3057–3062.
6. Rostovtsev, V.V., Green, L.G., Fokin, V.V., and Sharpless, K.B. (2002) *Angew. Chem. Int. Ed.*, **41**, 2596.
7. Zhang, L., Chen, X., Xue, P., Sun, H.H.Y., Williams, I.D., Sharpless, K.B., Fokin, V.V., and Jia, G. (2005) *J. Am. Chem. Soc.*, **127**, 15998–15999.
8. Kolb, H.C., Finn, M.G., and Sharpless, K.B. (2001) *Angew. Chem. Int. Ed.*, **40**, 2004–2021.
9. Meldal, M. and Tornoe, C.W. (2008) *Chem. Rev.*, **108**, 2952–3015.
10. Thirumurugan, P., Matosiuk, D., and Jozwiak, K. (2013) *Chem. Rev.*, **113**, 4905–4979.
11. Amblard, F., Cho, J.H., and Schinazi, R.F. (2009) *Chem. Rev.*, **109**, 4207–4220.
12. Dondoni, A. (2007) *Chem. Asian J.*, **2**, 700–708.
13. Leonet, V.A., Campo, V.L., Gomes, A.S., Field, R.A., and Carvalho, I. (2010) *Tetrahedron*, **66**, 9475–9492.
14. Kushwaha, D., Dwivedi, P., Kuanar, S.K., and Tiwari, V.K. (2013) *Curr. Org. Synth.*, **10**, 90–135.
15. Witczak, Z.J. and Bielski, R. (2013) *Click Chemistry in Glycoscience: New*

*Development and Strategies*, John Wiley & Sons, Inc., New York.

16. Tripathi, R.P., Dwivedi, P., Sharma, A., Kushwaha, D., and Tiwari, V.K. (2013) in *Click-Chemistry in Glycoscience* (eds Z.J. Witczak and R. Bielski), John Wiley & Sons, Inc., Hoboken, NJ, pp. 293–324.
17. Akula, R.A., Temelkoff, D.P., Artis, N.D., and Norris, P. (2004) *Heterocycles*, **63**, 2719–2725.
18. Mishra, K.B. and Tiwari, V.K. (2014) *J. Org. Chem.*, **79**, 5752–5762.
19. Kumar, D., Mishra, K.B., Mishra, B.B., Mondal, S., and Tiwari, V.K. (2014) *Steroids*, **80**, 71–79.
20. Mishra, K.B., Mishra, B.B., and Tiwari, V.K. (2014) *Carbohydr. Res.*, **399**, 2–7.
21. Mishra, K.B., Mishra, R.C., and Tiwari, V.K. (2015) *RSC Adv.*, **5**, 51779–51789.
22. Mishra, A. and Tiwari, V.K. (2015) *J. Org. Chem.*, **80**, 4869–4881.
23. Casas-Solvas, J.M., Vargas-Berenguel, A., Capitan-Vallvey, L.F., and Santoyo-Gonzalez, F. (2004) *Org. Lett.*, **6**, 3687–3690.
24. Yang, W., Wu, H.X., Xu, H.H., Hu, A.L., and Lu, M.L. (2011) *J. Agric. Food Chem.*, **59**, 12534–12542.
25. Dwivedi, P., Mishra, K.B., Mishra, B.B., Singh, N., Singh, R.K., and Tiwari, V.K. (2015) *Glycoconjugate J.*, **32**, 127–140.
26. Olomola, T.O., Klein, R., Mautsa, N., Sayed, Y., and Kaye, P.T. (2013) *Bioorg. Med. Chem.*, **21**, 1964–1971.
27. Palomo, C., Aizpurua, J.M., Balentova, E., Azcune, I., Santos, J.I., Nez-Barbero, J.J., Canada, J., and Miranda, J.I. (2008) *Org. Lett.*, **10**, 2227–2230.
28. Dedola, S., Hughes, D.L., Nepogodiev, S.A., Rejzek, M., and Field, R.A. (2010) *Carbohydr. Res.*, **345**, 1123–1134.
29. Carvalho, I., Andrade, P., Campo, V.L., Guedes, P.M.M., Sesti-Costa, R., Silva, J.S., Schenkman, S., Dedola, S., Hill, L., Rejzek, M., Nepogodiev, S.A., and Field, R.A. (2010) *Bioorg. Med. Chem.*, **18**, 2412–2427.
30. Harrison, J.A., Kartha, K.P.R., Fournier, E.J.L., Lowary, T.L., Malet, C., Nilsson, U.J., Hindsgaul, O., Schenkman, S., Naismith, J.H., and Field, R.A. (2011) *Org. Biomol. Chem.*, **9**, 1653–1660.
31. Campo, V.L., Ivanova, I.M., Carvalho, I., Lopes, C.D., Carneiro, Z.A., Saalbach, G., Schenkman, S., da Silva, J.S., Nepogodiev, S.A., and Field, R.A. (2015) *Tetrahedron*, **71**, 7344–7353.
32. Diot, J., Garcia-Moreno, M.I., Gouin, S.G., Mellet, C.O., Haupt, K., and Kovensky, J. (2009) *Org. Biomol. Chem.*, **7**, 357–363.
33. Yang, J.W., He, X.P., Li, C., Gao, L.X., Sheng, L., Xie, J., Shi, X.X., Tang, Y., Li, J., and Chen, G.R. (2011) *Bioorg. Med. Chem. Lett.*, **21**, 1092–1096.
34. Cheng, K., Liu, J., Liu, X., Li, H., Sun, H., and Xie, J. (2009) *Carbohydr. Res.*, **344**, 841–850.
35. Cheng, K., Liu, J., Sun, H., Bokor, E., Czifrak, K., Konya, B., Toth, M., Docsa, T., Gergely, P., and Somsak, L. (2010) *New J. Chem.*, **34**, 1450.
36. Alexacou, K.M., Hayes, J.M., Tiraidis, C., Zographos, S.E., Leonidas, D.D., Chrysina, E.D., Archontis, G., Oikonomakos, N.G., Paul, J.V., Varghese, B., and Loganathan, D. (2008) *Proteins Struct. Funct. Bioinf.*, **71**, 1307–1323.
37. Cheng, K., Liu, J., Sun, H., and Xie, J. (2010) *Synthesis*, **6**, 1046–1052.
38. Nandurkar, N., Zhang, J., Ye, Q., Ponomereva, L., She, Q., and Thorson, J.S. (2014) *J. Med. Chem.*, **57**, 7478–7484.
39. Hradilova, L., Polakova, M., Dvorakova, B., Hajduch, M., and Petruš, L. (2012) *Carbohydr. Res.*, **361**, 1–6.
40. Parida, P.K., Sau, A., Ghosh, T., Jana, K., Biswas, K., Raha, S., and Misra, A.K. (2014) *Bioorg. Med. Chem. Lett.*, **24**, 3865–3868.
41. Sahoo, L., Singhamahapatra, A., Kumar, K., and Loganathan, D. (2013) *Carbohydr. Res.*, **381**, 51–58.
42. Pal, R., Sarkar, S., Chatterjee, N., and Sen, A.K. (2013) *Tetrahedron Lett.*, **54**, 5642–5646.
43. Wilkinson, B.L., Bornaghi, L.F., Houston, T.A., Innocenti, A., Vullo, D., Supuran, C.T., and Poulsen, S.A. (2007) *J. Med. Chem.*, **50**, 1651–1657.
44. Morris, J.C., Chiche, J., Grellier, C., Lopez, M., Bornaghi, L.F., Maresca, A., Supuran, C.T., Pouyssegur, J., and

Poulsen, S.A. (2011) *J. Med. Chem.*, **54**, 6905–6918.

45. Li, J., Zheng, M., Tang, W., He, P.L., Zhu, W., Li, T., Zuo, J.P., Liu, H., and Jiang, H. (2006) *Bioorg. Med. Chem. Lett.*, **16**, 5009–5013.
46. Weiwer, M., Chen, C.C., Kemp, M.M., and Linhardt, R.J. (2009) *Eur. J. Org. Chem.*, **2009**, 2611–2620.
47. Kumar, R., Nasi, R., Bhasin, M., Khieu, N.H., Hsieh, M., Gilbert, M., Jarrell, H., Zou, W., and Jennings, H.J. (2013) *Carbohydr. Res.*, **378**, 45–55.
48. De Souza, T.B., Raimundo, P.O.B., Andrade, S.F., Hipolito, T.M.M., Silva, N.C., Dias, A.L.T., Ikegaki, M., Rocha, R.P., Coelho, L.F.L., Veloso, M.P., Carvalho, D.T., and Dias, D.F. (2015) *Carbohydr. Res.*, **410**, 1–8.
49. Sharma, S., Saquib, M., Verma, S., Mishra, N.N., Shukla, P.K., Srivastava, R., Prabhakar, Y.S., and Shaw, A.K. (2014) *Eur. J. Med. Chem.*, **83**, 474–489.
50. Feese, E., Sadeghifar, H., Gracz, H.S., Argyropoulos, D.S., and Ghiladi, R.A. (2011) *Biomacromolecules*, **12**, 3528–3539.
51. Lo Conte, M., Marra, A., Chambery, A., Gurcha, S.S., Besra, G.S., and Dondoni, A. (2010) *J. Org. Chem.*, **75**, 6326–6336.
52. Polakova, M., Stanton, R., Wilson, I.B.H., Holková, I., Sesak, S., Machova, E., Jandova, Z., and Kona, J. (2015) *Carbohydr. Res.*, **406**, 34–40.
53. Polakova, M., Bel, M., Miku, K., Lattov, E., and Perreault, H. (2011) *Bioconjugate Chem.*, **22**, 289–298.
54. Singh, A.K. and Gopu, K. (2010) *Tetrahedron Lett.*, **51**, 1180–1184.
55. Moni, L., Marra, A., Skotnicki, J.S., Koehn, F.E., Gharbia, M.A., and Dondoni, A. (2013) *Tetrahedron Lett.*, **54**, 6999–7003.
56. Kim, D.H., Choe, Y.S., Jung, K.H., Lee, K.H., Choi, J.Y., Choi, Y., and Kim, B.T. (2008) *Arch. Pharm. Res.*, **31**, 587–593.
57. Maschauer, S., Einsiedel, J., Haubner, R., Hocke, C., Ocker, M., Hübner, H., Kuwert, T., Gmeiner, P., and Prante, O. (2010) *Angew. Chem. Int. Ed.*, **49**, 976–979.
58. Fischer, C.R., Müller, C., Reber, J., Müller, A., Krämer, S.D., Ametamey, S.M., and Schibli, R. (2012) *Bioconjugate Chem.*, **23**, 805–813.
59. Vala, C., Chretien, F., Balentova, E., Lamande-Langle, S., and Chapleur, Y. (2011) *Tetrahedron Lett.*, **5**, 17–20.
60. Lang, C., Maschauer, S., Hübner, H., Gmeiner, P., and Prante, O. (2013) *J. Med. Chem.*, **56**, 9361–9365.
61. Maschauer, S., Haubner, R., Kuwert, T., and Prante, O. (2014) *Mol. Pharm.*, **11**, 505–515.
62. Mindt, T.L., Struthers, H., Brans, L., Anguelov, T., Schweinsberg, C., Maes, V., Tourwe, D., and Schibli, R. (2006) *J. Am. Chem. Soc.*, **9**, 15096–15097.
63. Fernández, S., Crócamo, N., Incerti, M., Giglio, J., Scarone, L., and Rey, A. (2012) *J. Labelled Compd. Radiopharm.*, **55**, 274–280.
64. Kizaka-Kondoh, S. and Konse-Nagasawa, H. (2009) *Cancer Sci.*, **100**, 1366–1373.
65. Fernández, S., Giglio, J., Rey, A.M., and Cerecetto, H. (2012) *Bioorg. Med. Chem.*, **20**, 4040–4048.
66. Gouin, S.G., Roger, M., Leygue, N., Deniaud, D., Julienne, K., Benoist, E., Picard, C., Kovensky, J., and Galaup, C. (2012) *Bioorg. Med. Chem. Lett.*, **22**, 2684–2688.
67. Mandal, S., Das, R., Gupta, P., and Mukhopadhyay, B. (2012) *Tetrahedron Lett.*, **53**, 3915–3918.
68. Benoist, E., Coulais, Y., Almant, M., Kovensky, J., Moreau, V., Lesur, D., Artigau, M., Picard, C., Galaup, C., and Gouin, S.G. (2011) *Carbohydr. Res.*, **346**, 26–34.
69. Khiar, N., Navas, R., and Fernandez, I. (2012) *Tetrahedron Lett.*, **53**, 395–398.
70. Wang, L., Liu, J., Miao, T., Zhou, W., Li, P., Ren, K., and Zhang, X. (2010) *Adv. Synth. Catal.*, **352**, 2571–2578.
71. Jha, A.K. and Jain, N. (2013) *Tetrahedron Lett.*, **54**, 4738–4741.
72. Ferlin, N., Gatard, S., Nguyen, A., Nhien, V., Courty, M., and Bouquillon, S. (2013) *Molecules*, **18**, 11512–11525.
73. Droumaguet, C.L., Wang, C., and Wang, Q. (2010) *Chem. Soc. Rev.*, **39**, 1233–1239.

74. Kushwaha, D., Singh, R.S., and Tiwari, V.K. (2014) *Tetrahedron Lett.*, **55**, 4532–4536.
75. Lei, Z., Wang, J., Tian, Y., Song, G., Zhao, G., and Xu, H. (2015) *Dyes Pigm.*, **113**, 627–633.
76. Tyagi, M. and Kartha, K.P.R. (2015) *Carbohydr. Res.* doi: 10.1016/j.carres.2015.04.008
77. Labrada, K.P., Brouard, I., Mendez, I., Perez, C.S., Gavín, J.A., and Rivera, D.G. (2014) *Eur. J. Org. Chem.*, **2014**, 3671–3683.
78. Lee, T., Cho, M., Ko, S.Y., Youn, H.J., Dong, J.B., Cho, W.J., Kang, C.Y., and Kim, S. (2007) *J. Med. Chem.*, **50**, 585–589.
79. Hassane, F.S., Frisch, B., and Schuber, F. (2006) *Bioconjugate Chem.*, **17**, 849–854.
80. Zerkowski, J.A., Nunez, A., Strahan, G.D., and Solaiman, D.K.Y. (2009) *J. Am. Oil Chem. Soc.*, **86**, 1115–1121.
81. Tang, S.C., Wong, Q.L., and Guo, Z.W. (2010) *Chem. Eur. J.*, **16**, 1319–1325.
82. Xie, J. and Bogliotti, N. (2014) *Chem. Rev.*, **114**, 7678–7739.
83. (a) Bodine, K.D., Gin, D.Y., and Gin, M.S. (2005) *Org. Lett.*, **7**, 4479–4482; (b) Bodine, K.D., Gin, D.Y., and Gin, M.S. (2004) *J. Am. Chem. Soc.*, **126**, 1638–1639.
84. Chandrasekhar, S., Rao, C.L., Nagesh, C., Reddya, C.R., and Sridhar, B. (2007) *Tetrahedron Lett.*, **48**, 5869–5872.
85. (a) Ghorai, A., Padmanaban, E., Mukhopadhyay, C., Acharia, B., and Chattopadhyay, P. (2012) *Chem. Commun.*, **48**, 11975–11977; (b) Ghorai, A., Gayen, A., Kulsi, G., Padmanaban, E., Laskar, A., Achari, B., Mukhopadhyay, C., and Chattopadhyay, P. (2011) *Org. Lett.*, **13**, 5512–5515.
86. Billing, J.F. and Nilsson, U.J. (2005) *J. Org. Chem.*, **70**, 4847–4850.
87. Argen, J.K.M., Grundberg, H.E., Billing, F., and Nilsson, U.J. (2006) *Synthesis*, **18**, 3141–3145.
88. Dorner, S. and Westermann, B. (2005) *Chem. Commun.*, 2852–2854.
89. Conte, M.L., Grotto, D., Chambery, A., Dondoni, A., and Marra, A. (2011) *Chem. Commun.*, **47**, 1240–1242.
90. (a) Lewandowski, B. and Jarosz, S. (2011) *Synth. Commun.*, **41**, 2161–2168; (b) Lewandowski, B. and Jarosz, S. (2010) *Org. Lett.*, **12**, 2532–2535.
91. Ray, A., Manoj, K., Bhadbhade, M.M., Mukhopadhyay, R., and Bhattacharjya, A. (2006) *Tetrahedron Lett.*, **47**, 2775–2778.
92. Leyden, R. and Murphy, P.V. (2009) *Synlett*, **12**, 1949–1950.
93. Campo, V.L., Carvalho, I., Silva, C.H.T.P., Schenkman, S., Hill, L., Nepogodiev, S., and Field, R.A. (2010) *Chem. Sci.*, **1**, 507–514.
94. Muthana, S., Yu, H., Cao, H., Cheng, J., and Chen, X. (2009) *J. Org. Chem.*, **74**, 2928–2936.
95. Tiwari, V.K., Kumar, A., and Schmidt, R.R. (2012) *Eur. J. Org. Chem.*, **2012**, 2945–2956.
96. Kumar, A., Geng, Y., and Schmidt, R.R. (2012) *Eur. J. Org. Chem.*, **2012** (35), 6846–6851.
97. Dondoni, A., Giovannini, P.P., and Massi, A. (2004) *Org. Lett.*, **6**, 2929–2932.
98. Kuijpers, B.H.M., Groothuys, S., Keereweer, A.R., Quaedflieg, P.J.L.M., Blaauw, R.H., van Delft, F.L., and Rutjes, F.P.J.T. (2004) *Org. Lett.*, **6**, 3123–3126.
99. Kuijpers, B.H.M., Groothuys, S., Soede, A.C., Laverman, P., Boerman, O.C., van Delft, F.L., and Rutjes, F.P.J.T. (2007) *Bioconjugate Chem.*, **18**, 1847–1854.
100. Fu, X., Albermann, C., Jiang, J., Liao, J., Zhang, C., and Thorson, J.S. (2003) *Nat. Biotechnol.*, **21**, 1467–1469.
101. Chun, C.K.Y. and Payne, R.J. (2009) *Aust. J. Chem.*, **62**, 1339–1343.
102. Bossu, I., Berthet, N., Dumy, P., and Renaudet, O. (2011) *J. Carbohydr. Chem.*, **30**, 458–468.
103. Lamandé-Langle, S., Collet, C., Hensienne, R., Vala, C., Chrétien, F., Chapleur, Y., Mohamadi, A., Lacolley, P., and Regnault, V. (2014) *Bioorg. Med. Chem.*, **22**, 6672–6683.
104. Miller, N., Williams, G.M., and Brimble, M.A. (2009) *Org. Lett.*, **11**, 2409–2412.
105. Lee, D.J., Yang, S.H., Williams, G.M., and Brimble, M.A. (2012) *J. Org. Chem.*, **77**, 7564–7571.

106. Cai, H., Huang, Z.-H., Shi, L., Zhao, Y.-F., Kunz, H., and Li, Y.-M. (2011) *Chem. Eur. J.*, **17**, 6396–6406.
107. Kempe, K., Krieg, A., Becer, C.R., and Schubert, U.S. (2011) *Chem. Soc. Rev.*, **41**, 176–191.
108. Geng, J., Mantovani, G., Tao, L., Nicolas, J., Chen, G., Wallis, R., Mitchell, D.A., Johnson, B.R.G., Evans, S.D., and Haddleton, D.M. (2007) *J. Am. Chem. Soc.*, **129**, 15156–15163.
109. Nurmi, L., Lindqvist, J., Randev, R., Syrett, J., and Haddleton, D.M. (2009) *Chem. Commun.*, **19**, 2727–2729.
110. Hetzer, M., Chen, G., Barner-Kowollik, C., and Stenzel, M.H. (2010) *Macromol. Biosci.*, **10**, 119–126.
111. Kumar, J., McDowall, L., Chen, G., and Stenzel, M.H. (2011) *Polym. Chem.*, **2**, 1879–1886.
112. Chen, Q., Cui, Y., Zhang, T.L., Cao, J., and Han, B.H. (2010) *Biomacromolecules*, **11**, 13–19.
113. Chen, Q. and Han, B.H. (2009) *J. Polym. Sci., Part A: Polym. Chem.*, **47**, 2948–2957.
114. Sletten, E.T., Svec, R.L., and Nguyen, H.M. (2015) *Tetrahedron Lett.* doi: org/10.1016/j.tetlet.2015.01.016
115. Ortega-Muñoz, M., Morales-Sanfrutos, J., Perez-Balderas, F., Hernandez-Mateo, F., Giron-Gonzalez, M.D., Sevillano-Tripero, N., Salto-Gonzalez, R., and Santoyo-Gonzalez, F. (2007) *Org. Biomol. Chem.*, **5**, 2291–2301.
116. Zhang, S. and Zhaob, Y. (2010) *Macromolecules*, **43**, 4020–4022.
117. Papp, I., Dernedde, J., Enders, S., Riese, S.B., Shiao, T.C., Roy, R., and Haag, R. (2011) *ChemBioChem*, **12**, 1075–1083.
118. Camponovo, J., Hadad, C., Ruiz, J., Cloutet, E., Gatard, S., Muzart, J., Bouquillon, S., and Astruc, D. (2009) *J. Org. Chem.*, **74**, 5071–5074.
119. Chabre, Y.M., Brisebois, P.P., Abbassi, L., Kerr, S.C., Fahy, J.V., Marcotte, I., and Roy, R. (2011) *J. Org. Chem.*, **76**, 724–727.
120. Mukosera, G.T., Adams, T.P., Rothbarth, R.F., Langat, H., Akanda, S., Barkley, R.G., Dolewski, R.D., Ruppel, J.V., and Snyder, N.L. (2015) *Tetrahedron Lett.*, **56**, 73–77.
121. Kushwaha, D. and Tiwari, V.K. (2013) *J. Org. Chem.*, **78**, 8184–8190.
122. Kannan, A. and Rajakumar, P. (2015) *RSC Adv.*, **5**, 46908–46915.
123. Thilgen, C. and Diederich, F. (2006) *Chem. Rev.*, **106**, 5049–5135.
124. Chen, Q., Yang, F., and Du, Y.G. (2005) *Carbohydr. Res.*, **340**, 2476–2482.
125. Joosten, A., Schneider, J.P., Lepage, M.L., Tarnus, C., Bodlenner, A., and Compain, P.A. (2014) *Eur. J. Org. Chem.*, **2014**, 1866–1872.
126. Isobe, H., Cho, K., Solin, N., Werz, D.B., Seeberger, P.H., and Nakamura, E. (2007) *Org. Lett.*, **9**, 4611–4614.
127. Papin, C., Doisneau, G., and Beau, J.M. (2008) *Chem. Eur. J.*, **15**, 53–57.
128. Ligeour, C., Vidal, O., Dupin, L., Casoni, F., Gillon, E., Meyer, A., Vidal, S., Vergoten, G., Lacroix, J.M., Souteyrand, E., Imberty, A., Vasseur, J.J., Chevolot, Y., and Morvan, F. (2015) *Org. Biomol. Chem.*, **13**, 8433–8444.
129. Rabuka, D., Hubbard, S.C., Laughlin, S.T., Argade, S.P., and Bertozzi, C.R. (2006) *J. Am. Chem. Soc.*, **128**, 12078–12079.
130. Webler, C.B., Jiang, H., Wang, W., Baughn, A.D., and Wu, P. (2011) *Bioorg. Med. Chem. Lett.*, **21**, 4989–4992.
131. Gupta, S.S., Kuzelka, J., Singh, P., Lewis, W.G., Manchester, M., and Finn, M.G. (2005) *Bioconjugate Chem.*, **16**, 1572–1579.
132. Gupta, S.S., Raja, K.S., Kaltgrad, E., Strable, E., and Finn, M.G. (2005) *Chem. Commun.*, 4315–4317.
133. Sawa, M., Hsu, T.L., Itoh, T., Sugiyama, M., Hanson, S.R., Vogt, P.K., and Wong, C.H. (2006) *Proc. Natl. Acad. Sci. U.S.A.*, **103**, 12371–12376.
134. Hong, V., Steinmetz, N.F., Manchester, M., and Finn, M.G. (2010) *Bioconjugate Chem.*, **21**, 1912–1916.
135. Astronomo, R.D., Kaltgrad, E., Udit, A.K., Wang, S.K., Doores, K.J., Huang, C.Y., Pantophlet, R., Paulson, J.C., Wong, C.H., Finn, M.G., and Burton, D.R. (2010) *Chem. Biol.*, **17**, 357–360.
136. Gerland, B., Goudot, A., Pourceau, G., Meyer, A., Dugas, V., Cecioni, S., Vidal, S., Souteyrand, E., Vasseur, J.J., Chevolot, Y., and Morvan, F. (2012) *Bioconjugate Chem.*, **23**, 1534–1547.

137. Dondoni, A. and Marra, A. (2010) *Chem. Rev.*, **110**, 4949–4977.
138. Dondoni, A. and Marra, A. (2006) *J. Org. Chem.*, **71**, 7546–7551.
139. Vecchi, A., Melai, B., Marra, A., Chiappe, C., and Dondoni, A. (2008) *J. Org. Chem.*, **73**, 6437–6440.
140. Moni, L., Rossetti, S., Scoponi, M., Marra, A., and Dondoni, A. (2010) *Chem. Commun.*, **46**, 475–477.
141. Fiore, M., Chambery, A., Marra, A., and Dondoni, A. (2009) *Org. Biomol. Chem.*, 7, 3910–3913.
142. Wang, K., Han, D., Cao, G., and Li, X. (2015) *RSC Adv.*, **5**, 47728–47731.
143. Chandrasekaran, V. and Lindhorst, T.K. (2012) *Chem. Commun.*, **8**, 7519–7521.

# 7
# Azide–Alkyne Click Reaction in Polymer Science

*Joydeb Mandal and S. Ramakrishnan*

## 7.1
## Introduction

Polymer science has been steadily enriched by the exciting developments in synthetic methodologies; new bond-forming reactions often provide rich avenues for exploitation in polymeric systems. One such bond-forming reaction is the Cu-catalyzed 1,3-dipolar cycloaddition between an organic azide and an alkyne (Scheme 7.1) to yield a triazole ring, which is popularly abbreviated as CuAAC; although this 1,3-dipolar cycloaddition reaction has been known for over half a century [1], it was only after the remarkable observation that Cu(I) catalyzes this process to speed it up by over a million fold [2], so as to enable it to be carried under ambient conditions with high regiospecificity, that this process catapulted to fame.

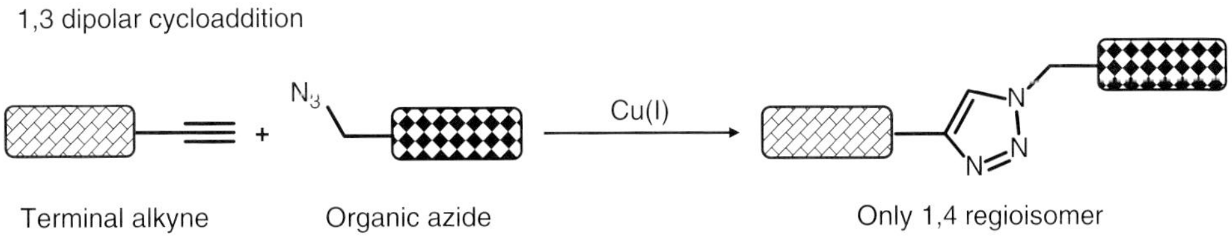


**Scheme 7.1** The azide–alkyne click reaction.

The primary reason for the popularity of the CuAAC reaction is its rapid, chemoselective and quantitative nature and, more importantly, the benign reaction conditions used; furthermore, the simple and straightforward strategies to install azide and alkyne groups in organic molecules have made this approach even more attractive. In the context of polymers, it is also important that both these functional groups tolerate several of the typical reaction conditions used for the preparation of polymers; specifically, both groups are reasonably stable to the frequently used standard free-radical-initiated polymerizations, although the terminal alkyne group could sometimes lead to a broadening of molecular

*Click Reactions in Organic Synthesis*, First Edition. Edited by Srinivasan Chandrasekaran.

weights. Similarly, several standard condensation polymerizations can also be readily conducted in the presence of these functional groups, thus, expanding the scope of the utility of this reaction even further.

Polymers are typically long string-like macromolecules with molecular weights ranging from a few thousands to over a million; they are often molecularly heterogeneous and consequently a pure sample of a synthetic polymer will have macromolecules with a distribution of molecular weights leading to high level of dispersity in chain lengths. During the past several decades, polymer chemists have constantly strived hard to reduce the various molecular-level structural heterogeneities; these include molecular weight, regio- and stereo-specificities, copolymer sequence, and topological (branching) variations.

Free-radical-initiated polymerization, which proceeds via a chain-growth process, is among the most important processes for the preparation of a variety of commercial polymers, such as low-density polyethylene (LDPE), polystyrene, polyvinyl chloride (PVC), polyvinyl acetate, polyacrylates, and many others; controlling the growth process in free-radical polymerizations to obtain "living" characteristics is probably one of the most important achievements of the past two decades. Although detailed elaboration of living polymerization here would be inappropriate, it is important to delineate some of the fundamental characteristics of living/controlled chain polymerizations, as these methods would be often referred to while describing the utility of CuAAC reaction. The three most important characteristics of living/controlled chain polymerizations are as follows:

1) All growing polymer chains remain unterminated (or active) and consequently can continue to grow if more monomer (either the same monomer or a comonomer) is added. If a second type of monomer is added, then the result is the formation of a diblock copolymer that contains a block of one polymer covalently linked to a block of second polymer (Scheme 7.2).
2) The molecular weight of the polymer is regulated by the number of moles of monomer consumed per mole of chain initiator.
3) The dispersity in chain lengths (often referred to as the polydispersity index or PDI) is very low, implying that the width of the molecular weight distribution is narrow.

The most important consequence of satisfying the aforementioned three conditions is that living/controlled polymerizations permit complete control over the nature of the functional group present at the chain end(s) and also allow for greater control of the molecular weight of the polymer and its distribution. As we shall illustrate in this chapter, using a combination of different controlled radical polymerizations (CRPs) and CuAAC reactions, a variety of structures depicted in Scheme 7.2 have be successfully prepared.

Step-growth polymerization is the other important class of polymerization processes; here an efficient bond-forming reaction between two different difunctional molecules, such as a diol and a dicarboxylic acid, is used to prepare polymers. Unlike chain-growth polymerizations, in step-growth polymerizations, very high

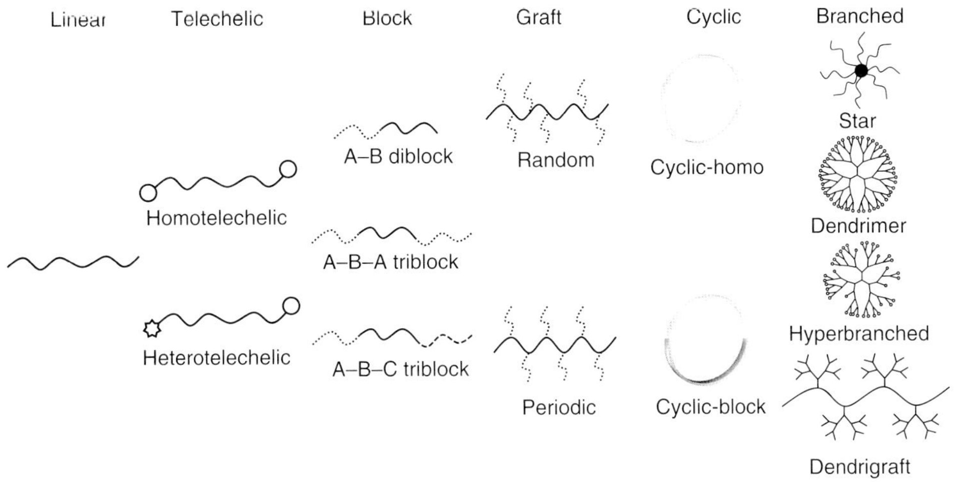


**Scheme 7.2** Schematic representation of the various types of polymers.

conversion, typically over 99%, is essential for the attainment of high-molecular-weight polymers. Therefore, not all bond-forming reactions can be used for the preparation of polymers; the most important characteristics that bond-forming reactions need to meet are as follows:

1) Rapid reaction rate and very high conversions
2) High chemoselectivity and absence of side reactions.

The CuAAC reaction clearly meets the aforementioned two criteria and does so with finesse; thus, it has quickly developed into an important reaction for the synthesis of polymers.

The purpose of this review is to illustrate the remarkably wide range of applications for which the azide–yne click reaction has been utilized in the context of polymer science. In reviewing the literature, we have selected pioneering examples that illustrate novel utilization of the click reaction to bring about a transformation or develop an application that best brings out the uniqueness of the strategy; thus, the intent is to be more illustrative than exhaustive.

## 7.2 Linear, Dendritic, and Hyperbranched Polymers

Using a step-growth process, simple linear polymers can be prepared by the self-reaction of a A-B-type monomer, where the functional groups A and B react only with each other and do not participate in any other side reaction; as depicted in Scheme 7.3, this would lead to the formation of a linear polymer molecule carrying A at one chain end and B at the other, provided that intramolecular cyclization does not occur.

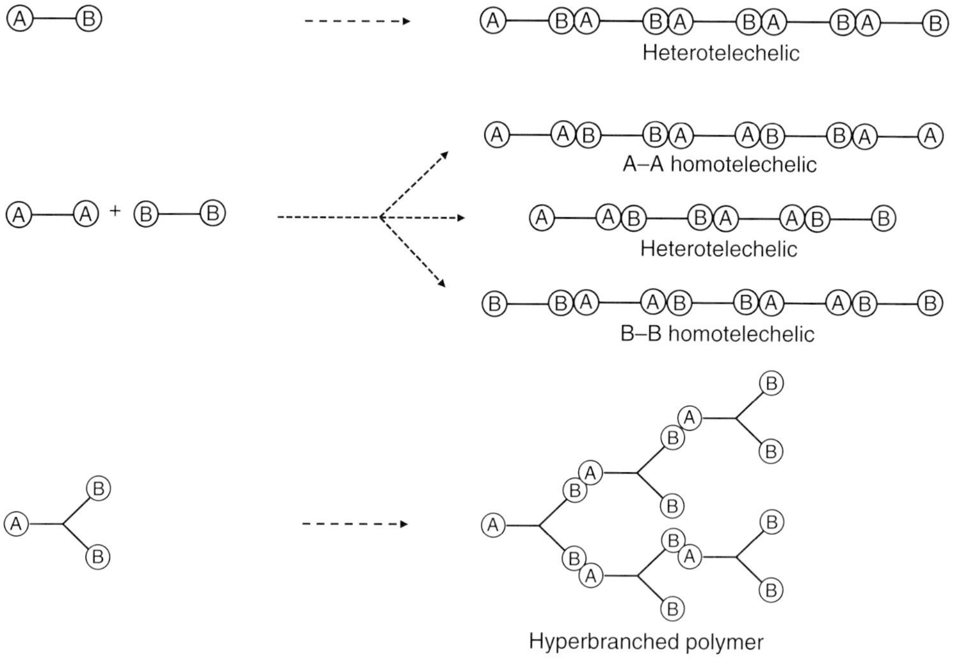


**Scheme 7.3** Types of step-growth polymerization processes.

Frequently, step-growth polymers are prepared by the reaction between two molecules, namely AA + BB; unlike in A-B-type self-condensation, this process would evidently leave chains with a statistical distribution of functional groups at the chain ends – some chains comprising either A or B groups at both ends while the others carrying A at one end and B at the other. The former are termed as *homotelechelic*, while the latter are termed as *heterotelechelic* polymers; further reaction using such systems would permit the preparation of different types of polymeric structures as depicted in Scheme 7.2.

One of the earliest reports describing the utilization of the CuAAC reaction for step-growth polymerization was by Diaz *et al.* over a decade ago [3]; they showed that a fairly simple di-azide and a di-alkyne can readily undergo reaction in the presence of added CuI or $CuSO_4$ in the presence of ascorbic acid as a reducing agent (Scheme 7.4). This first report also explored the use of a variety of other tri- and polyazides and trialkynes to prepare highly cross-linked polymeric materials; the primary objective was to utilize these polymers as structural adhesives, which were shown to perform just as well or even better than the best structural adhesives in use at that time.

One interesting observation was that when two copper plates needed to be stuck together, the mixture of triazides and trialkynes reacted even in the absence of added catalyst, suggesting the occurrence of oxidation–dissolution of Cu ions from the plates, which served as the catalyst; the triazole rings formed during

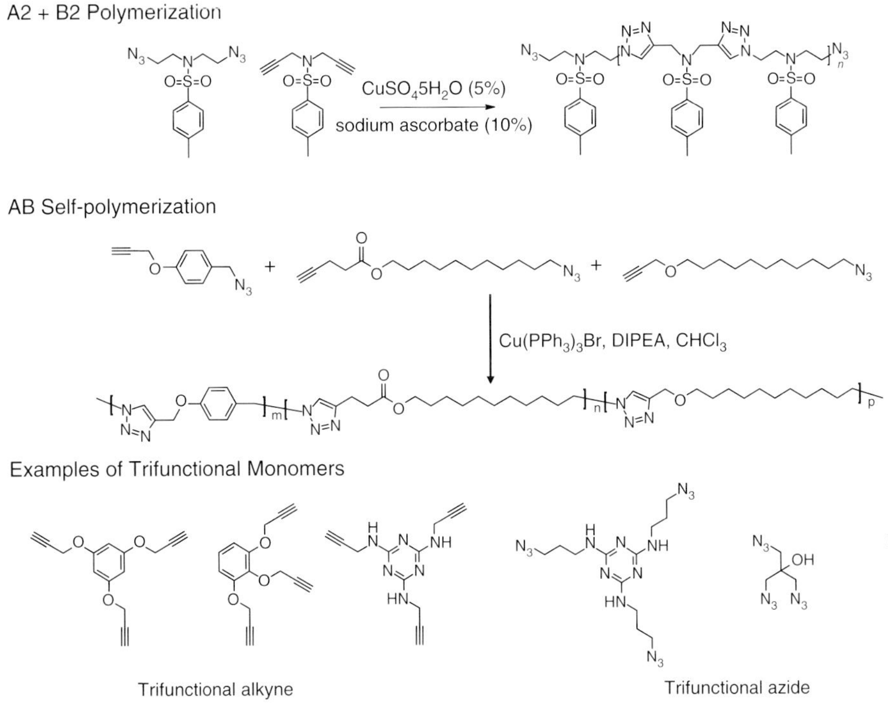


**Scheme 7.4** CuAAC-reaction-mediated synthesis of linear polymers; examples of trialkynes and triazides used for the preparation of cross-linked polymers.

the process are believed to coordinate with the metal plates, thereby substantially enhancing the adhesion characteristics.

Several simple A-B-type monomers (Scheme 7.4), bearing both alkyne and azide groups, can be readily prepared as was demonstrated by Binauld *et. al.* [4]; the simplicity of their preparation made them attractive targets for the preparation of linear polymers that bear triazole rings at varying periodicities along the polymer backbone and permitted the fine-tuning of their physical properties. One important demonstration of this study was the mutual compatibility of the azide and alkyne that renders these A-B-type monomers stable and storable, until exposed to the Cu catalyst that causes them to react, rapidly forming polymers. Furthermore, it also provided a potential to use these *heterotelechelic* polymers for purposeful generation of cyclic polymers; the idea of using *heterotelechelic* polymers has been used to generate cyclic (co)polymers using CRP as will be discussed later.

The triazole ring can also serve as a conjugated linker; in other words, the CuAAC reaction can be effectively used for the preparation of π-conjugated polymers by suitable design of the monomers. The first report describing the use of CuAAC for the preparation of conjugated polymers was by van Steenis *et al.* (Scheme 7.5) [5]; they prepared a variety of conjugated polymers based on fluorene as one of the building blocks, which permitted the installation of long

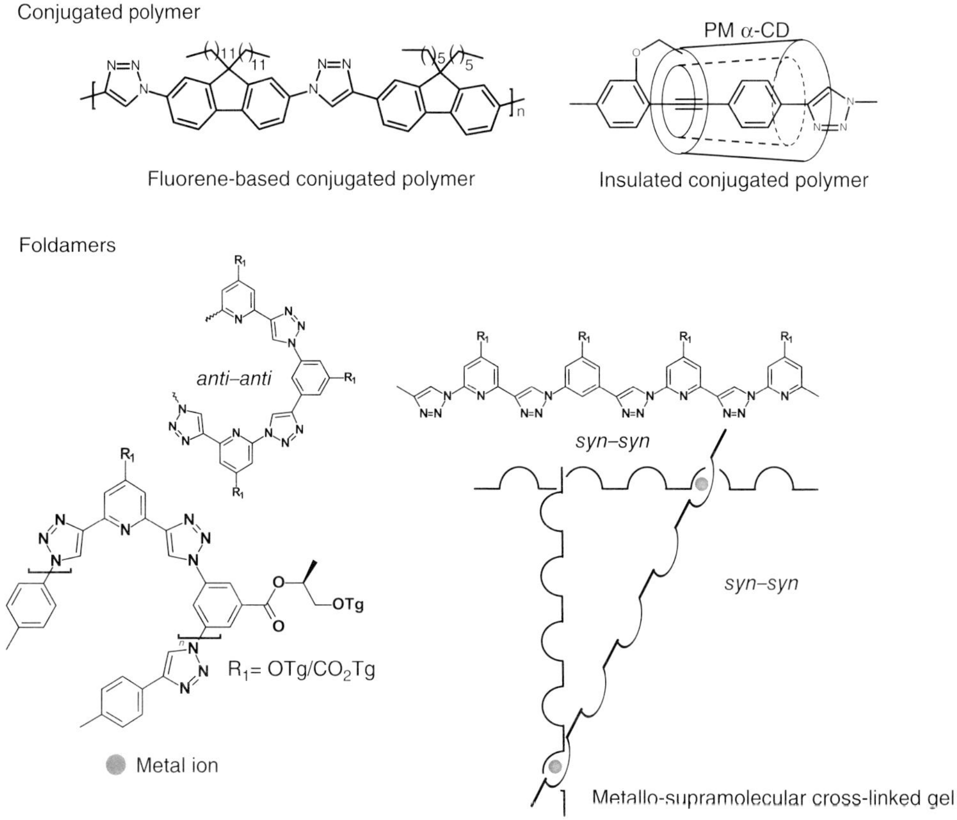


**Scheme 7.5** Examples of linear conjugated polymers and conjugated foldamers via CuAAC.

pendant alkyl chains (dodecyl) to enhance the polymer solubility. Moderately high-molecular-weight polymers with reasonable fluorescence quantum yields were obtained, which clearly demonstrated the efficacy of the CuAAC approach for the preparation of potentially useful conjugated polymers; removal of Cu catalyst had to be done using organic complexing agents, which is important should these materials be useful for device applications.

A more recent example utilizes an AB-type self-condensation to generate an *insulated conjugated polymer*; (Scheme 7.5) [6]; in this example, the tolane-based AB-type monomer, bearing an azide and an alkyne group, also carries a pendant permethylated cyclodextrin (PMCD), which undergoes self-inclusion in a polar solvent. The CuAAC reaction consequently leads to the formation of a conjugated polymer that is threaded through a series of PMCD rings – leading to what the authors term *insulated conjugated polymers*; the interesting feature is the substantially higher solution fluorescence quantum yields of the insulated polymer, when compared to the bare polymer due to the absence of aggregation.

The use of a *meta*-linkage in phenyl/pyridine containing stiff polymers was shown by Moore and coworkers to be an effective strategy to prepare polymers that can adopt a helical conformation under suitable conditions [7]; this first example utilized solvophobically driven folding by installing chiral PEG segments that line up along the periphery of the helically folded chains. Meudtner and Hecht utilized the CuAAC reaction to prepare similar systems carrying alternating pyridine, bearing pendant triethylene glycol (TrEG) units, and triazole units; the preponderance of the *anti–anti* conformation in the absence of a coordinating metal ion leads to the formation of a helical conformation. However, addition of a coordinating transition metal ion leads to a gel presumably due to the metal-coordination-induced formation of the *syn–syn* geometry that results in unfolding of the helix and formation of cross-links via interchain metal-complex formation [8], as depicted in Scheme 7.5.

Tetraphenyl ethylene (TPE) segment has been extensively investigated for its aggregation-induced-enhancement (AIE) fluorescence behavior [9], which arises because of severe restriction of the conformational mobility of the fluorophore upon aggregation. Qin *et al.* demonstrated that CuAAC reaction can be used to prepare polymers that carry the TPE units along the backbone (Scheme 7.6); these polymers exhibited the anticipated AIE leading to about 100-fold increase in fluorescence quantum yield upon solvent-induced precipitation [10]. Furthermore, these nanoaggregates in solution demonstrated efficient quenching when exposed to nitro-group-rich aromatics, such as TNT, making them potentially useful sensors for nitro-aromatic explosives.

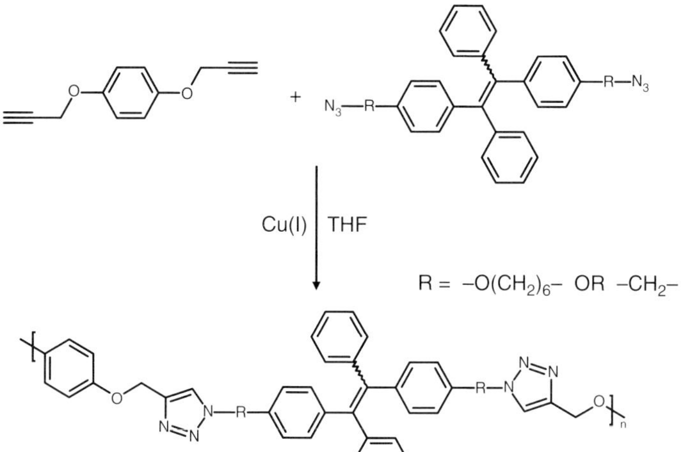


**Scheme 7.6** Tetraphenyl ethylene containing polymer via CuAAC; the polymer exhibits AIE.

Srinivasachari *et al.* utilized the click reaction between a diazido-trehalose derivative and dipropargyl amide of oligoethyleneimine (Scheme 7.7) [11]; the use of the CuAAC reaction enabled them to prepare polymers of significantly

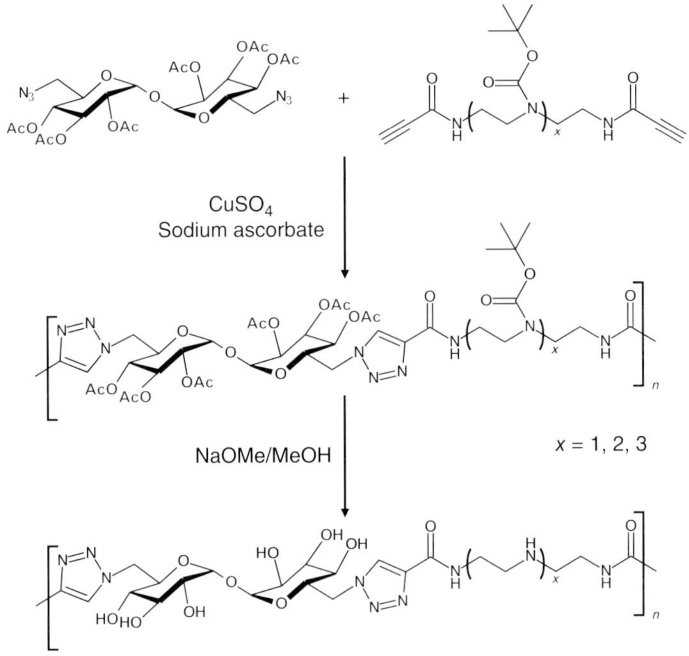


**Scheme 7.7** Biocompatible trehalose–oligoethyleneimine containing polymers for potential use as DNA delivery vehicles.

higher molecular weights than reported earlier. The presence of the disaccharide, the oligoethyleneimine, and the triazole ring in every repeat unit was shown to be critical for the formation of DNA polyplexes, which were stable to high salt and serum concentrations, thereby improving the transfection efficiency; they showed that polymers bearing longer oligoethyleneimine segments were more effective in gene transfection because of the larger number of backbone amine groups that can bind to DNA.

Small-molecule *organogelators* are an interesting class of molecules that organize in solution to form long enmeshed fibers because of which they are able to retain remarkably large quantities of an organic solvent; the self-assembly of the gelator molecules is driven by noncovalent interactions, such as H-bonding and aromatic π-stacking. Covalent linking of such preorganized assemblies without significantly affecting their morphology is an important step that could lead to a dramatic improvement in their mechanical properties.

Diaz *et al.* achieved this using an *organogelator* that carries two alkyne units at the termini of long alkyl chains linked to adjacent carbons on a cyclohexane ring via amide linkages; by using an organic diazide in the presence of a suitable Cu catalyst, they were able to "cross-link" the assemblies through the CuAAC reaction as shown in Scheme 7.8 [12]. Whereas the term *cross-linking* may not be appropriate to describe this process because the polymer formed would actually be a linear polymer, the interesting aspect is that the click reaction appears to

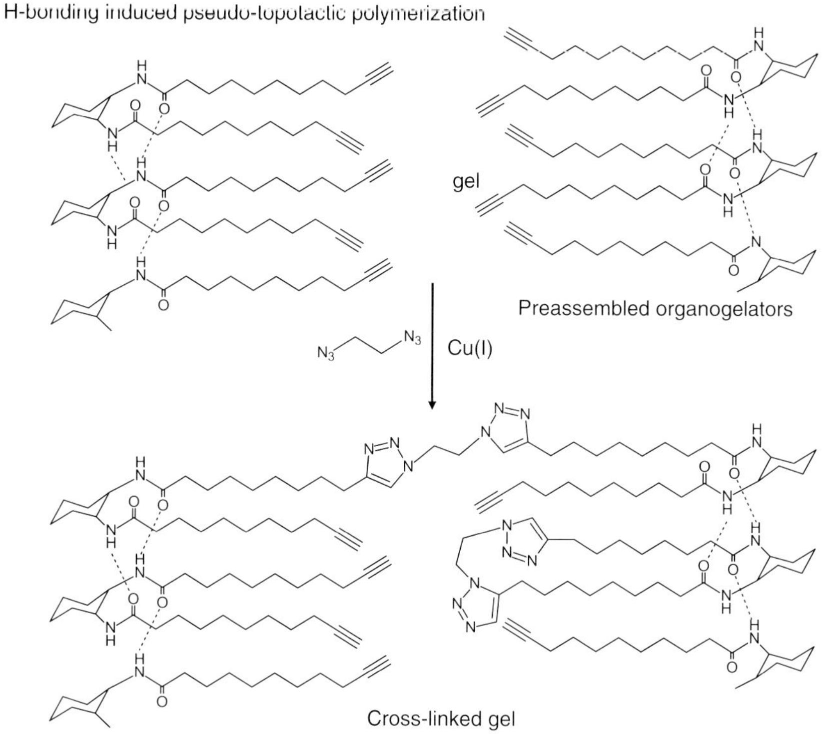


**Scheme 7.8** CuAAC reaction of preassembled organogelators – a case of pseudo-topotactic polymerization.

occur without significant disruption of the fibrillar morphology, which is an essential feature of gelators; this suggests that the assembly is adequately robust and therefore the linking between adjacent molecules occurs in a *pseudo-topotactic* manner.

Dendrimers are a fascinating class of highly symmetric branched polymers with very precise structure and molecular weight; they are typically prepared by a stepwise procedure that requires protection–deprotection (or functional group transformation) after each step; they also often need the use of an excess of one of the reactants and thereby require extensive purification after each step. Given the rapid and high conversion of the CuAAC reaction and the ease of preparation of the alkyne/azide, this strategy can be very effectively used for the synthesis of dendrimers. Wu *et al.* were the first to demonstrate the use of the CuAAC reaction for the preparation of dendrimers, using an approach shown in Scheme 7.9; they used 1,3-dipropargyloxy benzyl chloride (**1**) as one of the first starting reactants; after coupling this molecule with a monofunctional azide, the convergent strategy [13] was adopted to generate the first-generation dendron.

Transforming the chlorobenzyl to azidobenzyl unit was readily achieved in near quantitative yields using $NaN_3$; this azido-dendron is then reacted with

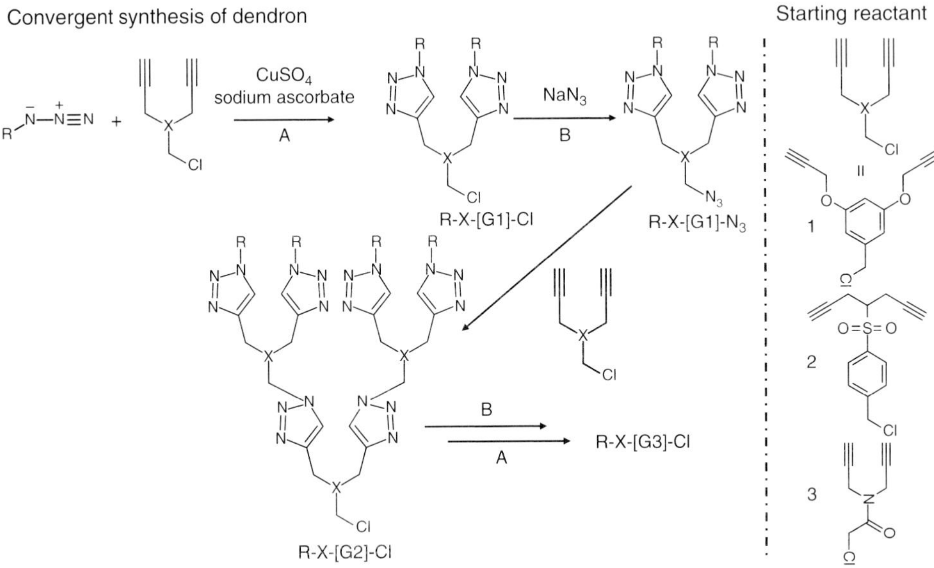


**Scheme 7.9** Convergent synthesis of dendron via CuAAC reaction.

the starting 3,5-dipropargyloxy benzyl chloride to yield the second-generation dendron. Repetition of this process of *functional-group transformation* and *coupling* leads to the formation of higher-generation dendrons; the quantitative and selective nature of the CuAAC reaction is key to the successful preparation of higher-generation dendrimers without structural defects that could arise due to incomplete reaction during any of the coupling steps. In a final step, the dendrons prepared via this convergent approach are coupled to a multifunctional core to generate the symmetric dendrimer.

In a later study, Wu *et al.*, used another interesting approach that began with a single starting molecule (**4**), which was transformed to either azide- or propargyl-bearing $AB_2$ monomers after deprotection of the acetonide; reacting the hydroxyl groups of these intermediate molecules with the starting molecule (**4**) iteratively permitted the generation of higher-generation dendrons as depicted in Scheme 7.10. Since each of these Dendrons is grown independently using the convergent approach, it allowed the investigators to choose the nature of the terminal groups in each dendron. In the final step, the CuAAC coupling of the azido- and propargyl-containing dendrons permitted an easy access to unsymmetrical dendrimers or *Janus* dendrimers that carry different types of terminal groups [14]. Similarly, the authors have expanded the repertoire of dendrimeric structures to generate sugar-decorated dendrimers with fluorophores tagged at the core, using CuAAC reactions to install both sugar residues and fluorophores; the general strategy developed in this study clearly points to the remarkable utility of the CuAAC reaction to develop a wide range of interesting multifunctional dendrimers.

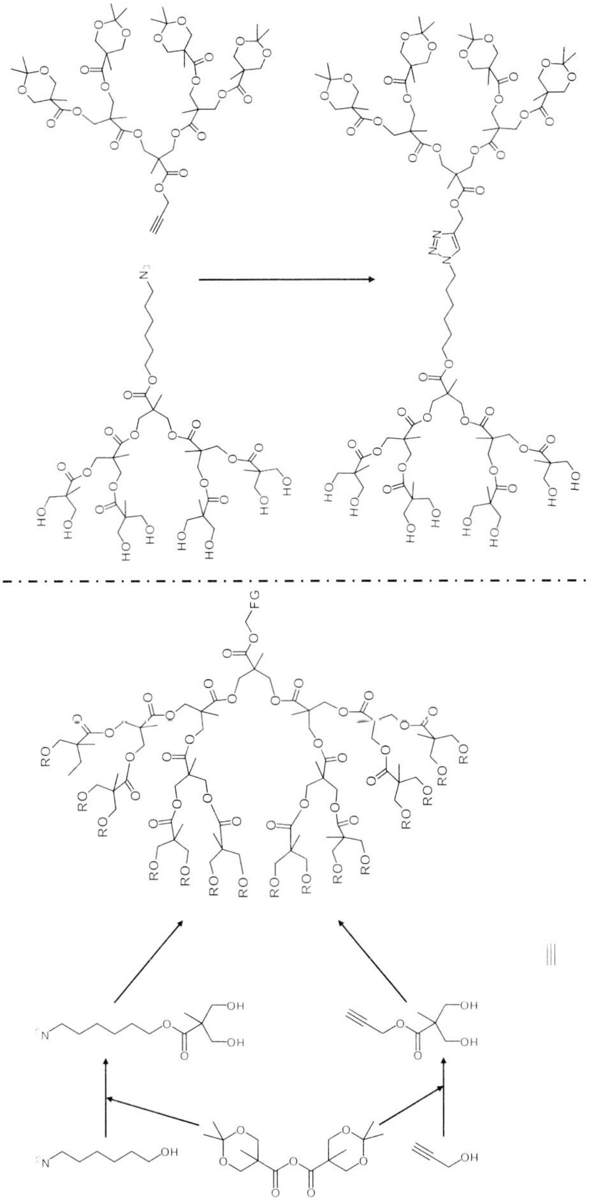


**Scheme 7.10** Synthesis of unsymmetrical dendrimer via CuAAC reaction of azide and alkyne containing dendrons.

Lee *et al.* used a similar strategy to prepare a diblock dendrimers; here they first prepared the Fréchet-type polyether dendron bearing an azide group at the focal point using the standard approach, while the PAMAM-type dendron with a propargyl group at the core was readily prepared by starting with propargyl amine and methyl acrylate. The two dendrons were linked together in the final step using the CuAAC reaction to generate the first diblock dendrimer carrying two structurally dissimilar halves as shown in Scheme 7.11 [15].

Dendronized linear polymers are very interesting macromolecules as these polymer chains carry pendant dendritic wedges on each repeat unit; as shown in Scheme 7.11, such polymers increase the molecular diameter of the chains substantially so as to enable one to directly visualize them using techniques, such as AFM [16]. One of the effective ways to prepare these dendronized polymers is by using a *graft-onto* approach wherein a dendron is installed on every repeat unit of a polymer using a suitable linking chemistry; CuAAC reaction has been very effectively used to achieve this by using a Fréchet-type dendron bearing an azide group at the focal point and clicking this on to poly(vinyl acetylene), which was first prepared by the free-radical polymerization of TMS-protected vinyl acetylene [17]. As expected, it was noted that higher-generation dendrons (>G3) lead to incomplete transformation of the pendant propargyl groups; clicking with the G3 dendron proceeded to approximately 98% conversion. This observation was readily rationalized as being due to steric reasons that impedes the access to adjacent pendant propargyl groups after a certain level of functionalization is achieved. To further increase the size of the pendant dendron, the same group used an improved strategy of using a combination of divergent–convergent approaches; here again, the CuAAC reaction was used for both stages [18].

More recently, in an effort to develop hydrophilic dendrimers, Wang *et al.* developed some interesting starting molecules bearing flexible TrEG segments (Scheme 7.12); iterative CuAAC reaction, followed by effective transformation of the chloro- to azido-groups after each coupling step, allowed the rapid development of hydrophilic dendrimers that are soluble in a variety of solvents [19].

An interesting aspect of this study is that the authors utilized the optimally located triazole pairs in each repeat unit to form a complex with Pd and subsequently used this dendrimeric catalyst for Suzuki–Miyaura-type coupling reaction. Unlike most of the other studies where the triazole ring is an unutilized consequence of the bond-forming process, this is a good example of using the triazole ring as a ligand for metal-promoted catalysis.

Hyperbranched polymers (HBPs), like dendrimers, are highly branched yet soluble polymers that have a similar *branch-upon-branch* topology; one primary difference is that HBPs are prepared via a single step by straightforward reaction between the A and B groups in an $AB_2$-type monomer (Scheme 7.3). In the first report of using CuAAC reaction to prepare HBPs, Scheel *et al.* used the 1,3-dipropargyloxy benzyl azide (derived from the chloride (**1**) in Scheme 7.9) to prepare a HBP by simple self-condensation, either thermally or using the CuAAC reaction [20]; the former approach leads to the formation of a soluble polymer, whereas the latter one leads to the formation of insoluble product. (Scheme 7.13)

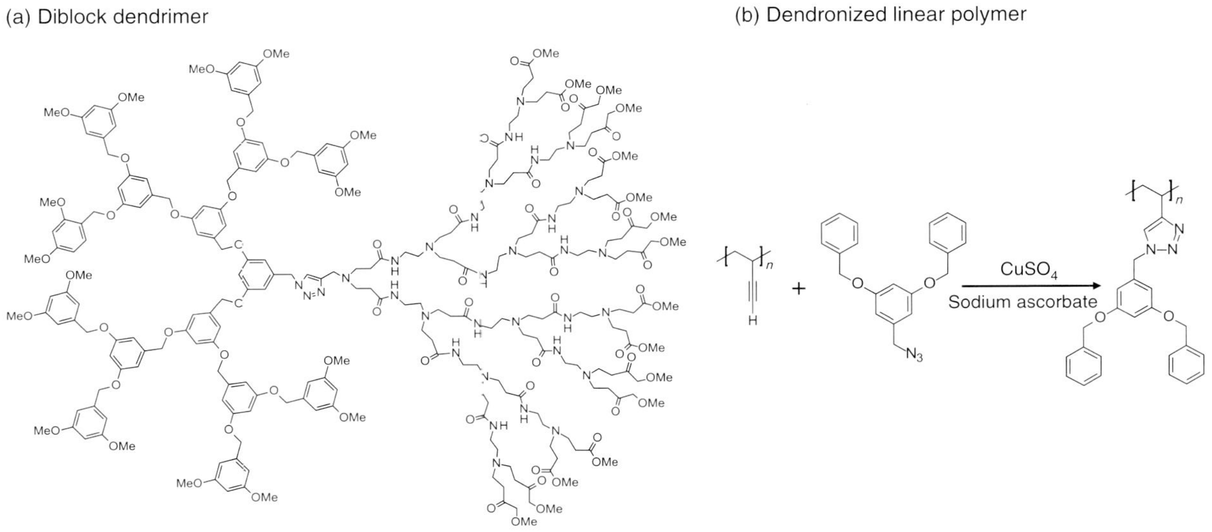


**Scheme 7.11** Synthesis of diblock dendrimer via CuAAC reaction (a) and a dendrinized linear polymer via CuAAC reaction of azido dendron with poly(vinylacetylene) (b).

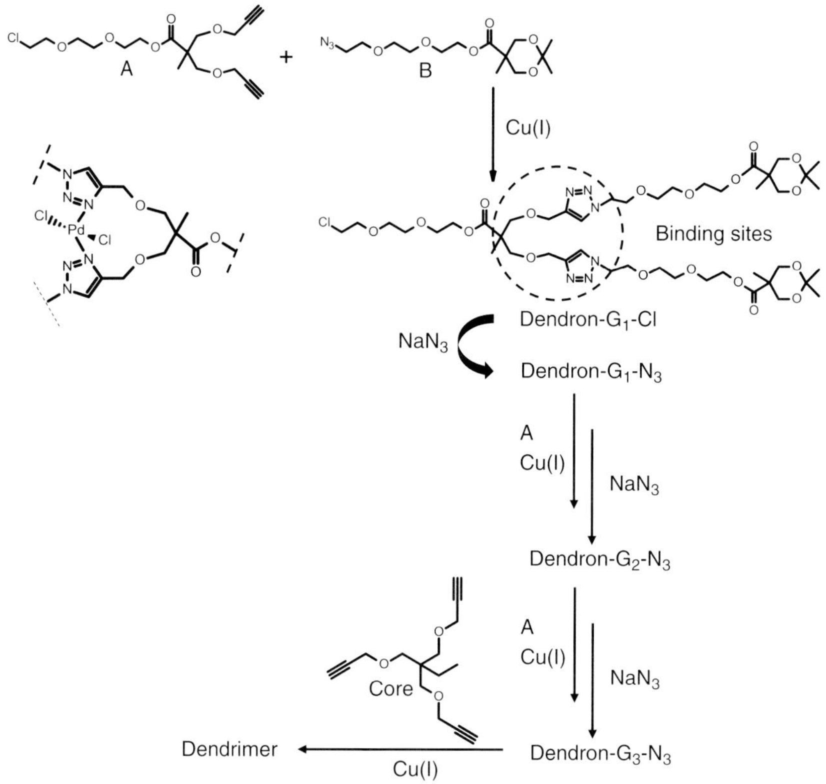


**Scheme 7.12** Convergent synthesis of hydrophilic dendrimer via iterative CuAAC reaction.

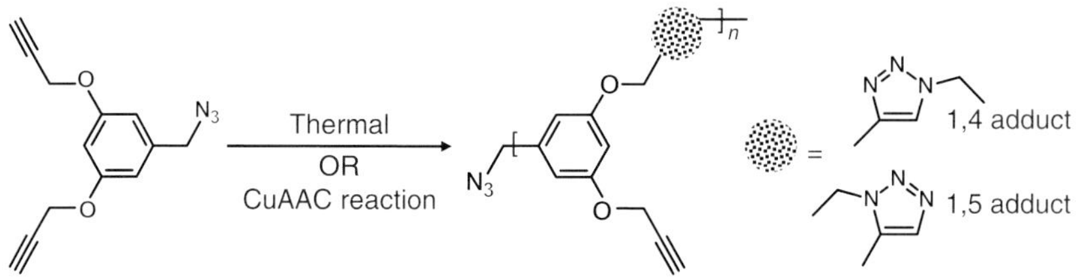


**Scheme 7.13** Direct synthesis of a hyperbranched polymer via CuAAC reaction of an $AB_2$ monomer.

Following this first report, Qin *et al.* showed that an $A_3 + B_2$ approach using a 1,3,5-tris(4-ethynylphenyl) amine and suitable diazide could be used to prepare HBPs (Scheme 7.14); this approach, of course, required careful control of both stoichiometry and conversion to preclude the formation of cross-linked products [21].

The authors also demonstrated that, in these systems, the presence of residual azide groups permits the photo-cross-linking of the polymer via the photolysis

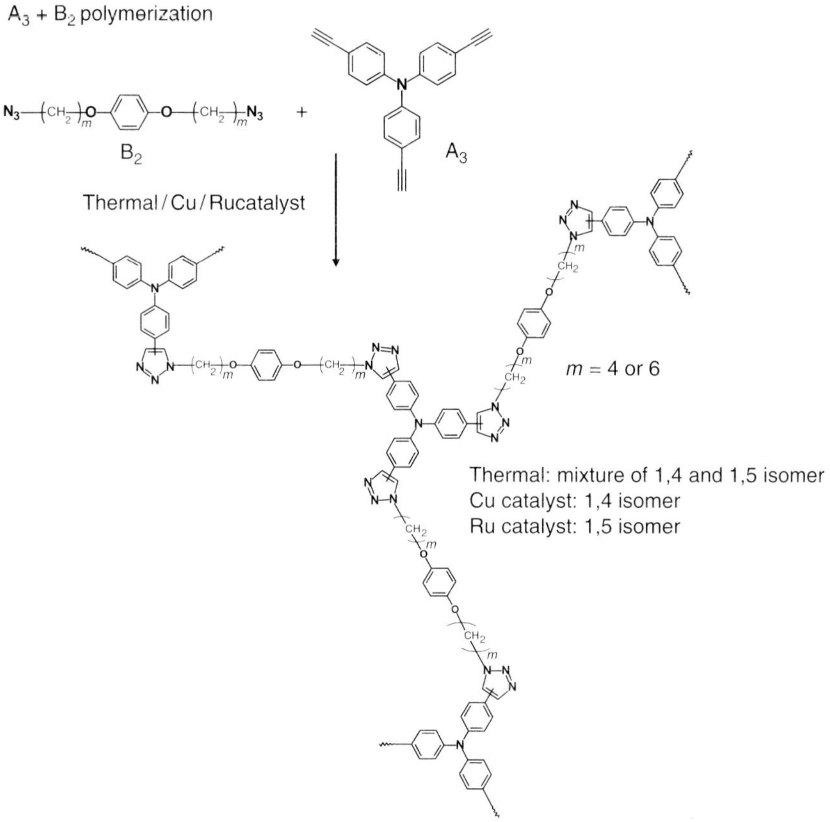


**Scheme 7.14** Synthesis of hyperbranched polymer via CuAAC reaction of trialkyne and diazide.

of the azide that forms reactive nitrenes that could lead to the formation of cross-links via a C-H insertion process; thus, these HBPs were used as photopatternable coatings.

In a very recent example, Shi *et al.* demonstrated an interesting and unusual approach to prepare a HBP using a simple $AB_2$ approach; unlike the earlier examples, they designed a monomer with two azido groups and a single propargyl group. Remarkably, polymerization of this $AB_2$ monomer in the presence of $CuSO_4$/ascorbic acid leads to the formation of high-molecular-weight polymer with low PDI and low level of linear defects [22]. It may be recalled here that, for a random statistical growth of an $AB_2$ monomer, the expected degree of branching (DB) is 0.5 [23]; they obtained a DB of 0.83 (the value of DB for a defect-free HBP would be 1.0). To explain this rather unusual observation, the authors postulated a very interesting mechanism (Scheme 7.15) wherein the Cu complexed with the triazole ring, formed by the first coupling reaction, facilitates the rapid and preferential coupling of the adjacent azide with the incoming propargyl-bearing

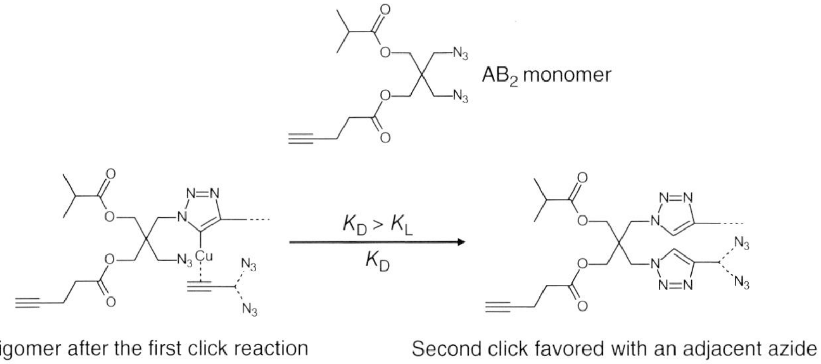


**Scheme 7.15** Unusual synthesis of a HBP with low PDI and high DB via pseudo-chain-growth CuAAC polymerization of an $AB_2$ monomer carrying two azides and one alkyne group.

$AB_2$ monomer; it is suggested that a transfer of the Cu atom from the triazole to the incoming monomer leads to a *chain-growth-type mechanism.* This postulate is also supported by the unprecedented linear increase in molecular weight with conversion.

This mechanism thus accounts for both the low levels of linear defects in the HBP and the chain-growth characteristics of the polymerization process. In short, the catalytic amount of Cu, taken initially, complexes with the triazole ring of the oligomeric species formed during the early stages of the polymerization and subsequent additions of the $AB_2$ monomer appear to occur only to those oligomers that have the Cu bound to them – making this a *pseudo-living-chain polymerization.* Furthermore, the authors also reveal that the molecular weight of the HBP increases linearly as the ratio of the monomer/Cu is increased, confirming the living chain-growth mechanism for this polymerization. This example clearly reveals the potential utility of the triazole rings formed during the CuAAC ligation, whose utility is still largely unexplored.

Dendrimers and HBPs form a unique class of polymers that bear numerous peripherally located terminal groups that could be further functionalized; extensive studies have utilized this approach to create core–shell type polymers, Janus structures, and so on [24]. Thus far, we have discussed the use of CuAAC reaction primarily in the context of a bond-forming process to link the monomers together to generate linear, dendritic, or HB polymers. An alternate strategy that was developed by us utilizes standard polycondensation processes for the preparation of HBPs, but by the suitable design of the $AB_2$-type monomers, the resulting HBPs bear a large number of clickable groups on the periphery. It is well known that an $AB_2$ monomer based on 5-hydroxy isophthalic acid (**5**) can readily polymerize under standard transesterification conditions to generate a HBP carrying numerous B-type groups at the periphery [25]; some years ago we showed that, just as the dimethyl ester, the dipropargyl ester of the $AB_2$ monomer also polymerizes similarly to yield a HB polyester with numerous propargyl groups at the

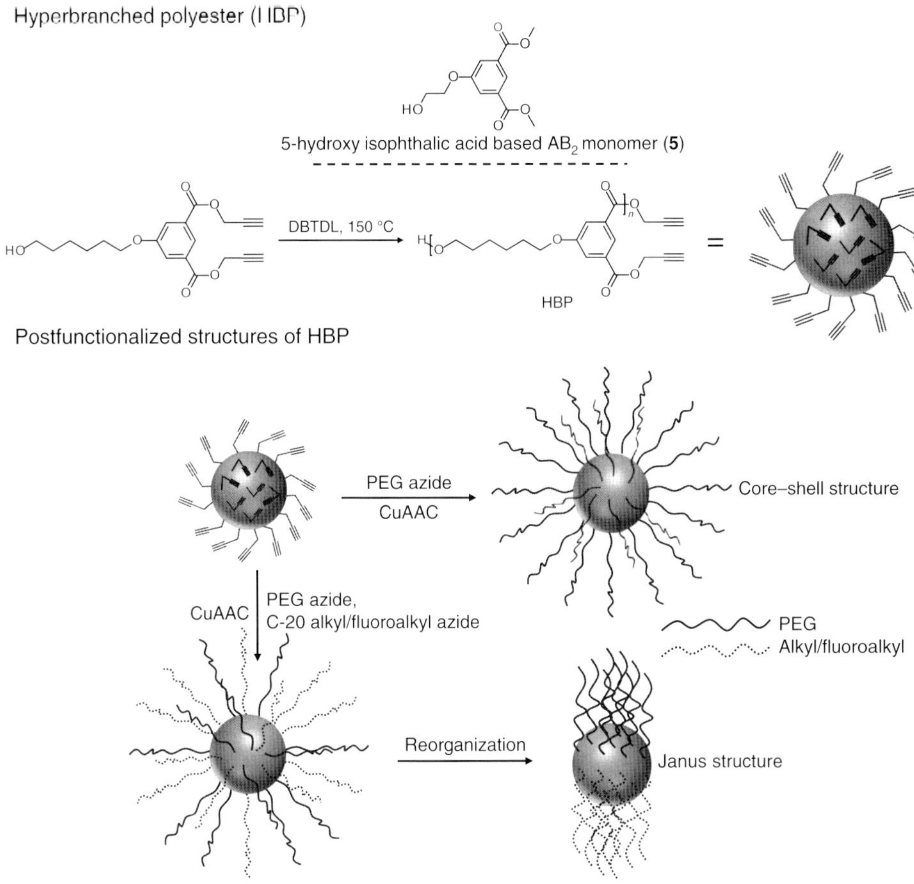


**Scheme 7.16** Synthesis of clickable hyperbranched polyesters and their postpolymerization modification to generate core–shell type and Janus type structures.

periphery (Scheme 7.16) [26]. These polymers could then serve as *clickable scaffolds* to install a variety of segments on their molecular periphery; whereas in this strategy, CuAAC reaction is not used to build the HBP, it is used to postfunctionalize the polymer and develop derivatives with interesting and potentially useful properties. For instance, core–shell-type HBPs wherein the core is hydrophobic and the shell is hydrophilic were prepared by the clicking the parent polyester with PEG azides [26]; similarly, clicking the periphery with two or more incompatible segments, such as PEG, C-20 alkyl chains, and/or perfluoroalkyl segments, yielded *Janus* and *tripodal* (Scheme 7.16) structures, wherein evidence for the self-segregation of the immiscible segments and its morphological consequences were studied [27]. These clickable HBPs could also serve as multifunctional cross-linkers for the preparation of amphiphilic gels – something that is clearly worth pursuing.

## 7.3 Telechelic and Block Copolymers

As mentioned earlier, controlling the nature of the end groups in linear polymers creates several interesting avenues for the preparation of a variety of polymeric systems; therefore, precise control of the end groups has constituted an important area of study. In the case of step-growth polymerization, one important strategy to control the end group(s) is to begin with an A-B-type monomer; the self-reaction between A and B would generate a polymer that has an A group at one chain end and a B group at the other, provided no chain closure has occurred, as depicted in Scheme 7.3. Some years ago, we designed an A-B-type monomer that carried one propargyl ester group and a hydroxyl group; the melt *trans*-esterification of this simple monomer generated a linear polymer carrying a clickable propargyl group at one end and a hydroxyl group at the other [26]. Clicking the single propargyl end group with a fluorophore permitted the determination of molecular weight using end-group analysis.

One of the most popular approaches to control the chain ends of polymer chains in a precise way is to use "living-chain polymerizations." Chain polymerizations, unlike step-growth polymerization, utilizes a chain initiator, such as a free radical or an ionic initiator, which reacts with a monomer to generate an oligomer with an active chain end; the oligomer chain end continues to further react with monomers leading to an enchainment process that effectively causes the length of the polymer chain to grow. As mentioned in the introduction, living polymerizations refer to a special class of chain polymerizations that is intrinsically devoid of chain termination and chain transfer processes; both these processes lead to the premature termination of a growing polymer chain and consequently the nature of the chain end is unregulated. The first "living-"chain polymerization used an anionic initiator, such as butyl lithium, to initiate the rapid formation of polystyrene from styrene monomer [28]; the growing polystyrene chain retained a carbanionic chain end that could be terminated using a variety of well-known reactions of carbanions leading to the generation of a desired chain end [29]. For instance, reaction of the polystyryl anion with ethylene oxide leads to the formation of a polystyrene chain with a hydroxyl group at one end; evidently, the other end of the chain would have the initiator fragment, which would be a *sec*-butyl group, if the chain was initiated by *sec*-BuLi (Scheme 7.17). The fidelity of the chain-end control clearly would depend on ensuring that all the chains initiated by the initiator remain active until terminated willfully: in this example, by using ethylene oxide. As mentioned earlier, the other most important utility of such living polymerizations is that a second monomer could be added after complete consumption of the first monomer; such a sequential addition leads to the formation of a block copolymer.

Free-radically initiated polymerizations typically do not possess living characteristics since termination process is unavoidable; this is because two free radicals very readily combine to form a stable covalent bond and, therefore, two growing chains would readily terminate, either by combination or by disproportionation,

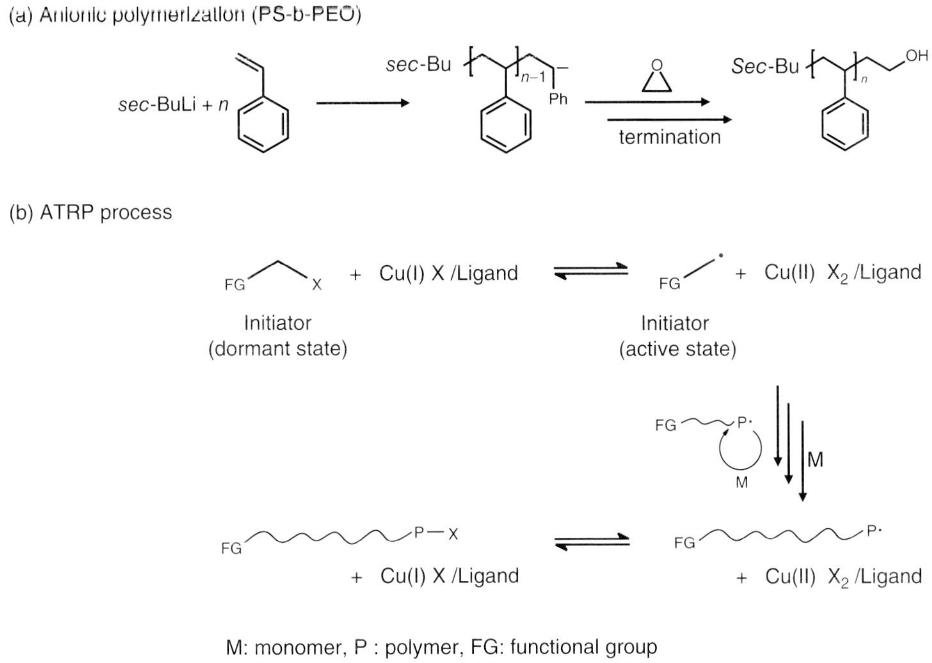


**Scheme 7.17** Anionic polymerization (a), ATRP process (b).

causing both the chains to lose activity [30]. It was recognized several decades ago that the only way to prevent or reduce the probability of chain termination in free-radical polymerizations is to reduce the concentration of growing polymeric radical chains. Several methods have been developed to prevent/reduce the termination process during free-radical polymerizations; all of them utilize a strategy that reversibly terminates (or effects reversible chain transfer of) the chains, which ensures that the steady-state concentration of radical chain ends is kept low [31]. Two of the most popular of such approaches are as follows: atom-transfer radical polymerization (ATRP) [32] and reversible addition–fragmentation chain transfer (RAFT) [33]; in the former, the chain initiator is typically an organic halide, which in the presence of a transition metal, in a lower oxidation state (Cu(I)), abstracts the halogen atom leading to the formation of a carbon-centered radical that can initiate the polymerization of an unsaturated monomer, such as styrene. Such a process has been termed *controlled radical polymerization* as it does not rigorously satisfy all the conditions of true "living" polymerization. The most important attribute of this approach is that most of the growing chain radicals reversibly terminate by reacquiring the halide atom from the Cu(II) species; this ensures that most chains are terminated by a halide in such a process (Scheme 7.17). As in other living-chain polymerizations, the initiator fragment (from the organic halide initiator) is installed at the other chain end. Thus, by a judicial choice of the starting initiator molecule, clickable functionality could be introduced at the initiator chain end and/or it can be introduced by modification of the alkyl halide chain end.

The first example of the use of CuAAC reaction in this context was the preparation of a diblock copolymer; Opsteen and van Hest used a trimethylsilyl blocked ATRP initiator (**6**) to initiate the CRP of styrene and methyl methacrylate (MMA), which after deprotection yielded a polymer chain with an alkyne unit at one end and a bromide at the other (Scheme 7.18) [34]. Similarly, using a simple α-bromo ethylbenzene as the ATRP initiator, polystyrene with a benzylic bromide at one end is readily synthesized; quantitative transformation of this bromide to an azide was achieved by treatment with a little excess of trimethylsilyl azide in the presence of tetrabutylammonium fluoride (TBAF).

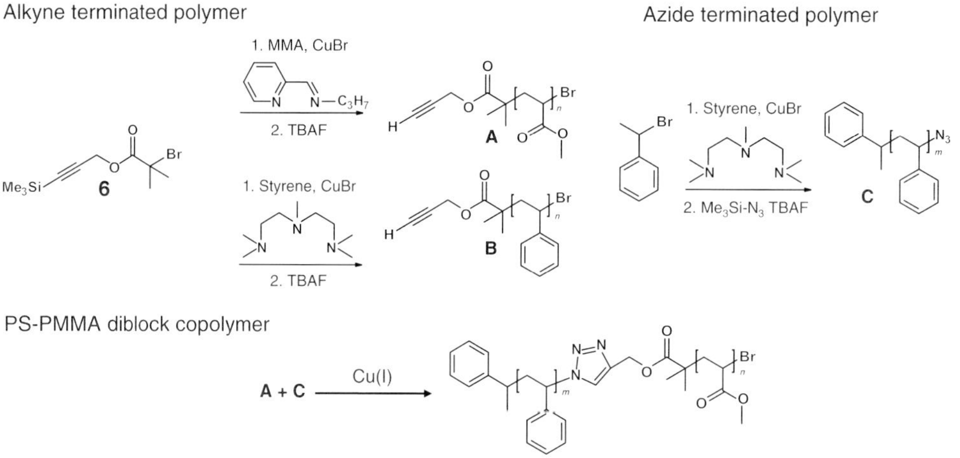


**Scheme 7.18** Synthesis of PS-PMMA diblock copolymer via CuAAC reaction.

Combining this azido-terminated polystyrene with the polymethyl methacrylate bearing an alkyne group under CuAAC conditions permitted the modular synthesis of a PS-PMMA diblock copolymer (Scheme 7.18) [34]; this modular approach allows for the independent synthesis of the two homopolymers with precise control over chain length. The rapid and quantitative nature of the CuAAC reaction is key to the success of this strategy. Several other diblock copolymers, such as PS-PEO and PMMA-PEO, have also been similarly synthesized using this CuAAC-promoted modular approach [35].

One of the major drawbacks of CRPs is the inability to prepare block copolymers of monomers with widely differing reactivities, such as styrene and vinyl acetate, by sequential monomer addition. Thus, the use of chemistries that permit effective coupling between suitably end-functionalized preformed polymer chains is clearly an attractive alternative. Quemener *et al.* designed two novel RAFT agents that could be used to polymerize styrene and vinyl acetate independently [36]; one RAFT agent installed an azide unit at the end of the polyvinyl acetate chains, whereas the other installed an alkyne unit at the polystyrene chain end. CuAAC reaction between these two polymers permitted the synthesis of a polystyrene-*block*-polyvinyl acetate, which is otherwise a difficult block copolymer to prepare because of the dramatic differences in the reactivities of the two monomers (Scheme 7.19).

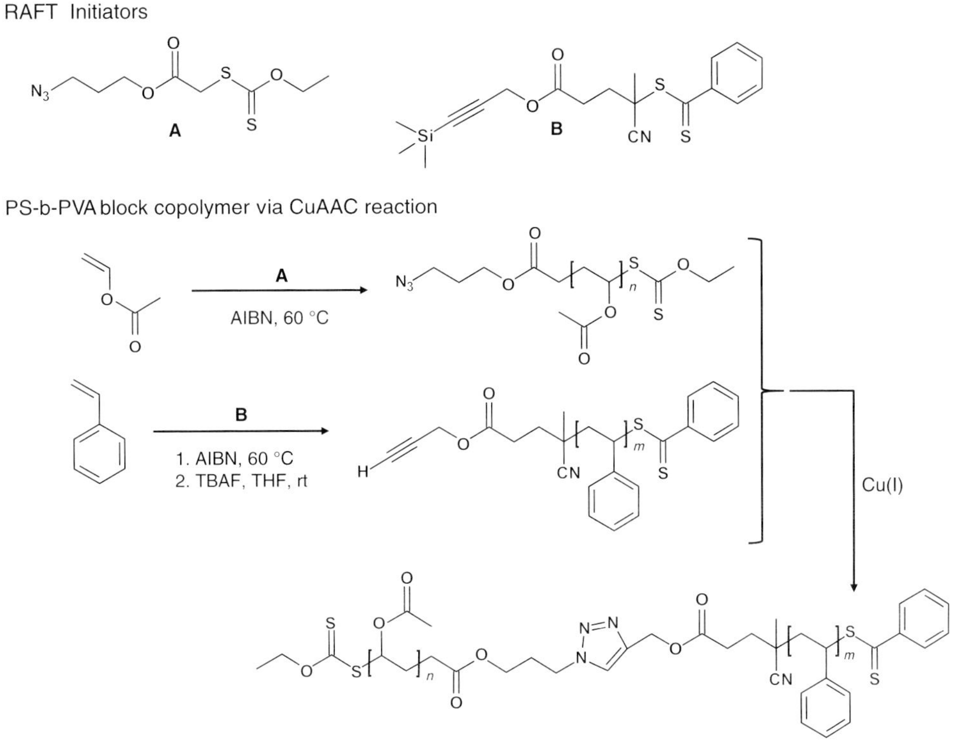


**Scheme 7.19** Synthesis of polystyrene-block-polyvinyl acetate copolymer via using RAFT polymerization in conjunction with CuAAC reaction.

The remarkable selectivity of the CuAAC reaction also provides an opportunity to utilize this reaction along with other click reactions, such as the Diels–Alder reaction; Durmaz *et al.* designed an elegant one-pot strategy to prepare an ABC-type triblock copolymer using a combination of CuAAC and Diels–Alder reaction (Scheme 7.20) [37]. To achieve this, they designed two ATRP initiators (**6** and **7**); one of them helped install a blocked maleimide unit at one end of a PMMA chain, whereas the other yielded a *heterotelechelic* polystyrene chain carrying an anthracene unit at one end and an azide at the other. Polyethylene glycol monomethyl ether of desired molecular weight was esterified using 4-pentynoic acid to yield the alkyne-terminated PEG block; heating all three polymers to 130 °C in the presence of Cu catalyst permitted the authors to *in situ* deprotect the maleimide, effect the Diels–Alder reaction between the

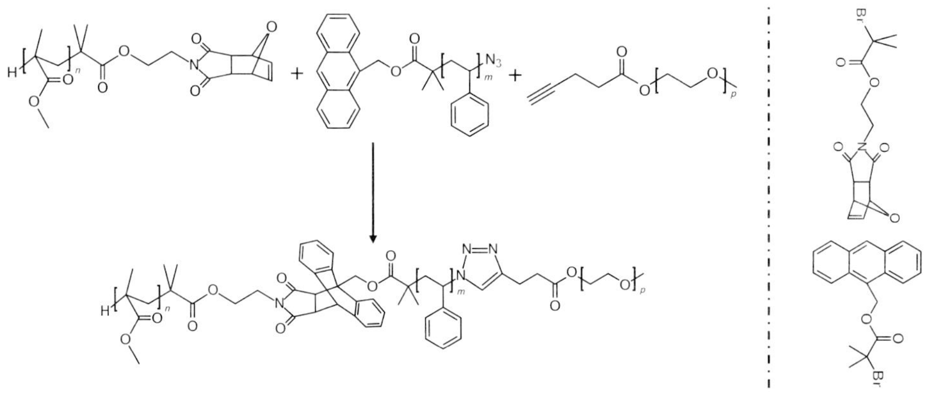

**Scheme 7.20** Synthesis of ABC triblock copolymer using a combination of CuAAC and Diels–Alder reaction.

maleimide and anthracene, at the same time also effect the click reaction between the alkyne–azide at the other end of the polystyrene chain. Thus, the authors demonstrate the power of the CuAAC reaction in combination with Diels–Alder reaction to generate a triblock copolymer simultaneously in one single final step.

Instead of using vinyl polymerization to prepare the different blocks, one can also use a different chain polymerization. In one of the early examples, Agut *et al.* synthesized a rod–coil block copolymer by ring-opening polymerization (ROP) of γ-benzyl-L-glutamate *N*-carboxyanhydride (NCA) using a suitably designed amine initiator (Scheme 7.21), which carries either an azide or an alkyne group, to prepare the first block, while the second block was prepared

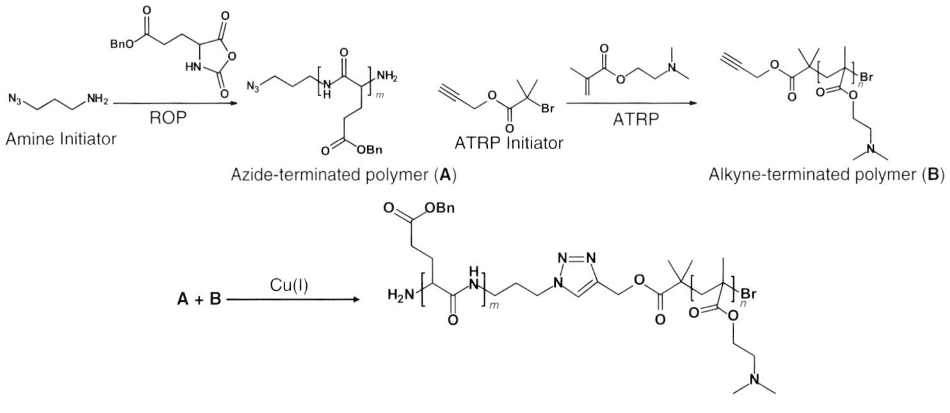


**Scheme 7.21** Synthesis of rod–coil block copolymers by a combination of ROP, ATRP, and CuAAC reaction.

by a copper-mediated ATRP of 2-(dimethylamino)ethyl methacrylate using an initiator bearing an alkyne/azide group [38].

The authors showed that the azide and alkyne functionalities remain unaffected during both the ROP of NCA by an amine and during the ATRP of the vinyl monomer in the presence of a Cu catalyst. The CuAAC reaction using either the azide/alkyne-bearing poly(benzyl glutamate) (PBLG) or the complementary alkyne/azide-bearing PDMAEMA yielded the diblock copolymer (Scheme 7.21) with excellent conversions, using a small excess of PDMAEMA component; the excess could be easily removed using an alumina column. PBLG is well known to adopt a rigid helical conformation, whereas PDMAEMA is known to adopt a random-coil conformation; importantly, the hydrolysis of the benzyl ester generates a water-soluble polyglutamic acid block, which in conjunction with PDMAEMA, renders both temperature- and pH-sensitive characteristics to this diblock copolymer.

Urien *et al.* demonstrated for the first time the preparation of a block copolymer containing a block of a conjugated polymer, poly(3-hexyl thiophene) (P3HT), and polystyrene blocks, as shown in Scheme 7.22 [39]. First, P3HT was prepared via Grignard metathesis (GRIM) [40] from 2,5-dibromo-3-hexyl thiophene, in the presence of 1 equiv. of *tert*-butylmagnesium bromide and a catalytic amount of Ni(dppp)$Cl_2$; terminating the pseudo-living polymerization with a slight excess of (5-chloro-1-pentynyl)trimethylsilane in the presence of additional Magnesium generated the required α,ω-pentynyl P3HT, after deprotection of the trimethylsilyl group using TBAF. The polystyrene block carrying the azide functionality at the chain was prepared by ATRP using the azido-initiator (**8**) instead of transforming the terminal bromide to an azide, as described previously; coupling of these two blocks under CuAAC conditions leads to the generation of a triblock copolymer, wherein the central block is the conjugated P3HT; by extending this approach,

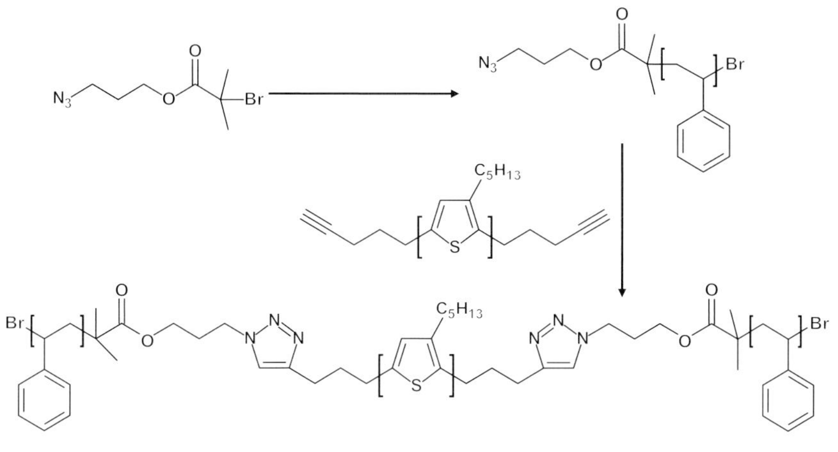


**Scheme 7.22** Synthesis of ABA triblock copolymer containing P3HT central block via CuAAC reaction of a dialkyne derivative of P3HT and azide-terminated polystyrene.

nanophase-separated conjugated block copolymers for potential optoelectronic applications were explored.

Smeets *et al.* developed a similar approach to prepare a diblock copolymer carrying two different poly(thiophene) derivatives, namely 3-alkyl and 3-alkoxy polythiophenes, using a CuAAC reaction as the final coupling reaction between the two different polythiophene blocks [41].

In another interesting variation, Zheng *et al.* used a rigid phenyl-capped bithiophene (Ph2TPh) functionalized at the two ends with PEG segments that carry either two azides or two propargyl groups; CuAAC reaction between these two yielded a multiblock copolymer bearing strongly interacting rigid Ph2TPh segments interspersed with long flexible hydrophilic PEG segments (Scheme 7.23) [42]. Careful study, using TEM, AFM, and X-ray scattering, demonstrated that in a polar solvent, such as DMSO, the periodic rigid–flexible amphiphilic multiblock copolymer folds to assist the collocation and interaction of the rigid Ph2TPh segments with the PEG segments looped above and below – 2D sheets with thickness corresponding to a single folded chain was realized.

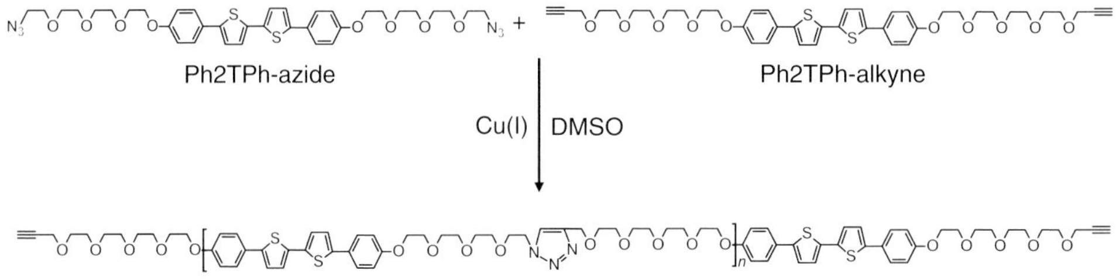


**Scheme 7.23** CuAAC-reaction-mediated synthesis of alternating multiblock copolymer poly(Ph2TPh-OEG).

## 7.4 Star and Star-Block Polymers

Star polymers are the simplest of the branched polymers; they represent polymers wherein several linear chains are linked to a single multivalent molecular core; 3-arm, 4-arm, and even 12-arm star polymers have been synthesized [43]. Typically, the most efficient process to prepare such polymers is to use a living polymerization, such as anionic polymerization, and terminate the process by a nucleophilic substitution reaction onto a multifunctional core. With the advent of CRPs and the CuAAC reaction, Gao and Matyjaszewski utilized, for the first time, a combination of ATRP and click reaction to prepare star-branched polymers [44]; as described earlier, ATRP can be readily used to prepare azide-terminated polystyrene by transforming the terminal bromide, which is typically installed at one chain end during ATRP, to an azide. The PS-azide was then reacted with different core molecules bearing multiple propargyl groups; an example is the tetra-pentynoyl ester of pentaerythritol, which yielded a tetra-arm polystyrene star polymer (Scheme 7.24) [44].

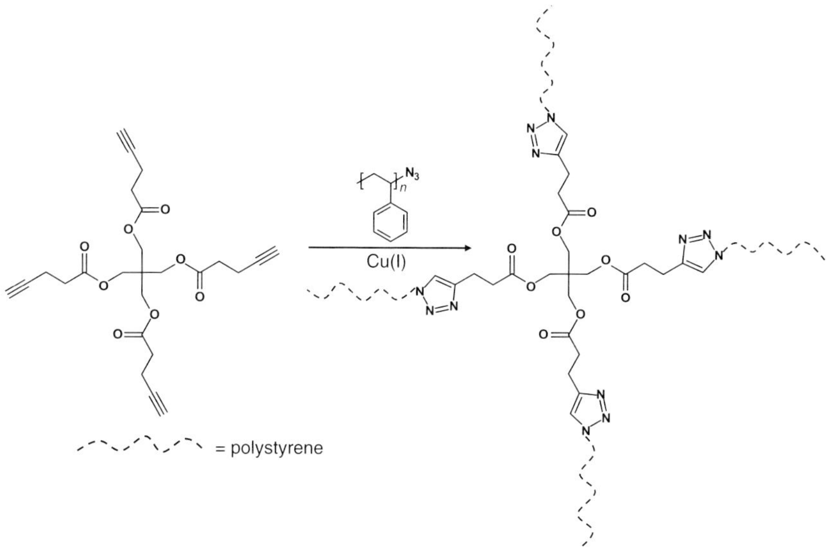


**Scheme 7.24** Synthesis of four-arm PS-star polymer via CuAAC reaction of PS-azide with a core bearing multiple propargyl groups.

Gao *et al.* extended their strategy further to prepare an amphiphilic three-arm star-block copolymer having a PS core and a PEO periphery; to achieve this, they used a trifunctional initiator (**9**) to initiate the ATRP polymerization of styrene, which led to the formation of a three-arm star polymer with bromide end groups; as in the earlier study, the bromo group was readily transformed to azide, which was then coupled with a pentynoyl ester of PEG-monomethyl ether using CuAAC reaction (Scheme 7.25) [45]. They were able to show that the retention of the terminal bromo-functionality at the three chain ends using ATRP, the transformation of the bromide to azide, and the CuAAC coupling reaction occurred with high fidelity to ensure the formation of the amphiphilic star-block copolymer.

In a different strategy, Durmaz *et al.* used a combination of CuAAC and Diels–Alder reaction *in situ* to create star-block copolymers in a single step [46]; to achieve this, first a *heterotelechelic* polystyrene was prepared using a suitably designed ATRP initiator that installs anthracene units on one end and a bromo group on the other, which is then transformed to an azide (**10**). PMMA was independently synthesized using a blocked maleimide-bearing ATRP initiator; in the final step, a tris(propargyl ether) core was reacted with 3 equiv. of the heterotelechelic anthracene-PS-$N_3$ and the PMMA bearing the thermally labile blocked maleimide leading to the formation of the star-block copolymer, as depicted in Scheme 7.26. The high fidelity and selectivity of both the CuAAC reaction and the Diels–Alder reaction are crucial for the generation

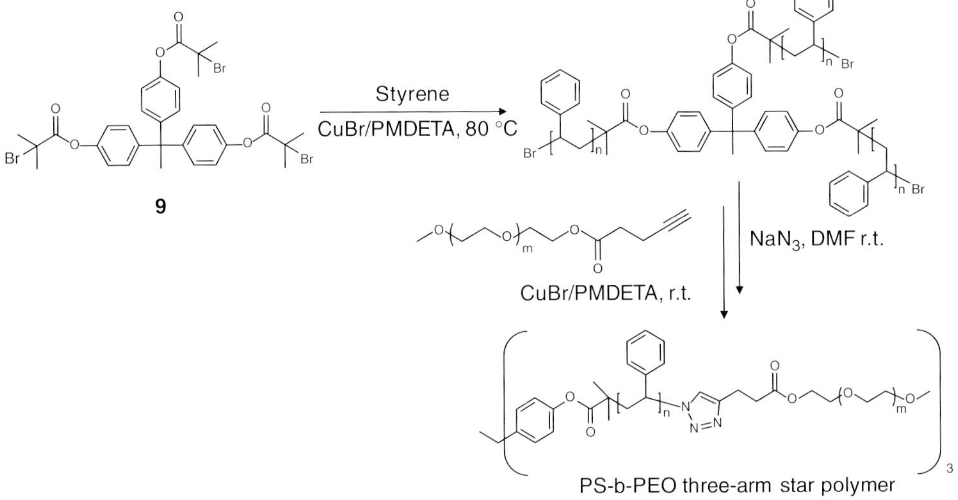


**Scheme 7.25** Synthesis of three-arm PS-PEO star-block copolymer grown from a trifunctional ATRP initiator by combination of ATRP and CuAAC reaction.

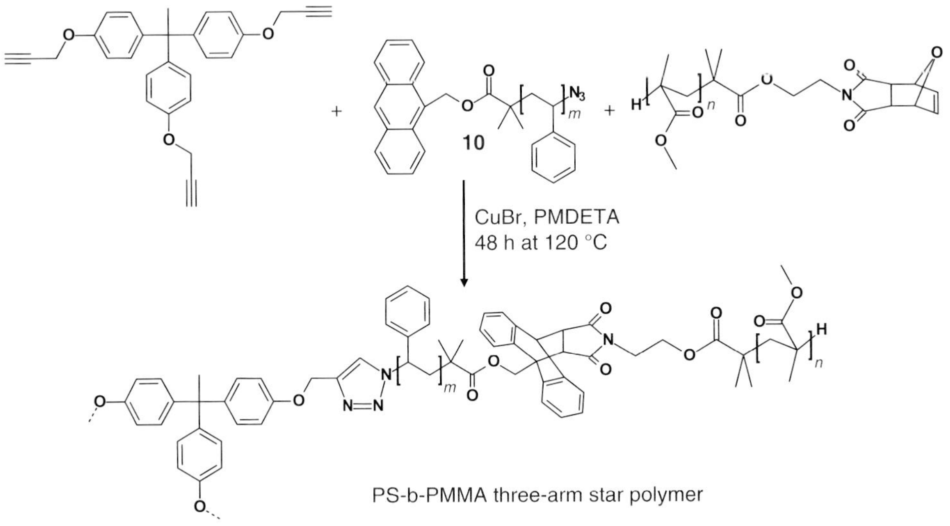


**Scheme 7.26** Single-step *in situ* formation of star-block copolymer using a combination of CuAAC and Diels–Alder reaction; three components, namely a heterotelechelic polystyrene bearing anthracene and azide end groups, maleimide-terminated PMMA, and a tris(propargyl ether) core, were used.

of the three-arm star-block copolymer in the final single step; the simultaneous retero-Diels–Alder deprotection to release the maleimide was clearly the icing on the cake.

Li *et al.* utilized a solvent-assisted collocation strategy to prepare star-like block copolymers; here an amphiphilic diblock copolymer was prepared by sequential monomer addition using a suitably designed RAFT initiator (**11**), which yielded diblock copolymers with both azide and alkyne groups at the same end of the polymer chain (Scheme 7.27). In this study, they have used either *N*-isopropylacrylamide (NIPAM) and dimethyl acrylamide (DMA) or styrene and NIPAM to first generate the diblock copolymer; the PNIPAM segment in both of them gives the system a temperature-responsive amphiphilic character due the lower critical solution temperature (LCST) behavior of PNIPAM block. In a suitable solvent, aggregates of the diblock copolymer are generated, wherein the azide and alkyne groups are collocated within the core; CuAAC reaction within the core leads to the formation of oligomers with 5–20 repeat units of the diblock copolymer, thus generating a star-like topology [47].

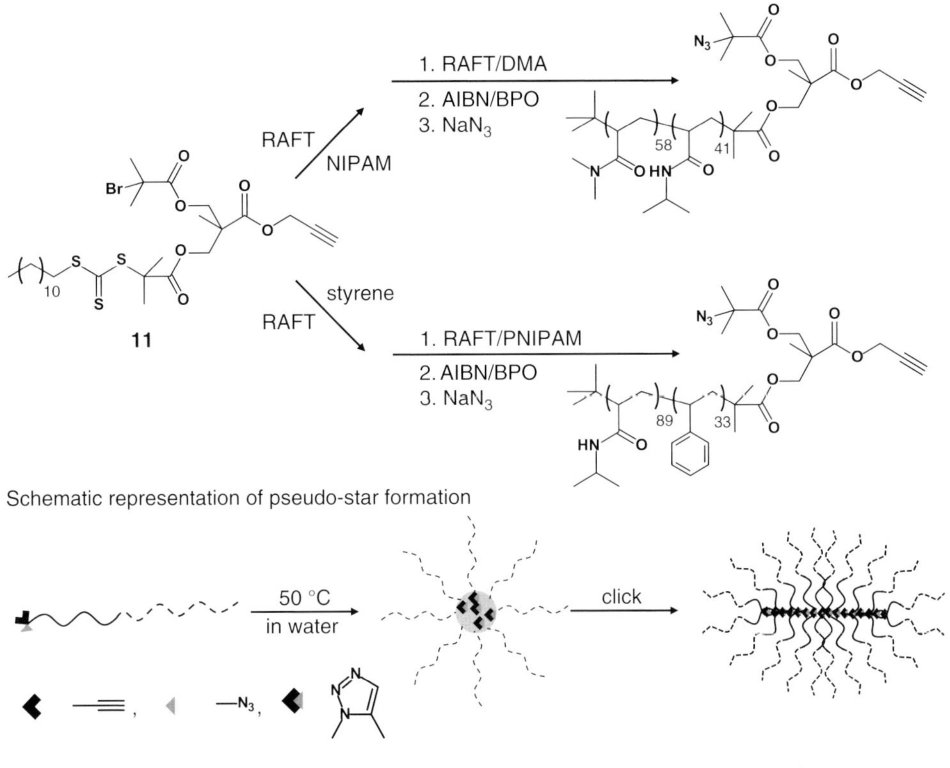


**Scheme 7.27** Synthesis of star-like diblock copolymer via CuAAC reaction using solvent-assisted collocation of the azide and an alkyne groups placed at the same end of the polymer chain; the bottom panel depicts the aggregation of PDMA-block-PNIPAM in water at 50 °C, which is above the LCST of PNIPAM.

## 7.5
## Cyclic Polymers

Cyclic polymers have long intrigued polymer scientists, as the absence of chain ends raises several interesting questions regarding their conformation, translocation, and diffusion characteristics [48]. As evident from the earlier sections, the use of CRP permits the generation of *heterotelechelic* polymers carrying different functional groups at the two chain ends; by proper choice of initiator structure, it would, therefore, be possible to install functional groups that could react with each other intramolecularly to generate cyclic structures. Although there are several methods that utilize high-dilution conditions to achieve cyclization, they often yield cyclic polymers with relatively low cyclization efficiency and/or poor control of molecular weight dispersity. Laurent and Grayson were the first to demonstrate the usefulness of CuAAC reaction in combination with ATRP for the generation of cyclic polymers [49]; using propargyl 2-bromoisobutyrate (**12**) as the ATRP initiator, they synthesized polystyrene bearing a propargyl group at one end and a benzylic bromide at the other, which was readily transformed quantitatively to an azide using sodium azide (Scheme 7.28).

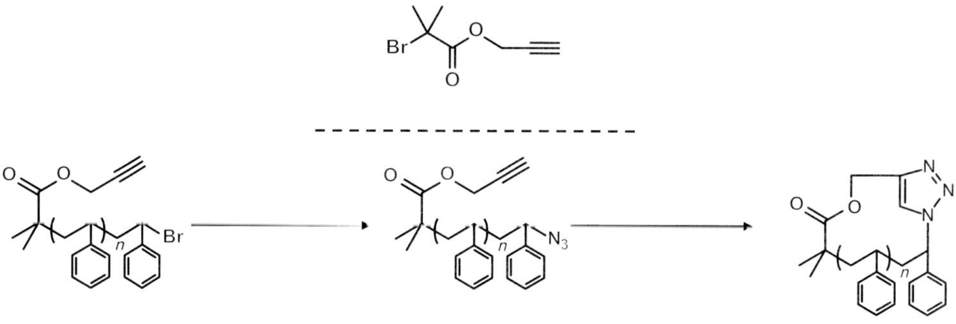

**Scheme 7.28** Synthesis of cyclic-PS from a heterotelechelic propargyl-PS-azide via CuAAC reaction.

Typically, such a *heterotelechelic* PS (propargyl-PS-azide) when subjected to CuAAC reaction would lead to the formation of both higher-molecular-weight polymers, by simple step-growth process via an interchain reaction, as well as cyclic polymers by an intrachain reaction; the usual strategy to enhance the formation of cyclic polymers is to carry out the reaction under very high dilution followed by a process of separation to isolate the cyclic. Laurent and Grayson used an alternative strategy to simulate high-dilution conditions by very slow addition of a relatively concentrated solution of propargyl-PS-azide to a solution containing Cu(I) catalyst, along with bipyridine; by choosing appropriate volumes of the two solutions and by regulating the addition rate, a very low instantaneous concentration of the linear polymers was ensured. Under these conditions, high yields of the cyclic polymer with minimal contamination of linear oligomers were obtained [49]. The success of this slow continuous addition method relies squarely on the high efficiency and fast reaction rate of the CuAAC reaction. Eugene and Grayson extended this approach for the preparation of a cyclic diblock copolymer of polymethylacrylate (PMA) and PS [50]; using the trimethylsilyl-protected ATRP initiator developed by Opsteen and van Hest [34] (Scheme 7.29), they first polymerized methyl acrylate followed by styrene to generate the TMS-protected PMA-PS-Br, which was then transformed to the required alkyne–PMA–PS–azide in two steps – first, the bromide was transformed to the azide and then the TMS group was deprotected using TBAF. Intrachain cyclization was done using the previously described process of slow continuous addition to yield high-purity samples of the *cyclic*-PMA-block-PS samples [50]. Several other cyclic polymers based on polyethylene oxide, polycaprolactone, and so on, were similarly prepared using the CuAAC reaction [51]. Although this process was demonstrated only with relatively low-molecular-weight polymers (up to 4 kD), the authors were later able to obtain cyclic polymer using CuAAC reaction of *heterotelechelic* polycaprolactone of reasonably high molecular weight (up to 25 kD) [51].

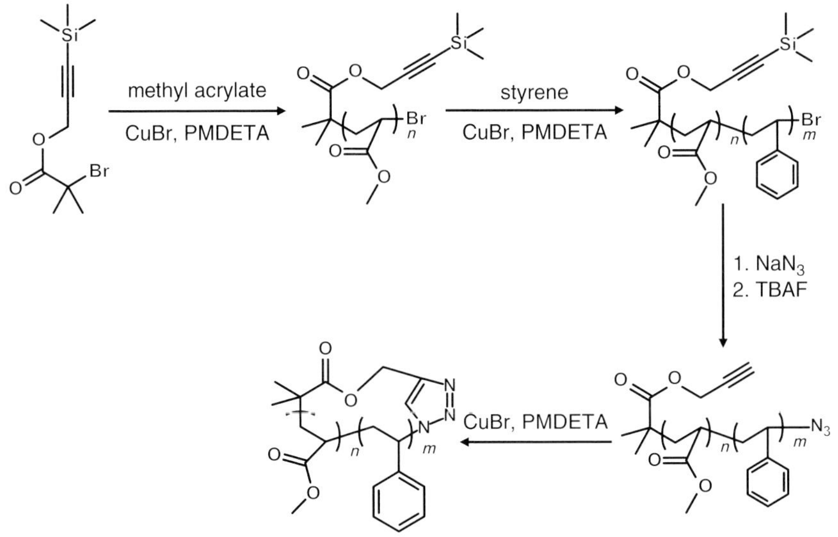


**Scheme 7.29** Synthesis of cyclic diblock copolymer of PMA and PS by a combination of ATRP and CuAAC reaction.

Pressly *et al.* developed an alternative approach using Cu nanoparticles in combination with microwave irradiation to significantly improve the efficiency of the click-based cyclization process [52]; this enabled the production of substantially larger quantities of almost pure cyclic polymers based on polystyrene and polyethylene oxide, in much shorter durations.

Poelma *et al.* devised an alternative strategy to prepare cyclic diblock copolymers via an intermolecular CuAAC reaction between two *homotelechelic* polymers (Scheme 7.30), namely PS with azide groups at both ends and

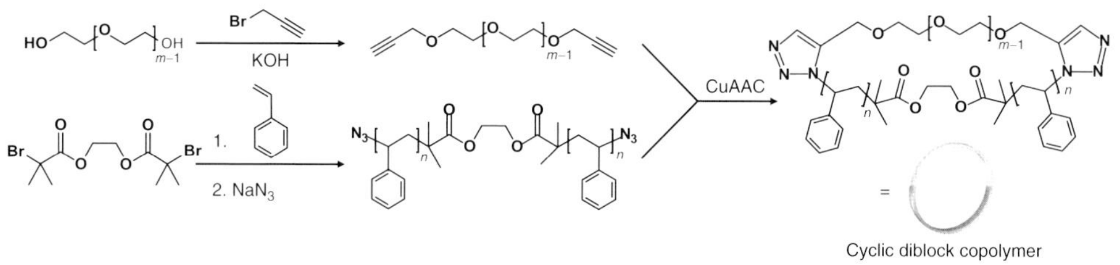


**Scheme 7.30** Synthesis of cyclic diblock copolymer of PS and PEO by intermolecular CuAAC reaction between a PS-diazide and dipropargylated PEO.

dipropargylated PEO [53]; this approach of utilizing an intermolecular reaction between two complementary polymer chains presents the advantage that the two blocks can be prepared by very different processes and is therefore not restricted to monomers that need to be polymerizable by sequential addition using a typically living polymerization process. This choice of the blocks is often very important from the point of view of the nanoscale phase-separated morphologies that result from block copolymers; controlling the length scale of the domains requires the immiscibility factor ($\chi$ value) to be maximized so that the phase separation can occur even when the block sizes are small.

The intermolecular cyclization between the two blocks were done under high-dilution conditions using a semibatch process, wherein after completion of the cyclization of the first batch, the second and subsequent charges are added and the reaction was allowed to proceed in the same reactor; this approach worked efficiently because the CuAAC reaction proceeded rapidly in the presence of a high concentration of the Cu catalyst, even though the polymer concentrations were low. The compact dimensions of the cyclic polymers when compared to their linear counterparts were evident from the significantly smaller domains sizes seen in the case of the cyclic diblock copolymer when compared to the linear diblock analog; periodicities of approximately 20 nm were seen for the cyclic diblock, whereas it was approximately 25 nm for the linear analog.

Qui *et al.* applied the CuAAC approach to prepare cyclic poly(*N*-isopropylacrylamide), *c*-PNIPAM, in an effort to examine the effect of cyclization on their solution behavior, specifically to probe the effect on the LCST [54]. They used an azido-derivatized RAFT agent (AIP) for the polymerization of NIPAM and subsequently transformed the trithiocarbonate end group to the desired propargylated derivative using a one-pot aminolysis followed by thiol–Michael addition to propargyl acrylate (Scheme 7.31). Cyclization of the *heterotelechelic* PNIPAM under high-dilution conditions yielded the desired *c*-PNIPAM; temperature-dependent solution behavior clearly revealed that *c*-PNIPAM exhibited broader collapse transition and at a higher temperature than the linear analog of the same molecular weight, thereby confirming the conformationally restricted nature of cyclic polymers.

Rather than form rings by linking the ends of the chains, it is conceivable that rings could be formed by linking the chain end to any point along the polymer

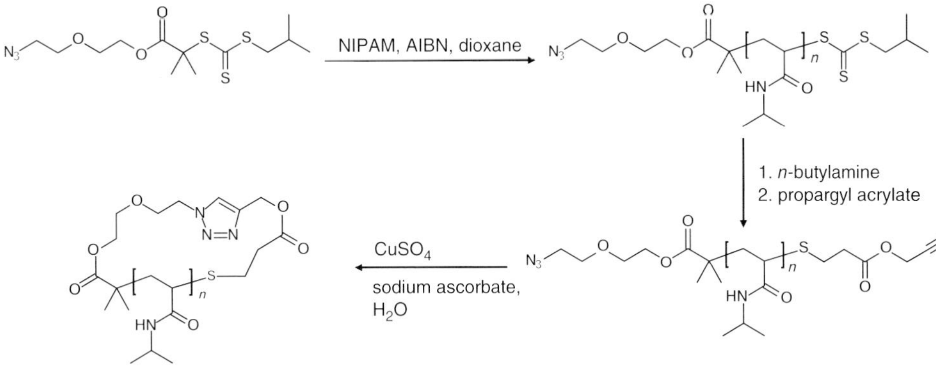


**Scheme 7.31** Preparation of heterotelechelic PNIPAM via CuAAC reaction.

chain or even by linking any two points along the chain; this would allow the generation of several other interesting topologies. Schmidt *et al.* [55] achieved precisely this by employing the strategy of controlled and time-regulated dosing of suitably chosen comonomer(s) during the ATRP of polystyrene; a maleimide-derivative was chosen as the comonomer, as it is well known that it does not homopolymerize but has a very strong tendency to form alternating copolymers with styrene [56]. Since ATRP exhibits "living" characteristics, implying that all chains grow linearly with conversion, a small dose (in proportion to number of growing chains) of maleimide comonomer during the polymerization installs a maleimide unit at a controlled location along the chain; the exact location depends on the timing of the comonomer addition. Using this approach, Schmidt *et al.* prepared different PS chains with the maleimide derivative at varied locations along the chain; the maleimide comonomer carried a trimethylsliyl-protected alkyne unit, which after deprotection yielded PS chains with a clickable alkyne unit at the desired locations [55]. As discussed earlier, since ATRP of styrene leaves a benzylic bromide at one chain end that can be converted to an azide, this process was utilized to affect an intramolecular CuAAC cyclization between the azide chain end and the alkyne group at predesignated locations along the chain. The authors have extended this idea to prepare a variety of other cyclic topologies; for instance, utilizing a difunctional ATRP initiator in combination with controlled dosing of the alkyne-bearing maleimide comonomer, they prepared a chain bearing two rings (Scheme 7.32). Needless to add, all the intramolecular cyclizations were carried out under high-dilution conditions.

In a remarkably interesting utilization of the CuAAC reaction, Leibfarth *et al.* very recently utilized the rapid and quantitative nature of key steps, namely the preparation of the azide from the halo precursor, deprotection of triisopropylsilyl-protected alkyne, and the CuAAC reaction, to prepare precise oligomers with exact control over structure and molecular weight [57]. To achieve this, they used flow synthesis in combination with iterative exponential growth (flow-IEG); the basic idea is to flow a single precursor (**13**) that bears both the protected-alkyne and the halo group through two independent channels, wherein they are subjected

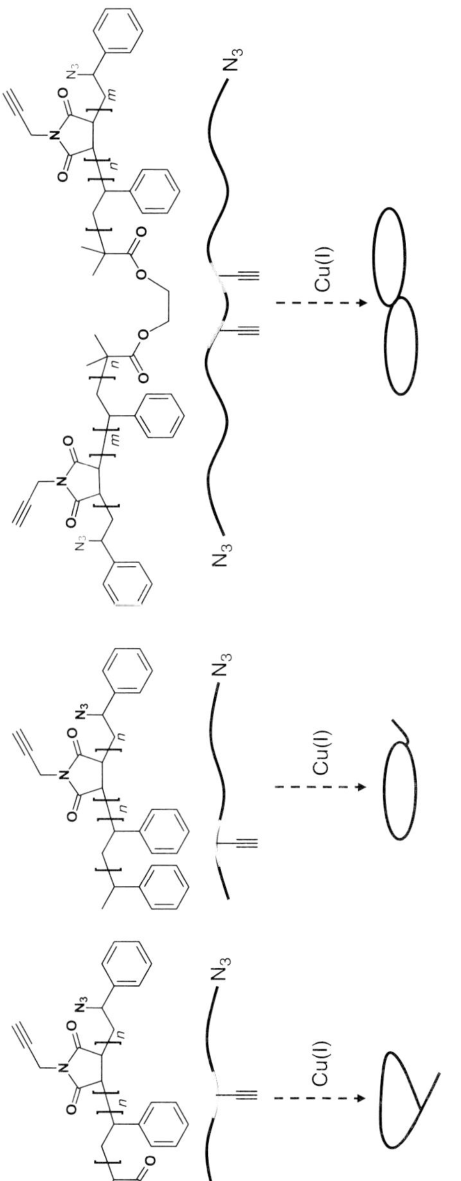


**Scheme 7.32** Preparation of different cyclic polymer topologies using a combination of time-regulated dosing of an alkyne-bearing comonomer during CRP and CuAAC reaction for generation of the ring.

to different reagents – one channel transforms the halo to azide, whereas the other deprotects the triisopropylsilyl group. The products from these two independent channels are then recombined and are allowed to react within a third channel in the presence of optimized Cu catalyst (Scheme 7.33) [57]; the product at the end of this channel is a pure dimer that carries at one end a halo group and at the other, the triisopropylsilyl-protected alkyne.

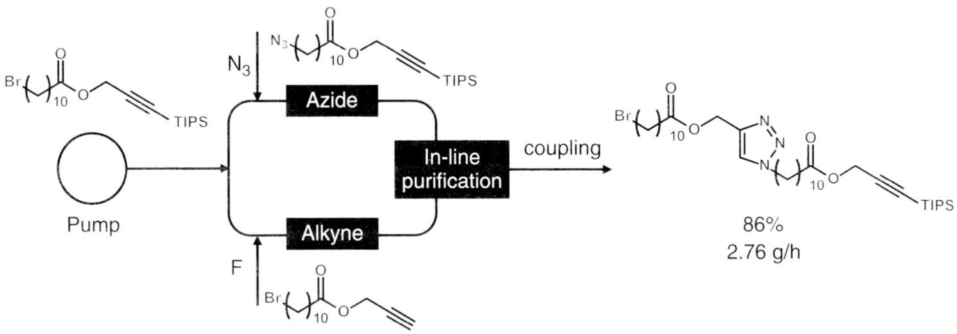


**Scheme 7.33** Iterative flow synthesis of precise oligomers from a TIPS-protected ω-bromoalkyne.

This dimeric product can be looped back and the dimers would similarly be independently deprotected and combined to form tetramers; this process when iteratively applied could result in an automated system that would produce precise oligomers of exact lengths. The fascinating aspect of this approach is that the species fed to the two channels need not be the same; a different halo-*R*-silylalkyne could be fed to one of the channels to yield an exact alternating oligomer sequence, or a second halo-*R*-silylalkyne derivative could be fed at any point in the iterative cyclic process to yield very precisely defined oligomers with the desired primary sequence. The success of this strategy clearly relies on the rapid and quantitative nature of the CuAAC reaction, and also on the rapid generation of the azide and the alkyne groups from suitable precursors; these characteristics in combination with flow synthesis holds the potential to automate the synthesis of a variety of precise oligomers in a very efficient and effective manner.

## 7.6 Side-Chain Clickable Polymers

Thus far, our focus has been on polymers that are prepared via the CuAAC reaction or those that bear clickable units primarily at the chain ends; in this section, we focus on another important application of CuAAC reaction, namely side-chain clickable polymers. These are polymers that bear either pendant propargyl or azide groups and can be clicked, post-polymerization, to access a wide variety of polymeric structures; one example that was earlier described was the dendronized linear polymers that permitted the direct observation of single polymer chains by using AFM.

Synthetic glycopolymers have elicited great interest in view of the multivalent interaction paradigm that has been proposed as the basis for the strong and specific protein–carbohydrate interactions. Of the several methods that have been developed to prepare synthetic glycopolymers, the approach by Ladmiral *et al.* utilizes the CuAAC reaction to install a variety of carbohydrate motifs as pendant units along a polymer backbone [58]; the CuAAC reaction is particularly attractive as it is compatible with unprotected sugar residues. In an effort to control the molecular weight and dispersity of the polymer, a trimethylsilyl-protected propargyl methacrylate was polymerized under standard ATRP conditions, which after deprotection using TBAF yielded the required parent polymer bearing a propargyl group in every repeat unit. This parent polymer was then clicked with various sugars, such as mannose- and galactose-based azides (Scheme 7.34), to provide a library of glycopolymers, whose interactions with lectin was examined. This approach is clearly a very general one and provides the opportunity to co-click different types of functional entities for a variety of biologically relevant applications.

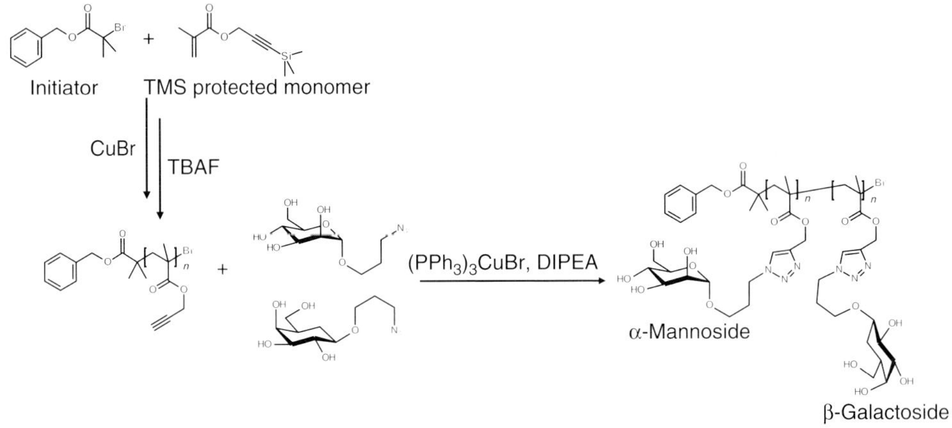


**Scheme 7.34** Synthesis of glycopolymers by a combination of ATRP and CuAAC reaction.

In an alternative approach, Sumerlin *et al.* utilized a methacrylate monomer carrying a pendant azido group; this was because the direct polymerization of the propargyl methacrylate leads to polymers with very high dispersity and sometimes even cross-linked gels at higher conversion, under standard ATRP conditions [59]. The azido groups were stable under the polymerization conditions and led to polymers with narrow molecular-weight distribution; the parent polymer was then clicked with a variety of propargylated molecules, such as propargyl alcohol and 4-pentynoic acid, to illustrate the potential of this polymer to yield a range of functional polymers starting from a single parent polymer (Scheme 7.35) [59]. Instead of having clickable units on every repeat unit, one can devise approaches

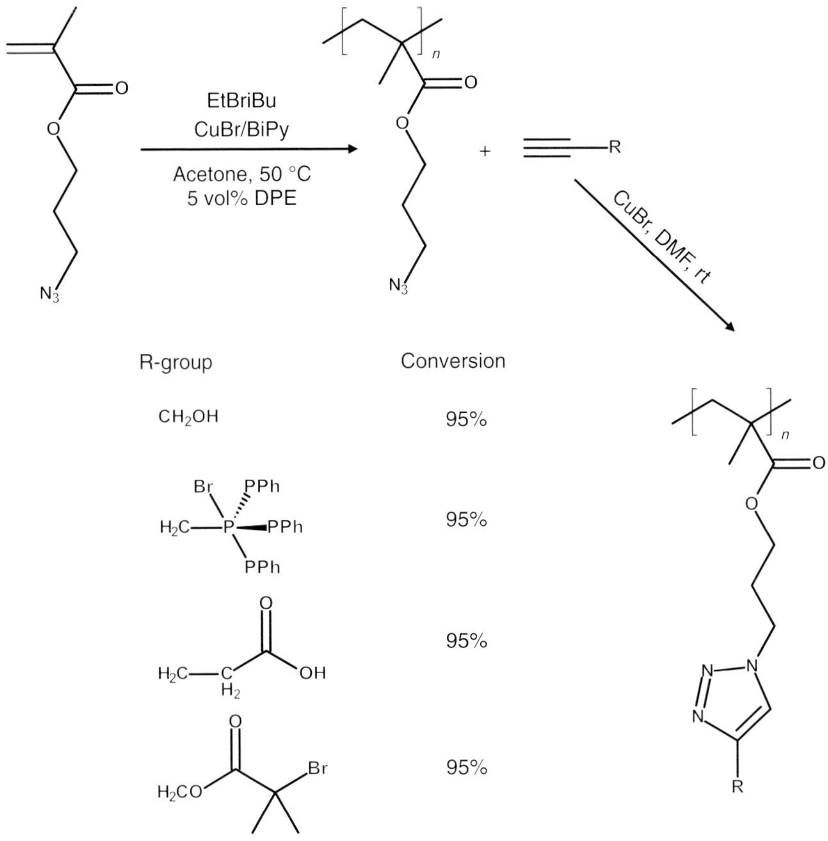


**Scheme 7.35** Synthesis of azide-containing clickable polymer by ATRP and subsequent post-polymerization modification with various alkynes via CuAAC reaction.

that would place clickable units at controlled locations along the polymer backbone. One such approach was taken by Lutz and coworkers for the preparation of interesting cyclic topologies (Scheme 7.32), as was described earlier [55].

In a different approach, we designed clickable monomers that could be condensed with a second monomer bearing either long hydrophobic alkylene or hydrophilic PEG segments to yield periodically clickable polymers via a step-growth polymerization [60, 61]; here propargylated 1,3-propane diol was condensed with a long-chain dicarboxylic acid, such as eicosanoic acid, to yield a periodically clickable polyester. This polyester was subsequently clicked with PEG-350 monomethyl ether monoazide (MPEG-350 azide) to yield periodically grafted amphiphilic copolymers (Scheme 7.36) [60]; these copolymers underwent an immiscibility-driven folding that enabled the collocation of the hydrophobic alkylene segments at the center and the PEG segments above and below the folded chains. The interesting feature of these polymers is that the central alkylene segments crystallize, providing further stability to the folded form; in bulk phase,

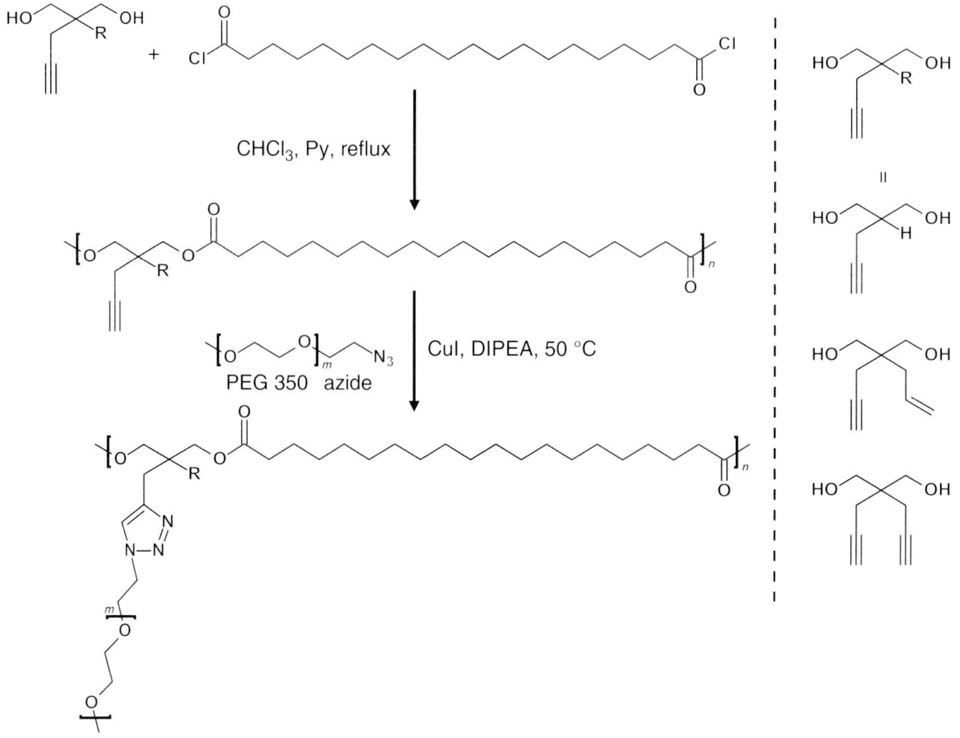


**Scheme 7.36** Synthesis of a series of periodically clickable polyesters and their postpolymerization modification with MPEG-350 azide via CuAAC reaction.

these systems form a lamellar morphology. Instead of PEG segments, pendant perfluoroalkyl moieties were also clicked onto the periodically clickable polyester; these polymers also fold similarly; however, here both the alkylene segments and the perfluoroalkyl segments crystallize independently, yielding a rather unique folded structure that exhibits a selective melting of only alternate layers in a lamellar structure leading to the formation of liquid crystalline systems [60]. Similarly, alternate polyesters where the PEG segments occupy the backbone and pendant long-chain alkyl groups that can undergo independent crystallization are installed as pendant units were also synthesized to generate other interesting structures [61].

## 7.7 Cross-linked Polymeric Systems

Cross-linked polymers are systems where individual polymer chains are covalently linked to form a network structure; such polymeric systems would not dissolve in any solvent but would swell to generate gels. The extent of swelling would strongly depend on the length of the chain segment linking the cross-link junctions; hydrophilic cross-linked polymers would form hydrogels when swollen in an aqueous medium, and such hydrogels have been extensively studied as they are potentially useful for drug delivery and tissue engineering applications. Traditional approaches to prepare cross-linked polymers typically utilize simple free-radical-initiated polymerization of a vinyl monomer in the presence of a divinyl cross-linker; this approach results in a statistical distribution of chain segment lengths between cross-link junctions. Other approaches utilize telechelic polymers of controlled chain lengths and react them with multifunctional cross-linkers to prepare gels; this approach, when carried out using very effective bond-forming reactions, can lead to well-defined gels with improved physical/mechanical properties.

Malkoch *et al.* were the first to recognize the utility of the CuAAC reaction to generate controlled hydrogels (Scheme 7.37); they utilized simple strategies from readily available starting materials to prepare a tetraazide molecules (**14**) carrying a central tetraethylene glycol segment and also a dipropargyl terminated PEG using a simple approach starting from commercially available PEG samples [62]. By carefully varying the ratio of the [Cu]:[ascorbate] with respect to the two PEG components, very high levels of cross-linking were achieved; consequently, the mechanical properties of some of these hydrogels with uniform distribution of cross-links were found to be substantially better than those prepared by the simple free-radical polymerization of PEG-diacrylate, wherein it is believed that clustering of cross-links occurs. Another important advantage of this approach is the possibility of tuning the reactant ratios to leave behind unreacted azide/alkyne groups that could be subsequently utilized to add further functionality to the hydrogels; this could be particularly useful for specific applications, such as tissue engineering. Strain-promoted azide–yne click (SPAAC) reaction utilizes a cyclooctyne that undergoes the cycloaddition with an azide even in the absence of the Cu catalyst because of the strain in the ring [63]; Jiang *et al.* utilized this idea to formulate an injectable hydrogel utilizing two PEG-based components – one a PEG-based tetraazide and the other a telechelic PEG carrying cyclooctyne at the

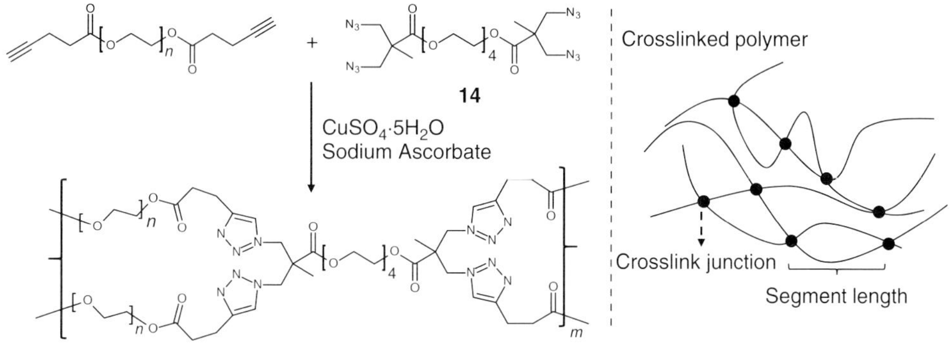


**Scheme 7.37** Synthesis of cross-linked hydrogels using the CuAAC reaction between a dialkyne and a tetraazide.

two chain ends – and mixing of the two components causes rapid gel formation. The authors have used linkages in their precursor polymers that undergo degradation under physiological conditions suggesting that such metal-free systems could be potentially useful for temporary embolizing agents [64].

Ossipov and Hilborn derivatized one of the very important commercially available water-soluble polymers, namely polyvinyl alcohol (PVA), in order to prepare cross-linked hydrogels; using an efficient carbonyldiimidazole (CDI)-mediated coupling, they introduced small amounts (1–5 mol%) of pendant azide and alkyne functionalities independently on to the polymer through a carbamate linkage. Cross-linked hydrogels were then prepared by reacting the PVA-azide and PVA-alkyne in the presence of Cu catalyst (Scheme 7.38) [65].

As expected, the swelling properties and mechanical behavior strongly depended on the extent of cross-linking, which in turn was readily controlled by varying the level of pendant functionalization of PVA; the low levels of derivatization of PVA, used in this study, ensured that the hydrophilicity of the gels remained high. This study provided the basis for a general strategy to derivatize a variety of synthetic and naturally occurring polymers bearing hydroxyl groups and cross-linking them using the CuAAC reaction to generate gels; hydrogels prepared under benign physiological conditions could have several potential applications in areas such as drug delivery, cell growth, and tissue engineering.

Altin *et al.* used an alternative strategy wherein a hydrophilic PEG chain was clicked with two dendrons on both chain ends [66]; using a slightly modified synthetic approach, earlier described by Wu *et al.*, [14] they generated the dendronized PEG that bore propargyl ester terminal groups at the periphery. Reacting this dendronized triblock copolymer with PEG-diazide of different molecular weights (Scheme 7.39), they prepared hydrogels [66]; two interesting aspects of this study are that both the size (generation) of the dendrons and the PEG lengths could be readily varied, which permitted the variation of the swelling behavior.

Importantly, by varying the ratio of the dendronized triblock and the PEG azide, they were able to show that not only the swelling characteristics can be modulated but also the residual propargyl groups could be utilized to click important functional segments on to the hydrogels. More recently, they showed that the residual propargyl can be functionalized with azide-derivatized biotin, which in turn was able to efficiently bind to a fluorescently tagged streptavidin, thus opening up the possibilities of ligating other biologically relevant entities [67].

Liquid crystalline elastomeric systems exhibit several interesting properties due to the presence of ordered domains; Xia *et al.* utilized a combination of living ring-opening metathesis polymerization (ROMP) in the presence of a chain transfer agent (CTA) to prepare a telechelic low $T_g$ block with pendant mesogens and bromo groups at both the chain ends [68]. The complete transformation of the terminal bromo groups to azides was readily accomplished; this liquid crystalline block was then cross-linked using tripropargyl amine to yield a liquid crystalline elastomer (Scheme 7.40). The homogeneous distribution of the cross-link junctions in this polymer and its swellability in standard liquid crystals permitted the

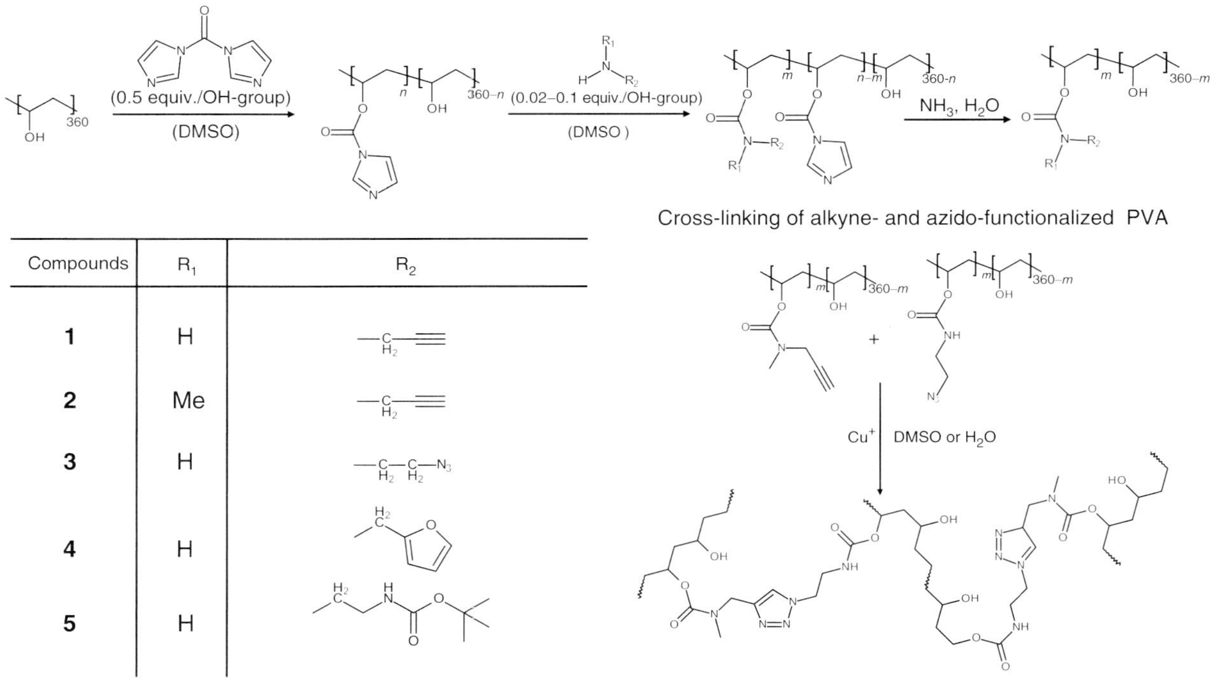


**Scheme 7.38** Synthesis of PVA-based hydrogels via CuAAC reaction between azide- and alkyne-functionalized PVA.

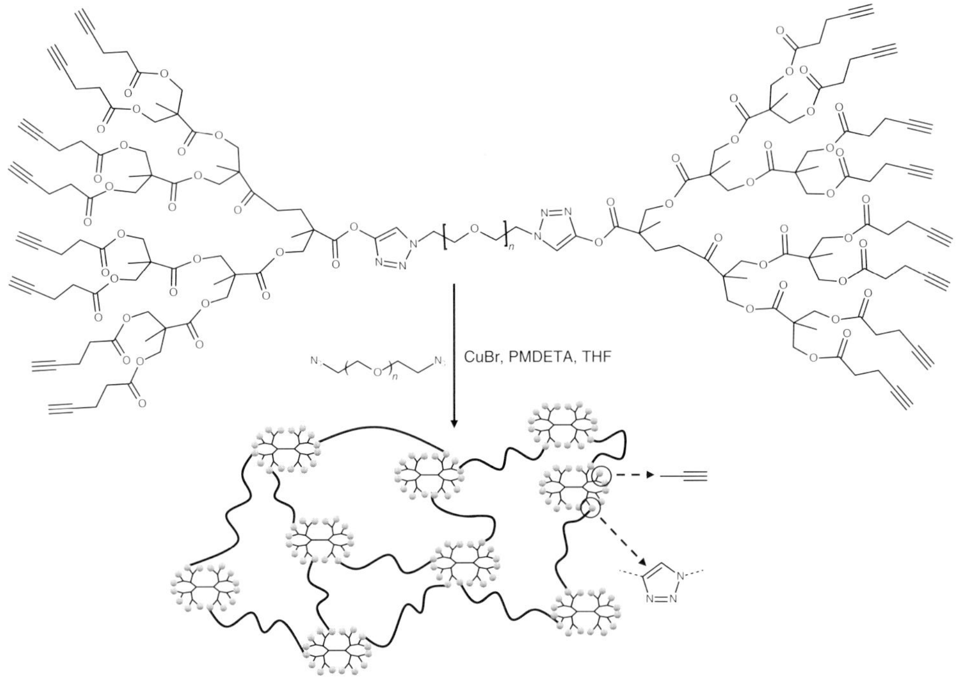


Scheme 7.39 Synthesis of hydrogels via CuAAC reaction between a PEG-diazide and a dendronized triblock copolymer carrying propargyl ester terminal groups and a central PEG segment.

examination of its electro-optic switching characteristics; it was noted that the uniform distribution of cross-links leads to significantly faster switching times.

## 7.8
## Porous Organic Polymers

Porous organic polymers have caught the imagination of chemists, triggered by the remarkable range of metal–organic framework solids (MOFs) and covalent organic frameworks (COFs) structures that have been studied during the past decades, specifically in terms of their potential for gas adsorption and catalysis. Several bond-forming reactions have been explored for the creation of porous organic polymers and conjugated microporous polymers (CMPs); these are essentially highly cross-linked polymers that are built from rigid multifunctional monomers. Holst *et al.* were the first to explore the utility of the CuAAC reaction to create porous polymeric materials; they used the reaction between tetrakis(4-azidophenyl)methane and tetrakis(4-ethynylphenyl)methane (Scheme 7.41) in the presence of $CuSO_4$ and ascorbic acid to generate three-dimensionally cross-linked polymeric materials [69]. The interesting feature of this material

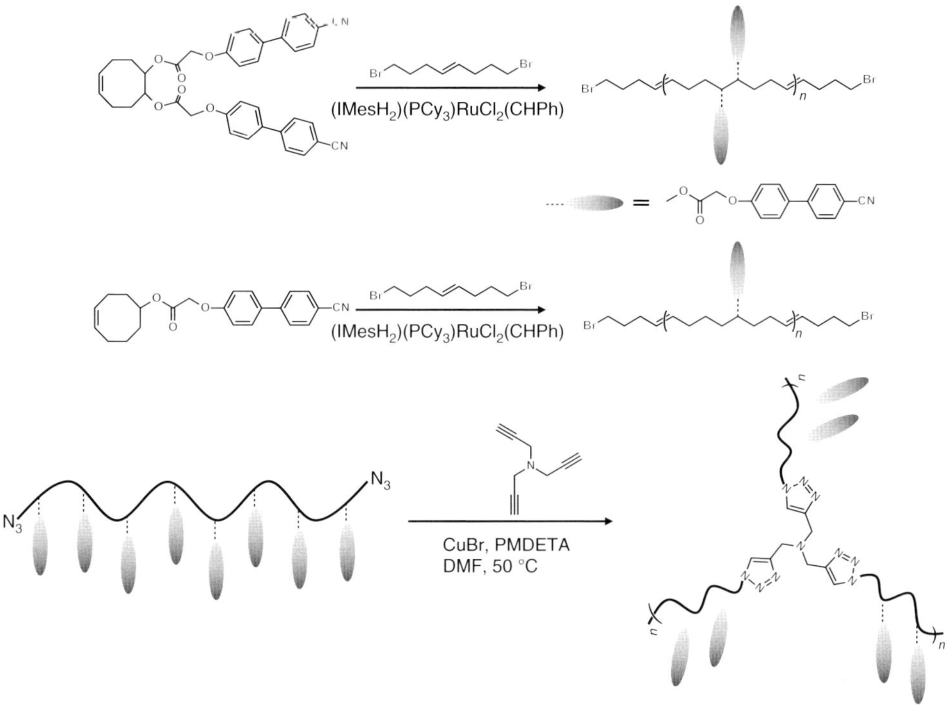


**Scheme 7.40** Synthesis of liquid-crystalline elastomers via CuAAC reaction between a diazide liquid crystalline block and tripropargyl amine.

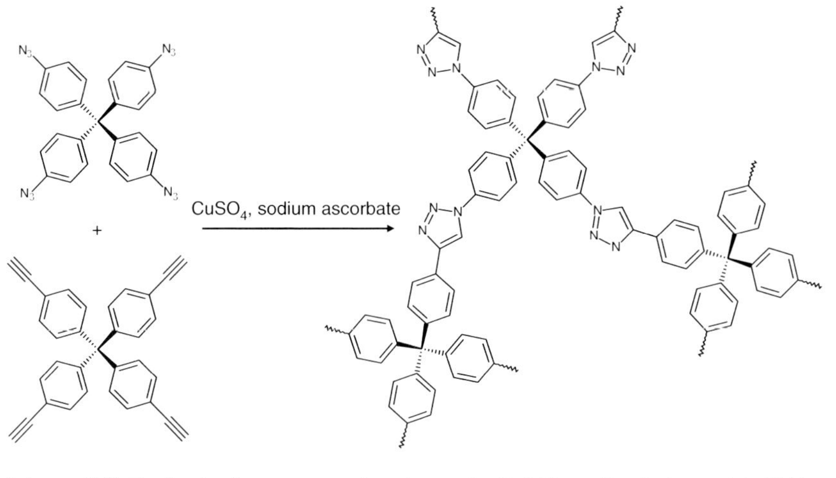


**Scheme 7.41** Synthesis of porous organic polymers by CuAAC reaction between tetrakis(4-azidophenyl)methane and tetrakis(4-ethynylphenyl)methane.

was that, despite the relatively smaller surface area (1128 $m^2/g$), it exhibited significantly larger $CO_2$ uptake when compared with similar polymers prepared via alternate routes that had substantially higher surface areas, suggesting the possible presence of some $CO_2$-philic units in the polymer. Very recently, Lindemann *et al.* [70] extended this approach using the same two monomers to prepare free-standing thin films by utilizing a layer-by-layer (LBL) deposition strategy on to a sacrificial substrate, which allowed the facile transfer of the thin membrane on to a second substrate to conduct gas permeability studies.

Zhong *et al.* utilized a similar approach but used 2,6-diethynyl pyridine along with the tetrakis(4-azidophenyl)methane to generate the porous polymeric network (Scheme 7.42); the interesting feature of this approach is that it generates 2,6-bis(1,2,3-triazol-4-yl)pyridine (BTP) linking units bearing two triazolyl moieties bridged by a central pyridine ring [71].

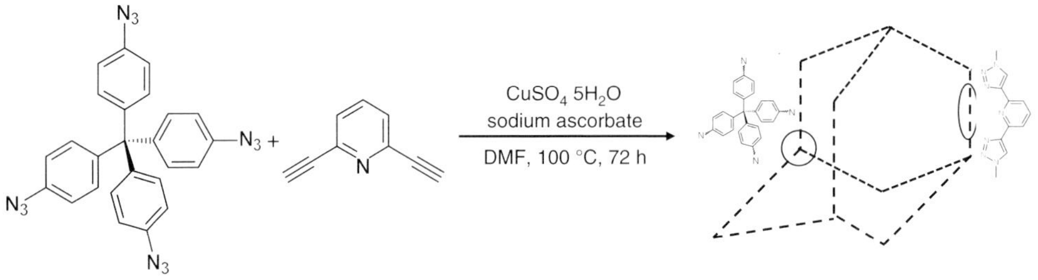


**Scheme 7.42** Porous organic polymers bearing BTP ligands for efficient anchoring of Pd nanoparticles.

These BTP units serve as excellent ligands for metal ions; using these cross-linked polymers, the authors prepared Pd nanoparticles that are anchored to the framework by the direct reduction of $Pd(OAc)_2$ in the presence of the porous polymeric scaffold. The highly porous network structure with Pd nanoparticles anchored on the inner walls served as an efficient hydrogenation catalyst for olefins with high turnover numbers and high recyclability.

## 7.9
## Surface Modification using CuAAC Reaction

Another important applications of CuAAC reaction is its utility in surface modifications of a variety of substrates using small molecules and for the preparation of densely grafted polymeric brushes. The first demonstration of surface modification using click chemistry was by Collman *et al.*; they utilized a ω-azido alkanethiol, along with a simple alkanethiol, to first modify the surface of a gold-coated electrode by the formation of a self-assembled monolayer (SAM) and subsequently used CuAAC reaction with a suitable Ferrocene derivative bearing an alkyne group [72]. The authors demonstrated that the CuAAC reaction proceeded quantitatively using grazing angle IR, and they examined the electroactivity of the modified surface using cyclic voltammetry. This first study demonstrated that CuAAC reaction can also be adapted heterogeneously to modify interfaces. Grafting polymer chains onto solid substrates is a useful approach to change its surface properties, such as wetting, friction, adhesion, and biocompatibility. One of the most effective ways to create a dense polymer layer, often referred to as *polymer brushes*, is to anchor initiators on the substrate surface and grow polymer chains directly from such modified substrates; it has been demonstrated that such an approach leads to the formation of densely packed extended polymer chains [73].

Lee *et al.* first anchored ATRP initiators onto gold-coated substrates using a suitably designed α,ω-difunctionalized disulfide (**15**) to create a densely formed SAM bearing ATRP initiators at the exposed chain ends; subsequently, they polymerized PEG-methacrylate using these anchored initiators in the presence of the Cu catalyst. The terminal bromo groups at the chain ends were then transformed quantitatively to azide groups; this now permits the clicking of desired functional entities onto the surface to elicit desired properties (Scheme 7.43) [74]. PEG-methacrylate was selected as the monomer to ensure biocompatibility, as PEGylated surfaces are known to inhibit protein binding. To demonstrate the accessibility of the surface azide groups, an azido-derivative of biotin was clicked to the polymer-brush-modified surface; subsequently, surface plasmon resonance (SPR) was used to confirm the selective binding with streptavidin in preference to other proteins.

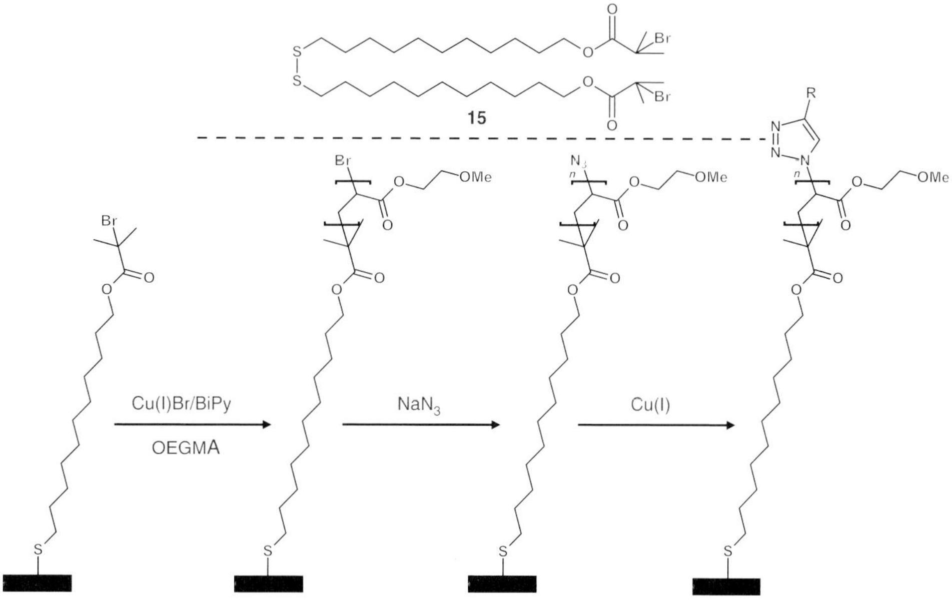


**Scheme 7.43** Surface modification via a combination of SAM, ATRP and CuAAC reaction.

Microcontact printing (μCP) is simple yet powerful approach to create patterned substrates [75]; typically, a soft polysiloxane-based stamp is used to transfer the required pattern onto the desired substrate. Several studies have demonstrated the ability to conduct reactions rapidly at the interface using suitable chemistries; in other words, rather than merely transfer material from the stamp onto the surface, covalent reactions can be made to occur between reactive sites on the substrate and molecules delivered from the stamp [76]. Rozkiewicz *et al.* modified silicon wafer surface by a standard strategy, first using a ω-bromoalkyl triethoxysilane, then transforming the bromo to an azide using $NaN_3$ to yield a high density of azide functionality [77]; the stamp was then treated with a simple long-chain alkyne and was pressed against the azide-functionalized surface for approximately 5 min. The high local concentration of the azide and alkyne at the contact interface caused the click reaction to proceed rapidly at room temperature even in the absence of a Cu catalyst; this was verified using AFM, XPS, and also by clicking a fluorescent alkyne for direct visualization [77]. A particularly revealing aspect of this study was the demonstration that the click reaction can be made more facile when the concentrations of the reactants are enhanced at the reaction loci.

In another interesting demonstration, Nandivada *et al.* utilized an ethynyl-derivatized polyxylylene (**16**), which was coated by a chemical vapor deposition (CVD) process using ethynyl paracyclophane as the precursor [78]; CVD of paracyclophane has been extensively used to prepare uniform conformal coatings on a variety of substrates with strong substrate-adhesion characteristics [79]. The authors were able to demonstrate that under controlled conditions of monomer vaporization/pyrolysis, the polymer coatings formed underwent minimal inadvertent cross-linking and the ethyne groups were accessible to CuAAC reaction with organic azides; they utilized the μCP protocol to create patterned substrates wherein azido-derivatized biotin was covalently linked only in selected regions of the substrate. Subsequently, using a fluorescently labeled streptavidin, the authors were able to show that selective noncovalent binding occurred only to the biotin-derivatized regions on the substrate (Scheme 7.44).

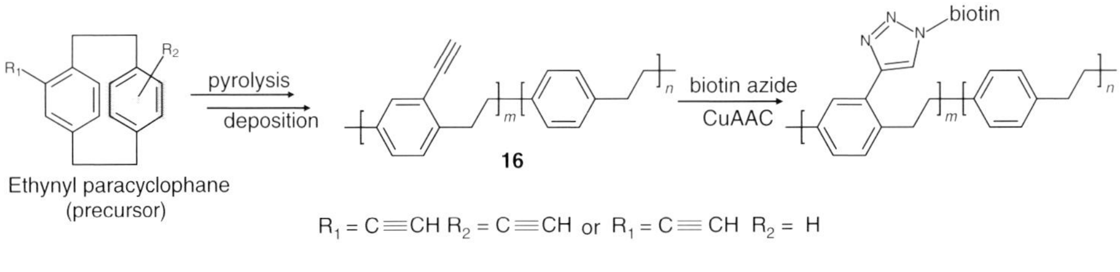


**Scheme 7.44** Modification of CVD-coated surface via CuAAC reaction.

LBL deposition is a technique that has been effectively used to deposit thin films of polymers onto surfaces [80]; most often this method uses electrostatic interactions between oppositely charged polyelectrolytes to build the films, although other noncovalent interactions such as H-bonding have also been used. The Cu-catalyzed click reaction has several unique attributes, such as rapid reaction rates and mild reaction conditions that are compatible with water; Such *et al.* used the CuAAC reaction for the first time to covalently link alternate polymer layers deposited in an LBL manner [81]. Importantly, they achieved this using a single polymer, namely polyacrylic acid (PAA), which was derivatized with either azide or alkyne groups to the extent of approximately 15 mol%; this was achieved by alternate immersion of the silicon/gold substrate into an aqueous solution containing either the azido- or alkyne-PAA, along with $CuSO_4$ and ascorbic acid [81]. Multiple cyclic immersions, with intermediate pure-water washes, permitted the generation of a cross-linked multilayer of a polyelectrolyte (PAA); unlike PAA films deposited without azide–yne cross-linking, these films were stable under high pH conditions, wherein the un-cross-linked PAA films would be completely removed by dissolution. Measurement of thickness variation as a function of number of deposition cycles clearly revealed that essentially single layer of a polymer was deposited during each immersion; the average thickness of each layer was estimated to be about 4.6 nm. Similar studies using dendrimers that carried numerous terminal azide or alkyne groups, carried out by Vestberg *et al.*, [82] revealed that, as anticipated, the thickness of each dendrimer layer was significantly smaller compared to their linear analog, thus reconfirming the compact nature of these highly branched Dendritic macromolecules. These studies reveal the remarkable efficacy of the CuAAC reaction for surface modification; clearly, having multiple azide or alkyne groups in the polymer chains used for the LBL deposition plays a key role, as a single covalent ligation event first anchors the chain onto the substrate and consequently further reaction between the two components is made a lot more facile. Thus, simple immersion into a solution of the macromolecule is shown to be adequate for covalent deposition of a single polymer layer via the CuAAC reaction.

## 7.10 Strain-Promoted Click Reaction

Although the CuAAC reaction has been an extremely valuable strategy, often the complete removal of Cu is hampered by its strong binding to the triazole ring and/or other coordinating ligands that could be present in the polymer structure; this poses a challenge especially when biological applications are targeted, because of the cytotoxic nature of Cu ions. In an effort to remedy this situation, Bertozzi and coworkers developed the copper-free strain-promoted alkyne-azide click (SPAAC) reaction [83] based on the early observation by Wittig that cyclooctyne reacts highly exothermically with organic azides [84]. Canalle *et al.* developed an interesting strategy that utilized this approach to prepare cross-linked polymer coatings using azidoethyl-methacrylate-based polymer; a thin film of the polymer, along with the cross-linker that carried two cyclooctyne units linked to the ends of tetraethylene glycol (**17**), was coated onto the substrate [85]. To ensure that the cross-linked coating is covalently anchored to the glass substrate, the substrate was first functionalized to install alkyne units and then treated with the azide-containing polymer along with the cross-linker (**17**); thermal treatment at 120 °C led to the formation of a covalently anchored cross-linked coating (Scheme 7.45).

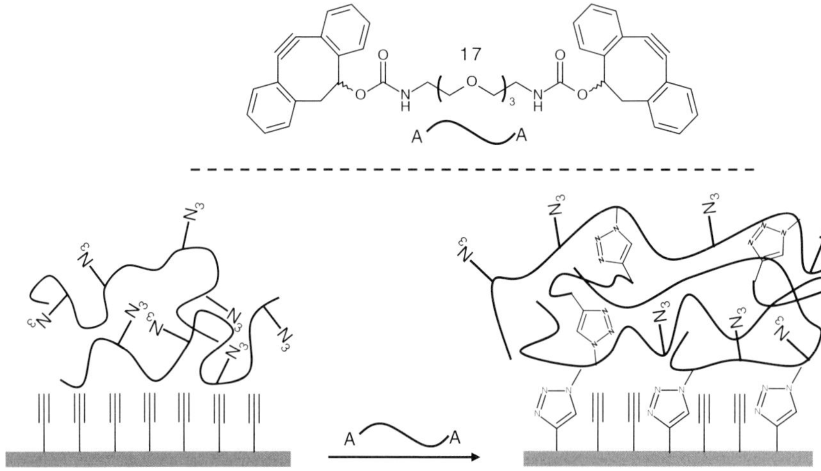

**Scheme 7.45** Formation of a cross-linked polymer coating on an alkyne-functionalized substrate using strain-promoted click reaction.

The useful feature of this coating is the controllable excess of azide functionality that remained stable under the thermal curing process, which was confirmed by IR spectroscopic measurements; these azide groups were subsequently treated with several clickable partners, based on suitably functionalized coumarin derivatives, which permitted the authors to compare and conclude that the strain-promoted click partner was the most effective in the postcoating treatment. The authors suggest that such coatings could be effectively used to ligate a variety of biologically active units. The SPAAC process was later successfully used for the production of enzyme-linked immunosorbent assay (ELISA) by immobilization of azide-functionalized peptides on microtiter plates carrying bicyclo[6.1.0]non-4-yne (BCN) [86].

Ornelas *et al.* utilized the strain-promoted alkyne–azide click reaction to develop PEGylated dendrimers (Scheme 7.46); here, the authors demonstrate that polyamido dendrimers bearing peripheral azide groups can be readily clicked using a PEGylated cyclooctyne derivative (**18**) [87]. The authors made a careful comparison of this SPAAC process with the standard CuAAC process and show that while in the latter case, the complete removal of the Cu presents a difficult problem, the SPAAC process goes to completion within 30 min at room temperature in a ethanol–water solvent mixture; a 1.5-fold excess of the cyclooctyne was used to effect rapid conversion. This cycloaddition reaction yields a mixture of both regio-isomers, unlike the CuAAC reaction. However,

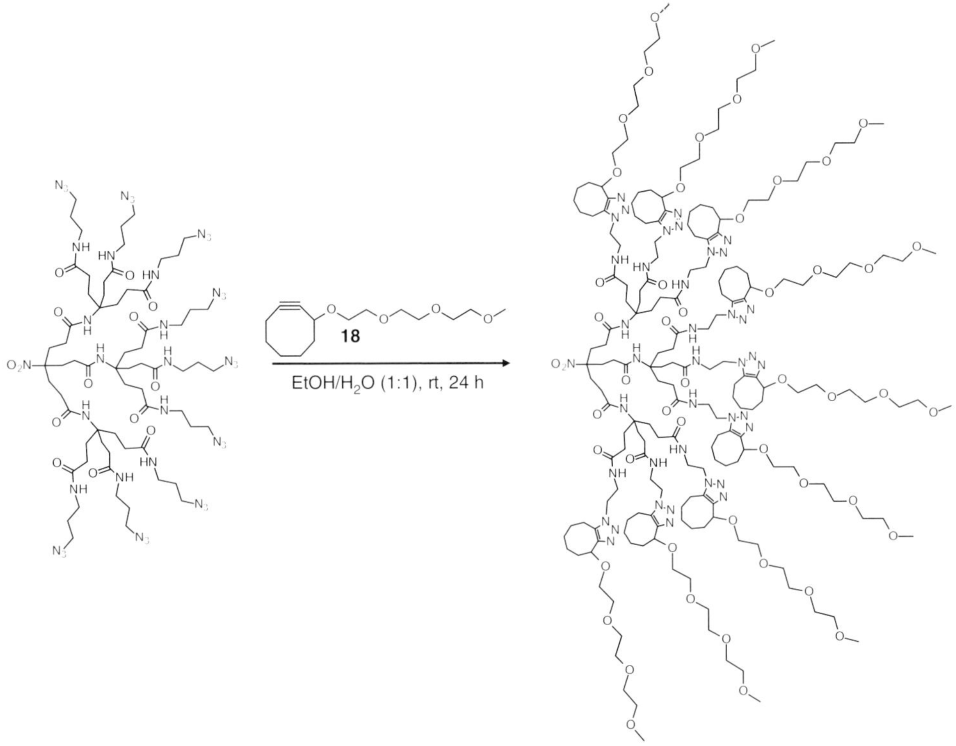


**Scheme 7.46** Synthesis of PEGylated polyamido dendrimers via SPAAC.

the purity of the final product was demonstrated to be far superior to the conventional Cu-catalyzed approach, rendering this a more suitable method for developing scaffolds for biological applications. Thus, in situations wherein the regiospecificity is not crucial, but it is crucial to ensure the absence of Cu, the SPAAC serves as an excellent alternative to covalently link desired moieties onto a variety of scaffolds. Furthermore, the simple three-step strategy to prepare the PEGylated cyclooctyne derivative in good yields, starting from commercially available cycloheptene, enhances the scope of the strategy considerably.

## 7.11 Topochemical Azide–Alkyne Cycloaddition (TAAC) Reactions

Whereas strain-promoted click reaction relied on enhancing the reactivity of the alkyne by introducing ring strain, Sureshan and coworkers explored the azide–yne click reaction in suitably designed crystalline solids [88, 89]; their strategy was to bring the two reactants together in the right orientation and distance to lower the barrier to reaction considerably and thereby effect a Topochemical Azide–Alkyne Cycloaddition (TAAC) reactions. This solid-state catalyst-free azide–alkyne cycloaddition was shown to be regiospecific and led to the selective formation of either of the two isomeric triazoles.

By adopting crystal engineering approaches, the monomers having azide and alkyne motifs are designed in such a way that they crystallize with proximal placement of azide and alkyne motifs in an orientation close to that of the transition state for their cycloaddition reaction. Mild activation of such crystals produces triazole-linked polymers under topochemical control and often in a single-crystal-to-single-crystal (SCSC) fashion. By changing the protecting groups, the regiospecificity of triazole formation could be tuned; while the crystals of 1-deoxy-β-1-azido-4-*O*-propargyl-D-galactose react topochemically to give the polymer with 1,4-triazole linkage exclusively in an SCSC fashion [88], the crystals of its tribenzyl ether derivative give the polymer with 1,5-triazole linkages [89]. More recently, another interesting demonstration of the TAAC involved the SCSC synthesis of a 1,5-triazolyl linked polynucleoside from its monomer [90], as depicted in Scheme 7.47. Thus, bringing the reactants in close proximity greatly enhances the reaction rates; this was also evident in an earlier example wherein simple stamping in the μCP process enabled the reaction to occur even in the absence of a catalyst [77].

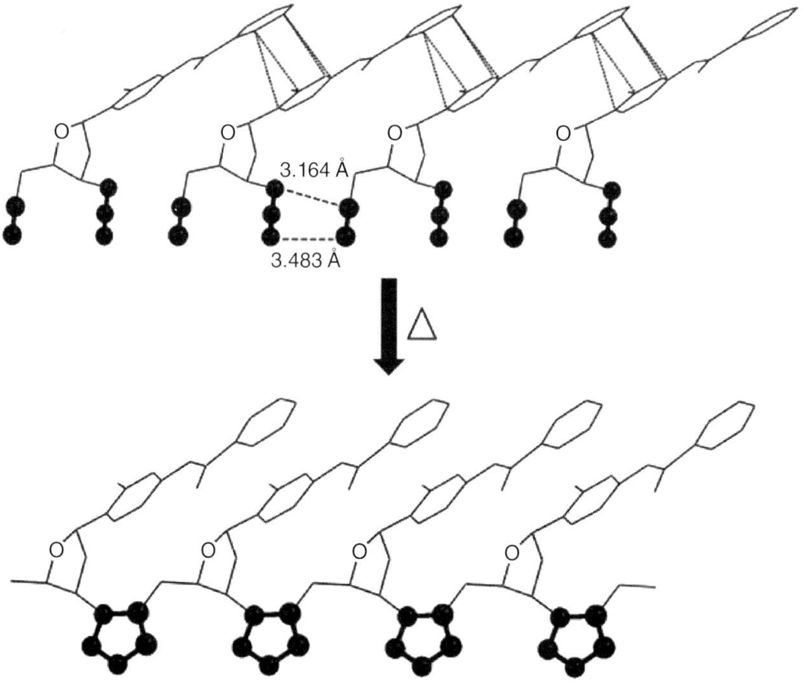


**Scheme 7.47** Topochemical azide–alkyne click reaction for the synthesis of a polynucleoside from its monomer in the crystalline solid state.

## 7.12
## Summary and Outlook

The use of CuAAC reaction in polymer science has clearly provided very rich dividends both in terms of creating very novel structures and formulating innovative strategies to improve the synthesis of previously studied systems. The rapid and chemoselective nature of the CuAAC reaction, along with its remarkable compatibility to a range of reaction conditions and its tolerance to a variety of functional groups ensured its successful utilization in the exceptionally wide range of situations that were described in this chapter. The simple strategies to prepare polymers/functional agents/substrates containing either the alkyne or azide functionality are clearly crucial to the success of the CuAAC reaction. As evident from the foregone discussions, polymer scientists have explored a very rich range of applications of the CuAAC reaction – ranging from creating novel cyclic structures, cross-linked hydrogels, cross-linked functional coatings, biocompatible substrates for tissue engineering, porous polymeric systems for catalysis and many others. The potential for automated synthesis, using a combination of CuAAC reaction and flow synthesis, is clearly one of the most creative demonstrations of the utility of the CuAAC reaction, which exploits all the most important and unique characteristics of this reaction. While in itself, the CuAAC reaction has created and will continue to create rich avenues for further exploitation; we believe, it is the philosophical outlook, namely the *search for the nearly perfect chemical reaction* that would create even greater ripples in the field of polymers and material science. The CuAAC reaction has kindled hope and expectation that this search would be a worthwhile one.

## References


1. Huisgen, R. (1963) *Angew. Chem. Int. Ed. Engl.*, **2**, 565–632.
2. (a) Tornøe, C.W., Christensen, C., and Meldal, M. (2002) *J. Org. Chem.*, **67**, 3057; (b) Rostovtsev, V.V., Luke, V., Green, G., Fokin, V.V., and Sharpless, K.B. (2002) *Angew. Chem. Int. Ed.*, **41**, 2596–2599; (c) Wang, Q., Chan, T.R., Hilgraf, R., Fokin, V.V., Sharpless, K.B., and Finn, M.G. (2003) *J. Am. Chem. Soc.*, **125**, 3192–3193.
3. Diaz, D.D., Punna, S., Holzer, P., Mcpherson, A.K., Sharpless, K.B., Fokin, V.V., and Finn, M.G. (2004) *J. Polym. Sci., Part A: Polym. Chem.*, **42**, 4392–4403.
4. Binauld, S., Damiron, D., Hamaide, T., Pascault, J.P., Fleurya, E., and Drockenmuller, E. (2008) *Chem. Commun.*, 4138–4140.
5. van Steenis, D.J.V.C., David, O.R.P., van Strijdonck, G.P.F., van Maarseveen, J.H., and Reek, J.N.H. (2005) *Chem. Commun.*, 4333–4335.
6. Terao, J., Kimura, K., Seki, S., Fujihara, T., and Tsuji, Y. (2012) *Chem. Commun.*, **48**, 1577–1579.
7. Nelson, J.C., Saven, J.G., Moore, J.S., and Wolynes, P.G. (1997) *Science*, **277**, 1793–1796.
8. Meudtner, R.M. and Hecht, S. (2008) *Macromol. Rapid Commun.*, **29**, 347–351.
9. Dong, Y., Lam, J.W.Y., Qin, A., Liu, J., Li, Z., Tang, B.Z., Sun, J., and Kwok, H.S. (2007) *Appl. Phys. Lett.*, **91**, 011111.

10. Qin, A., Lam, J.W.Y., Tang, L., Jim, C.K.W., Zhao, H., Sun, J., and Tang, B.Z. (2009) *Macromolecules*, **42**, 1421–1424.
11. Srinivasachari, S., Liu, Y., Zhang, G., Prevette, L., and Reineke, T.M. (2006) *J. Am. Chem. Soc.*, **128**, 8176–8184.
12. Dıaz, D.D., Rajagopal, K., Strable, E., Schneider, J., and Finn, M.G. (2006) *J. Am. Chem. Soc.*, **128**, 6056–6057.
13. Wu, P., Feldman, A.K., Nugent, A.K., Hawker, C.J., Scheel, A., Voit, B., Pyun, J., Frechet, J.M.J., Sharpless, K.B., and Fokin, V.V. (2004) *Angew. Chem. Int. Ed.*, **43**, 3928–3932.
14. Wu, P., Malkoch, M., Hunt, J.N., Vestberg, R., Kaltgrad, E., Finn, M.G., Fokin, V.V., Sharpless, K.B., and Hawker, C.J. (2005) *Chem. Commun.*, 5775–5777.
15. Lee, J.W., Kim, B.K., Kim, J.H., Shin, W.S., and Jin, S.H. (2006) *J. Org. Chem.*, **71**, 4988–4991.
16. Shu, L., Schluter, A.D., Ecker, C., Severin, N., and Rabe, J.P. (2001) *Angew. Chem. Int. Ed.*, **40**, 4666–4669.
17. Helms, B., Mynar, J.L., Hawker, C.J., and Frechet, J.M.J. (2004) *J. Am. Chem. Soc.*, **126**, 15020–15021.
18. Mynar, J.L., Choi, T.L., Yoshida, M., Kim, V., Hawker, C.J., and Frechet, J.M.J. (2005) *Chem. Commun.*, 5169–5171.
19. Wang, L., Kiemle, D.J., Boyle, C.J., Connors, E.L., and Gitsov, I. (2014) *Macromolecules*, **47**, 2199–2213.
20. Scheel, A.J., Komber, H., and Voit, B.I. (2004) *Macromol. Rapid Commun.*, **25**, 1175–1180.
21. Qin, A., Lam, J.W.Y., Jim, C.K.W., Zhang, L., Yan, J., Haussler, M., Liu, J., Dong, Y., Liang, D., Chen, E., Jia, G., and Tang, B.Z. (2008) *Macromolecules*, **41**, 3808–3822.
22. Shi, Y., Graff, R.W., Cao, X., Wang, X., and Gao, H. (2015) *Angew. Chem. Int. Ed.*, **54**, 7631–7635.
23. Frey, H. and Holter, D. (1999) *Acta Polym.*, **50**, 67–76.
24. (a) Caminade, A.M., Laurent, R., Nicota, B.D., and Majoral, J.P. (2012) *New J. Chem.*, **36**, 217–226; (b) Satoh, T. (2009) *Soft Matter*, **5**, 1972–1982; (c) Wang, D., Deraedt, C., Ruiz, J., and Astruc, D. (2015) *Acc. Chem. Res.*, **48**, 1871–1880.
25. (a) Feast, W.J. and Stainton, N.M. (1995) *J. Mater. Chem.*, **5**, 405–411; (b) Parker, D. and Feast, W.J. (2001) *Macromolecules*, **34**, 2048–2059.
26. (a) Ramkumar, S.G., Rose, K.A.A., and Ramakrishnan, S. (2010) *J. Polym. Sci., Part A: Polym. Chem.*, **48**, 3200–3208; (b) Saha, A. and Ramakrishnan, S. (2009) *Macromolecules*, **42**, 4956–4959.
27. (a) Samuel, A.Z. and Ramakrishnan, S. (2012) *Macromolecules*, **45**, 2348–2358; (b) Samuel, A.Z. and Ramakrishnan, S. (2013) *Langmuir*, **29**, 1245–1257.
28. Szwarc, M. (1956) *Nature*, **178**, 1168–1169.
29. Higashihara, T., Hayashib, M., and Hirao, A. (2011) *Prog. Polym. Sci.*, **36**, 323–375.
30. Buback, M., Gunzler, F., Russell, G.T., and Vana, P. (2009) *Macromolecules*, **42**, 652–662.
31. Braunecker, W.A. and Matyjaszewski, K. (2007) *Prog. Polym. Sci.*, **32**, 93–146.
32. Matyjaszewski, K. and Xia, J. (2001) *Chem. Rev.*, **101**, 2921–2990.
33. Lowea, A.B. and McCormick, C.L. (2007) *Prog. Polym. Sci.*, **32**, 283–351.
34. Opsteen, J.A. and van Hest, J.C.M. (2005) *Chem. Commun.*, 57–59.
35. (a) Ko, Y.S., Circu, M.V., Geiger, T., Dunki, S., Nueschab, F.A., and Opris, D.M. (2014) *RSC Adv.*, **4**, 35027–35034; (b) Wu, Z.Q., Ono, R.J., Chen, Z., Lia, Z., and Bielawski, C.W. (2011) *Polym. Chem.*, **2**, 300–302; (c) Greene, A.C., Zhu, J., Pochan, D.J., Jia, X., and Kiick, K.L. (2011) *Macromolecules*, **44**, 1942–1951.
36. Quemener, D., Davis, T.P., Kowollik, C.B., and Stenzel, M.H. (2006) *Chem. Commun.*, 5051–5053.
37. Durmaz, H., Dag, A., Altintas, O., Erdogan, T., Hizal, G., and Tunca, U. (2007) *Macromolecules*, **40**, 191–198.
38. Agut, W., Taton, D., and Lecommandoux, S. (2007) *Macromolecules*, **40**, 5653–5661.
39. Urien, M., Erothu, H., Cloutet, E., Hiorns, R.C., Vignau, L., and Cramail, H. (2008) *Macromolecules*, **41**, 7033–7040.
40. Loewe, R.S., Khersonsky, S.M., and McCullough, R.D. (1999) *Adv. Mater.*, **11**, 250–253.
41. Smeets, A., Willot, P., Winter, J.D., Gerbaux, P., Verbiest, T., and

Koeckelberghs, G. (2011) *Macro molecules*, **44**, 6017–6025.
42. Zheng, Y., Zhou, H., Liu, D., Floudas, G., Wagner, M., Koynov, K., Mezger, M., Butt, H.J., and Ikeda, T. (2013) *Angew. Chem. Int. Ed.*, **52**, 4845–4848.
43. Hadjichristidis, N. (1999) *J. Polym. Sci., Part A: Polym. Chem.*, **37**, 857–871.
44. Gao, H. and Matyjaszewski, K. (2006) *Macromolecules*, **39**, 4960–4965.
45. Gao, H., Min, K., and Matyjaszewski, K. (2007) *Macromol. Chem. Phys.*, **208**, 1370–1378.
46. Durmaz, H., Dag, A., Hizal, A., Hizal, G., and Tunca, U. (2008) *J. Polym. Sci., Part A: Polym. Chem.*, **46**, 7091–7100.
47. Li, L., Yan, M., Zhang, G., and Wu, C. (2013) *Macromolecules*, **46**, 8152–8160.
48. (a) Semlyen, E.R. (2000) *Cyclic Polymers*, 2nd edn, Kluwer Academic Publishers; (b) Tezuka, Y. (2013) *Topological Polymer Chemistry: Progress of Cyclic Polymers in Syntheses, Properties and Functions*, World Scientific.
49. Laurent, B.A. and Grayson, S.M. (2006) *J. Am. Chem. Soc.*, **128**, 4238–4239.
50. Eugene, D.M. and Grayson, S.M. (2008) *Macromolecules*, **41**, 5082–5084.
51. (a) Hoskins, J.N. and Grayson, S.M. (2009) *Macromolecules*, **42**, 6406–6413; (b) Zhang, B., Zhang, H., Li, Y., Hoskins, J.N., and Grayson, S.M. (2013) *ACS Macro Lett.*, **2**, 845–848.
52. Pressly, E.D., Amir, R.J., and Hawker, C.J. (2011) *J. Polym. Sci., Part A: Polym. Chem.*, **49**, 814–819.
53. Poelma, J.E., Ono, K., Miyajima, D., Aida, T., Satoh, K., and Hawker, C.J. (2012) *ACS Nano*, **6**, 10845–10854.
54. Qiu, X.P., Tanaka, F., and Winnik, F.M. (2007) *Macromolecules*, **40**, 7069–7071.
55. Schmidt, B.V.K.J., Fechler, N., Falkenhagen, J., and Lutz, J.F. (2011) *Nat. Chem.*, **3**, 234–238.
56. Cowie, J.M.G. (1985) *Alternating Copolymers*, Plenum Press, New York.
57. Leibfarth, F.A., Johnson, J.A., and Jamison, T.F. (2015) *Proc. Natl. Acad. Sci. U.S.A.*, **112**, 10617–10622.
58. Ladmiral, V., Mantovani, G., Clarkson, G.J., Cauet, S., Irwin, J.L., and Haddleton, D.M. (2006) *J. Am. Chem. Soc.*, **128**, 4823–4830.
59. Sumerlin, B.S., Tsarevsky, N.V., Louche, G., Lee, R.Y., and Matyjaszewski, K. (2005) *Macromolecules*, **38**, 7540–7545.
60. Mandal, J., Prasad, S.K., Rao, D.S.S., and Ramakrishnan, S. (2014) *J. Am. Chem. Soc.*, **136**, 2538–2545.
61. Mandal, J. and Ramakrishnan, S. (2015) *Langmuir*, **31**, 6035–6044.
62. Malkoch, M., Vestberg, R., Gupta, N., Mespouille, L., Dubois, P., Mason, A.F., Hedrick, J.L., Liao, Q., Frank, C.W., Kingsburye, K., and Hawker, C.J. (2006) *Chem. Commun.*, 2774–2776.
63. Agard, N.J., Prescher, J.A., and Bertozzi, C.R. (2004) *J. Am. Chem. Soc.*, **126**, 15046–15047.
64. Jiang, H., Qin, S., Dong, H., Lei, Q., Su, X., Zhuo, R., and Zhong, Z. (2015) *Soft Matter*, **11**, 6029–6036.
65. Ossipov, D.A. and Hilborn, J. (2006) *Macromolecules*, **39**, 1709–1718.
66. Altin, H., Kosif, I., and Sanyal, R. (2010) *Macromolecules*, **43**, 3801–3808.
67. Kaga, S., Yapar, S., Gecici, E.M., and Sanyal, R. (2015) *Macromolecules*, **48**, 5106–5115.
68. Xia, Y., Verduzco, R., Grubbs, R.H., and Kornfield, J.A. (2008) *J. Am. Chem. Soc.*, **130**, 1735–1740.
69. Holst, J.R., Stockel, E., Adams, D.J., and Cooper, A.I. (2010) *Macromolecules*, **43**, 8531–8538.
70. Lindemann, P., Tsotsalas, M., Shishatskiy, S., Abetz, V., Sidenstein, P.K., Azucena, C., Monnereau, L., Beyer, A., Gölzhäuser, A., Mugnaini, V., Gliemann, H., Bräse, S., and Woll, C. (2014) *Chem. Mater.*, **26**, 7189–7193.
71. Zhong, H., Gong, Y., Zhang, F., Liab, L., and Wang, R. (2014) *J. Mater. Chem. A*, **2**, 7502–7508.
72. Collman, J.P., Devaraj, N.K., and Chidsey, C.E.D. (2004) *Langmuir*, **20**, 1051–1053.
73. Edmondson, S., Osborne, V.L., and Huck, W.T.S. (2004) *Chem. Soc. Rev.*, **33**, 14–22.
74. Lee, B.S., Lee, J.K., Kim, W.J., Jung, Y.H., Sim, S.J., Lee, J., and Choi, I.S. (2007) *Biomacromolecules*, **8**, 744–749.
75. (a) Quist, A.P., Pavlovic, E., and Oscarsson, S. (2005) *Anal. Bioanal. Chem.*, **381**, 591–600; (b) Xia, Y. and

Whitesides, G.M. (1998) *Angew. Chem. Int. Ed. Engl.*, **37**, 550–575.

76. Sullivan, T.P., van Poll, M.L., Dankers, P.Y.W., and Huck, W.T.S. (2004) *Angew. Chem. Int. Ed.*, **43**, 4190–4193.
77. Rozkiewicz, D.I., Janczewski, D., Verboom, W., Ravoo, B.J., and Reinhoudt, D.N. (2006) *Angew. Chem. Int. Ed.*, **45**, 5292–5296.
78. Nandivada, H., Chen, H.Y., Bondarenko, L., and Lahann, J. (2006) *Angew. Chem. Int. Ed.*, **45**, 3360–3363.
79. Alf, M.E., Asatekin, A., Barr, M.C., Baxamusa, S.H., Chelawat, H., Ozaydin-Ince, G., Petruczok, C.D., Sreenivasan, R., Tenhaeff, W.E., Trujillo, N.J., Vaddiraju, S., Xu, J., and Gleason, K.K. (2010) *Adv. Mater.*, **22**, 1993–2027.
80. (a) Decher, G. (1997) *Science*, **277**, 1232–1237; (b) Peyratout, C.S. and Dähne, L. (2004) *Angew. Chem. Int. Ed.*, **43**, 3762–3783.
81. Such, G.K., Quinn, J.F., Quinn, A., Tjipto, E., and Caruso, F. (2006) *J. Am. Chem. Soc.*, **128**, 9318–9319.
82. Vestberg, R., Malkoch, M., Kade, M., Wu, P., Fokin, V.V., Sharpless, K.B., Drockenmuller, E., and Hawker, C.J. (2007) *J. Polym. Sci., Part A: Polym. Chem.*, **45**, 2835–2846.
83. Johnson, J.A., Baskin, J.M., Bertozzi, C.R., Koberstein, J.T., and Turro, N.J. (2008) *Chem. Commun.*, 3064–3066.
84. Wittig, G. and Krebs, A. (1961) *Chem. Ber.*, **94**, 3260–3275.
85. Canalle, L.A., van Berkel, S.S., de Haan, L.T., and van Hest, J.C.M. (2009) *Adv. Funct. Mater.*, **19**, 3464–3470.
86. Canalle, L.A., Vong, T., Adams, P.H.H.M., van Delft, F.L., Raats, J.M.H., Chirivi, R.G.S., and van Hest, J.C.M. (2011) *Biomacromolecules*, **12**, 3692–3697.
87. Ornelas, C., Broichhagen, J., and Weck, M. (2010) *J. Am. Chem. Soc.*, **132**, 3923–3931.
88. Pathigoolla, A. and Sureshan, K.M. (2013) *Angew. Chem. Int. Ed.*, **52**, 8671–8675.
89. Atchutarao, P., Gonnade, R.G., and Sureshan, K.M. (2012) *Angew. Chem. Int. Ed.*, **51**, 4362–4366.
90. Pathigoolla, A. and Sureshan, K.M. (2014) *Angew. Chem. Int. Ed.*, **53**, 9522–9525.

# 8
# Thiol-Based “Click” Chemistry for Macromolecular Architecture Design

*Weidong Zhang, Kui Chen, and Gaojian Chen*

## 8.1
## Introduction

“Click chemistry” introduced by Sharpless [1] and Hartmuth and coworkers [2] has been considered as an extraordinarily powerful tool for the preparation of new materials due to the mild reaction conditions, regioselectivity, high reaction rate, functionality tolerance, and so on. With the development of the most popular Cu(I)-mediated azide–alkyne click reactions [3], the thiol-based “click” reaction has recently attracted researchers due to its facile and versatile metal-free process that possesses the basic characteristics of a click reaction. It should be noted that some of the thiol-based reactions do not strictly follow the click philosophy. However, most of the reactions can be carried out in reasonable reaction times with high yields and chromatographic separation is not required for purification. In this chapter, the thiol-based “click” reaction is used to denote the thiol–ene [4], thiol–yne [5], thiol–isocyanate [6], thiol–epoxy [7], thiol–bromo reactions [8], and so on (Figure 8.1) carried out in polymer synthesis. A variety of reaction types such as radical addition, nucleophilic addition, and nucleophilic substitution are involved in these reactions. Herein the focus is on the fabrication of different topological architectures by thiol-based “click” reactions, the reaction mechanism and the application in materials synthesis (photo-click hydrogels) [9] and modification will not be discussed in this chapter. Several excellent reviews have appeared detailing its mechanism [10] and the application [11]. There are generally two approaches in synthesizing target polymers: one is using the thiol–ene or thiol–yne functionalized small-molecule compound as monomer to fabricate the topological polymers, whose major drawback is that the molecular weight and molecular weight distribution of polymer obtained are hard to control. However, this strategy is a facile and convenient tool for the preparation of functionalized network/gel [12], especially for the self-healing materials [13]. An alternative strategy is the postpolymerization modification of reactive precursor (co)polymers for the construction of complex (macro)molecules. It is valuable to note that the great advantages of this approach are owed to the

*Click Reactions in Organic Synthesis*, First Edition. Edited by Srinivasan Chandrasekaran.

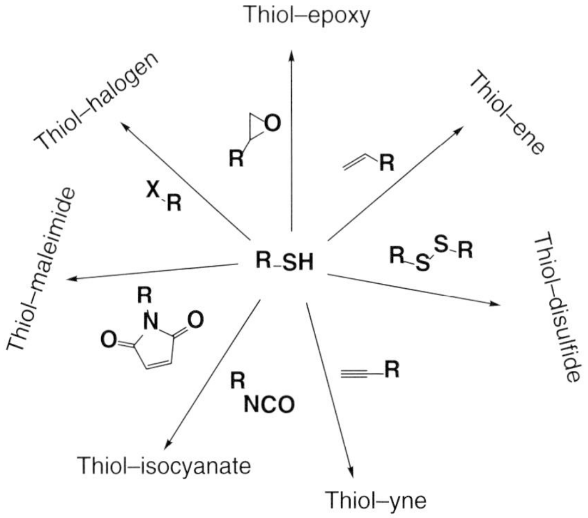


**Figure 8.1** Example of reactions carried out in thiol-based "click" for the fabrication of polymeric materials.

availability of different controlled polymerization (CLP) techniques, such as living radical polymerization (LRP) [14], which is tolerant to many functional groups, a large choice of polymerizable monomers, enabling the preparation of polymers with well-defined chain length, various architectures, and specific end-group functionalities. Indeed, the thiol-terminated polymers can be easily obtained by reversible addition–fragmentation chain-transfer (RAFT) polymerization [15] of a wide range of monomers and a subsequent cleavage of the chain-transfer agent. The combination of CLP techniques and thiol-based "click" chemistry provides an efficient platform for the preparation of polymers in various architectures.

## 8.2 Thiol Chemistry for Macromolecular Architecture Design

### 8.2.1 Linear Polymers

To synthesize linear polymers, a feasible strategy is to "click" one molecule with dithiol functionalities and another molecule with two thiol-reactive functionalities such as diene together as shown in Figure 8.2.

In a typical procedure, the thiol–epoxy "click" reaction near quantitative yields can be employed as a polymerization reaction for preparing linear polymer chains substituted with free hydroxyl groups. Khan *et al.* [16] reported that the linear polymer can be easily constructed by step-growth polymerization of the commercially available diepoxide (AA-type) and bis-sulfur (BB-type) monomers using lithium hydroxide (base catalysts) as the polymerization catalyst. (Figure 8.3)

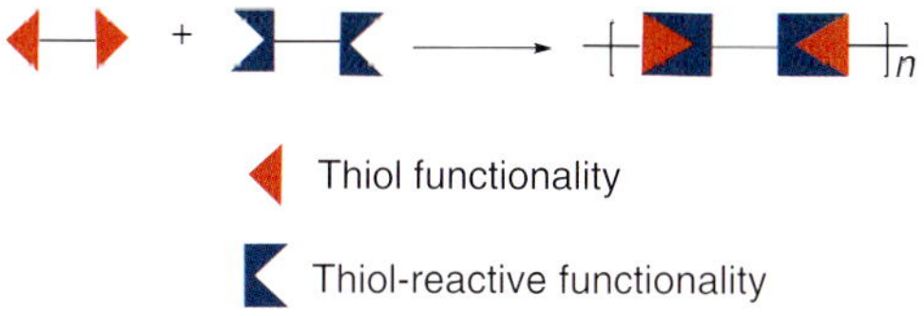


**Figure 8.2** Synthesis of linear polymers by polymerizing dithiol and molecules with two thiol-reactive functionalities.

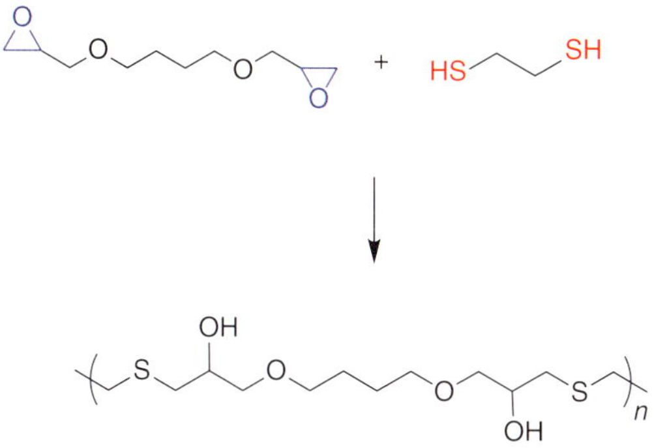

**Figure 8.3** Synthesis of linear polymers by thiol–epoxy "click" polymerization.

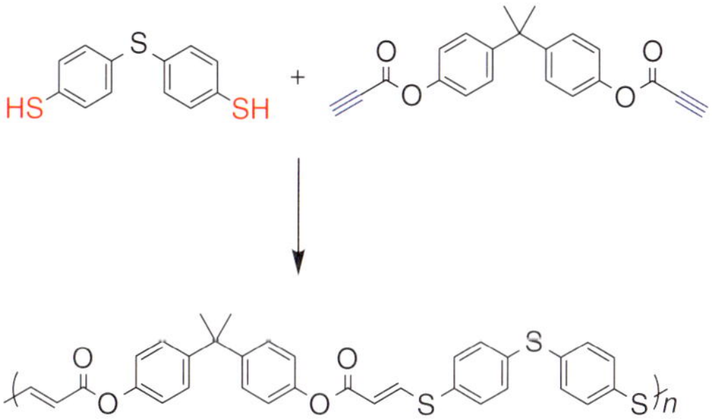

**Figure 8.4** Synthesis of linear polymers by polyhydrothiolation of diynes with dithiols.

Here, it should be pointed out that the thiol–epoxy "click" reaction is a feasibility implemented in many important biosynthetic and biomedical applications because the base-catalyzed reactions can be easily carried out in water and in solvent-free conditions with high yield.

Recently, Tang and coworkers also succeeded in developing the polyhydrothiolation reaction using Diynes (AA-type) with Dithiols (BB-type) (Figure 8.4) [17]. Moreover, the alkyne hydrothiolation reaction proceeded in a linear manner, and no branched isomer was obtained at all. More importantly, they found that both regioselectivity and stereoselectivity of the polyhydrothiolation reaction can be well controlled in the presence of the Rh catalyst ($Rh(PPh_3)_3Cl$) [17].

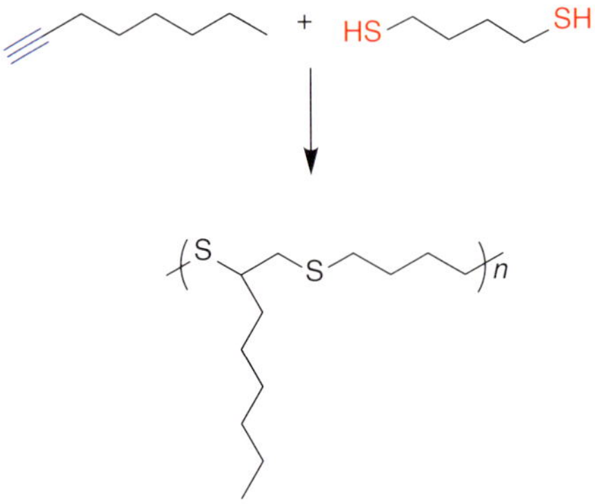


**Figure 8.5** Synthesis of linear polymers by polymerization of monoalkyne and a dithiol compound.

Another strategy for the synthesis of linear polymers has been proposed by Meier and coworkers via rapid step-growth radical-mediated thiol–yne polymerization of a monoalkyne (A-type) and a dithiol compound (BB-type) (Figure 8.5) [18]. It is confirmed that the rate of the polymerization strongly depends on the reactivity order of alkynes as reported by Du Prez and coworkers [19]: 1-octyne > propargyl acetate > methyl propargyl ether > 2-octyne > ethyl propiolate, methyl propargyl amine. It is important to point out that this strategy can be employed to design the characteristics of main chain as well as the density of the grafted functional groups.

### 8.2.2 Graft and Comb Polymers

Generally, the comb or graft polymers can be successfully prepared by the direct polymerization of the linear functionalized monomer or macromonomer. However, the monomer is hard to synthesize and the polymerization is relatively difficult to be carried out in a controlled manner. In contrast to the traditional strategy (monomer-first), the thiol-based click reaction has been widely used for the preparation of graft and comb polymers by postmodification strategy, which enables a modular approach for introducing diverse functionalities onto the polymer chain and tuning of the targeted chemical and physical properties. Typically, as illustrated in Figure 8.6, the functionalized graft or comb polymers could be easily fabricated by introducing thiol functional groups onto the reactive backbone.

A good example of using the modular approach is the construction of glycopolymers from alkene backbone-functional polymers via metal-free thiol–ene click reaction as reported by Stenzel and coworkers (Figure 8.7) [20]. This biologically friendly nature of thiol–ene click reaction is a powerful synthetic tool for preparing the biofunctional polymer and investigating bioprocesses. Another example can be found in the work of McCormick in which an attractive strategy

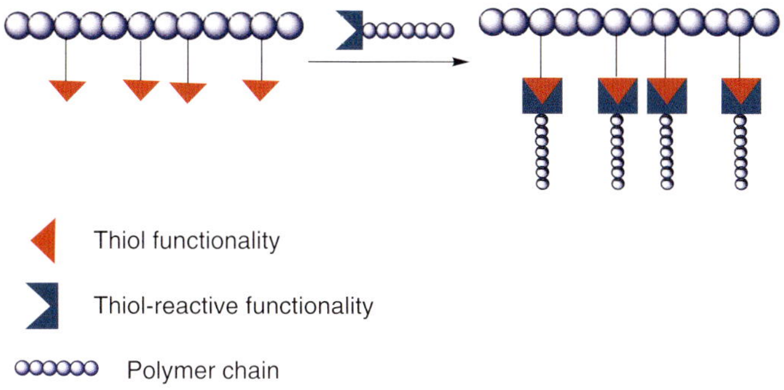


**Figure 8.6** Synthesis of the graft or comb polymers by thiol-based click reactions.

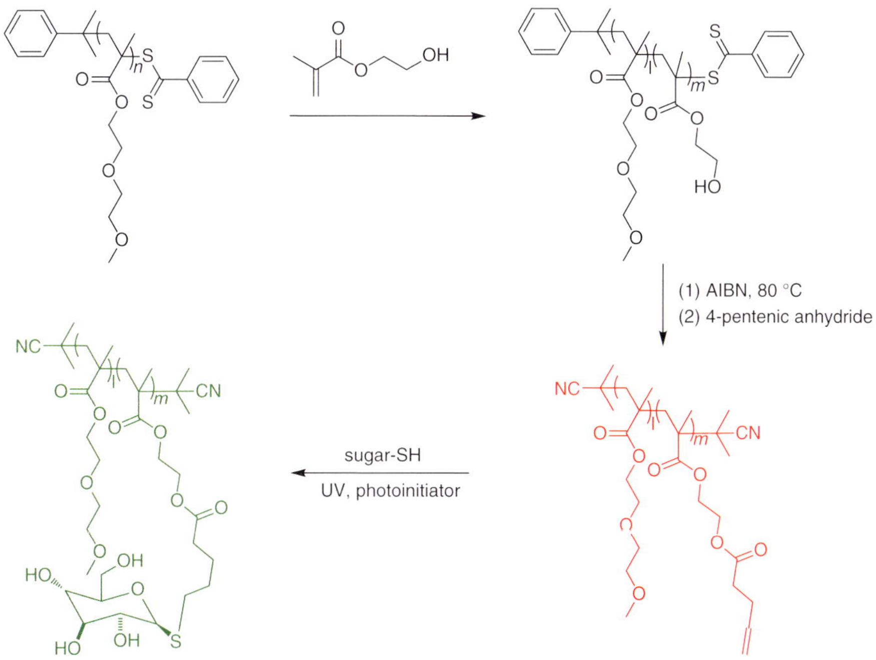


**Figure 8.7** Synthetic strategies for the preparation of glucose-functionalized polymers [20].

utilizing RAFT polymerization, isocyanate–hydroxyl coupling, and thiol–ene addition were used to design well-defined functional block copolymers from a single copolymer scaffold [21].

In addition to the alkene-functionalized polymer approach of making side-chain functionalized polymer, other approaches have also been reported. For example, based on the protection–deprotection chemistry of ortho-nitrobenzyl moiety, Barner-Kowollik and coworkers investigated light-triggered release of thiols at ambient temperature and developed a novel method to modify functionalization

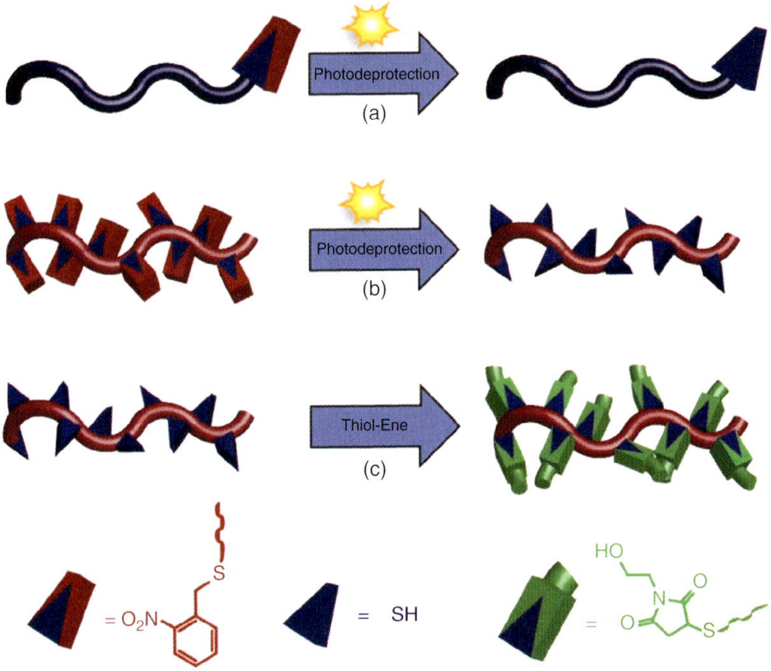


**Figure 8.8** Synthesis of graft polymers by the photo-triggered deprotection of the 2-nitrobenzyl thioether moiety on a polymer backbone, followed by the highly efficient thiol–maleimide chemistry. (Reprinted from Pauloehrl *et al.* [22], with permission of Royal Society of Chemistry.)

of polymeric backbones by *in situ* phototriggered deprotection and thiol–ene chemistry (Figure 8.8) [22], where the photodeprotection of the 2-nitrobenzyl thioether moiety with UV light and the photoinduced free-radical thiol–ene coupling reaction were carried out in one-pot. It is worth mentioning that the reaction is readily controlled by manipulating the light intensity, the dose or the duration of the exposure, when photoinduced thiol-based click reaction is used.

The development of photocatalytic systems for the thiol-based "click" reactions, such as photoredox catalyst $Ru(bpz)_3(PF6)_2$ [23] and semiconducting titania ($TiO_2$) [24], leads to unprecedented opportunities in materials design. Boyer *et al.* developed a visible-light photocatalytic thiol–ene reaction to prepare the linear polymers by step-growth addition polymerization of dithiol with diene compounds using $Ru(bpy)_3Cl_2$ as photoredox catalyst and *p*-toluidine as redox mediator (Figure 8.9) [25]. Moreover, this thiol–ene reaction could also be switched "ON" and "OFF." More importantly, this highly efficient reaction can be performed in the presence of air, water, and visible light. This approach could be successfully employed to synthesize comb polymers by postfunctionalization of poly(ethyl methacrylate-*r*-allyl methacrylate), synthesized via RAFT polymerization.

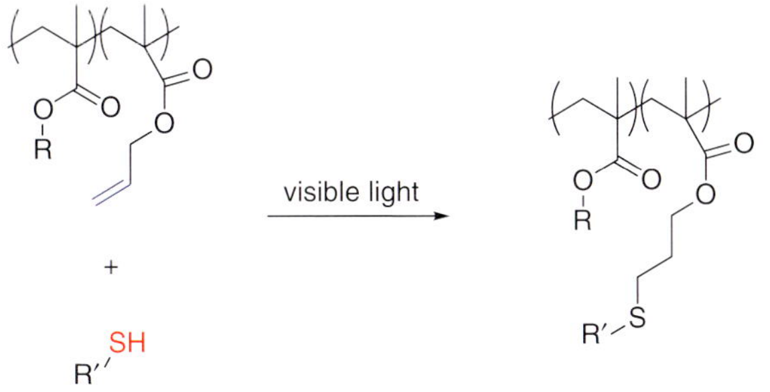


**Figure 8.9** Visible light photocatalytic thiol–ene reaction [25].

### 8.2.3 Star Polymers

Star polymers have gained great attention over the past decades due to their unique three-dimensional shape and properties. Compared with linear polymers, star polymers have many advantages because of their compact structures. For example, star polymers have lower solution and melt viscosities than linear polymers at the same molecular weights. Until now, star polymers can be prepared by two main methods, that is, "core-first" and "arm-first." Compared with the "core-first" technique, the "arm-first" technique was much easier to employ in the preparation of the star polymers. As presented in Figure 8.10, the thiol-based click reaction is a good strategy to fabricate multiarm star polymers. Moreover, the unique advantage of this strategy is that the thiol-terminated polymers can be easily prepared by transforming end-functionalization of polymers formed via RAFT polymerizations. As is well known, RAFT agent can be easily reduced into thiol-functionalized polymer in the presence of RAFT-reducing agents.

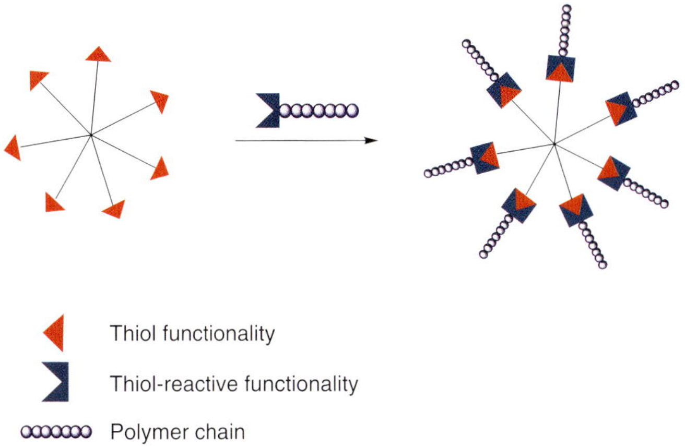


**Figure 8.10** The "arm-first" technique used to fabricate star polymers via thiol-based "click" reaction.

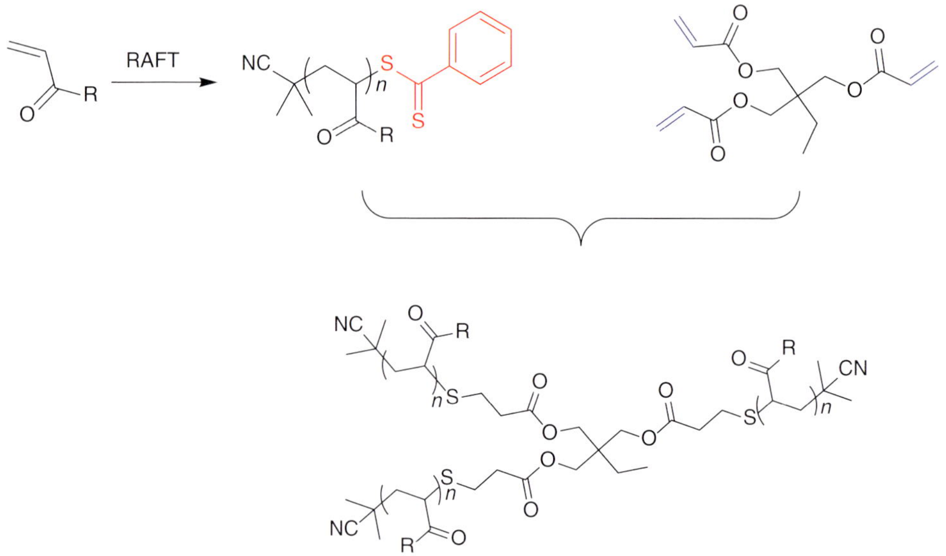


**Figure 8.11** Synthesis of three-arm star polymers via a thiol–vinyl Michael "click" reaction.

Once the thiol-functional polymer is obtained, it can be easily "clicked" to a multiarm core-bearing thiol reactive groups via the thiol-based "click" reaction. In 2008, Lowe *et al.* demonstrated this synthetic pathway using the nucleophilic, phosphine-catalyzed thiol–Michael addition reaction to successfully synthesize three-arm star polymer with narrow polydispersity [26]. (Figure 8.11) It is worth noting that the dimethylphenylphosphine (DMPP) is employed to effectively suppress the presence of any polymeric disulfide. Lowe and coworker further developed another strategy for one-pot preparation of the star polymer by the arm-first technique. Firstly, RAFT polymerization of *N*,*N*-diethylacrylamide using 1-cyano-1-methylethyl dithiobenzoate as RAFT agent was carried out, and then reduction and thiol–ene click reaction in the presence of hexylamine, radical initiator, and core were used to fabricate three-arm star polymer via one-pot. The NMR spectroscopy, GPC, and MALDI-TOF MS were used to confirm the reaction [26]. This approach provides a synthetic pathway where the well-defined polymer architectures can be successfully fabricated via combination of RAFT and thiol-based "click" reactions.

Cyclodextrin (CD) is a natural molecule made up of multisugar units in a ring. It is widely used in host–guest chemistry and has attracted great interest for various applications. Haddleton and coworker firstly synthesized the thiol-functionalized β-CD, and then β-CD was modified via base-catalyzed thiol–ene Michael addition using vinyl-terminated polymer precursors (Figure 8.12). As the hydroxy groups on the secondary face of CD remained unreacted, ring-opening polymerization (ROP) of ϵ-CL was then conducted to design the miktoarm star polymers [27].

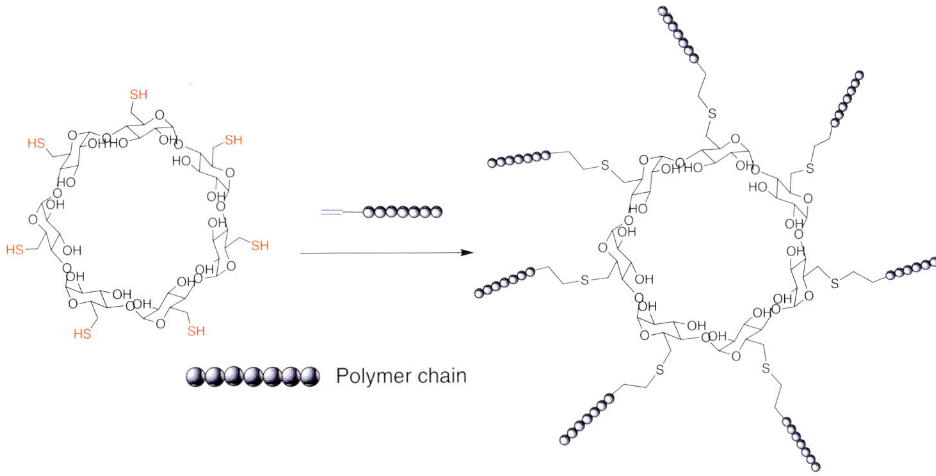


**Figure 8.12** Synthesis of cyclodextrin-centered star polymers via thiol-based "click" reaction.

### 8.2.4 Cyclic Polymers

It is well known that cyclic polymers possess higher density, lower intrinsic viscosity, and higher glass transition temperature because of the absence of chain ends compared to their linear counterparts of the same molecular weight [28]. Despite the fact that the cyclic polymers exhibit different physical properties under dilute and bulk conditions, the synthesis of the cyclic polymers is still relatively difficult. The two main types of cyclization synthetic methodologies are as follows: (i) ring-closure reactions and (ii) ring-expansion polymerizations (REPs) [29]. Compared with the REPs, to avoid the competing step-growth side reaction, the cyclization should be efficiently performed at lower concentrations. With the discovery and development of the "click" reactions in combination with LRP, the synthesis of cyclic polymers has again received ever-increasing interest of researchers. The extremely efficient click reactions have been widely applied to cyclize the precursors and to use these precursors as blocks to build more complex architectures [30]. Compared with the copper-catalyzed azide/alkyne cycloaddition (CuAAC) "click" reaction, the "thiol–ene" reactions have been well studied to provide alternative strategies for polymer cyclization without the need for toxic copper catalysts (Figure 8.13). For example, Dove and coworkers developed the A–A + B–B approach to synthesize and functionalize cyclic poly(lactide)s using the thiol–ene click chemistry (Figure 8.14), where the α,ω-maleimido-functional stereoregular poly(lactide)s could be successfully cyclized using bifunctional thiols. Moreover, the mild nature of these reactions has no significant effect on the stereoregularity [31].

Winnik *et al.* demonstrated that the α,ω-functional linear precursor can be successfully prepared via combination of RAFT polymerization using

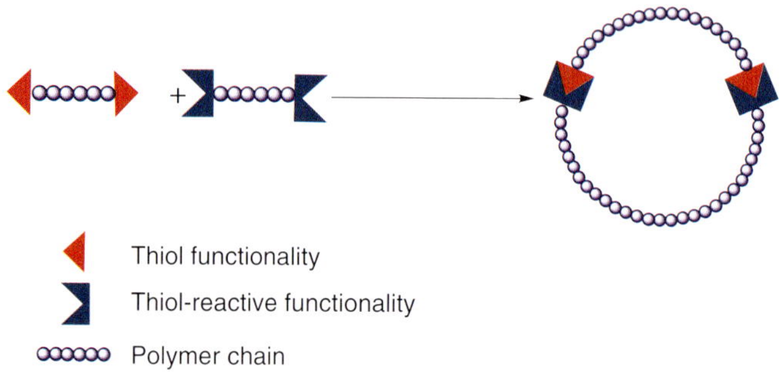


**Figure 8.13** Cyclization of linear polymers by thiol-based click reaction to form cyclic polymer.

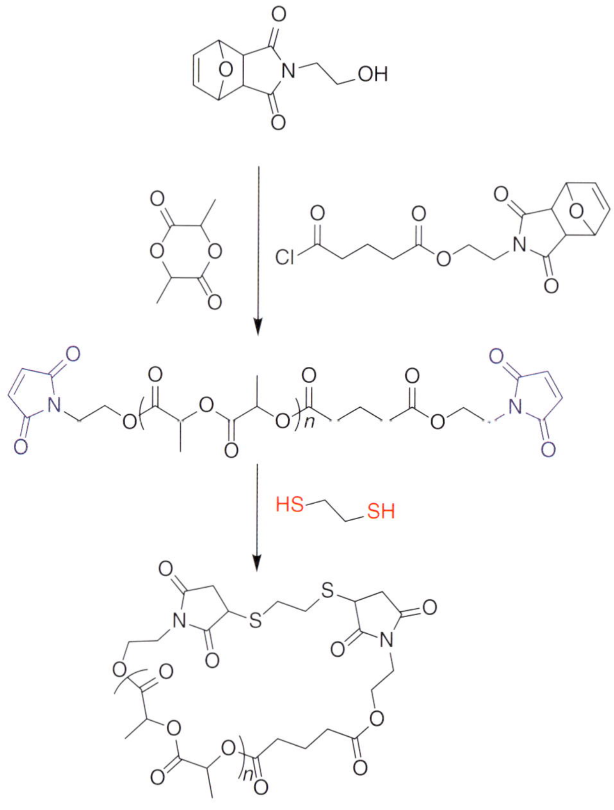

**Figure 8.14** Synthesis of cyclic poly(lactide) by thiol-based click reaction.

azide-containing trithiocarbonate as RAFT agent and thiol-based click reaction. Firstly, the homopolymers of poly(*N*-isopropylacrylamide) (PNIPAM) can be converted to the propargyl-terminated linear precursors by one-pot aminolysis/Michael addition reaction in the presence of the *n*-butylamine and propargyl

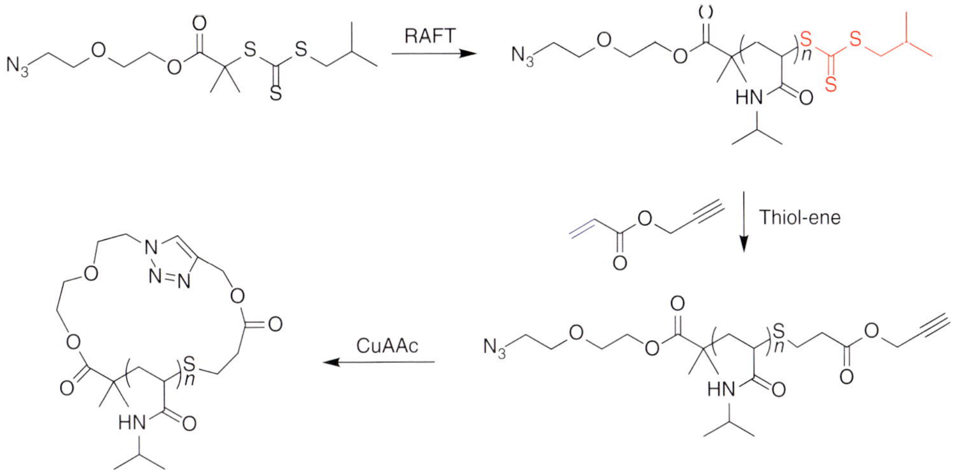


**Figure 8.15** Synthesis of cyclic PNIPAM via thiol–ene in combination with CuAAc chemistry [32].

acrylate. Then, the linear precursor was subsequently employed to synthesize temperature-responsive cyclic PNIPAM via the copper(I)-catalyzed alkyne–azide reaction [32]. (Figure 8.15) This strategy of combining LRP, thiol-based click reaction, and copper(I)-catalyzed alkyne–azide click reaction provides a general route to prepare various well-defined cyclic polymers.

Similarly, Wang and coworkers developed a simple and versatile route to synthesizing functional cyclic polyesters with two consecutive click reactions, including the cyclization of the linear precursor with CuAAC and side-chain functionalization with thiol–ene coupling reaction [33] (Figure 8.16). Note that the cyclic polymers can lead to a significantly longer elimination time in blood circulation and a greater tumor accumulation than their linear polymer counterparts [34]. It is believed that cyclic materials prepared by thiol-based click reaction would display inherent advantages in tumor therapy.

## 8.2.5 Dendritic and Hyperbranched Polymers

Physical properties of highly branched polymers are different from their linear analogs and have many potential applications in medicine and nanoengineering. Dendritic polymers with monodispersed composition and well-defined tree-like structures are even attractive. Efficient reactions are required, and thiol–ene was used by Hawker and coworkers to prepare dendrimers starting from a tris-alkene core 2,4,6-triallyloxy-1,3,5-triazine. 1-Thioglycerol was used to introduce two hydroxy groups via UV-irradiated thiol–ene reaction, which was carried out at room temperature for 30 min. Due to the efficient nature of thiol–ene chemistry and less byproducts, hydroxy-terminated dendrimers

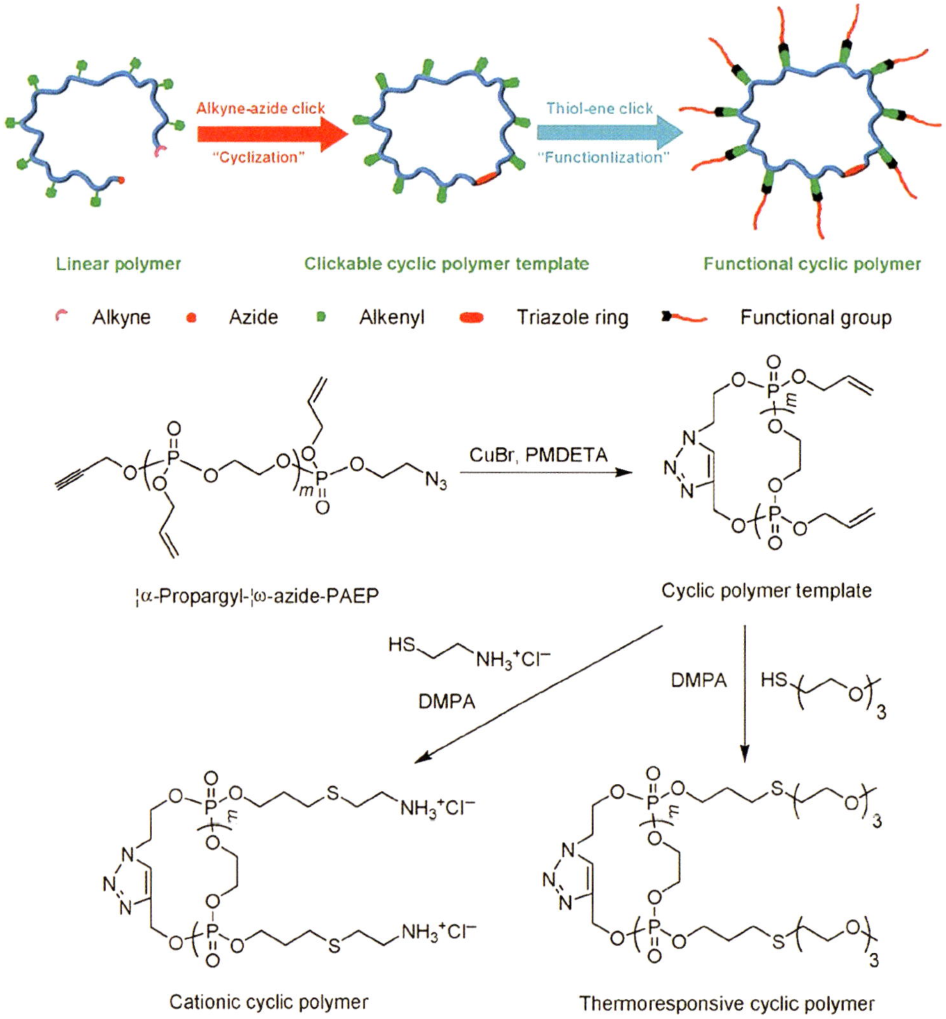


**Figure 8.16** Schematic illustration of the syntheses of cyclic polymer template, functionalized cationic and thermoresponsive cyclic polymers. (Reproduced from Yuan *et al.* [33], with permission of Royal Society of Chemistry.)

can be purified by simple precipitation into diethyl ether [35] (Figure 8.17). By combing thiol–ene and Grignard reactions with tetravinylsilane as a core, Rissing and Son synthesized carbosilane–thioether dendrimers up to the fifth generation [36].

Akin to thiol–ene reaction, thiol–yne was used by Stenzel and coworkers for the synthesis of dendrimers, where thiol–yne reaction was fast (10 min UV irradiation) and proceeded by reacting 2 equiv. of thiol with alkyne; therefore, a limited number of steps were needed to produce a high number of terminal functional

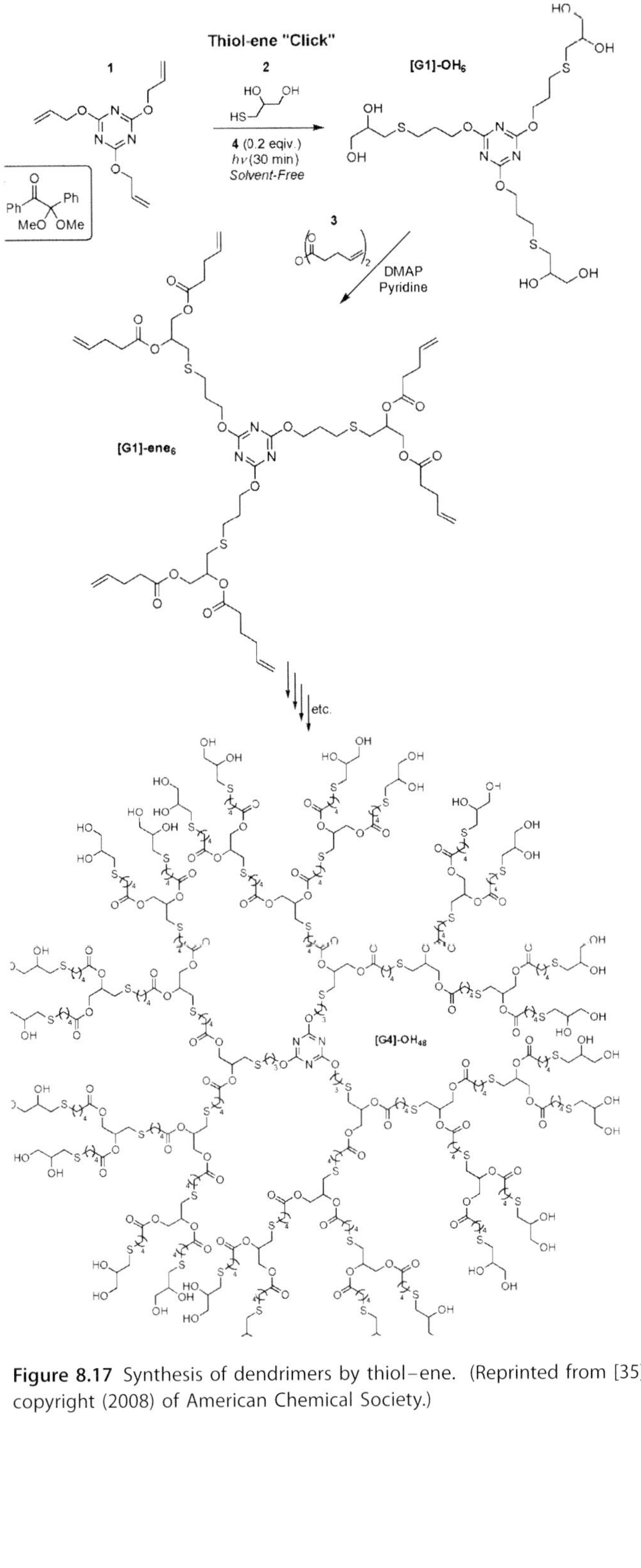


**Figure 8.17** Synthesis of dendrimers by thiol–ene. (Reprinted from [35], with permission copyright (2008) of American Chemical Society.)

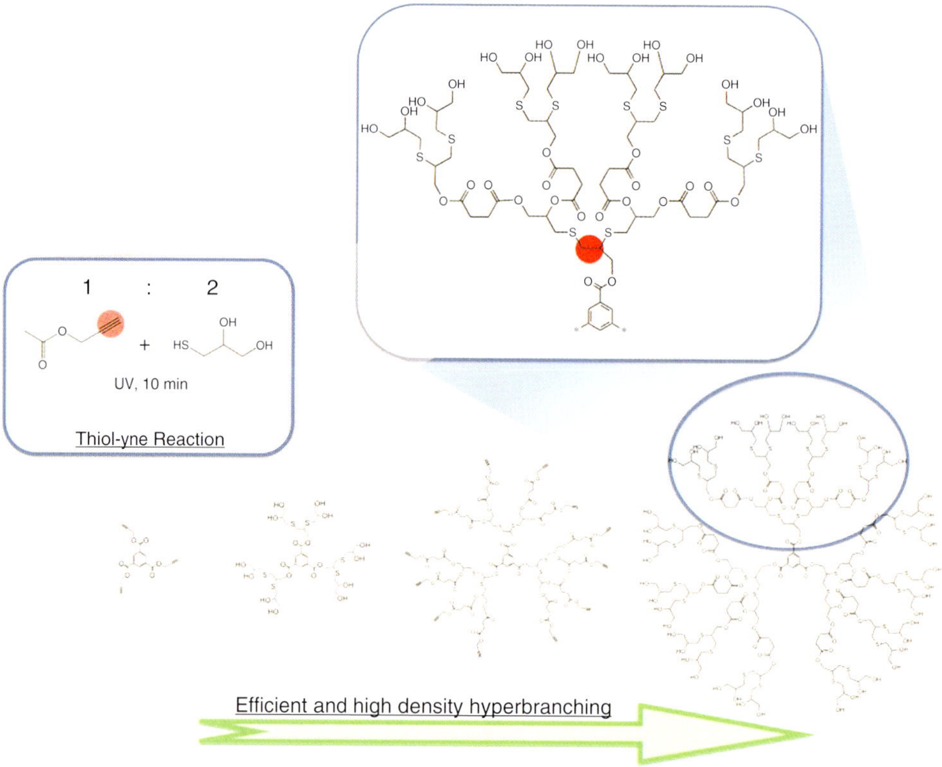


**Figure 8.18** Synthesis of dendrimers via thiol–yne chemistry and esterification reactions. (Reproduced from Chen *et al.* [37], with permission of Royal Society of Chemistry.)

groups. As shown in Figure 8.18, a dendrimer with 48 end-group functionalities was prepared in three steps [37].

Thiol–yne is widely used to prepare hyperbranched polymers (HPs), which are attractive alternatives to dendrimers and are easily prepared with less reaction and purification steps. In Perrier's lab, hyperbranched polymers were easily prepared by using molecules that have a thiol and an alkyne at each of the two ends, which is an $AB_2$-type monomer where one alkyne can react with two thiols (Figure 8.19). The molecules used here can be small molecules, homopolymers, or block copolymers [38].

Using a molecule with two thiols at each end ($A_2$) and another molecule with alkene and alkyne at each end ($CB_2$), Gao and coworkers synthesized hyperbranched polymers via sequential thiol–ene and thiol–yne chemistry, as shown in Figure 8.20 [39].

Still using dithiols, and a molecule with halogen and alkyne at each end, they prepared HPs by a combination of thiol–halogen and thiol–yne chemistry (Figure 8.21) [40].

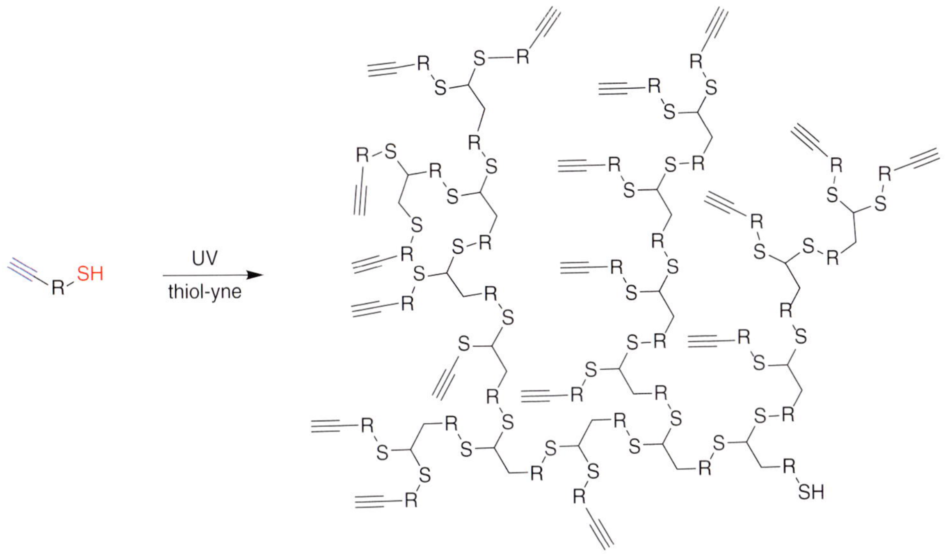


**Figure 8.19** Synthesis of hyperbranched polymer by polymerization of a molecule bearing an alkyne and a thiol.

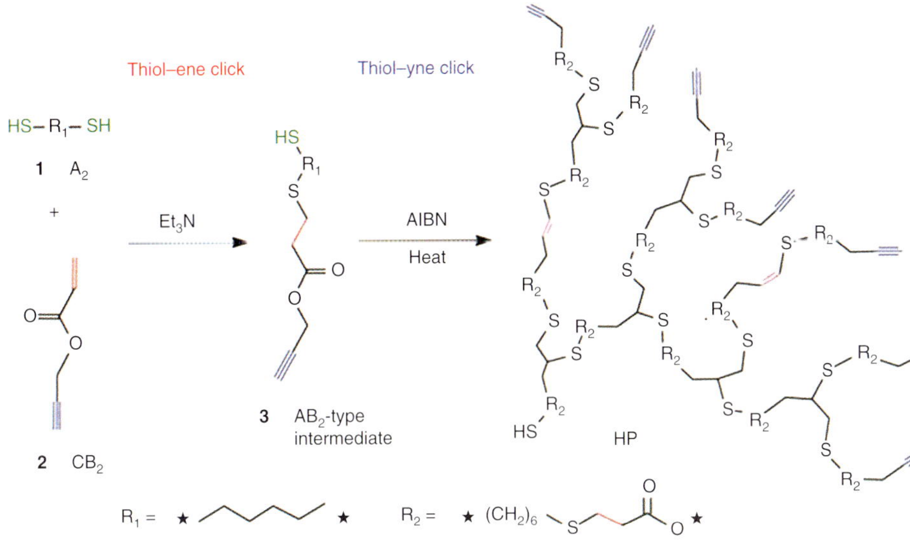


**Figure 8.20** Synthesis of hyperbranched polymer by sequential thiol–ene and thiol–ene click chemistry. (Reprinted from Han *et al.* [39], with permission of Royal Society of Chemistry.)

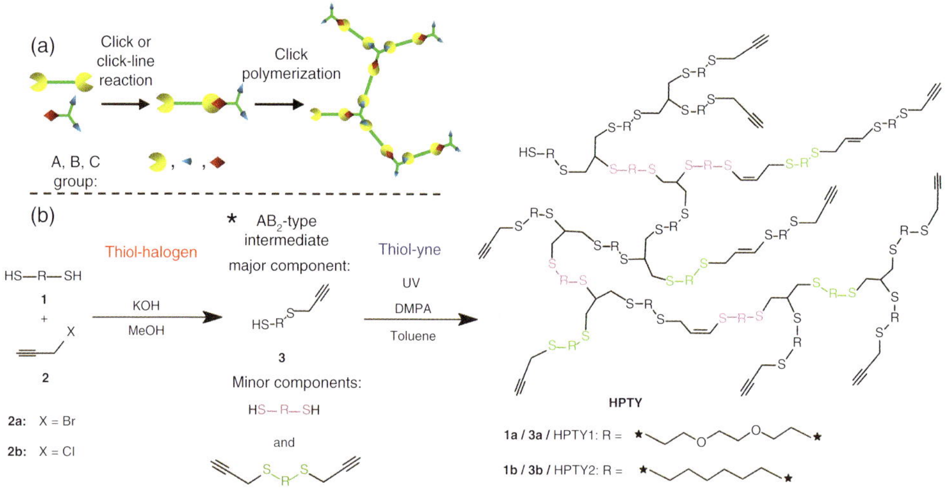


**Figure 8.21** Synthesis of hyperbranched polymer by sequential thiol–halogen and thiol–yne chemistry. (Reproduced from Han *et al.* [40], with permission of Royal Society of Chemistry.)

Xu *et al.* designed RAFT agents bearing two bromoesters in the R group fragment for the synthesis of polymers. The thiocarbonylthio end groups of RAFT polymers can be converted to thiols via aminolysis that then react intermolecularly with the bromoester terminal groups on other polymer chains to yield HPs via a thiol–bromo "click" chemistry (Figure 8.22). In addition, there are excess bromoester functionalities in the HPs obtained, and these bromoester groups could be further modified to introduce other functionalities via metal-catalyzed LRP [41].

### 8.2.6 Conjugated and Hybrid Polymers

In nature, sugar plays an important role in biological processes in the form of carbohydrate chain and also conjugates with protein and other biomolecules such as glycoprotein and peptidoglycan. To prepare the mimic of natural glycoconjugates and other functional conjugated materials, thiol-based "click" is a good choice due to its high efficiency and without the need of a metal catalyst. Typically, as shown in Figure 8.23, polymers are synthesized via LRP and further modified to possess vinyl groups, and protein with thiol functionality is then conjugated to the semitelechelic vinyl functional polymer via thiol–ene chemistry.

Maynard and coworkers synthesized Poly(ethylene glycol) methyl ether acrylate (PEGA) via RAFT polymerization and dithioester functional groups at the chain end, which were reduced via aminolysis and reacted with divinyl sulfone to afford semitelechelic vinyl sulfone polymers. Thiol–ene reaction was then used to prepare conjugates between the vinyl-terminated polymer and a

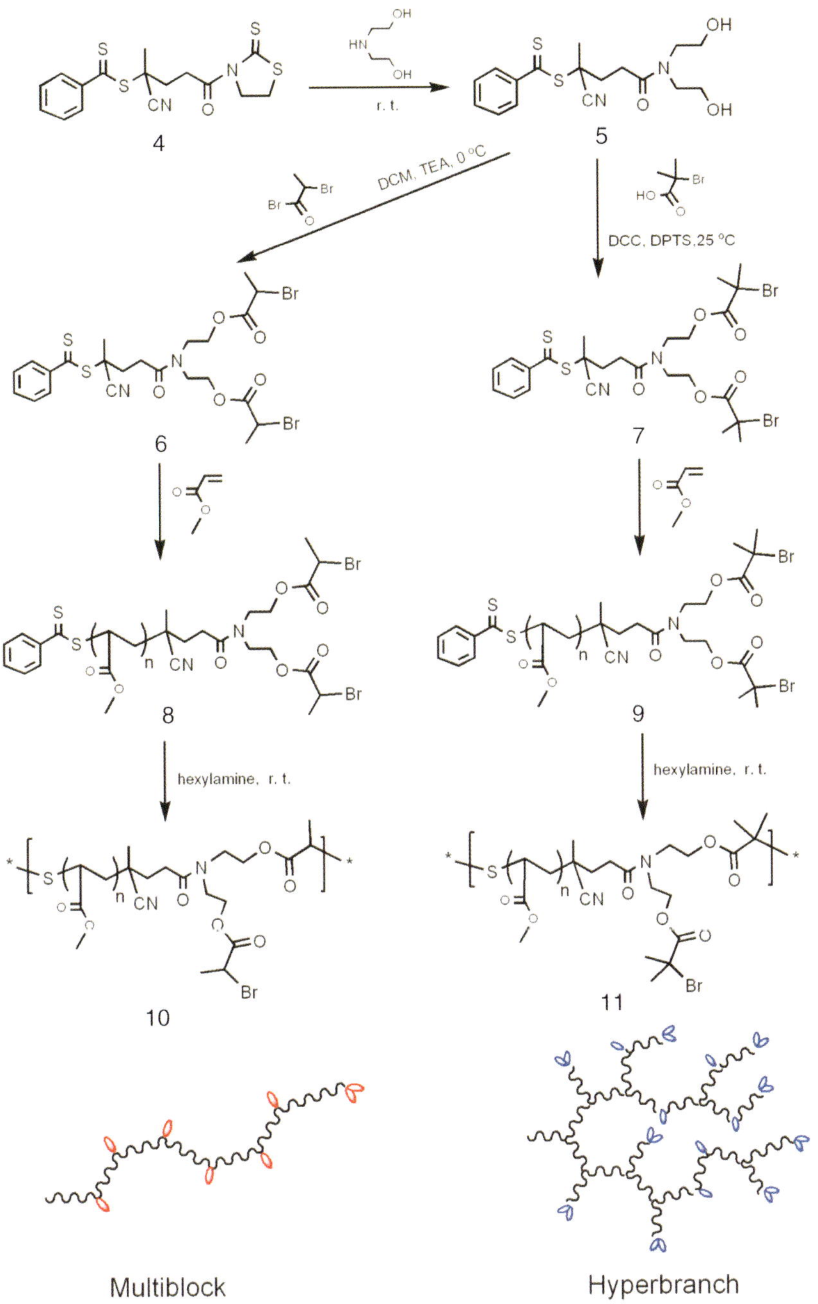


**Figure 8.22** One-pot preparation of multiblock and hyperbranched polymers. (Reproduced from Xu *et al.* [41], with permission of American Chemical Society.)

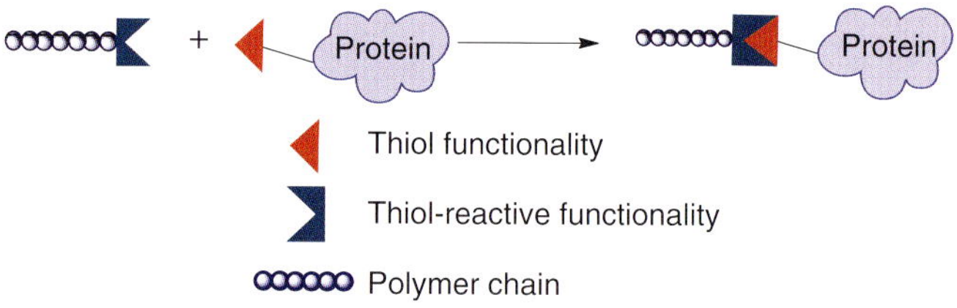


**Figure 8.23** Synthesis of protein–polymer conjugate via thiol–ene chemistry.

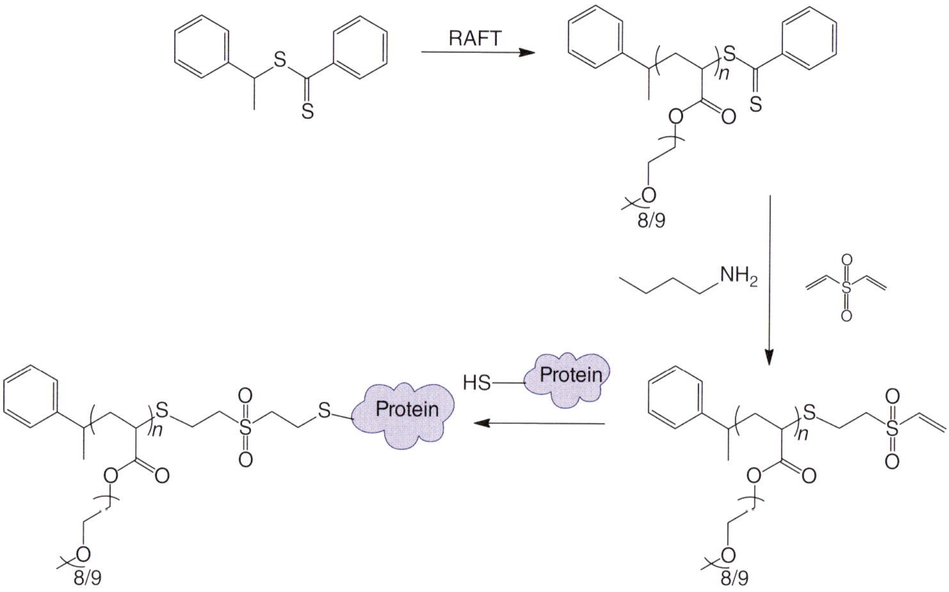


**Figure 8.24** Polymer synthesis and protein conjugation [42].

protein with a free cysteine (Figure 8.24). It should be noted that the reaction was performed with 20 equiv. of polymer to protein in mixed solvent (pH 9.0 phosphate buffer with 10% dimethyl sulfoxide), indicating a large excess of polymer is needed, which may be not ideal as "click" reaction. However, the reaction is relatively fast (2 h) and mild (at 23 °C) and the conjugate is stable without complication due to hydrolysis of chain ends [42]. In another example, a bismaleimide, 1,8-bis-maleimidodiethyleneglycol, was used to connect RAFT polymerized thermoresponsive PNIPAM and ovalbumin by two consecutive thiol–ene reactions [43]. Reaction was carried out in nitrogen-purged PB (pH 7.2) at room temperature for 24 h with about 8 equiv. of polymer to protein.

Similarly, RAFT polymer poly[*N*-(2-hydroxypropyl)methacrylamide], after being converted to semitelechelic thiol-terminated polymers, can be conjugated to maleimide-modified resin-bound peptide via thiol–maleimide reaction. The formed peptide–polymer conjugate is cleaved from the resin and the ability of the peptide to form β-sheet structures is investigated [44].

Not only conjugated with biomolecules, polymeric nanoparticles can be further modified to extend applications. For example, thiol–ene click chemistry was adopted to stabilize and functionalize polymer multilayers and capsules, in which PEG was introduced to the capsules via thiol–ene to afford protein-resistant particles or capsules that are reactive to protein with amine functionalities [45], microspheres prepared by suspension polymerization can be further modified with sugar moieties via thiol–ene chemistry for the preparation of microspheres with high bioactivity [46], and thiol–halogen coupling in combination with seeded-emulsion polymerization is used to prepare Janus particles that show region-selective binding of protein [47]. Vinyl benzyl chloride (VBC) was used as swelling monomer for the seed particles to introduce chloride groups at one side of the particle, and biomolecules with thiol functionality were then clicked onto the PVBC lobe of particle via thiol–chloride chemistry (Figure 8.25).

The hybrid inorganic/polymeric materials are attractive that they incorporate the properties of both. In this category, it is essential to find a molecule that can

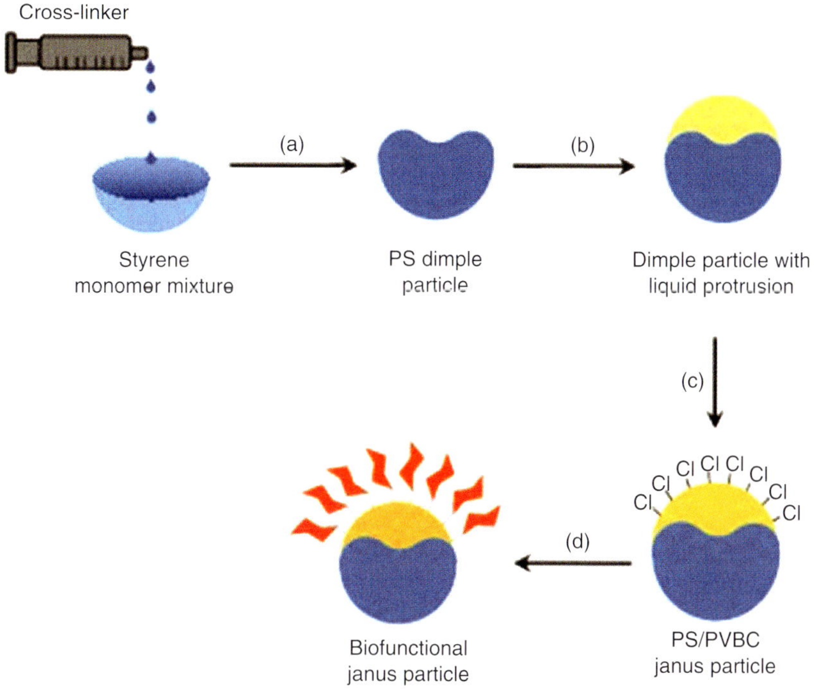


**Figure 8.25** Synthesis of bioactive Janus particles by SEP and thiol–halogen chemistry. (Reproduced from Li *et al.* [47], with permission of Wiley.)

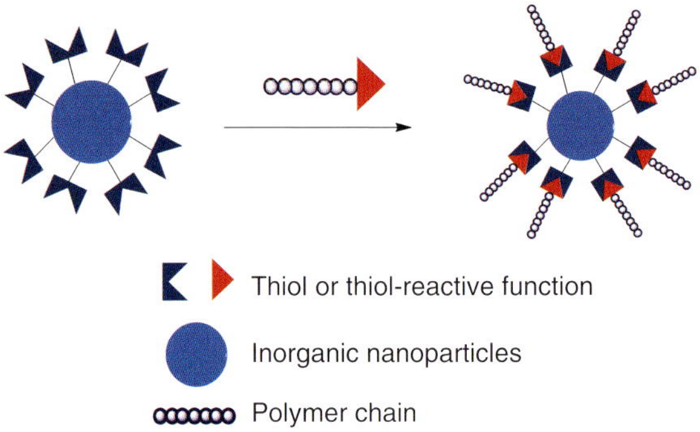


**Figure 8.26** Modified inorganic nanoparticles via thiol–ene chemistry.

effectively attach to the inorganic part and is ready to be clicked via thiol-based chemistry, as shown in Figure 8.26.

For example, polymerizing inorganic nanoparticles with cross-linker divinylbenzene yields composite polymer–inorganic latex particles functionalized with vinyl groups. In the process of composite miniemulsion polymerization, vinyl groups were easily introduced to the particle surface by using cross-linker divinylbenzene, thiol-terminated polymers can be further clicked on the particle surface via a thiol–ene approach catalyzed by V-50 as a free-radical source (Figure 8.27) [48].

Organosilane is widely used in surface modification, which can form stable condensation products with silanol groups on siliceous surface and other oxides

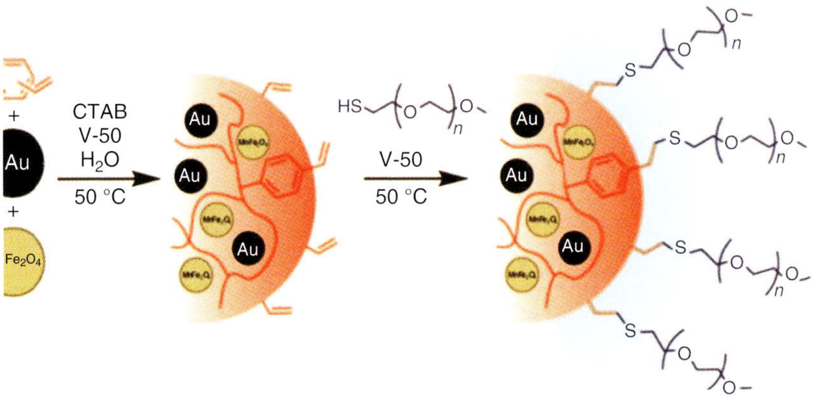


**Figure 8.27** Preparation of multimodal latex particles by composite miniemulsion polymerization, followed by attachment of PEG chains to the surface of composite particles using thiol–ene chemistry. (Reproduced from van Berkel *et al.* [48], with permission of American Chemical Society.)

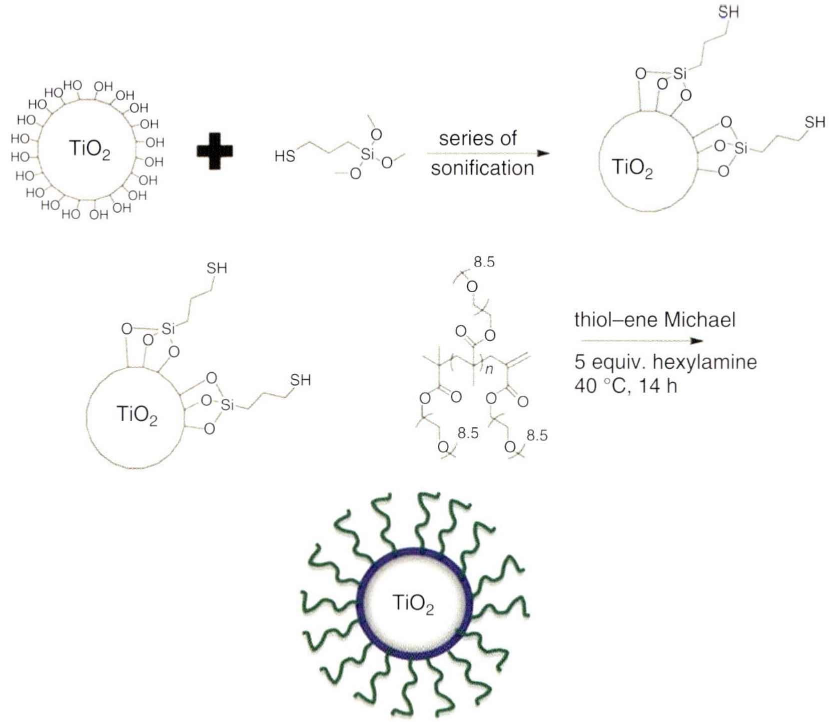


**Figure 8.28** Overall synthetic approach for the surface modification of $TiO_2$ nanoparticles with POEGMA. (Reproduced from Tedja *et al.* [50], with permission of Royal Society of Chemistry.)

of aluminum, tin, titanium, and so on [49]. Using (3-mercaptopropyl) trimethoxy silane, $TiO_2$ nanoparticles are thiolated and polymers with vinyl end groups made by the catalyst chain-transfer polymerization (CCTP) technique are clicked to the nanoparticles via thiol–ene Michael addition reactions, as shown in Figure 8.28 [50].

Inspired by the adhesive proteins secreted by marine mussels, dopamine has strong interactions with different surfaces from the catechol groups due to hydrogen bonding, pi-pi stacking, or bidentate bonding through the hydroxyl groups. Nonspherical iron oxide nanoparticles are firstly modified to incorporate desired vinyl groups by catecholic chemistry, and then well-defined glycopolymers synthesized by RAFT polymerization are conjugated with iron oxide nanoparticles via thiol–ene reaction [51]. Catecholic chemistry has been extensively used to create virtually all types of material surfaces, regardless of their chemical functionality or surface energy. The approach of combining thiol-based click and catecholic chemistry provides a versatile tool to prepare functional hybrid inorganic/organic nanoparticles (Figure 8.29).

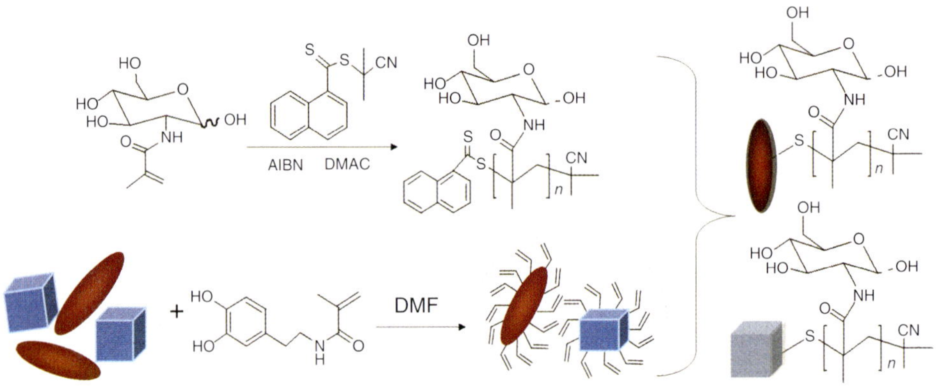


**Figure 8.29** Synthesis of glycopolymer-coated iron oxide nanoparticles. (Reproduced from Li *et al.* [51], with permission of Royal Society of Chemistry.)

## 8.3 Conclusion

This chapter summarizes the design and preparation of polymers with various architectures via thiol-based "click" chemistry (Table 8.1). With the efficient and metal-free nature of the reactions, in combination with LRP and other click techniques, variable polymer materials have been synthesized including polymers, conjugates, and inorganic/organic hybrids. As demonstrated by many examples, thiol-terminated polymer can be easily obtained, by RAFT polymerization of a wide range of monomers. This guarantees the efficient synthesis of various functional polymers via thiol-based click chemistry. Most of the reactions are carried out in room temperature and proceed in fast pace, especially those UV-initiated thiol–ene reactions. These polymers with diverse architectures guarantee their further applications. We believe more is coming with more complex architectures and more applications.

The versatility of thiol group allows thiol-based click reaction to be efficiently functionalized via many different pathways, including thermal, photochemical, and redox processes. Recent developments in visible-light catalysis have expanded the scope and application of polymer chemistry. The combination of visible-light and thiol-based chemistry has received more attention and is expected to develop quickly. New catalytic systems that are cheap and nontoxic such as titania ($TiO_2$) with visible light have been reported recently for thiol–ene reaction. In addition, light can be easily turned "on" and "off." It can be controlled both in the timescale and geographic scale, which provides potential for programmed and patterned designing of novel materials. Thiol group is available in biomolecules, which guarantee the wide application of thiol-based click chemistry in bio-related synthesis. In this regard, visible-light-assisted click is promising.

It should be noted here that for thiol-based click higher efficiency is still in demand. It is hard to couple polymers with thiol–ene reaction, compared to

**Table 8.1** Example of polymers synthesized via thiol-based click chemistry.

| Entry | Architecture | Building blocks | Reaction type | Reaction conditions | References |
|---|---|---|---|---|---|
| 1 | Linear | 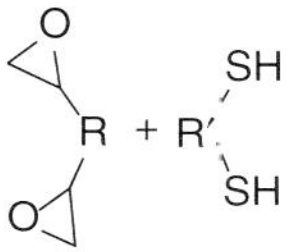 | Thiol–epoxy | LiOH, THF, 0 °C, 6 h | [16] |
| 2 | Linear | 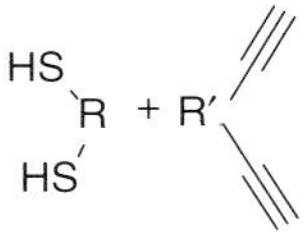 | Thiol–yne | $(Ph)_2NH$,DMF, 25 °C, 24 h | [17] |
| 3 | Linear | 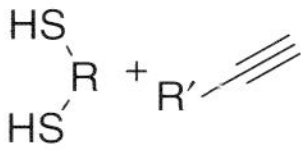 | Thiol–yne | UV-initiation, RT, 1–2 h | [18] |
| 4 | Graft | 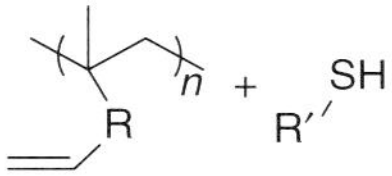 | Thiol–ene | UV-initiation, DMF, RT, 2 h | [20] |
| 5 | Graft | 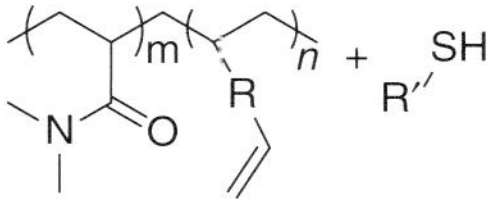 | Thiol–ene | TEA, DMSO, 20 °C, 12 h | [21] |

(continued overleaf)

**Table 8.1** (*Continued*).

| Entry | Architecture | Building blocks | Reaction type | Reaction conditions | References |
|---|---|---|---|---|---|
| 6 | Graft | 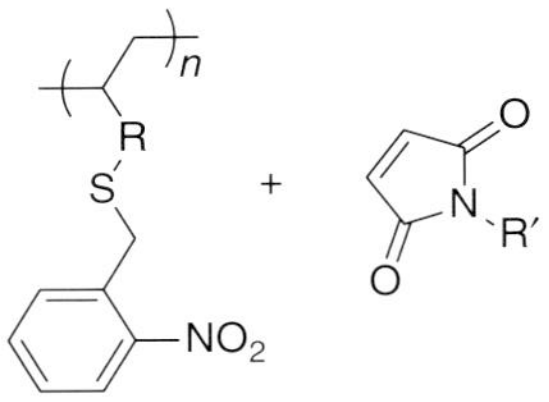 | One-pot photodeprotection and thiol–maleimide | UV-initiation, DCM, RT, 16 h | [22] |
| 7 | Graft | 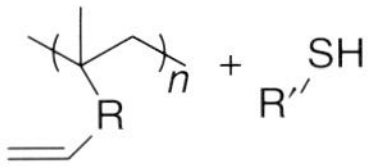 | Thiol–ene | Visible light, RT, 20 min | [25] |
| 8 | Star | 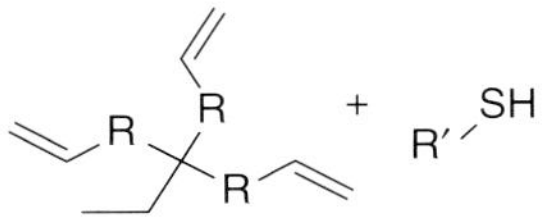 | Thiol–ene | RT, THF, DMPP, 12 h | [26] |
| 9 | Star | 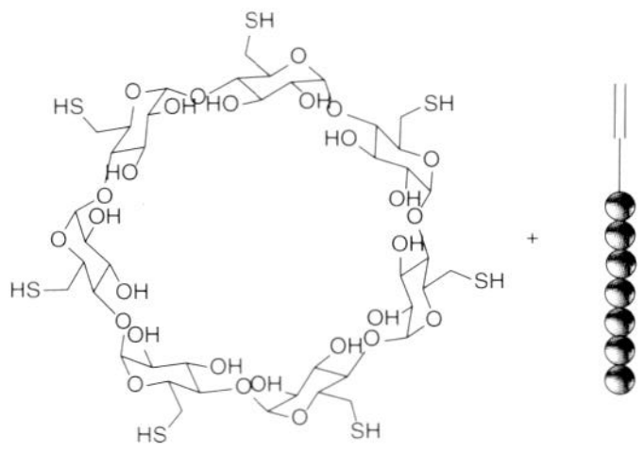 | Thiol–ene | RT, DMSO, DMPP, 24 h | [27] |

| | | | | | |
|---|---|---|---|---|---|
| 10 | Cyclic | 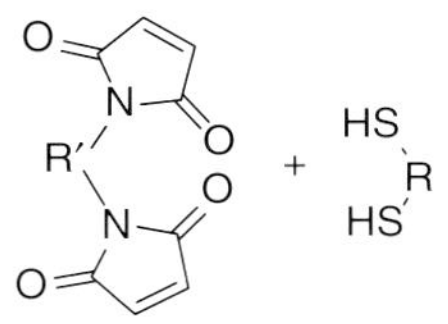 | Thiol–maleimide | $NEt_3$, $Na_2S_2O_5$, RT | [31] |
| 11 | Cyclic | 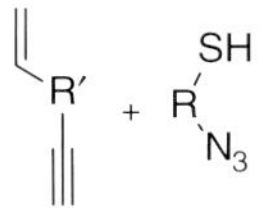 | Thiol–ene | *n*-butylamine, THF, RT | [32] |
| 12 | Dendritic | 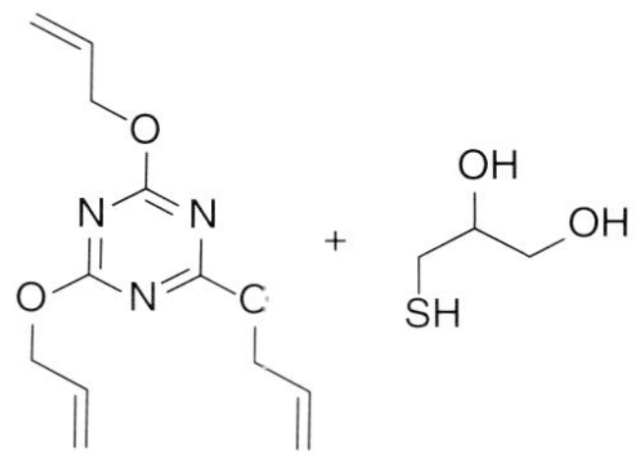 | Thiol–ene | UV-initiation, RT, 30 min | [35] |
| 13 | Dendritic | 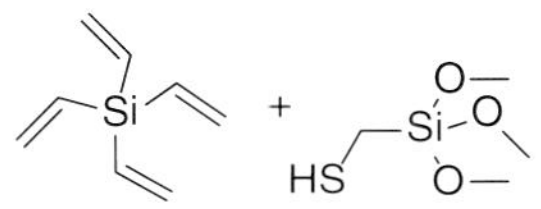 | Thiol–ene | UV-initiation, DMF, RT, 3 h | [36] |

*(continued overleaf)*

**Table 8.1** *(Continued).*

| Entry | Architecture | Building blocks | Reaction type | Reaction conditions | References |
|---|---|---|---|---|---|
| 14 | Dendritic | 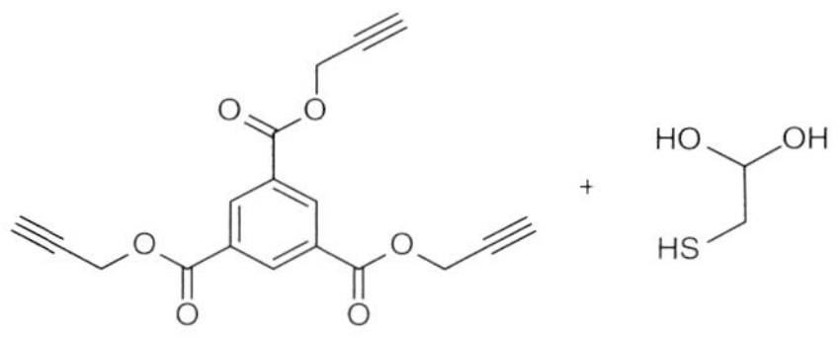 | Thiol–yne | UV-initiation, RT, 10 min | [37] |
| 15 | Hyperbranched | 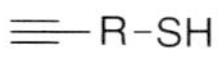 | Thiol–yne | UV-initiation, RT, DMF, DMPA | [38] |
| 16 | Hyperbranched | 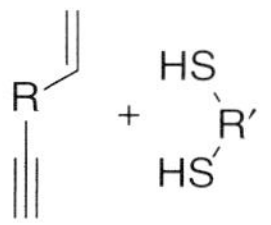 | Thiol–ene and thiol–yne | Thiol–ene: $NEt_3$, 35 °C, 6 h; thiol–yne AIBN, 65 °C, 12 h | [39] |
| 17 | Hyperbranched | 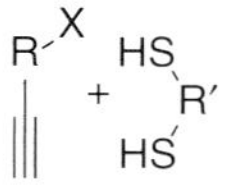 | Thiol–yne and thiol–halogen | UV-initiation, toluene, DMPA, 3 h | [40] |
| 18 | Hyperbranched | 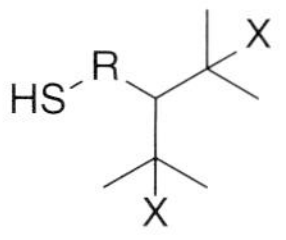 | Thiol–halogen | Hexylamine, RT, 30 min | [41] |

| | | | | | |
|---|---|---|---|---|---|
| 19 | Protein conjugate | 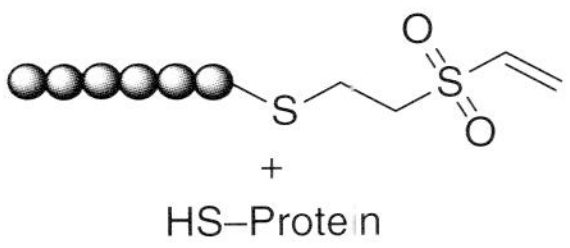 | Thiol–ene | TCEP, PB, 23 °C, 2 h | [42] |
| 20 | Protein conjugate | 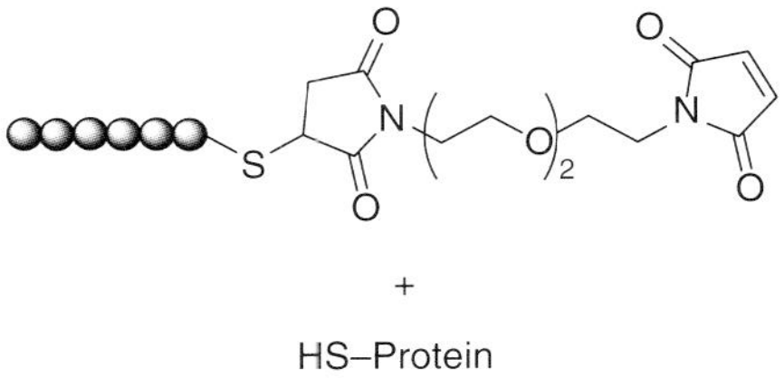 | Thiol–ene | PBS pH 7.2, RT, 24 h | [43] |
| 21 | Peptide conjugate | 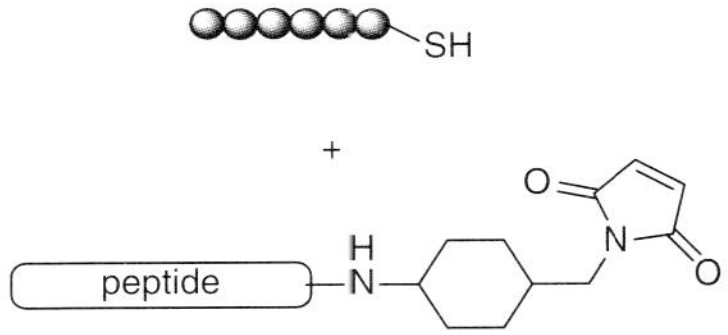 | Thiol–maleimide | DMF, RT, 24 h | [44] |
| 22 | Polymeric capsule | 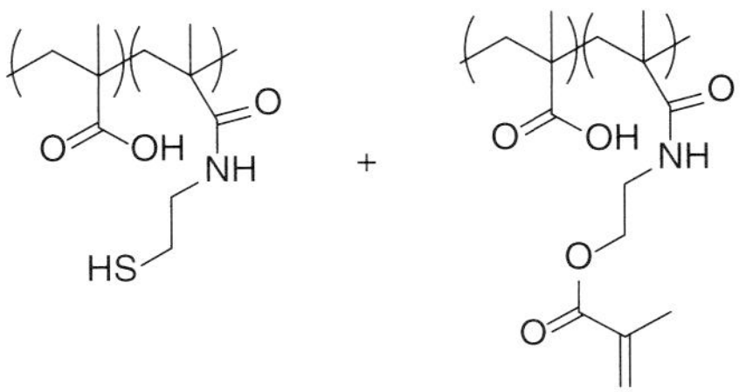 | Thiol–ene | PBS pH 4, UV-initiation RT,2 h | [45] |

(continued overleaf)

**Table 8.1** (*Continued*).

| Entry | Architecture | Building blocks | Reaction type | Reaction conditions | References |
|---|---|---|---|---|---|
| 23 | Polymeric particles | 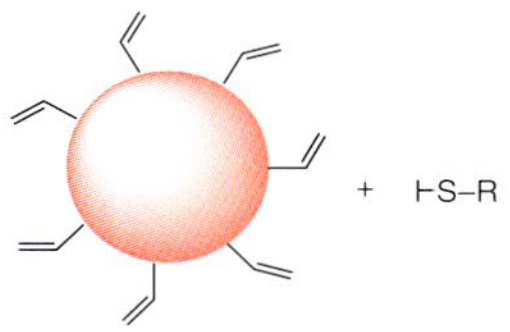 | Thiol–ene | DMSO,TCEP, RT, 72 h | [46] |
| 24 | Polymeric particles | 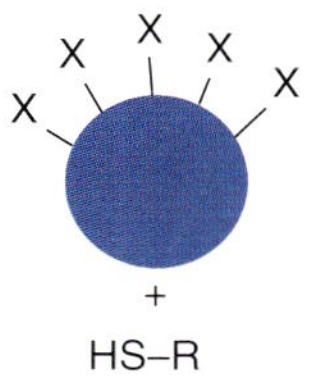 | Thiol–halogen | 40 °C, 110 h | [47] |
| 25 | Inorganic/organic hybrid | 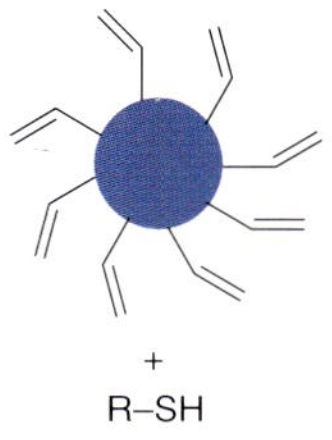 | Thiol–ene | 50 °C, V-50 | [48] |

| | | | | | |
|---|---|---|---|---|---|
| 26 | Inorganic/organic hybrid | 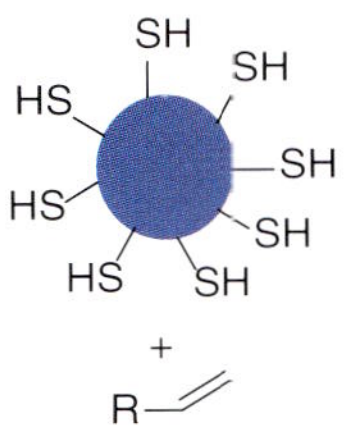 | Thiol–ene | Hexylamine, 40 °C, 14 h | [50] |
| 27 | Inorganic/organic hybrid | 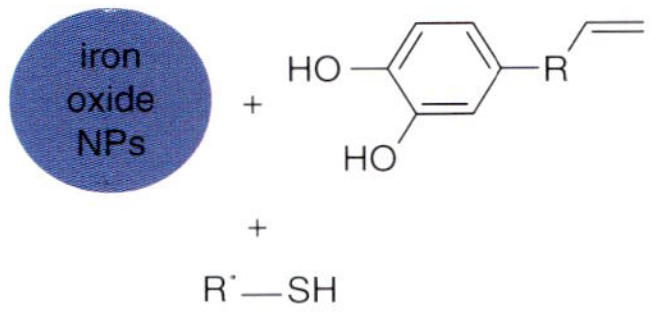 | Thiol–ene | Ultrasound, RT, DMF, 10 h | [51] |

CuAAc click. To increase yield, excess of thiols is normally needed. These drawbacks limit the application of thiol-based click in some areas. Overall, thiol-based click is versatile in polymer synthesis, and we believe they are promising in designing materials with bioapplications and beyond.

## Acknowledgments

We thank National Natural Science Foundation of China (No. 21374069, 21004042, 21104051) and Scientific Research Foundation for the Returned Overseas Chinese Scholars, State Education Ministry for financial support.

## References

1. Rostovtsev, V.V., Green, L.G., Fokin, V.V., and Sharpless, K.B. (2002) *Angew. Chem. Int. Ed.*, **41**, 2596–2599.
2. Hartmuth, M.G.F., Kolb, C., and Barry Sharpless, K. (2001) *Angew. Chem. Int. Ed.*, **40**, 2004–2021.
3. (a) Binder, W.H. and Sachsenhofer, R. (2007) *Macromol. Rapid Commun.*, **28**, 15–54; (b) Malkoch, M., Thibault, R.J., Drockenmuller, E., Messerschmidt, M., Voit, B., Russell, T.P., and Hawker, C.J. (2005) *J. Am. Chem. Soc.*, **127**, 14942–14949; (c) Helms, B., Mynar, J.L., Hawker, C.J., and Fréchet, J.M.J. (2004) *J Am. Chem. Soc.*, **126**, 15020–15021; (d) Sumerlin, B.S., Tsarevsky, N.V., Louche, G., Lee, R.Y., and Matyjaszewski, K. (2005) *Macromolecules*, **38**, 7540–7545; (e) Johnson, J.A., Finn, M.G., Koberstein, J.T., and Turro, N.J. (2007) *Macromolecules*, **40**, 3589–3598; (f) Zhang, W., Zhang, W., Zhang, Z., Zhu, J., and Zhu, X. (2010) *Macromol. Rapid Commun.*, **31**, 1354–1358.
4. (a) Lowe, A.B. (2010) *Polym. Chem.*, **1**, 17–36; (b) Dondoni, A. and Marra, A. (2012) *Chem. Soc. Rev.*, **41**, 573–586; (c) Scanlan, E.M., Corce, V., and Malone, A. (2014) *Molecules*, **19**, 19137–19151.
5. Lowe, A.B., Hoyle, C.E., and Bowman, C.N. (2010) *J. Mater. Chem.*, **20**, 4745–4750.
6. Hensarling, R.M., Rahane, S.B., LeBlanc, A.P., Sparks, B.J., White, E.M., Locklin, J., and Patton, D.L. (2011) *Polym. Chem.*, **2**, 88–90.
7. Jian, Y., He, Y., Sun, Y., Yang, H., Yang, W., and Nie, J. (2013) *J. Mater. Chem. C*, **1**, 4481–4489.
8. Zhang, Q., Anastasaki, A., Li, G.-Z., Haddleton, A.J., Wilson, P., and Haddleton, D.M. (2014) *Polym. Chem.*, **5**, 3876–3883.
9. Shih, H. and Lin, C.C. (2015) *Biomacromolecules*, **16**, 1915–1923.
10. Nair, D.P., Podgórski, M., Chatani, S., Gong, T., Xi, W., Fenoli, C.R., and Bowman, C.N. (2014) *Chem. Mater.*, **26**, 724–744.
11. Lowe, A.B. (2014) *Polymer*, **55**, 5517–5549.
12. Shih, H., Fraser, A.K., and Lin, C.C. (2013) *ACS Appl. Mater. Interfaces*, **5**, 1673–1680.
13. Hillewaere, X.K.D., Teixeira, R.F.A., Nguyen, L.-T.T., Ramos, J.A., Rahier, H., and Du Prez, F.E. (2014) *Adv. Funct. Mater.*, **24**, 5575-5583.
14. (a) Kamigaito, M., Ando, T., and Sawamoto, M. (2001) *Chem. Rev.*, **101**, 3689–3745; (b) Chiefari, J., (Bill) Chong, Y.K., Ercole, F., Krstina, J., Jeffery, J., Le, T.P.T., Mayadunne, R.T.A., Meijs, G.F., Moad, C.L., Moad, G., Rizzardo, E., and Thang, S.H. (1998) *Macromolecules*, **31**, 5559–5562; (c) Matyjaszewski, K. and Xia, J. (2001) *Chem. Rev.*, **101**, 2921–2990.
15. Roth, P.J., Boyer, C., Lowe, A.B., and Davis, T.P. (2011) *Macromol. Rapid Commun.*, **32**, 1123–1143.
16. Brändle, A. and Khan, A. (2012) *Polym. Chem.*, **3**, 3224–3227.

17. Liu, J., Lam, J.W.Y., and Tang, B.Z. (2009) *Chem. Rev.*, **109**, 5799–5867.
18. Türünç, O. and Meier, M.A.R. (2012) *J. Polym. Sci., Part A: Polym. Chem.*, **50**, 1689–1695.
19. Gokmen, M.T., Brassinne, J., Prasath, R.A., and Du Prez, F.E. (2011) *Chem. Commun.*, **47**, 4652–4654.
20. Chen, G., Amajjahe, S., and Stenzel, M.H. (2009) *Chem. Commun.*, 1198–1200.
21. Flores, J.D., Treat, N.J., York, A.W., and McCormick, C.L. (2011) *Polym. Chem.*, **2**, 1976.
22. Pauloehrl, T., Delaittre, G., Bastmeyer, M., and Barner-Kowollik, C. (2012) *Polym. Chem.*, **3**, 1740–1749.
23. (a) Tyson, E.L., Ament, M.S., and Yoon, T.P. (2013) *J. Org. Chem.*, **78**, 2046–2050; (b) Tyson, E.L., Niemeyer, Z.L., and Yoon, T.P. (2014) *J. Org. Chem.*, **79**, 1427–1436.
24. Bhat, V.T., Duspara, P.A., Seo, S., Abu Bakar, N.S., and Greaney, M.F. (2015) *Chem. Commun.*, **51**, 4383–4385.
25. Xu, J. and Boyer, C. (2015) *Macromolecules*, **48**, 520–529.
26. Chan, J.W., Yu, B., Hoyle, C.E., and Lowe, A.B. (2009) *Polymer*, **50**, 3158–3168.
27. Zhang, Q., Li, G.Z., Becer, C.R., and Haddleton, D.M. (2012) *Chem. Commun.*, **48**, 8063–8065.
28. Zhu, X., Zhou, N., Zhang, Z., Sun, B., Yang, Y., Zhu, J., and Zhu, X. (2011) *Angew. Chem. Int. Ed.*, **50**, 6615–6618.
29. Jia, Z. and Monteiro, M.J. (2012) *J. Polym. Sci., Part A: Polym. Chem.*, **50**, 2085–2097.
30. Lonsdale, D.E. and Monteiro, M.J. (2011) *J. Polym. Sci., Part A: Polym. Chem.*, **49**, 4603–4612.
31. Stanford, M.J., Pflughaupt, R.L., and Dove, A.P. (2010) *Macromolecules*, **43**, 6538–6541.
32. Qiu, X.-P., Tanaka, F., and Winnik, F.M. (2007) *Macromolecules*, **40**, 7069–7071.
33. Yuan, Y.Y., Du, J.Z., and Wang, J. (2012) *Chem. Commun.*, **48**, 570–572.
34. Chen, B., Jerger, K., Frechet, J.M., and Szoka, F.C. Jr., (2009) *J. Controlled Release*, **140**, 203–209.
35. Killops, K.L., Campos, L.M., and Hawker, C.J. (2008) *J. Am. Chem. Soc.*, **130**, 5062–5064.
36. Rissing, C. and Son, D.Y. (2009) *Organometallics*, **28**, 3167–3172.
37. Chen, G., Kumar, J., Gregory, A., and Stenzel, M.H. (2009) *Chem. Commun.*, 6291–6293.
38. (a) Konkolewicz, D., Gray-Weale, A., and Perrier, S.B. (2009) *J. Am. Chem. Soc.*, **131**, 18075–18077; (b) Konkolewicz, D., Poon, C.K., Gray-Weale, A., and Perrier, S. (2011) *Chem. Commun.*, **47**, 239–241.
39. Han, J., Zhao, B., Gao, Y., Tang, A., and Gao, C. (2011) *Polym. Chem.*, **2**, 2175–2178.
40. Han, J., Zhao, B., Tang, A., Gao, Y., and Gao, C. (2012) *Polym. Chem.*, **3**, 1918–1925.
41. Xu, J., Tao, L., Boyer, C., Lowe, A.B., and Davis, T.P. (2010) *Macromolecules*, **43**, 20–24.
42. Grover, G.N., Alconcel, S.N., Matsumoto, N.M., and Maynard, H.D. (2009) *Macromolecules*, **42**, 7657–7663.
43. Li, M., De, P., Li, H., and Sumerlin, B.S. (2010) *Polym. Chem.*, **1**, 854.
44. Radu, L.C., Yang, J., and Kopecek, J. (2009) *Macromol. Biosci.*, **9**, 36–44.
45. Connal, L.A., Kinnane, C.R., Zelikin, A.N., and Caruso, F. (2009) *Chem. Mater.*, **21**, 576–578.
46. Gu, W., Chen, G., and Stenzel, M.H. (2009) *J. Polym. Sci., Part A: Polym. Chem.*, **47**, 5550–5556.
47. Li, B., Wang, M., Chen, K., Cheng, Z., Chen, G., and Zhang, Z. (2015) *Macromol. Rapid Commun.*, **36**, 1200–1204.
48. van Berkel, K.Y., Piekarski, A.M., Kierstead, P.H., Pressly, E.D., Ray, P.C., and Hawker, C.J. (2009) *Macromolecules*, **42**, 1425–1427.
49. Arkles, B. (1977) *ChemTech*, **7**, 766–778.
50. Tedja, R., Soeriyadi, A.H., Whittaker, M.R., Lim, M., Marquis, C., Boyer, C., Davis, T.P., and Amal, R. (2012) *Polym. Chem.*, **3**, 2743.
51. Li, X., Bao, M., Weng, Y., Yang, K., Zhang, W., and Chen, G. (2014) *J. Mater. Chem. B*, **2**, 5569–5575.

# 9
# Synthesis of Macrocycles and Click Chemistry

*Dario Pasini*

## 9.1
## Introduction

The chemistry of macrocycles goes back a long way, but it is safe to say that the field has greatly expanded in breadth and scope since the advent of the "chemistry beyond the molecule," also defined as supramolecular chemistry [1, 2]. Macrocycles are defined by IUPAC as "cyclic macromolecules or macromolecular cyclic portions of a molecule" [3]. The birth of supramolecular chemistry is traditionally fixed at the publication of the synthesis of crown ethers by Charles Pedersen almost 50 years ago [4]. Since then, macrocycles have been used as hosts capable of recognition of specific guests, and, in conjunction with suitable assembling strategies, have been used for the construction of a large variety of nanoscale structures [5, 6].

The synthesis of macrocycles has evolved in recent years by using concepts developed in different areas of chemistry. We can summarize three key topics as follows:

1) The use of high-dilution conditions has improved yields in the cyclization process; this tool statistically favors ring closing versus polymerization reaction, either in single chains containing complementarily reactive chain-end functionalities (AB; A reacting with B) or in chains containing complementary reactive functionalities (AA with BB) [5–7].
2) Coherently designed guest systems, recognized by the forming host, can effectively act as templates during the reaction and preorganize the host–guest system for the macrocyclization reaction (forming one or more new covalent bonds), lowering the activation energies for the desired chemistry [5, 6]. These are usually considered template-directed syntheses.
3) Dynamic combinatorial chemistry, as a method for the generation of new molecules by means of reversible reactions between simple building blocks under thermodynamic control. In a dynamic combinatorial library (DCL), all constituents are in equilibrium, and their distribution is determined by their

*Click Reactions in Organic Synthesis*, First Edition. Edited by Srinivasan Chandrasekaran.

thermodynamic stability within the DCL. This tool has been used with success for the construction, among others, of macrocyclic structures [8].

The key issue, whatever of the previous concepts is used, is to utilize efficient, high-yielding, functional-group-tolerant, orthogonal [9] organic chemistry for the ring-closing step (the macrocyclization reaction).

Click reactions have emerged in recent years as a particular class of molecular transformations, whose requirements have been precisely codified. They are gaining popularity especially in those fields, such as biomaterials synthesis or bioconjugations, requiring specific characteristics, which cannot be found in simply high-yielding organic reactions [10]. In fact, Sharpless [11] defined the term *click chemistry* as follows: "The reaction must be modular, wide in scope, give very high yields, generate only inoffensive byproducts that can be removed by nonchromatographic methods [ … ]. The required process characteristics include simple reaction conditions (ideally, the process should be insensitive to oxygen and water), readily available starting materials and reagents, the use of no solvent or a solvent that is benign (such as water) or easily removed, and simple product isolation [ … ]. Click processes proceed rapidly to completion and also tend to be highly selective for a single product: we think of these reactions as being springloaded for a single trajectory."

By virtue of its simplicity, click chemistry has quickly become an established tool for the construction of complex molecular architectures. The prototypical click chemistry reaction is the Cu(I)-catalyzed azide–alkyne cycloaddition (CuAAC). In fact, 1,3-dipolar cycloadditions between azides and terminal alkynes have been known for a long time and were studied in detail by Huisgen and coworkers almost 50 years ago [12]. The reaction is thermodynamically favored, but it requires heating, and it does not show regioselectivity (both 1,4-disubstitued and 1,5-disubstituted-1,2,3-triazoles are formed) [13]. The catalyzed version of this synthetic methodology was introduced in the early 2000s. Seminal papers by Sharpless and Meldal [14, 15] showed that catalysis by Cu(I) species greatly enhanced the reactivity (up to $10^7$ rate enhancements), and the reaction can be carried out at room temperature; furthermore, the catalyzed reaction is extremely regioselective toward the formation of the 1,4-disubstituted triazole (over the 1,5-disubstituted isomer).

Sharpless initially utilized the generation of catalytic Cu(I) by the addition *in situ* of the reducing sodium ascorbate to Cu(II) salts, in a mixture of solvents capable of dissolving both the organic reactants and the inorganic counterparts (Scheme 9.1). The functionalization of macromolecular structures, the synthesis of dendrimers, and the conjugation between natural and artificial macromolecules have been targeted using CuAAC [16–22]. A copper-free variant of CuAAC has also been successfully developed, and it is gaining increasing popularity [23].

Another molecular transformation classified as "click chemistry" is the thiol–ene reaction (or alkene hydrothiolation) (Scheme 9.1b), which occurs between a thiol and an alkene to form an alkyl sulfide. It has all the desirable features of a click reaction, being highly efficient, simple to execute with no side

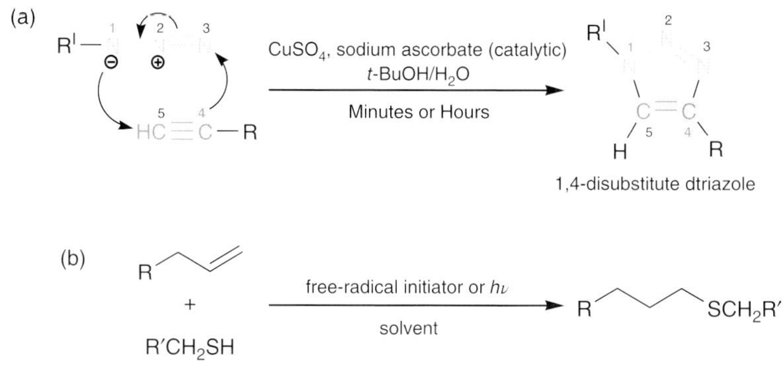


**Scheme 9.1** Prototypical conditions for: the Cu(I)-catalyzed azide–alkyne cycloaddition (CuAAC) reaction, (a) 1,3-dipolar cycloaddition and (b) the thiol–ene click reaction.

products, and proceeding rapidly to high yields. It is most frequently photoinitiated, promoting unique capabilities related to spatial and temporal control of the click reaction [24, 25].

In this chapter, we highlight recent examples of the use of the mentioned click reactions as the macrocyclization reaction in several contexts. We exclude the following from the current review: (i) cyclic structures in which part of the cyclic connection is formed by noncovalent bonds [26]; (ii) the postmodification with CuAAC click chemistry of preformed macrocycles in order to introduce triazoles for functional applications [27, 28]; and (iii) coordination polymers, or metal-organic frameworks, formed by "clicked" struts or ligands [29]. This chapter is structured in several subchapters according to the different classes of macrocycles obtained.

### 9.1.1 Peptide- and Sugar-Containing Click Macrocycles

Macrocyclic peptides are attractive for the development of bioactive compounds, with the aim to modulate biomolecular interactions. In these cyclic frameworks, the high degree of functional complexity, which is usually achievable within simpler oligopeptides, is combined with restricted conformational flexibility, making them suited to achieve selective and tight binding to extended biomolecular interfaces, such as those mediating protein–protein and protein–nucleic acid complex formation [30]. Some of the enhanced features and properties exhibited by conformationally constrained peptide-based ligands, compared to linear peptides, include enhanced cell permeability [31] and higher affinity toward the target biomolecule.

Recent developments in the use of the CuAAC reactions in the synthesis, modification, and conformational control of peptidomimetic oligomers were reviewed in 2010 [32]. Both synthetic and biosynthetic methods have been implemented to afford peptides in cyclic or conformationally constrained configurations.

Fasan's group has reported a new method for constructing conformationally constrained organopeptide hybrids by combining a genetically encoded polypeptide and a synthetic precursor (Figure 9.1) [33]. In their strategy, an alkyne-bearing unnatural amino acid was incorporated within the N-terminal portion of the linear hybrid **1**, formed by an intein protein (a protein segment able to excise itself) fused to a chosen polypeptide ($AA_1 \ldots AA_n$). The synthetic precursors (SPs, structure **2**) were aromatic units containing suitable complementary functionalities (to alkyne and thioesters of the hybrid strands) in the form of azide and hydrazide moieties. The bioorthogonal functionality is a thioester bond formed *in situ* by action of the thiol group of the N-termini cysteine residue of the intein protein on its amide connectivity (see structure **3a** in Figure 9.1). The CuAAC reaction between the azide of the synthetic precursor and the alkyne moiety of the unnatural aminoacid, coupled with the hydrazide thioester reaction, afforded the organopeptide macrocycle **4** with the concomitant release of the intein protein. The CuAAC protocol used proceeded quantitatively with the various biosynthetic precursors within minutes; prototypical conditions for the click chemistry ($CuSO_4$, sodium ascorbate in buffered aqueous solutions) were used, with the addition of EDTA (ethylenediaminetetraacetic acid) to facilitate copper removal. The authors could easily monitor the reactions using MALDI-TOF spectra, acquired after the coupling reaction.

The efficiency and applicability of CuAAC methodologies to peptide chemistry have recently been stressed by Liu's group [34]. They have implemented a chemical system in which DNA templates the generation of sequence-defined synthetic polymers (modified cyclic and acyclic oligopeptides), with no structural similarity to the corresponding templating nucleic acid moieties. The building blocks can be structurally varied, resulting in different backbones (polyethylene glycols, α-peptides, and β-peptides). The authors identified the CuAAC reaction as the key factor to achieve efficient translation. Prototypical conditions for the click chemistry ($CuSO_4$, sodium ascorbate) were used, with overnight incubation at 4 °C and salt removal with short-column chromatography. The CuAAC reaction has also been reported for cyclization reactions of oligopeptides carried out directly on the resins, thus under the heterogeneous conditions, which always characterize solid-phase synthesis [35–37]. In all these reports, the CuAAC reaction yields are high.

Anseth *et al.* have recently reported a rapid synthetic strategy for the on-resin cyclization of peptides using thiol–ene photochemistry. The method takes advantage of the thiol group of natural cysteine amino acids inserted into the peptide sequence and allows for various alkenes, as the second component to be incorporated into the peptide backbone [38]. A ligand of an integrin protein was synthesized as the model peptide (Figure 9.2).

Although cysteine thiol functionalities are used in the cyclization reaction, it is in principle possible to incorporate other inert cysteine residues by exploiting highly selective and orthogonal cysteine protecting groups. Two different alkenes were incorporated and cyclized: an allyl ester, introduced as a Fmoc-protecting group within the peptide sequence, and a bicyclic alkene, incorporated through

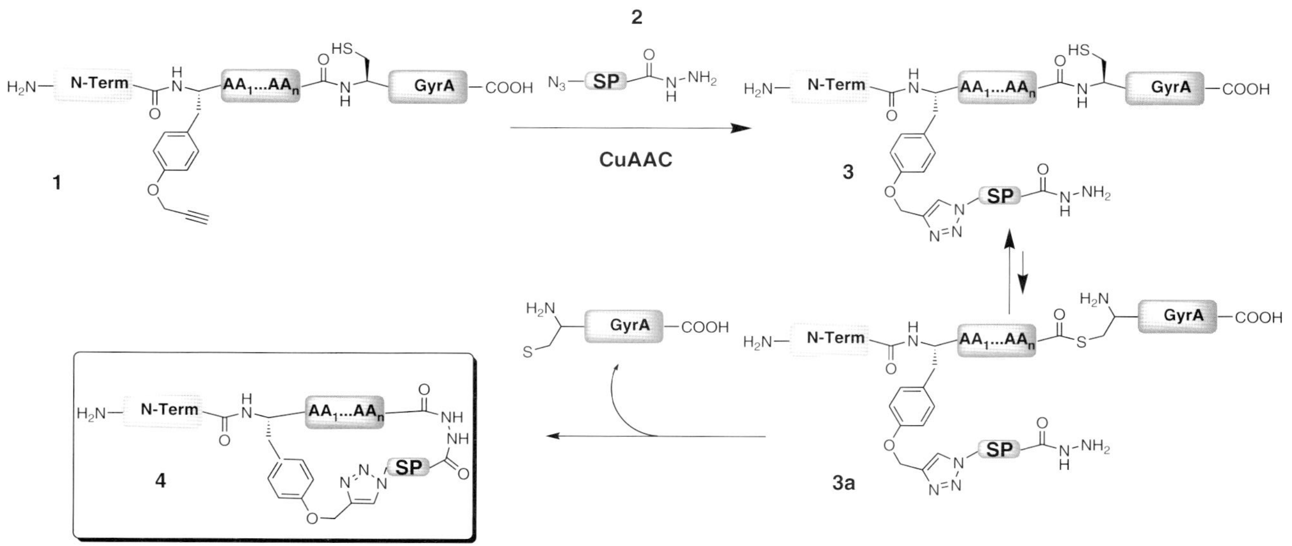


**Figure 9.1** Strategy for the modular synthesis of macrocyclic organopeptide hybrids **4**. The starting linear polypeptides comprise an N-terminal tail (light gray), *O*-propargyl tyrosine, a target sequence ($AA_1 \ldots AA_n$), and a GyrA intein segment. Macrocyclization occurs upon coupling of this protein to a synthetic precursor (SP) by concomitant CuAAC and thioester–hydrazide coupling [33].

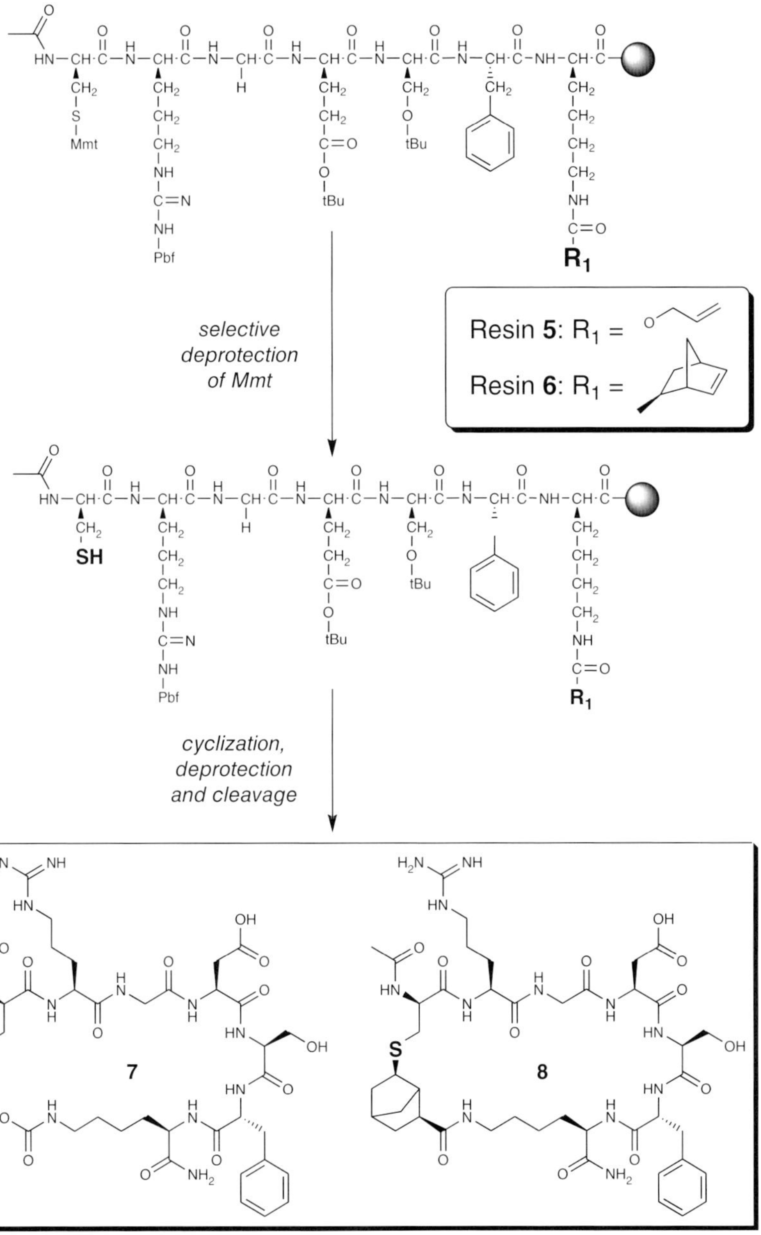


**Figure 9.2** Synthetic route to on-resin peptide macrocyclization using thiol–ene photochemistry [38].

an on-resin modification using 5-norbornene 2 carboxylic acid, since strained alkenes are known to exhibit higher reactivity.

Linear peptides were built on solid phase to afford functionalized resins **5** and **6**, and the monomethoxytrityl (Mmt)-protecting group for the thiol was selectively removed. 2,2-Dimethoxy-2-phenylacetophenone (DMPA) was used as the photoinitiator for the thiol–ene cyclization. The reaction with the allyl ester reached completion at 1 h, whereas with the strained norbornene, the reaction reached completion after just 20 min. After cleavage from the resin, **8** and **7** were recovered in 37% and 24% yield, respectively. Competitive binding ELISA assays confirmed that the thiol–ene reaction did not exhibit negative effects on the cyclic peptides' activity, with **7** and **8** exhibiting $IC_{50}$ values comparable to reported literature values for analogous systems.

Kolmar's group demonstrated the utility of a 1,2,3-triazole bridge as a disulfide replacement [39]. The efficient *in vitro* generation of disulfide bonds is still a challenge in contemporary biochemistry; in fact, it is usually achieved in peptides postsynthetically by means of the use of mediators such as air oxygen. It is particularly challenging in cysteine-rich oligopeptides, for which the regiospecific formation of one out of several disulfide bonds is not trivial to control. 1,2,3-Triazole can be expected to act as efficient disulfide surrogates as they are redox stable, and dissimilar to common building blocks of nature, so that improved pharmacokinetic properties can be anticipated. In their communication, the authors designed a series of triazole analogs of a monocyclic variant of the sunflower trypsin inhibitor-I (cyclic peptide **9** in Figure 9.3). The analogs **10–12** have been obtained through the cyclization of linear peptides, bearing sequences of aminoacids, which include unnatural azide- and alkyne-functionalized ones (precursors **P**). The authors used both Cu(I)- and the Ru(II)-catalyzed versions of the azide–alkyne cycloaddition reaction. This latter version, introduced by Sharpless [40] subsequent to CuAAC, allows for the regioselective formation of 1,5-disubstituted triazole derivatives from azides and terminal alkynes. Although the RuAAC variant gathers much less popularity than CuAAC, it has been used with success in the context of different scenarios, as demonstrated in recent examples [41–46]. CuAAC could be carried out using solution-phase synthesis techniques in solution: cyclic peptides **11** and **12** (Figure 9.3) were obtained using classic CuAAC conditions ($Cu(SO_4)_2$, sodium ascorbate) and high dilution and isolated in modest yields (10–20%) after purification by HPLC. RuAAC was carried out after anchoring the linear precursors on a solid phase, eventually achieving the 1,5-disubstituted triazole-containing cyclic peptide **10**, although in low yields (2%). The authors stated that RuAAC did not work satisfactorily using solution-phase synthesis techniques. The low yields are likely the consequence of the high degree of complexity and of conformational flexibility of the precursors to be cyclized, rather than of the inefficient protocols/conditions utilized for the CuAAC reaction.

Energy-minimized 3D models of all variants of **9** were able to substantiate the design considerations. The 1,5-disubstituted triazole bridge of **10** resulted as more capable, when compared to the 1,4-sustitution pattern on the triazole present in

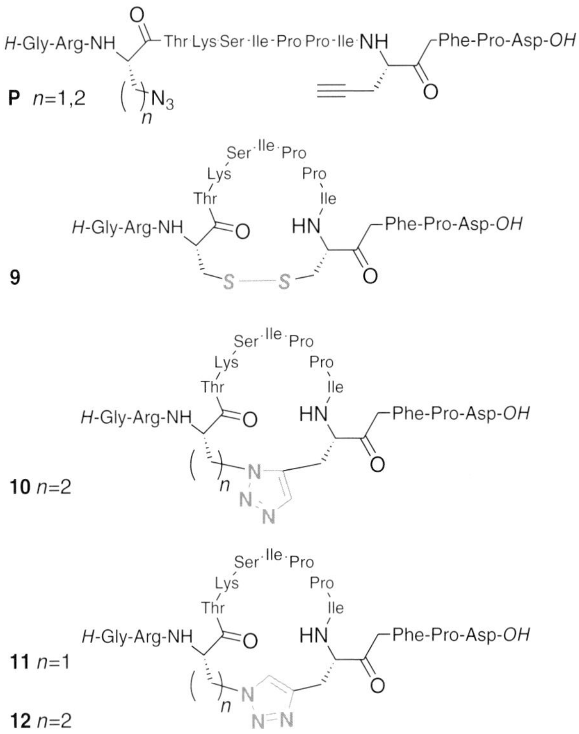


**Figure 9.3** Cyclic peptides obtained by CuAAC or RuAAC click chemistries incorporating 1,4- or 1,5-disubtituted triazole linkages as trypsin inhibitors analog of **9** [39].

substrates **11** and **12**, of optimal structural similarity and almost perfect superposition with cyclic peptide skeleton of **9**. Indeed, inhibitory activity studies showed that, whereas analog **10** essentially retained the activity of **9**, both **11** and **12** are at least 2 orders of magnitude less effective than **9**. This work demonstrates the feasibility of the substitution of a disulfide bridge by a triazole linkage and, in general, the broad applicability of AAC reactions for cyclopeptide mimic.

One of the most appealing areas of research related to macrocyclic chemistry is their use as building blocks for the assembly of nanotubes [47–49]. Cyclic D,L-α-peptides, and more recently cyclic $\beta_3$-peptides, have been shown to assemble into tubular structures through the association of complementary and suitably positioned amide hydrogen-bonding units within the cyclic backbone [50]. Triazoles can be considered as amide bond analogs; they are, in fact, suitable mimics of the amide bond in terms of polarity, as well as hydrogen-bond-donating and -accepting ability (Figure 9.4).

In recent work, Chattoparday's group has reported on cyclic peptides that are able to assemble in oriented nanotubes [51–53]. The triazoles embedded within the cyclic structures were obtained through CuAAC reactions between alkyne and azide moieties of elaborated precursors, containing cis-β-furanoid sugars and

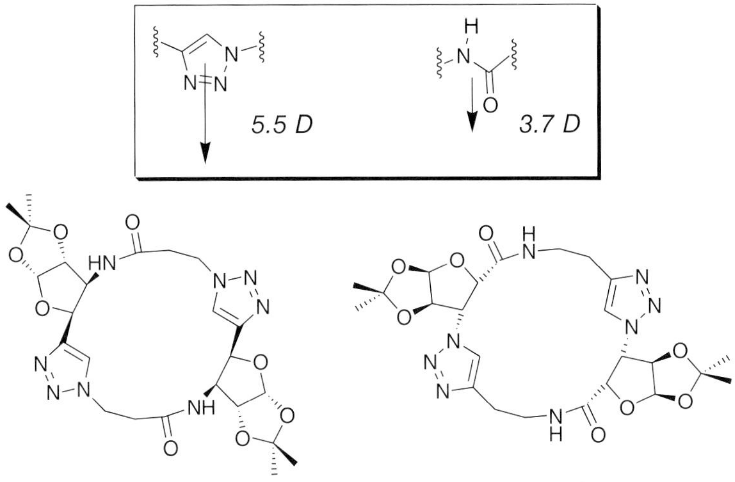


**11a** **11b**

**Figure 9.4** Regioisomeric triazole-containing cyclic peptides composed of *cis*-β-furanoids and β-alanines [51].

β-alanine moieties, although the click reaction was not used in the cyclization step. The reactions were carried out in classical CuAAC conditions ($Cu(SO_4)_2$, sodium ascorbate) in mixtures of organic solvents and water. Two regioisomeric products (**11a**,**b** in Figure 9.4) were designed and separately obtained. Changes in the packing and self-assembly of these macrocycles, as a consequence of the subtle structural changes in their regioisomeric structures, resulted in different polarity of the nanotubes due to different orientation of functional groups. These macrocycles are effective model systems for artificial ion channels, and their unidirectionality in terms of dipole moments augurs well for functional applications of this new class of peptidomimetic macrocycles. Work in this area has also been recently reported by the groups of Abell and James [54, 55].

The use of CuAAC click chemistry for the synthesis of sugar-containing macrocycles has recently been reported by several groups, and very interesting properties have been demonstrated for these derivatives [56–60], such as metal complexation ability, the possibility to address simply stereochemical diversity, and multivalent binding to biological systems. Chen's group [61] reported the synthesis and characterization of structurally well-defined macrocyclic oligosaccharides of various dimensions. Their strategy involves a series of chemical, but also highly efficient chemoenzymatic methods, for the preparation of the precursor oligosaccharides **12** (Figure 9.5), incorporating an azido-containing sialic acid at the nonreducing end and a propargyl group at the reducing end. The CuAAC macrocyclization reaction in a mixture of organic solvent/$H_2O$ was carried out using "nonclassical" conditions, by means of directly providing Cu(I) as the catalyst, instead of generating it *in situ*. Yields in the range 30–80% could be obtained: such yields have to be considered outstanding, given the relative

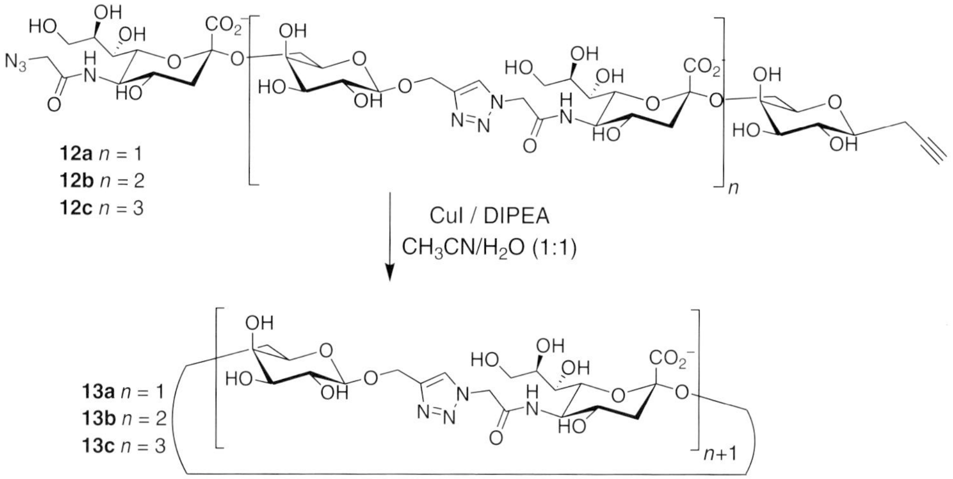


**Figure 9.5** Synthesis of click macrocycles through the cyclization of oligosaccharide linear monomers of varying lengths [61].

flexibility and poor preorganization of the substrates. The ionic macrocycles **13** possess high solubility in water and can encapsulate hydrophobic aromatic small molecules in a size-dependent manner, similarly to cyclodextrins.

Xie *et al.* have obtained carbohydrate-based BODIPY(4,4-difluoro-4-bora-3*a*,4*a*-diaza-*s*-indacene)-containing macrocycles **14–16** (Figure 9.6) with dimensions varying from 23- to 27-membered rings [62]. BODIPY dyes are gaining increasing popularity given their excellent photophysical properties such as high fluorescence quantum yield, photochemical stability, and strong absorbance. The macrocycles were efficiently synthesized employing CuAAC

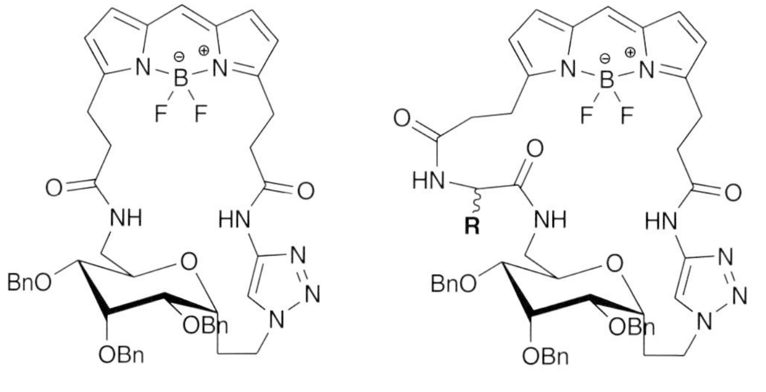


**Figure 9.6** Structures of BODIPY-based sugar-containing macrocycles [62].

in an intramolecular cyclization step, starting from a precursor built from alkyne-functionalized BODIPY acid, *C*-glucosyl azido amine, and amino acids. The reaction proceeded in high optimized yields (47% for **14**), using classical conditions, $Cu(SO_4)_2$, sodium ascorbate, and an unusual combination of solvents ($CH_2Cl_2$, $H_2O$). BF(OMe) and $B(OMe)_2$ derivatives have also been obtained by fluoride exchange with MeOH. The excellent photophysical properties of BODIPY remain intact in the macrocyclic structure. Macrocycles **14** and **15** showed receptor-like properties toward selected anions ($F^-$ and $CN^-$) and cations ($Cu^{2+}$ and $Fe^{3+}$) in MeCN.

### 9.1.2
### Click Macrocycles for Anion Binding and Supramolecular Recognition

The synthesis of rigid macrocycles using CuAAC click chemistry and the characterization of their host–guest binding properties have only recently been addressed [63, 64]. It is the work of Flood and coworkers that has demonstrated the possibility of using CuAAC click chemistry not only for the rapid, orthogonal, and high-yielding construction of shape-persistent macrocycles, but also for the utilization of the embedded triazoles resulting from the click process as amide surrogates for the binding of anions. The prototypical example is the highly preorganized, shape-persistent macrocycle **17** (Figure 9.7), in which the acidic C–H hydrogens of the four triazole units are perfectly positioned for the recognition of a spherical anion. Indeed, macrocycle **17** showed a very high binding affinity toward chloride anions in organic solvents [65]. Flood and coworkers have recently summarized the state of the art in this emerging subfield of research related to CuAAC click chemistry [69]. Other triazole macrocycles (compounds **18** and **19** in Figure 9.7), in which more flexible $sp^3$ carbon atoms have been introduced within the rigid cyclic backbone, revealed an oblique–pillar cavity with the chair-like conformation in the solid state, with substantially reduced binding capabilities toward anions such as $HSO_4^-$ and $H_2PO_4^-$ to form 1 : 1 and 1 : 2 host–guest complexes in $CDCl_3$ [66].

More recent reports have dealt with the possibility of increasing the hydrogen bonding capability of the neutral CH hydrogen-bond-donor motif (i.e., triazole) by making it cationic (i.e., triazolium moieties). Beer's group has recently reported the synthesis and characterization of a clicked ferrocene-containing bis(triazole) macrocycle [67]. Alkylation gives the corresponding triazolium macrocycle (**20** in Figure 9.7), which strongly binds anions such as chloride and benzoate even in polar organic solvents. The authors presented good evidence ($^1H$ NMR titration experiment) that the high binding constant is favored by charge-assisted C–H• anion interactions, and they use the redox-active macrocycle **20** for electrochemical sensing of chloride ions in $CH_3CN$ solution.

Sessler *et al.* have described rigid, shape-persistent click macrocycle **21** designed to bind tetrahedral oxyanions (Figure 9.7). The macrocycle was obtained from triazole-containing precursors (obtained using CuAAC click

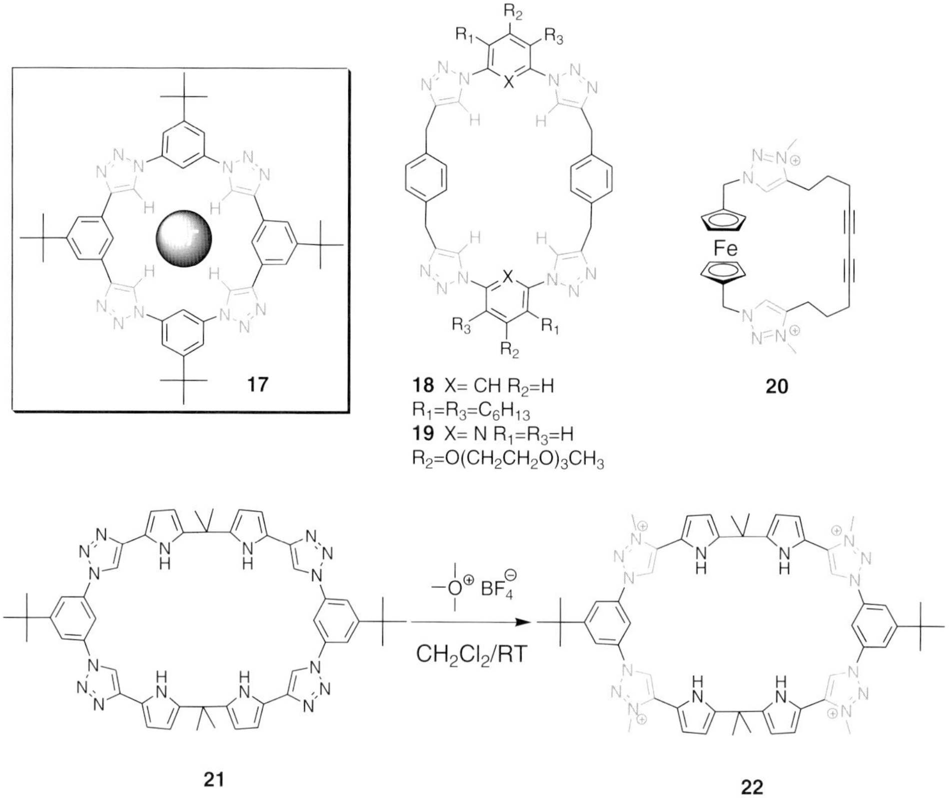


**Figure 9.7** Triazole- and triazolium-containing macrocycles for the binding of anions. In gray: triazole or triazolium amide surrogates as the key functionalities responsible for binding [65–68].

chemistry), which were then cyclized by a condensation reaction with acetone to form the quaternary carbon atoms within the macrocyclic backbone. Postmodification of macrocycle **21** by reaction with the methylating agent trimethyloxonium tetrafluoroborate afforded the macrocycle **22**. The system incorporates neutral NH hydrogen-bond-donor groups (pyrrole), as well as cationic CH hydrogen-bond-donor motifs (triazolium moieties) [68]. The system displays a high selectivity for tetrahedral oxyanions, relative to monoanions and trigonal planar anions, in mixed polar organic–aqueous solvents. Theoretical calculations were carried out in an effort to understand the influence of solvent on the intrinsic hydrogen-bonding ability of the donor groups (pyrrole N–H and triazolium C–H). Macrocycle **22** can bind pyrophosphate and phosphate anions in the solid state as confirmed by X-ray diffraction analyses.

Pasini *et al.* have recently reported the design and synthesis of a chiral click macrocycle behaving as a chiroptical sensor for the detection of halide anions

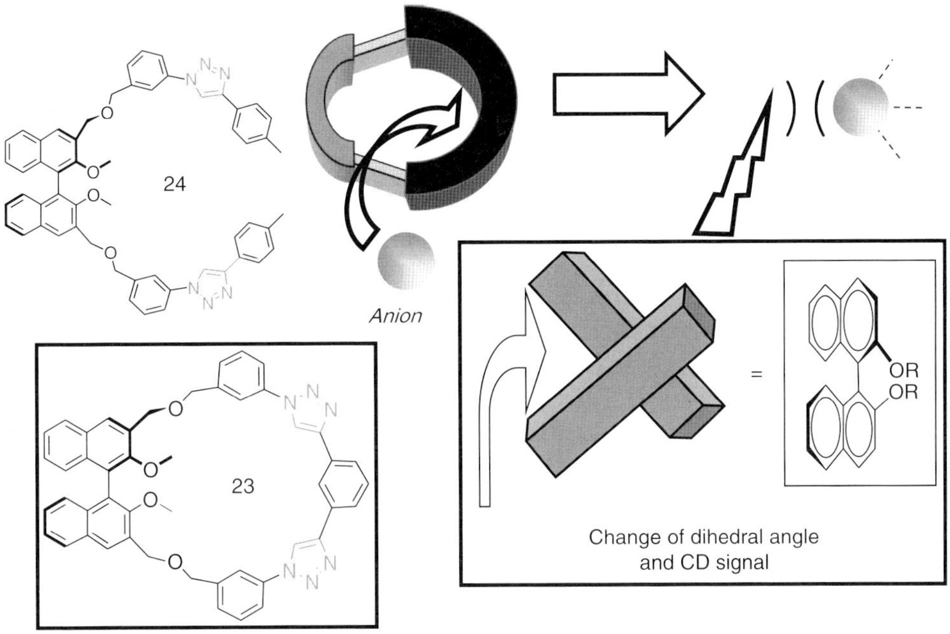


**Figure 9.8** Structures of macrocycle **23** and control compound **24**, and (bottom) operating principles of the macrocyclic chiroptical sensor **23** in which the variation of the CD response of the Binol unit is the key sensing output. (Reproduced from Caricato *et al.* [70], with permission of Elsevier.)

(Figure 9.8) [70]. This work is part of a broader activity in the field of binaphthyl-based chiral macrocycles for chiroptical sensing and chiral nanostructuring [71–87]. The macrocycle **23** incorporates a Binol unit. In order to counterbalance the inherent distortion brought about by the binaphthyl units into the macrocyclic framework, $sp^3$ carbon atoms, imparting a higher conformational freedom with respect to sp- or $sp^2$-hybridized carbon atoms, have been introduced.

Open-control compound **24** was synthesized through classical CuAAC conditions. Macrocycle **23**, instead, needed to be synthesized by the use of nonclassical CuAAC conditions, by directly introducing the Cu(I) catalytic source. As in previous examples [65], the CH hydrogen atoms of the triazole moiety are able to undergo hydrogen bonds with spherical anions, stabilizing the host–guest complex. The binding constants, measured in apolar organic solvents, are not remarkable, either for the macrocycle or for the model compound. The system shows selectivity toward iodide, bromide, and chloride over fluoride and carboxylate anions. The macrocycle (and not the control compound) is a "truly chiroptical" sensor for halide anions: strong signaling is generated only when CD spectroscopy is used as the detection tool. The signaling mechanism is not a result of a specific recognition of the analyte with the CD reporter, the Binol moiety, but rather a secondary effect. The anions, held into the macrocyclic cavity, are in close proximity to the Binol moiety, and its conformational change generates the variation of the exciton-coupled signal classical signature of the Binol moiety.

### 9.1.3
### Clicking Macrocycles to form Mechanical Bonds

The CuAAC click reaction has been widely used for the preparation of interlocked molecules, such as catenanes and rotaxanes [88]. In the case of rotaxanes, the CuAAC reaction has been used mainly for the synthesis of the axle; subsequent threading, clipping, or slipping procedures were utilized to produce the mechanically interlocked product.

In selected recent examples, the CuAAC reaction has been used for the ring-closing step of suitably positioned and entangled precursors. Megiatto and Schuster [89] have recently reported the synthesis of "Sauvage-type" catenanes, using click chemistry. "Sauvage-type" catenanes (Figure 9.9) are synthesized by making use of a preorganized coordination complex in which two substituted phenanthroline units are optimally arranged around a Cu(I) template (structure **25** in Figure 9.9). One of the challenges often encountered in the ring closure step of the preoriented phenanthroline moieties is that the complex is inherently unstable, and it can dissociate when subjected to temperature or solvent polarity changes, or to the presence of competing complexation ligands. The authors describe two different protocols based on Cu(I) template synthesis and "click" reactions for the synthesis of functionalized [2]catenanes. A straightforward procedure, involving high-dilution conditions at high temperatures (70 °C), was developed for the synthesis of [2]catenanes bearing two identical peripheral groups in high yields (Figure 9.9). The authors used "nonclassical" click chemistry conditions for the cyclization of precatenane **25**, by means of the direct introduction in the reaction mixture of a Cu(I) source and a base (DBU), but they also introduced sodium ascorbate. After demetalation, serendipitously occurring during the workup of the reaction, catenane **26** was obtained in an astonishing 90% isolated yield. The same authors reported the formation using click chemistry of a complex unsymmetrical material that would otherwise be impossible to synthesize, such as the porphyrin-$C_{60}$ [2]catenane [90]. The introduction of peripheral functional groups into the

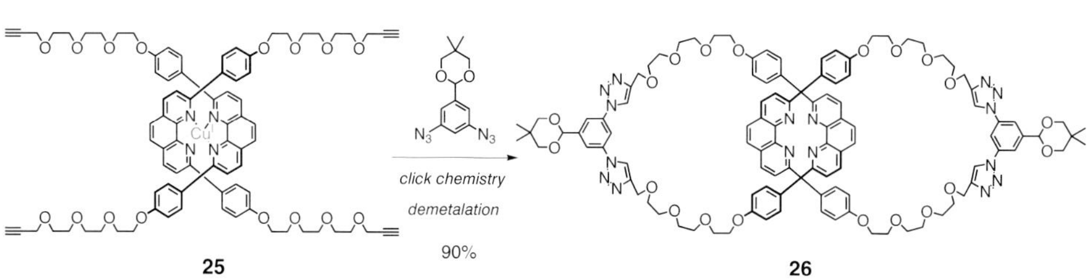


**25** **26**

**Figure 9.9** High-yielding synthesis of "Sauvage-type" symmetrical catenane **26** [89].

catenane structure, using such high-yielding methodology, paves the way to the use of [2]catenanes for the preparation of even more complex structures.

In a recent article, Leigh and coworkers have applied CuAAC in a very elegant way for the synthesis of a trefoil knot [91, 92]. The strategy has been dubbed as "the active-metal template": in fact, Cu(I) ions both preorganize the knot precursor and act as the cyclization catalyst (Figure 9.10). The open, flexible precursor **27** needs to be folded, before being cyclized into the trefoil knot **29.** The design of the molecule was chosen so that the arrangement of the ligands around the metal center forms the necessary entanglement to create the knot by coordination in a tetrahedral way to the bipyridine units (structure **28**). At the same time, a catalytic amount of the copper(I) ions complexes the two unsymmetrical alkyne and azide extremities of the thread and the pyridine moiety, catalyzing the CuAAC click ring-closing reaction between the two terminal moieties. After demetalation, a racemic mixture of left- and right-handed knots **29** was isolated in a yield of 24%, a very reasonable yield considering the synthetic challenge.

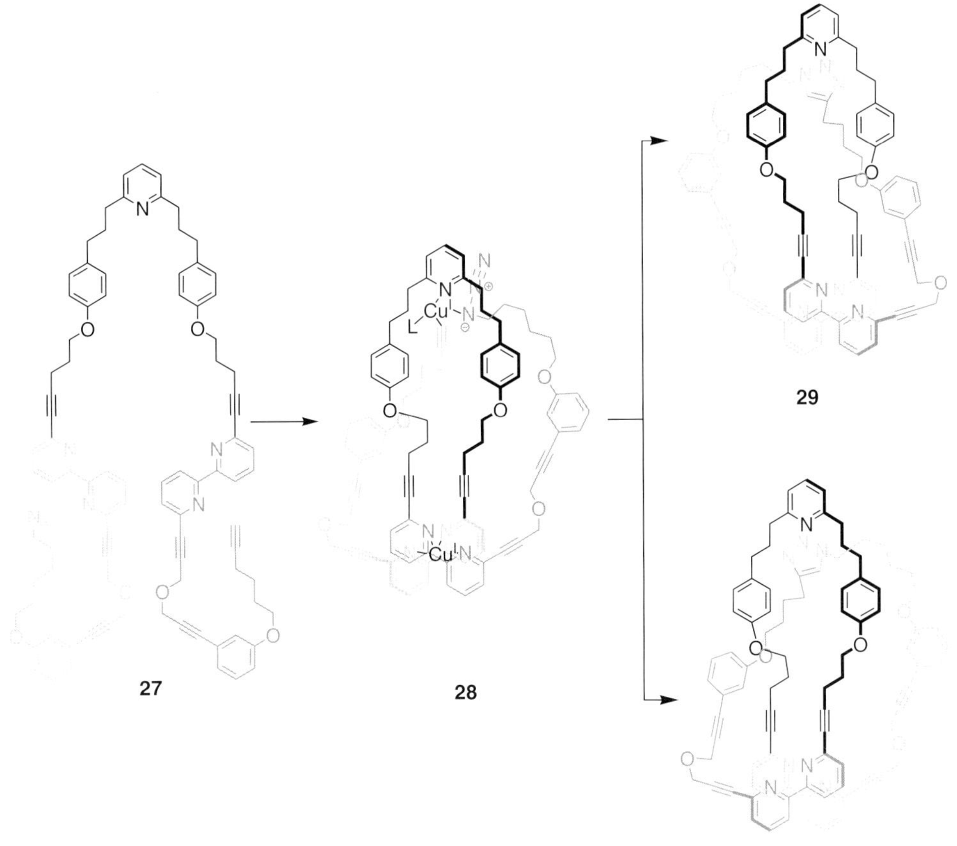


**Figure 9.10** Synthesis of trefoil knot **29** by Leigh and coworkers [92].

A similar active-template strategy was applied by Goldup *et al.* for the formation of doubly interlocked [3]rotaxanes [93]. A photochemical thiol–yne click reaction (a variant of thiol–ene click) was used for the formation of rotaxanes, proceeding mildly at ambient temperature and humidity under an air atmosphere [94].

### 9.1.4
### Cyclic Polymers Obtained by the CuAAC Click Reaction

Cyclic polymers represent a special class of macromolecules among the various possible polymeric architectures. Because of the absence of chain ends in cyclic polymers, different chain dynamics can strongly influence physical properties compared with linear analogs of the same mass [95]. Recently, therefore, large cyclic polymers have emerged from the realm of "scientific curiosities" to that of potentially useful macromolecular architectures for a variety of functional applications [96–99]. By virtue of the IUPAC definition [1], cyclic polymers can be considered as macrocycles. Polymers with (small or large) cyclic repeating units within the macromolecular backbone, cyclopolymers, have been known for some time [100–104], and they must not be assimilated to cyclic polymers.

Grayson's group has pioneered the application of CuAAC click chemistry to the realm of cyclic polymers. They prepared polystyrene macrocycles from ATRP (Atom Transfer Radical Polymerization) precursors containing a terminal alkyne (see Figure 9.11) [105]. Polymer samples **30** were reliably obtained with low polydispersities (<1.2) and varying degrees of polymerization by virtue of the use of the controlled polymerization technique (ATRP). Further, the use of the ATRP methodology can afford polymers with tailored end-capping chain functionalities. The authors were able to transform the terminal bromide to an azide, to afford intermediate polymer **31**. Polymer **32** was then obtained by "click" cyclization with the pendant alkyne from the initiator; the concentration of the polymer samples has been carefully controlled to avoid polymer **31** to prefer condensation instead of cyclization. This route offers outstanding control over the size and polydispersity of the macrocyclic polymers, as well as being tolerant of a number of functional groups. More recently, the same group has reported the synthesis of well-defined toroidal macromolecules through a similar design [106].

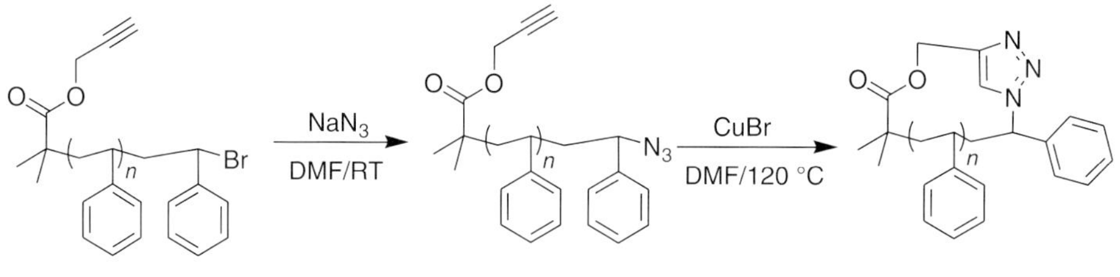


**30** **31** **32**

**Figure 9.11** Terminal insertion of azide and "Click" cyclization of polystyrene prepared via ATRP [105].

The cyclization of linear polymers by coupling end groups relies on the very slow feed of the uncyclized, linear polymer into a highly dilute reaction mixture of the CuAAC catalyst and solvents, leading to monocyclic polymer in very low concentrations over long periods of time (>10 h) and at high temperatures (>100 °C). Montero's group has recently reported an alternative approach, in which the Jacobson–Stockmayer theory is utilized to predict the ratio of monocyclic polystyrene (c-PSTY) in a one-pot reaction at 25 °C [107]. Using the same substrates shown in Figure 9.11, they found empirically, from experimental diffusion-controlled rate coefficients for cyclization and condensation of α,ω-polymer chains, that the Jacobson–Stockmayer theory is applicable for the CuAAC reaction. Therefore, the percentage of monocyclic polymer is independent of reaction rate parameters (such as catalysts concentration and temperature) and is only dependent on polymer concentration.

The groups of Li and Zhu reported an elegant example of photosensitive cyclic amphiphilic homopolymers, in which biphenyl azobenzene chromophores in the cyclic main chain are tethered with hydrophilic ethylene glycol chains [108]. The synthetic approach is particularly interesting, and it involves uncatalyzed, thermally activated alkyne–azide cycloadditions as the step-growth polymerization mechanism, occurring in bulk, to afford, starting from monomer **33** (Figure 9.12), polymer **34**. This is followed by a CuAAC intramolecular cyclization from the α-alkyne/ω-azide linear precursors. The cyclic amphiphilic homopolymers **35** thus obtained are able to self-assemble into spherical nanoparticles showing photoresponsive behavior, as a consequence of the *cis*–*trans* isomerization of the azobenzene moieties embedded in the backbone. The cyclic topology greatly influences the physical properties and self-assembly behavior of these materials.

Sequential "click" strategies were utilized by Zhang, Cheng, and coworkers to precisely construct new surfactants based on polyhedral oligomeric silsesquioxane (POSS), tethering cyclic polymers [109]. The general synthetic approach involves different types of click chemistries: sequentially performed strain-promoted azide–alkyne cycloaddition (SPAAC), CuAAC for POSS–polymer conjugation, and thiol–ene reactions for POSS surface functionalization.

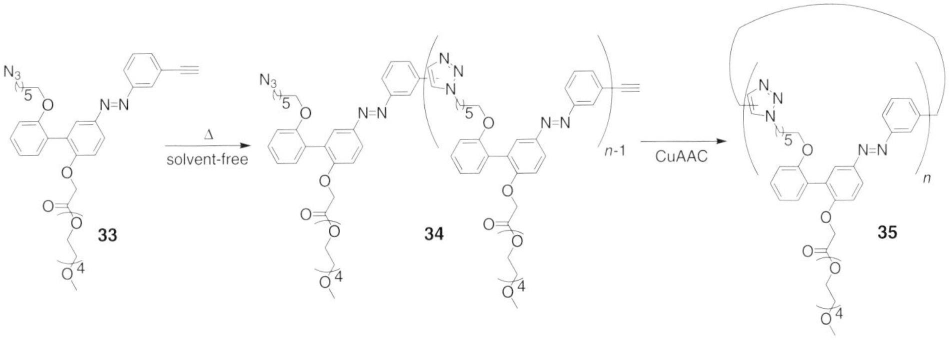


**Figure 9.12** The synthetic routes of the linear and cyclic polymers [108].

## 9.2 Summary and Conclusions

This chapter highlights recent achievements in the synthesis of cyclic structures by the use of the click reaction, both CuAAC and thiol–ene. Perhaps the most useful demonstration of the utility and versatility of this reaction is given by the breadth of different chemical structures illustrated in this contribution. Cross-fertilization among different areas of chemistry has flourished owing to the utilization of the click reaction technique. In some of the selected cases, especially in the untemplated cyclization of peptides in solution or in the solid phase (Section 9.1.1), CuAAC represents the option of choice. We have also highlighted how sometimes CuAAC is now traditionally associated not only to triazoles, the organic functionality resulting from its application, but also from its transformation into more useful moieties for functional applications (triazolium salts). It is safe to anticipate a long life to these versatile reactions, in association with the synthesis of complex and functional molecular architectures. Given the low toxicity of triazoles [110], applications of CuAAC click chemistry in the field of functionalization of biopolymers toward innovative biomaterials and bio-based plastics can be envisaged.

## References

1. Lehn, J.-M. (1988) *Angew. Chem. Int. Ed. Engl.*, **27**, 89–112.
2. Cram, D.J. (1988) *Angew. Chem. Int. Ed. Engl.*, **27**, 1009–1020.
3. IUPAC http://goldbook.iupac.org/M03662.html (accessed on June 2015).
4. Pedersen, C.J. (1967) *J. Am. Chem. Soc.*, **89**, 7017–7036.
5. Weber, E., Vögtle, F., and Burrell, A.K. (1992) *Macrocycles*, Springer-Verlag.
6. Lehn, J.-M. (1995) *Supramolecular Chemistry. Concepts and Perspectives*, Wiley-VCH Verlag GmbH.
7. Fastrez, J. (1989) *J. Phys. Chem.*, **93**, 2635–2642.
8. Cougnon, F.B.L. and Sanders, J.K.M. (2012) *Acc. Chem. Res.*, **45**, 2211–2221.
9. Wong, C.-H. and Zimmerman, S.C. (2013) *Chem. Commun.*, 1679–1695.
10. Barner-Kowollik, C., Du Prez, F.E., Espeel, P., Hawker, C.J., Junkers, T., Schlaad, H., and Van Camp, W. (2011) *Angew. Chem. Int. Ed.*, **50**, 60–62.
11. Kolb, H.C., Finn, M.G., and Sharpless, K.B. (2001) *Angew. Chem. Int. Ed.*, **40**, 2004–2021.
12. Huisgen, R., Szeimies, G., and Möbius, L. (1967) *Chem. Ber.*, **100**, 2494–2507.
13. Meldal, M. and Tornøe, C.W. (2008) *Chem. Rev.*, **108**, 2952–3015.
14. Tornøe, C.W., Christensen, C., and Meldal, M. (2002) *J. Org. Chem.*, **67**, 3057–3064.
15. Rostovtsev, V.V., Green, L.G., Fokin, V.V., and Sharpless, K.B. (2002) *Angew. Chem. Int. Ed.*, **41**, 2596–2599.
16. Sen, G.S., Kuzelka, J., Singh, P., Lewis, W.G., Manchester, M., and Finn, M.G. (2005) *Bioconjugate Chem.*, **16**, 1572–1579.
17. Touaibia, M., Shiao, T.C., Papadopoulos, A., Vaucher, J., Wang, Q., Behamioud, K., and Roy, R. (2007) *Chem. Commun.*, 380–382.
18. Pacini, A., Caricato, M., Ferrari, S., Capsoni, D., Martínez de Ilarduya, A., Muñoz-Guerra, S., and Pasini, D. (2012) *J. Polym. Sci., Part A: Polym. Chem.*, **50**, 4790–4799.
19. Sharma, A.K., Caricato, M., Quartarone, E., Edizer, S., Schieroni, A.G., Mendichi, R., and Pasini, D. (2012) *Polym. Bull.*, **69**, 911–923.
20. Brennan, J.L., Hatzakis, N.S., Tshikhudo, T.R., Dirvianskyte, N., Razumas, V., Patkar, S., Vind, J.,

Svendsen, A., Nolte, R.J.M., Rowan, A.E., and Brust, M. (2006) *Bioconjugate Chem.*, **17**, 1373–1375.

21. Mynar, J.L., Choi, T.-L., Yoshida, M., Kim, V., Hawker, C.J., and Fréchet, J.M.J. (2005) *Chem. Commun.*, 5169–5171.
22. Wu, P., Feldman, A.K., Nugent, A.K., Hawker, C.J., Scheel, A., Voit, B., Pyun, J., Fréchet, J.M.J., Sharpless, K.B., and Fokin, V.V. (2004) *Angew. Chem. Int. Ed.*, **43**, 3928–3932.
23. Sletten, E.M. and Bertozzi, C.R. (2011) *Acc. Chem. Res.*, **44**, 666–676.
24. Hoyle, C.E. and Bowman, C.N. (2010) *Angew. Chem. Int. Ed.*, **49**, 1540–1573.
25. Northrop, B.H. and Coffey, R.N. (2012) *J. Am. Chem. Soc.*, **134**, 13804–13817.
26. Romuald, C., Cazals, G., Enjalbal, C., and Coutrot, F. (2013) *Org. Lett.*, **15**, 184–187.
27. Lau, Y.H., Price, J.R., Todd, M.H., and Rutledge, P.J. (2011) *Chem. Eur. J.*, **17**, 2850–2858.
28. Nierengarten, I., Guerra, S., Holler, M., Nierengarten, J.-F., and Deschenaux, R. (2012) *Chem. Commun.*, 8072–8074.
29. Savonnet, M., Kockrick, E., Camarata, A., Bazer-Bachi, D., Bats, N., Lecocq, V., Pinela, C., and Farrusseng, D. (2011) *New J. Chem.*, **35**, 1892–1897.
30. Driggers, E.M., Hale, S.P., Lee, J., and Terrett, N.K. (2008) *Nat. Rev. Drug Discovery*, **7**, 608–624.
31. Walensky, L.D., Kung, A.L., Escher, I., Malia, T.J., Barbuto, S., Wright, R.D., Wagner, G., Verdine, G.L., and Korsmeyer, S.J. (2004) *Science*, **305**, 1466–1470.
32. Holub, J.M. and Kirshenbaum, K. (2010) *Chem. Soc. Rev.*, **39**, 1325–1337.
33. Smith, J.M., Vitali, F., Archer, S.A., and Fasan, R. (2011) *Angew. Chem. Int. Ed.*, **50**, 5075–5080.
34. Niu, J., Hili, R., and Liu, D.R. (2013) *Nat. Chem.*, **5**, 282–292.
35. Turner, R.A., Oliver, A.G., and Lokey, R.S. (2007) *Org. Lett.*, **9**, 5011–5014.
36. Ingale, S. and Dawson, P.E. (2011) *Org. Lett.*, **13**, 2822–2825.
37. Saludes, J.P., Morton, L.A., Ghosh, N., Beninson, L.A., Chapman, E.R., Fleshner, M., and Yin, H. (2012) *ACS Chem. Biol.*, **7**, 1629–1635.
38. Aimetti, A.A., Shoemaker, R.K., Lin, C.-C., and Anseth, K.S. (2010) *Chem. Commun.*, **46**, 4061–4063.
39. Empting, M., Avrutina, O., Meusinger, R., Fabritz, S., Reinwarth, M., Biesalski, M., Voigt, S., Buntkowsky, G., and Kolmar, H. (2011) *Angew. Chem. Int. Ed.*, **50**, 5207–5211.
40. Zhang, L., Chen, X., Xue, P., Sun, H.H.Y., Williams, I.D., Sharpless, K.B., Fokin, V.V., and Jia, G. (2005) *J. Am. Chem. Soc.*, **127**, 15998–15999.
41. Boz, E. and Tüzün, N.Ş. (2013) *J. Organomet. Chem.*, **724**, 167–176.
42. Boren, B.C., Narayan, S., Rasmussen, L.K., Zhang, L., Zhao, H.T., Lin, Z.Y., Jia, G.C., and Fokin, V.V. (2008) *J. Am. Chem. Soc.*, **130**, 8923–8930.
43. Majireck, M.M. and Weinreb, S.M. (2006) *J. Org. Chem.*, **71**, 8680–8683.
44. Ruan, Y.-B., Yu, Y., Li, C., Bogliotti, N., Tang, J., and Xie, J. (2013) *Tetrahedron*, **69**, 4603–4609.
45. Zhang, J., Kemmink, J., Rijkers, D.T.S., and Liskamp, R.M.J. (2013) *Chem. Commun.*, 4498–4500.
46. Arisawa, M., Fujii, Y., Kato, H., Fukuda, H., Matsumoto, T., Ito, M., Abe, H., Ito, Y., and Shuto, S. (2013) *Angew. Chem. Int. Ed.*, **52**, 1003–1007.
47. Pasini, D. and Ricci, M. (2007) *Curr. Org. Synth.*, **4**, 59–80.
48. Baudry, Y., Bollot, G., Gorteau, V., Litvinchuk, S., Mareda, J., Nishihara, M., Pasini, D., Perret, F., Ronan, D., Sakai, N., Shah, M.R., Som, A., Sorde, N., Talukdar, P., Tran, D.-H., and Matile, S. (2006) *Adv. Funct. Mater.*, **16**, 169–179.
49. (a) Shimizu, L.S. (2007) *Polym. Int.*, **56**, 444–452; (b) Shimizu, L.S., Salpage, S.R., and Korous, A.A. (2014) *Acc. Chem. Res.*, **47**, 2116–2127.
50. Bong, D.T., Clark, T.D., Granja, J.R., and Ghadiri, M.R. (2001) *Angew. Chem. Int. Ed.*, **40**, 988–1011.
51. Ghorai, A., Padmanaban, E., Mukhopadhyay, C., Acharia, B., and Chattopadhyay, P. (2012) *Chem. Commun.*, 11975–11977.
52. Ghorai, A., Gayen, A., Kulsi, G., Padmanaban, E., Laskar, A., Achari, B., Mukhopadhyay, C., and Chattopadhyay, P. (2011) *Org. Lett.*, **13**, 5512–5515.

53. Kulsi, G., Ghorai, A., Achari, B., and Chattopadhyay, P. (2015) *RSC Adv.*, **5**, 64675–64681.
54. Pehere, A.D., Sumby, C.J., and Abell, A.D. (2013) *Org. Biomol. Chem.*, **11**, 425–429.
55. Chouhan, G. and James, K. (2013) *Org. Lett.*, **15**, 1206–1209.
56. Ajay, A., Sharma, S., Gupt, M.P., Bajpai, V., Kumar, H.B., Kaushik, M.P., Konwar, R., Ampapathi, R.S., and Tripathi, R.P. (2012) *Org. Lett.*, **14**, 4306–4309.
57. Allam, A., Dupont, L., Behr, J.-B., and Plantier-Royon, R. (2012) *Eur. J. Org. Chem.*, **2012**, 817–823.
58. Kumar, A., Geng, Y., and Schmidt, R.R. (2012) *Eur. J. Org. Chem.*, **2012**, 6846–6851.
59. Lewandowski, B. and Jarosz, S. (2010) *Org. Lett.*, **12**, 2532–2535.
60. Campo, V.L., Ivanova, I.M., Carvalho, I., Lopes, C.D., Carneiro, Z.A., Saalbach, G., Schenkman, S., Santana da Silva, J., Nepogodiev, S.A., and Field, R.A. (2015) *Tetrahedron*, **71** (39), 7344–7353.
61. Muthana, S., Yu, H., Cao, H., Cheng, J., and Chen, X. (2009) *J. Org. Chem.*, **74**, 2928–2936.
62. Yu, Y., Bogliotti, N., Tang, J., and Xie, J. (2013) *Eur. J. Org. Chem.*, **2013**, 7749–7760.
63. Schulz, M. and Christoffers, J. (2013) *Tetrahedron*, **69**, 802–809.
64. Hradilová, L., Grepl, M., Dvoráková, B., and Hradil, P. (2013) *Tetrahedron Lett.*, **54**, 1218–1221.
65. Li, Y. and Flood, A.H. (2008) *Angew. Chem. Int. Ed.*, **47**, 2649–2652.
66. Duan, Q., Xia, W., Lin, C., Pan, Y., and Wang, L. (2015) *Tetrahedron Lett.*, **56**, 4002–4006.
67. White, N.G. and Beer, P.D. (2012) *J. Org. Chem.*, **8**, 246–252.
68. Cai, J., Hay, B.P., Young, N.J., Yang, X., and Sessler, J.L. (2013) *Chem. Sci.*, **4**, 1560–1567.
69. Hua, Y. and Flood, A.H. (2010) *Chem. Soc. Rev.*, **39**, 1262–1271.
70. Caricato, M., Olmo, A., Gargiulli, C., Gattuso, G., and Pasini, D. (2012) *Tetrahedron*, **68**, 7861–7866.
71. Moletti, A., Coluccini, C., Pasini, D., and Taglietti, A. (2007) *Dalton Trans.*, **16**, 1588–1592.
72. Boiocchi, M., Bonizzoni, M., Moletti, A., Pasini, D., and Taglietti, A. (2007) *New J. Chem.*, **31**, 352–356.
73. Bencini, A., Coluccini, C., Garau, A., Giorgi, C., Lippolis, V., Messori, L., Pasini, D., and Puccioni, S. (2012) *Chem. Commun.*, **48**, 10428–10430.
74. Coluccini, C., Mazzanti, A., and Pasini, D. (2010) *Org. Biomol. Chem.*, **8**, 1807–1815.
75. Colombo, S., Coluccini, C., Caricato, M., Gargiulli, C., Gattuso, G., and Pasini, D. (2010) *Tetrahedron*, **66**, 4206–4211.
76. Coluccini, C., Dondi, D., Caricato, M., Taglietti, A., Boiocchi, M., and Pasini, D. (2010) *Org. Biomol. Chem.*, **8**, 1640–1649.
77. Caricato, M., Leza, N.J., Gargiulli, C., Gattuso, G., Dondi, D., and Pasini, D. (2012) *Beilstein J. Org. Chem.*, **8**, 967–976.
78. Caricato, M., Coluccini, C., Dondi, D., Vander Griend, D.A., and Pasini, D. (2010) *Org. Biomol. Chem.*, **8**, 3272–3280.
79. Caricato, M., Deforge, A., Bonifazi, D., Dondi, D., Mazzanti, A., and Pasini, D. (2015) *Org. Biomol. Chem.*, **13**, 3593–3601.
80. Coluccini, C., Castelluccio, A., and Pasini, D. (2008) *J. Org. Chem.*, **73**, 4237–4240.
81. Agnes, M., Sorrenti, A., Pasini, D., Wurst, K., and Amabilino, D.B. (2014) *CrystEngComm*, **16**, 10131–10138.
82. Caricato, M., Leza, N.J., Roy, K., Dondi, D., Gattuso, G., Shimizu, L.S., Vander Griend, D.A., and Pasini, D. (2013) *Eur. J. Org. Chem.*, **2013**, 6078–6083.
83. Asakawa, M., Ashton, P.R., Hayes, W., Janssen, H.M., Meijer, E.W., Menzer, S., Pasini, D., Stoddart, J.F., White, A.J.P., and Williams, D.J. (1998) *J. Am. Chem. Soc.*, **120**, 920–931.
84. Asakawa, M., Ashton, P.R., Boyd, S.E., Brown, C.L., Menzer, S., Pasini, D., Stoddart, J.F., Tolley, M.S., White, A.J.P., Williams, D.J., and Wyatt, P.G. (1997) *Chem. Eur. J.*, **3**, 463–481.

85. Crespo Alonso, M., Arca, M., Isaia, F., Lai, R., Lippolis, V., Callear, S.K., Caricato, M., Pasini, D., Coles, S.J., and Aragoni, M.C. (2014) *CrystEngComm*, **16**, 8582–8590.
86. Caricato, M., Sharma, A.K., Coluccini, C., and Pasini, D. (2014) *Nanoscale*, **6**, 7165–7174.
87. Coluccini, C., Caricato, M., Cariati, E., Righetto, S., Forni, A., and Pasini, D. (2015) *RSC Adv.*, **5**, 21495–21503.
88. Miljanić, O.Š., Dichtel, W.R., Aprahamian, I., Rohde, R.D., Agnew, H.D., Heath, J.R., and Stoddart, J.F. (2007) *QSAR Comb. Sci.*, **26**, 1165–1174.
89. Megiatto, J.D. Jr., and Schuster, D.I. (2010) *New J. Chem.*, **34**, 276–286.
90. Megiatto, J.D. Jr.,, Schuster, D.I., Abwandner, S., de Miguel, G., and Guldi, D.M. (2010) *J. Am. Chem. Soc.*, **132**, 3847–3861.
91. Romuald, C. and Coutrot, F. (2012) *Angew. Chem. Int. Ed.*, **51**, 2544–2545.
92. Barran, P.E., Cole, H.L., Goldup, S.M., Leigh, D.A., McGonigal, P.R., Symes, M.D., Wu, J., and Zengerle, M. (2011) *Angew. Chem. Int. Ed.*, **50**, 12280–12284.
93. Neal, E.A. and Goldup, S.M. (2015) *Chem. Sci.*, **6**, 2398–2404.
94. Zhou, W., Zheng, H., Li, Y., Liu, H., and Li, Y. (2010) *Org. Lett.*, **12**, 4078–4081.
95. Fox, M.E., Szoka, F.C., and Frechet, J.M.J. (2009) *Acc. Chem. Res.*, **42**, 1141–1151.
96. Binauld, S., Hawker, C.J., Fleury, E., and Drockenmuller, E. (2009) *Angew. Chem. Int. Ed.*, **48**, 6654–6658.
97. Schulz, M., Tanner, S., Barqawi, H., and Binder, W.H. (2010) *J. Polym. Sci., Part A: Polym. Chem.*, **48**, 671–680.
98. Narumi, A., Zeidler, H.S., Barqawi, H., Enders, C., and Binder, W.H. (2010) *J. Polym. Sci., Part A: Polym. Chem.*, **48**, 3402–3416.
99. Wan, X., Liu, T., and Liu, S. (2011) *Biomacromolecules*, **12**, 1146–1154.
100. Sharma, A.K., Cornaggia, C., and Pasini, D. (2010) *Macromol. Chem. Phys.*, **211**, 2254–2259.
101. Edizer, S., Veronesi, B., Karahan, O., Aviyente, V., Değirmenci, I., Galbiati, A., and Pasini, D. (2009) *Macromolecules*, **49**, 1860–1866.
102. Coluccini, C., Metrangolo, P., Parachini, M., Pasini, D., Resnati, G., and Righetti, P. (2008) *J. Polym. Sci., Part A: Polym. Chem.*, **46**, 5202–5213.
103. Cagnoni, E., Pasini, D., Galbiati, A., Ricci, M., and Righetti, P.P. (2003) *Macromolecules*, **36**, 8894–8897.
104. Blazquez, E., Mustarelli, P., Pasini, D., Righetti, P.P., and Tomasi, C. (2004) *J. Mater. Chem.*, **14**, 2524–2529.
105. Laurent, B.A. and Grayson, S.M. (2006) *J. Am. Chem. Soc.*, **128**, 4238–4239.
106. Laurent, B.A. and Grayson, S.M. (2011) *J. Am. Chem. Soc.*, **133**, 13421–13429.
107. Lonsdale, D.E., Bell, C.A., and Monteiro, M.J. (2010) *Macromolecules*, **43**, 3331–3339.
108. Sun, Y., Wang, Z., Li, Y., Zhang, Z., Zhang, W., Pan, X., Zhou, N., and Zhu, X. (2015) *Macromol. Rapid Commun.*, **36**, 1341–1347.
109. Li, Y., Su, H., Feng, X., Yue, K., Wang, Z., Lin, Z., Zhu, X., Fu, Q., Zhang, Z., Cheng, S.Z.D., and Zhang, W.-B. (2015) *Polym. Chem.*, **6**, 827–837.
110. Ulloora, S., Shabaraya, R., and Adhikari, A.V. (2013) *Bioorg. Med. Chem. Lett.*, **23**, 3368–3372.

# 10
# Modifications of Nucleosides, Nucleotides, and Nucleic Acids using Huisgen's [3+2] Azide–Alkyne Cycloaddition: Opening Pandora's Box

*Franck Amblard, Ozkan Sari, Sebastien Boucle, Ahmed Khalil, and Raymond F. Schinazi*

## 10.1
## Introduction

The Cu(I)-catalyzed Huisgen [3+2] Azide–Alkyne cycloaddition (CuAAC), popularized by Sharpless in the early 2000s as part of his "click chemistry" concept, is now a widely used reaction in the field of organic chemistry. This trend, which can also be observed in the field of nucleosides and nucleotides synthesis, has led to the publication of hundreds of articles reporting the use of CuAAC in drug discovery for the synthesis of new nucleosides, but also for the preparation of nucleoside bioconjugates with novel biological and physical properties. This reaction has also been successfully utilized in the formation of artificial oligonucleotides and applied to DNA derivatization, conjugation, and labeling. The purpose of this review is not to be exhaustive, but rather to focus on the extraordinary diversity of applications related to the use of the CuAAC in the field of nucleoside, nucleotide, and oligonucleotide chemistry.

### 10.1.1
### Nucleoside Modifications

#### 10.1.1.1 Nucleoside Analogs as Potential Drugs

The development of nucleoside analogs for medicinal uses has had a marked impact on clinical chemotherapy, and numerous nucleoside analogs were successfully developed for the treatment of human immunodeficiency viruses (HIV), hepatitis B virus (HBV), hepatitis C virus (HCV), herpes simplex virus (HSV), cytomegalovirus (CMV), varicella zoster virus (VZV), and various cancers [1].

Based on these successes, several base- and sugar-modified nucleoside analogs bearing a triazole group at various positions have been prepared and evaluated.

**Nucleobase Modifications:** In the search for new therapeutic derivatives, substituted triazole rings have been introduced on both pyrimidines [2] and purine nucleosides [3–5](Figure 10.1). Among all the synthesized compounds, some

*Click Reactions in Organic Synthesis*, First Edition. Edited by Srinivasan Chandrasekaran.

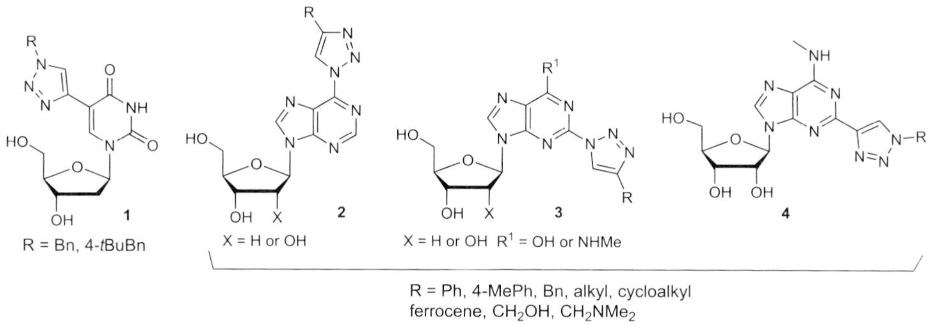


**Figure 10.1** Base-modified nucleosides **1–4**.

of them displayed anticancer activity (compounds **1**, R = Bn, 4-phenoxyphenyl) [6] or moderately inhibited vesicular stomatitis virus and CMV replication (compounds **3**, X = H, R = *n*-Bu, $R^1$ = OH; X = OH, R = Ph).

The discovery of clinically relevant nucleoside analogs, such as Ribavirin [7], Viramidine [8], or AICAR [9], containing a 1,2,4-triazole or an imidazole ring as nucleobase has led to an intensive search for new five-membered ring base-modified nucleoside analogs including 1,2,3-triazole derivatives (Figure 10.2)

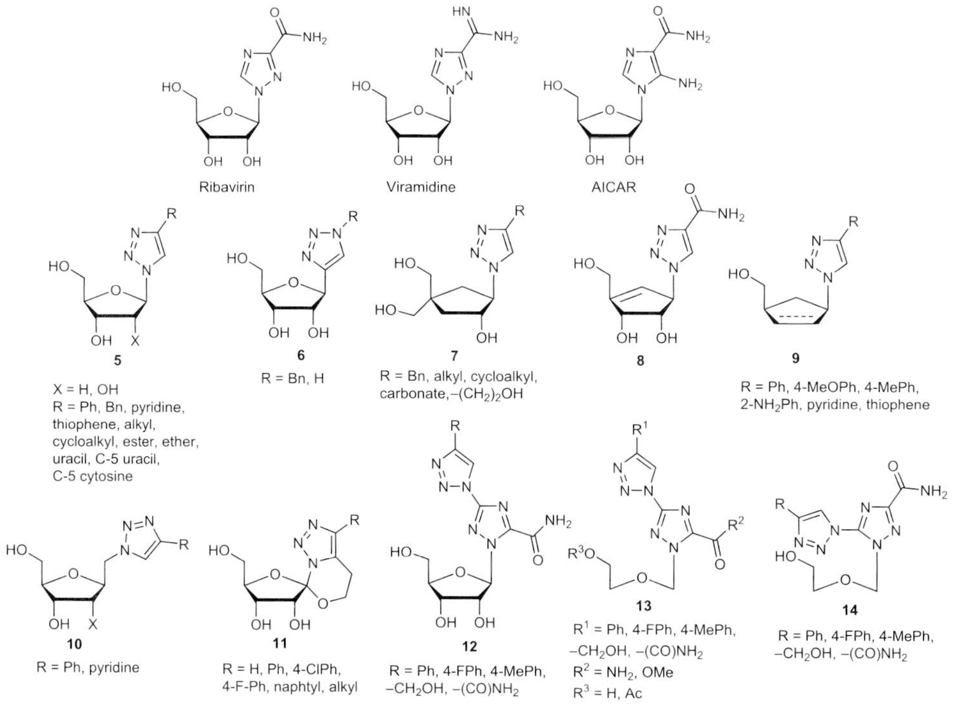


**Figure 10.2** Base-modified nucleoside analogs.

[10]. Interestingly, although the majority of these compounds were synthesized using classical CuAAC conditions, in certain cases, the use of ultrasound [11] or microwave irradiation [10f, 12–14] was shown to significantly improve the yield and speed of the cycloaddition.

Among all the compounds prepared using a CuAAC, compound **8** exhibited moderate activity against cowpox virus and vaccinia virus [10b] and compounds **5** (X = OH, R = 6-methoxy-naphthyl and 2-pyridyl) were more potent against the cancer cell line Chronic Myeloid Leukemia (CML), K562 than the reference compound AICAR (Figure 10.2). Compounds **11** (R = 4-Chlorophenyl) and **14** (R = $CH_2OH$, $CONH_2$) displayed anti-MHV (Mouse Hepatitis Virus) [15] and anti-TMV (Tobacco Mosaic Virus) activities, respectively [16, 17] while compound **13** ($R^1$ = 1-methylpyrenyl, $R^2$ = OMe, and $R^3$ = Ac) acted as an antiproliferative agent and an immunomodulator [18]. Although most of the base-modified nucleosides are prepared through the reaction of a β-1′-azido sugar with an alkyne, β-1′-alkynyl sugar moieties can also react with azides in order to form C-nucleosides bearing a substituted 1,2,3-triazole base (compound **6**) (Figure 10.2) [19]. Finally, "fleximers," where the nucleobase is splinted but still retains the key recognition of DNA bases, have been studied [20, 21]. For instance, St Amant *et al.* [22] prepared several C-5 connected uracil and cytosine analogs (compound **5**), and Parmenopoulou *et al.* [23] reported compound **5** (X = OH, R = uracil) as a potent competitive inhibitor of RNAase A (Figure 10.2).

**Sugar-Modified Nucleosides:** 1,2,3-Triazole rings have been introduced at the 2′- [24, 25], 3′- [26–29], 4′- [30, 31], and 5′- [32, 33] positions of the nucleoside sugar moiety (Figure 10.3). Among all the synthesized compounds, compounds **16** (R = Me, $R^1$ = H, $R^2$ = Naphthyl), **18** (R = H, Ph, *t*-Bu), and **19** (R = Ph, 4-Br-Ph, 4-NHAc-Ph, 4-phenoxyphenyl, $R^1$ = Me) displayed anti-HIV, HBV, and/or influenza A activity while **21** (R = 1-cyclopentyl-1-hydroxyethyl) inhibited the α-2,3-sialyltransferase. In addition, several highly modified triazolo nucleoside analogs such as compounds **22** and **23** have been prepared and evaluated as potential antiviral agents (Figure 10.3) [34–38].

#### 10.1.1.2 Nucleoside Bioconjugates

The high efficiency and simplicity of CuAAC methodology made this reaction suitable for the development of new nucleoside bioconjugates with novel or increased biological properties, altered hydrophobicity, increased bioaffinity, or ability to carry metal ions.

Thus, Lesnikowski's group [39] described the coupling of nucleoside analogs **24–26** with carborane or *closo*-dodecarborane to form conjugate complexes for boron neutron capture therapy (BNCT) (Figure 10.4).

The CuAAc methodology was used to prepare a variety of nucleoside bioconjugates bearing spin labels or fluorescent probes (compounds **27–30**) [40–43] and was also used for the $^{18}F$-labeling of nucleosides (compound **27G**) (Figure 10.5) [44].

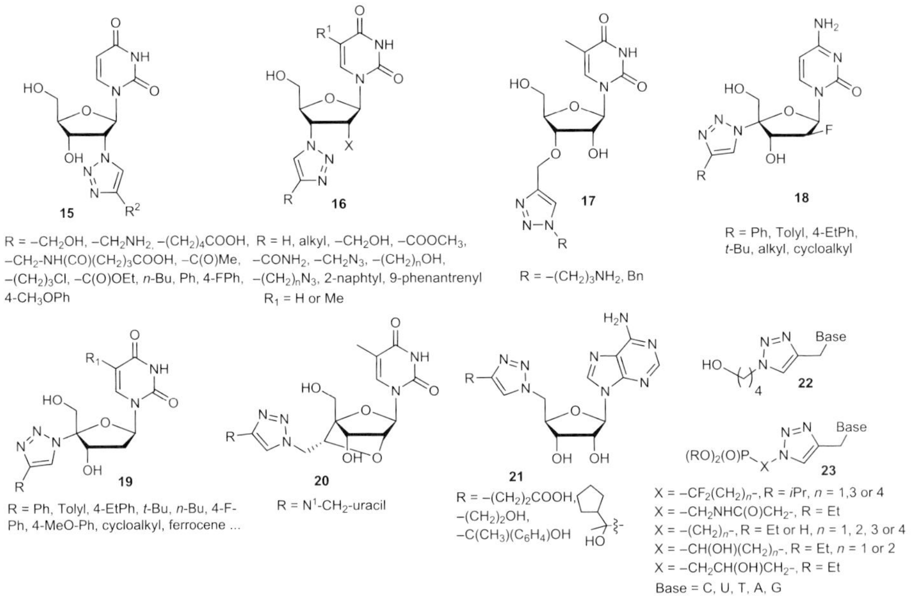


**Figure 10.3** Structures of sugar-modified nucleosides **15–23**.

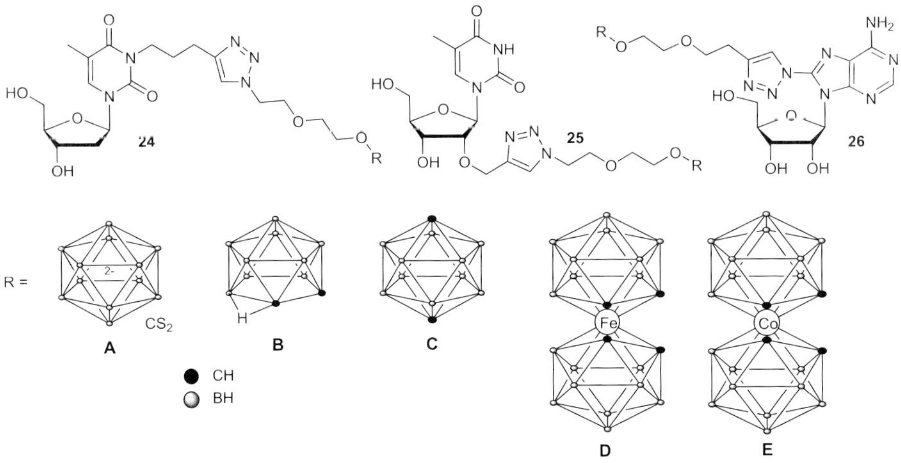


**Figure 10.4** Nucleoside bioconjugates bearing carboranes and *closo*-dodecarborane.

For dynamic measurement and tracking of signaling events inside single living cells using fluorescence lifetime imaging (FLIM) of living MCF-7 cancer cells, Zayas *et al.* prepared triazole adducts **31–33** through a Strain-Promoted Alkyne–Azide Cycloaddition (SPAAC) between a 2-azidopurine or 5-azido pyrimidine nucleoside and highly reactive cyclooctyne derivatives (Figure 10.6) [45].

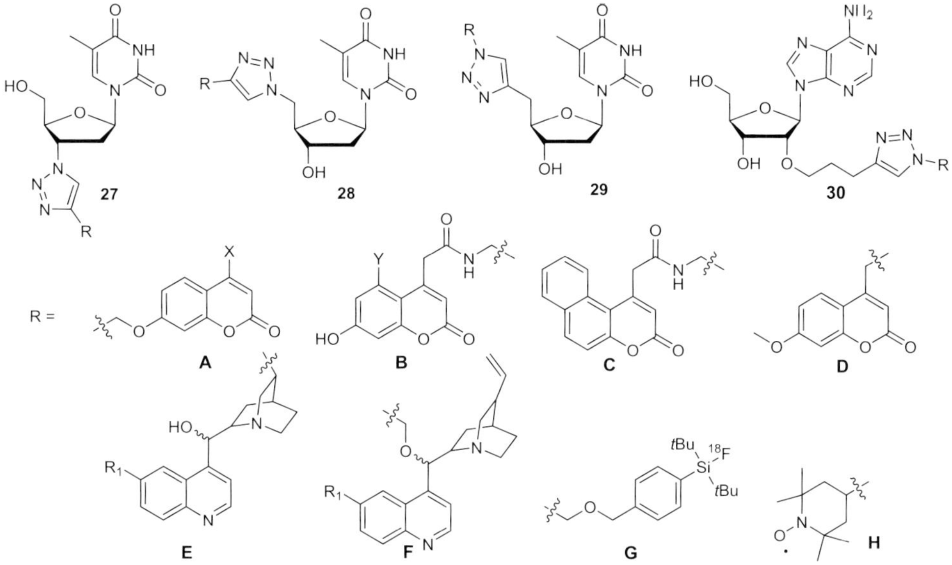

Figure 10.5 Structures of nucleoside bioconjugates **27–30**.

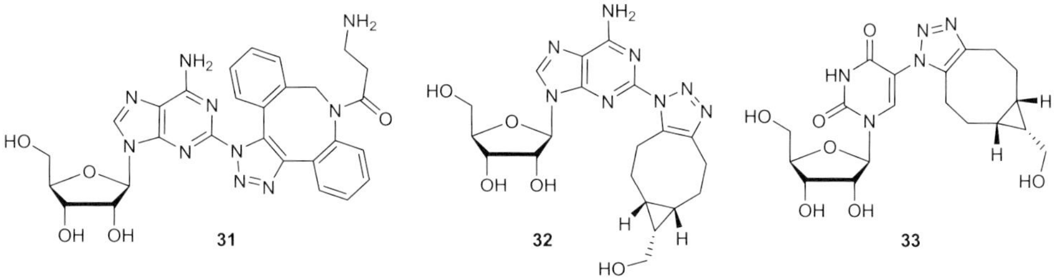

Figure 10.6 Structure of bioconjugates nucleosides **31–33**.

Taking advantage of the versatility of the Cu(I)-catalyzed 1,3-cycloaddition, Lee *et al.* developed an approach to synthesize and screen fucosyltransferase (Fuc-T) inhibitors *in situ*. Among the 85 compounds prepared, (**34**) was found to be a selective inhibitor of α-1,3-Fuc-T VI ($IC_{50} = 0.15\ \mu M$) (Figure 10.7) [46].

In the search for more potent HIV-1 derivatives, Danel *et al.* prepared compound **35A** (Figure 10.8) [47] by reacting AZT and a very potent HIV-1 inhibitor, 6-(3,5-dimethylbenzyl)-5-ethyl-1-((prop-2-yn-1-yloxy)methyl)pyrimidine-2,4(1*H*,3*H*)-dione, through "click" chemistry. Interestingly, this "chimera" showed antiviral activity against wild-type HIV-1 and mutant strains comparable to those observed for the parent compound (assuming no residual AZT was present during the biological testing). Similarly, AZT was also coupled with Bevirimat, an anti-HIV drug, or various triterpenoid acids to generate compounds **35B** and **35C** (Figure 10.8), which displayed, respectively, anti HIV-1 [48] and antitumoral activities [49]. Using the same AZT as a starting material,

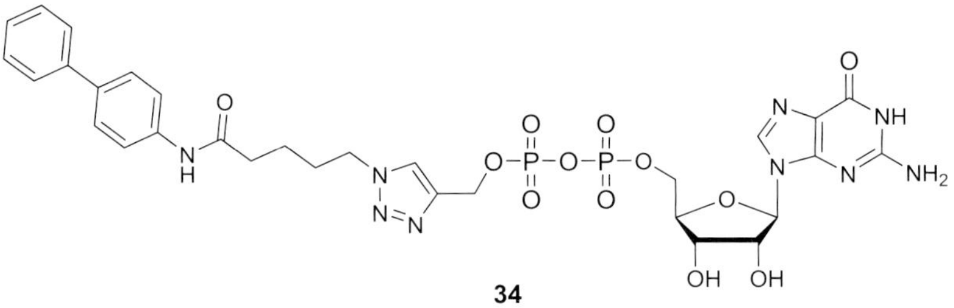

**34**

**Figure 10.7** Fucosyltransferase inhibitor **34**.

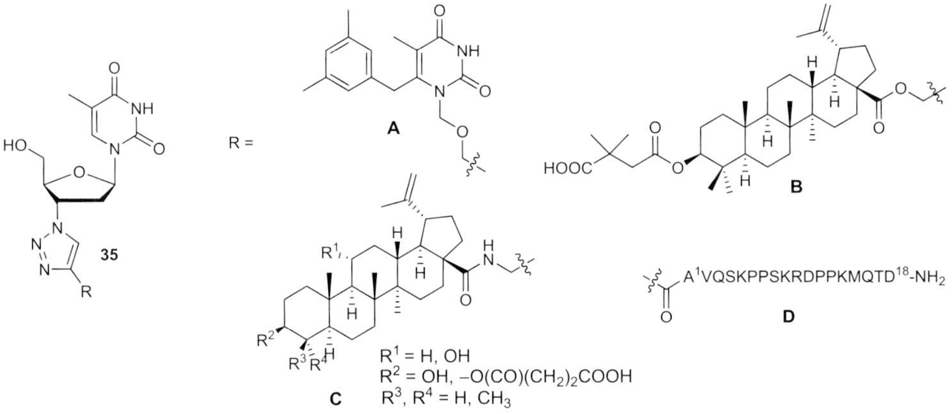


**Figure 10.8** Structures of bioconjugate nucleosides **35**.

Dobkowsky *et al.* prepared compound **35D** and demonstrated that a peptidic palindromic sequence (Sys) could be effectively used to transport low-weight molecules in tomato plants (Figure 10.8) [50].

Bioconjugates of nucleosides and Pleuromutilin, a naturally occurring antibacterial agent, have also been prepared and found to display antibacterial activity (compounds **36–41**) (Figure 10.9). Interestingly, some of these analogs showed improved affinity to the peptidyl transferase center in the ribosome compared to the natural Pleuromutilin [51].

In order to potentially improve solubility and molecular recognition of anti-VZV bicyclic furo[2,3-*d*]pyrimidine nucleosides, Jin *et al.* [52] prepared new carbohydrate conjugated analogs **42** (Figure 10.10). Evaluation of these compounds has not been reported yet.

Mindt *et al.* [53] developed the "click-to-chelate" approach, which allowed them to synthesize metal-labeled nucleoside conjugate **43** in a one-pot procedure (Figure 10.11). The authors showed that the 1,4-disubstituted triazole moiety forms an integral part of the metal-chelating system and facilitates the incorporation of labeled complexes into biomolecules. Nikan *et al.* [54] were also able to generate compound **44**, the first example of a water-soluble cavitand

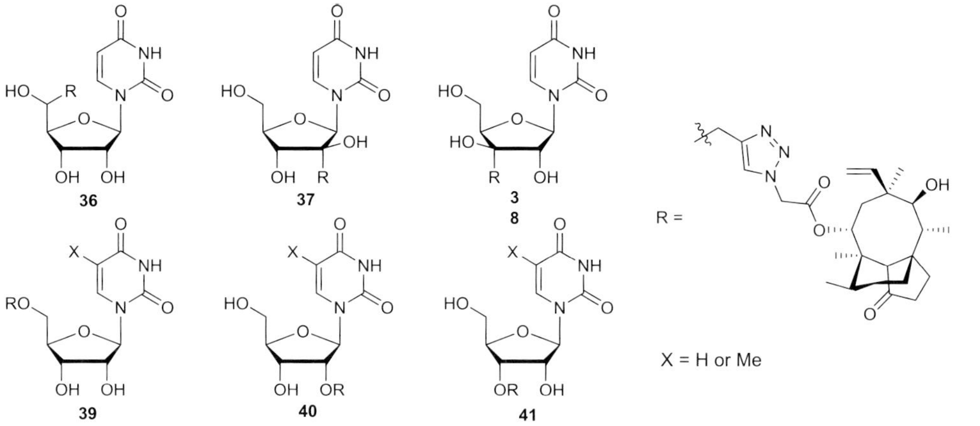


Figure 10.9 Structures of nucleoside bioconjugates 36–41.

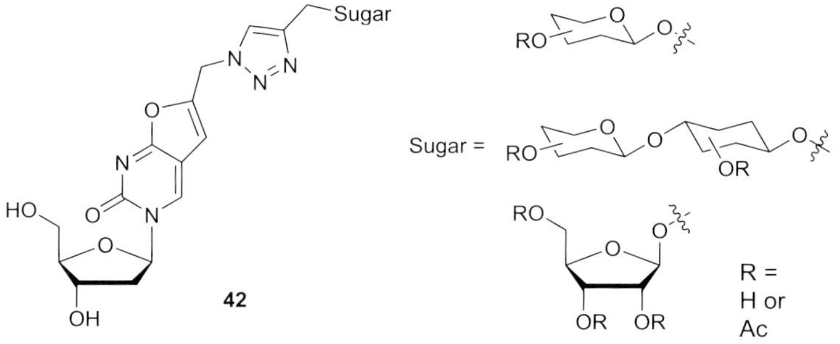


Figure 10.10 Structure of nucleoside bioconjugates 42.

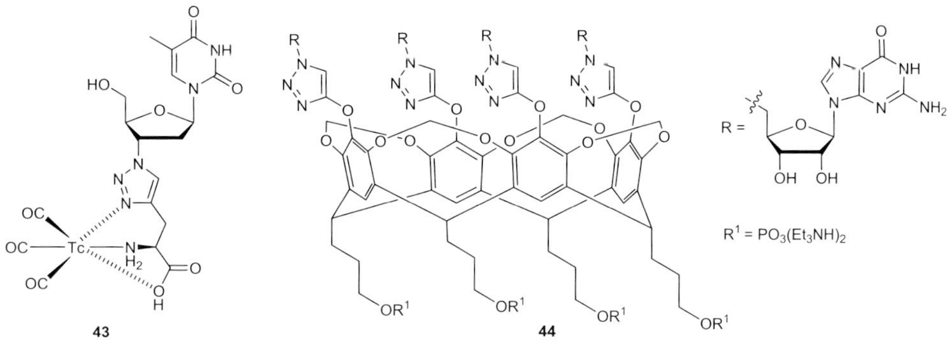


Figure 10.11 Structure of radiolabeled bioconjugate 43 and cavitand 44.

bearing four guanine moieties. Although further characterization of this unique compound is still under process, it could potentially be used already as a screening tool for binding ligands or proteins or as components of nanoscale structures (Figure 10.11).

## 10.2 Nucleotide and Nucleic Acid Modifications

### 10.2.1 "Artificial" DNA

The synthesis of DNA containing an unnatural triazole linkage instead of a regular phosphate moiety has also been investigated (Figure 10.12). These DNA analogs can be formed by reaction of nucleotides containing either a 3′-azido group or a 5′-alkyne in the presence of a complementary template in order to facilitate the CuAAC. Thus, click-ligated DNA **45** (Figure 10.12) was used as a template in Polymerase Chain Reaction (PCR) and proved to be readable except for the fact that the progeny strand only had one thymidine whereas the original template had two at the triazole linkage [55]. On the other hand, the more flexible triazole-linked DNA **46** produced perfect amplicons, in which the bases around the triazole linkage were copied correctly [56]. This specific type of linker also appeared biocompatible and was accurately replicated in *E. coli* and more recently in human cells [57].

Triazole-linked analogs of deoxyribonucleic acid or ribonucleic acids ($^{TL}$DNA and $^{TL}$RNA), in which every phosphate unit is replaced by a 1,2,3-triazole moiety, were also prepared, either in solution or using solid-phase synthesis (compounds **47** and **48**) (Figure 10.12). Interestingly, the duplex of oligo dT with dA $^{TL}$DNA appeared to be stable [58].

Another example of "artificial" DNA preparation was the incorporation into DNA of 1,2,3-triazole residues with nucleobase tethers (compound **49**) (Figure 10.12). However, nucleobases of these "fleximers" failed to pair with the opposite bases, which resulted in duplex instability [59].

### 10.2.2 Presynthetic Modification DNA

In order to prepare modified Oligodeoxynucleotide (ODNs), two main strategies coexist: Pre- and postsynthetic modifications. The term premodification is used

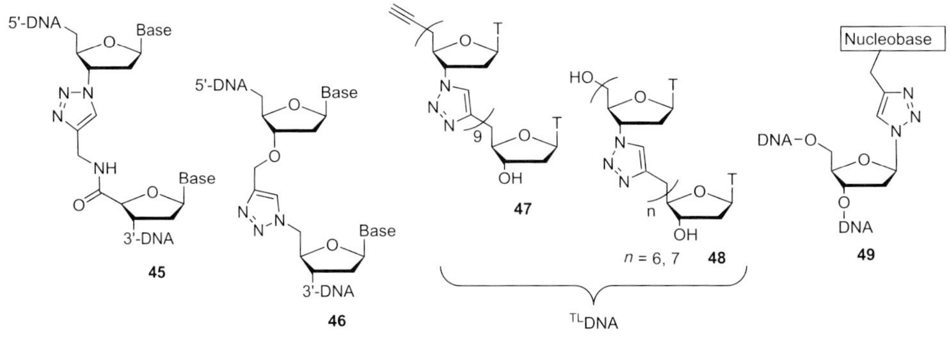


**Figure 10.12** Examples of "artificial" DNAs.

when an already modified nucleoside is incorporated into the DNA strand, while the term postmodification means that the modification is achieved afterward on the strand of DNA itself. Although Seela and coworkers frequently used a postsynthetic approach (See Section 10.2.3.) to label alkyne-containing oligonucleotides, they recently described the synthesis of fluorescent "clicked" nucleosides. These compounds were functionalized with pyrene and were later incorporated into a DNA strand (Figure 10.13). Interestingly, this group showed that compound **51** possessing a short linker destabilized DNA, while compound **50** including a longer linker improved the stability of the duplex. [60]

In the same manner, boronic-acid-labeled thymidine-5′-triphosphate **52** was synthesized by Lin *et al.* [61] using a CuAAC approach (Figure 10.14), and these compounds were successfully incorporated into DNA for aptamer selection.

A presynthetic approach has also been used to prepare lipid-conjugated oligonucleotides (LONs) in order to improve cellular uptake and intracellular delivery of various ODNs. Thus, Grijalvo *et al.* [62] discovered lipid-siRNA with anti-inflammatory properties while Godeau *et al.* [63] synthesized nontoxic lipid-ON conjugates, which efficiently inhibit HCV internal ribosome entry site-dependent translation in human Huh7 cells.

Finally, Carell's group successfully synthesized gamma-labeled alkyne nucleoside triphosphate analogs that can be coupled with fluorophore azide groups via

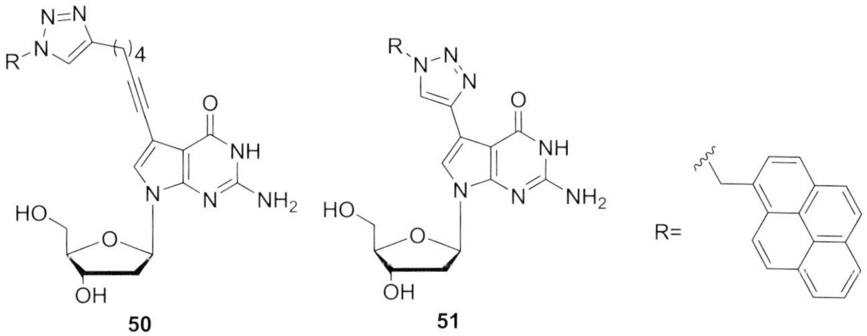


**50** **51**

**Figure 10.13** Pyrene-containing nucleosides **50** and **51**.

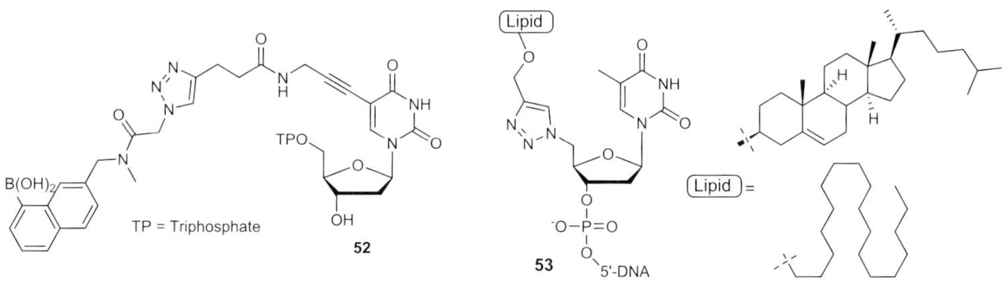


**52** **53**

**Figure 10.14** Structures of boronic-acid-modified thymidine-5′-triphosphate **52** and lipid–oligonucleotide conjugates **53**.

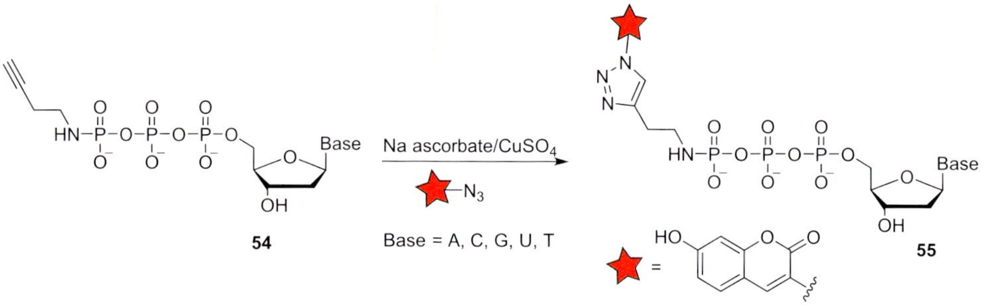


**Figure 10.15** Synthesis of fluorophore gamma-labeled nucleoside 5′-triphosphates. (Reproduced from Serdjukow *et al.* [64], with permission of Royal Society of Chemistry.)

CuAAC [64]. Interestingly, these fluorescent triphosphates were shown to be substrates for DNA polymerases (Figure 10.15).

### 10.2.3 Postsynthetic Modification

The derivation of complex biomolecules in order to enhance the understanding of various biological mechanisms *in vitro* and *in vivo* has been thoroughly studied over the years. However, the wide variety of functionalities present on these systems has been a significant barrier toward an effective selective derivation. Until the advent of the CuACC, the probes were usually prepared by coupling amino-containing nucleotides with activated esters of fluorescent tags followed by enzymatic incorporation of their triphosphate form into the DNA (presynthetic modification). Unfortunately, the incorporation of sterically hindered probes has proven to be extremely difficult to achieve, especially since numerous fluorophores (i) decompose under the basic conditions used during the solid-phase oligonucleotide synthesis (SPOS) and (ii) are unstable to the deprotection conditions at the end of the PCR process. In this context, the specificity of CuAAC, related to the absence of azide or alkynyl groups in natural systems, enabled the derivation of DNAs within a complex biological medium (postsynthetic modification) [65]. Despite these unique features, there are several limitations related to the use of CuAAC to modify DNA and/or RNA strands. First of all, copper ions are known to cause strand breaks [66] and to potentially degrade DNA within few minutes. Earliest examples of click-chemistry-mediated tagging of a single-stranded DNA [67] were thus performed without copper catalyst, but the cycloaddition would be hindered by long reaction time and high temperature as a consequence. Since then, Cu(I)-stabilizing ligands such as tris(benzyltriazolyl)methylamine (TBTA) have been developed and are commonly used to both enhance reaction rate and protect DNA from degradation. Another solution to this problem was the development of catalyst-free SPAAC reactions between azides and superreactive cyclooctynes derivatives. It is noteworthy though that SPAAC is not completely bioorthogonal since cyclooctynes can react

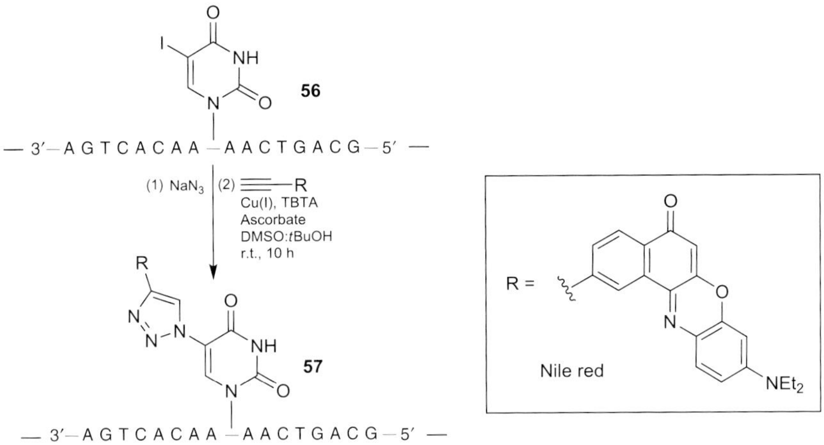


**Figure 10.16** Preparation of the nile-red-modified DNA1.

with strong nucleophiles such as thiols, which are omnipresent in cells [68] and can lead to the formation of two to four diastereoisomers. The final limitation of the CuAAC is the challenging synthesis of azide containing DNA using a solid-phase approach since azides are easily reduced by the phosphorous(III) atom inevitably present in the most popular phosphoramidite strategy (Staudinger reaction). Nowadays, DNA derivation is most commonly achieved by "clicking" an alkynyl containing nucleotide with an azide-functionalized dye. Several groups tried to circumvent this issue. For instance, Beyer *et al.* reported the late-stage introduction of an azide to the DNA by reaction of a halogenated nucleotide (containing 5-iodo-2′-deoxyuridine) with sodium azide before reaction with the ethynyl-modified nile red (Figure 10.16) [69].

As mentioned earlier, the labeling of ODNs is usually achieved by the incorporation of nucleoside bearing an alkyne on the base before reaction with azide-labeled dyes [70, 71]. Alkynyl functional groups are commonly attached to the 5-position of pyrimidines and 7-position of purines (Figure 10.17). Such alkyne-containing bases are well tolerated by polymerase (PCR) and are also compatible with solid-phase oligonucleotide synthesis. Overall, they only slightly increase the melting temperature of the DNA and show a negligible impact on the duplex [72].

Seela also showed that tripropargylamine-dU containing oligonucleotide can easily be clicked twice with two initially nonfluorescent 3-azido-7-hydroxycoumarins (Figure 10.18). Interestingly, the resulting bis-coumarin-labeled oligonucleotide **61** was less fluorescent than its single triazolide coumarin counterpart due to the self-quenching effect between the fluorophores [73].

The modification of DNA including differently protected alkyne-containing bases, which can be sequentially/selectively cleaved in order to "click" various tags on the same strand, has also been explored [74]. Thus, a DNA strand containing a 5-alkynyl uridine derivative along with TMS- and TIPS-protected 5-alkyne-2′-deoxycytidine analogs was synthesized, and after a series of

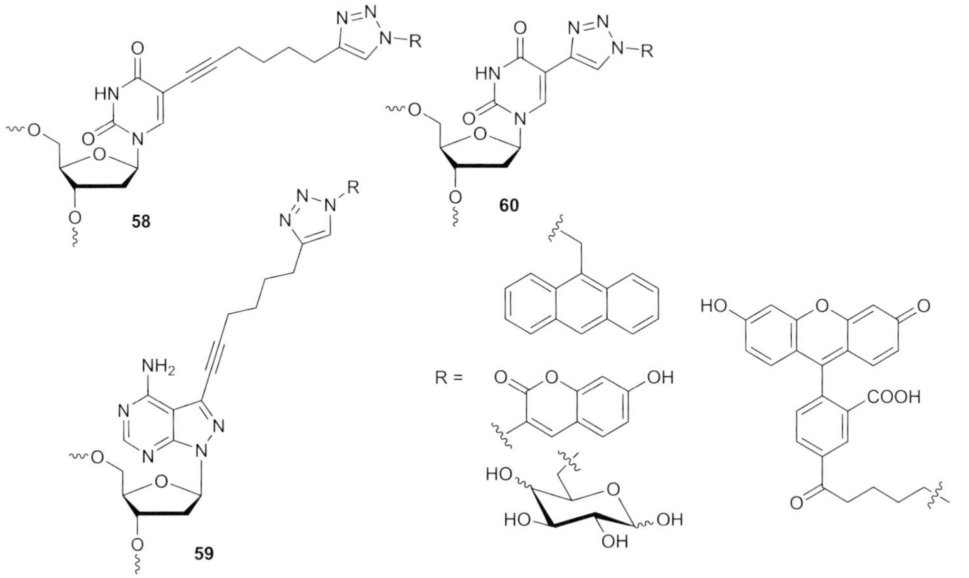

Figure 10.17 Internal DNA labeling between 8-aza-7-deaza-2′-deoxyadenosine, 5′-modified dU alkyne precursors and azide-labeled fluorogenic dyes.

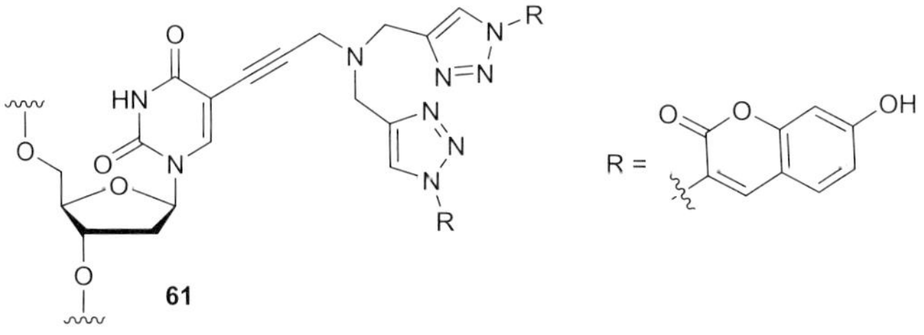

Figure 10.18 Labeling of the 5-position of uridine analog with 3-azido-7-hydroxycoumarins.

selective deprotection, each of them was "clicked" with an azido-modified tag (Figure 10.19).

((5-(Octa-1,7-diynyl)-2′-deoxyuridine was also used in the preparation of peptide–DNA constructs **69** by reaction of an alkynyl containing DNA strand and different azido-containing peptides (Figure 10.20) [75]. Interestingly, these peptide–DNA hybrid constructs showed retention of their peptide and DNA properties, opening the way to the preparation of orthogonally tunable scaffolds having various functions.

Sasaki *et al.* reported an approach to specifically detect mutagenic nucleotide $O^6$-Me-dG in DNA using a sequential functionality transfer–CuAAC strategy (Figure 10.21) [76]. $O^6$-Me-dG is formed by reaction with alkylating agents [77] and is known to induce GC to AC transition mutation during DNA replication both *in vitro* [78] and *in vivo* [79]. It is assumed that its genotoxicity is involved

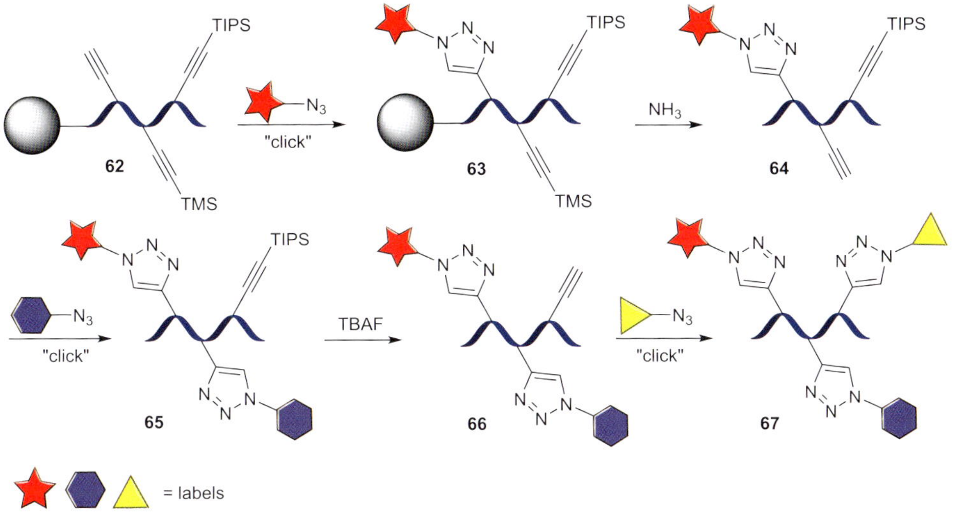


**Figure 10.19** Sequential modification of DNA by three consecutive CuAAC reactions.

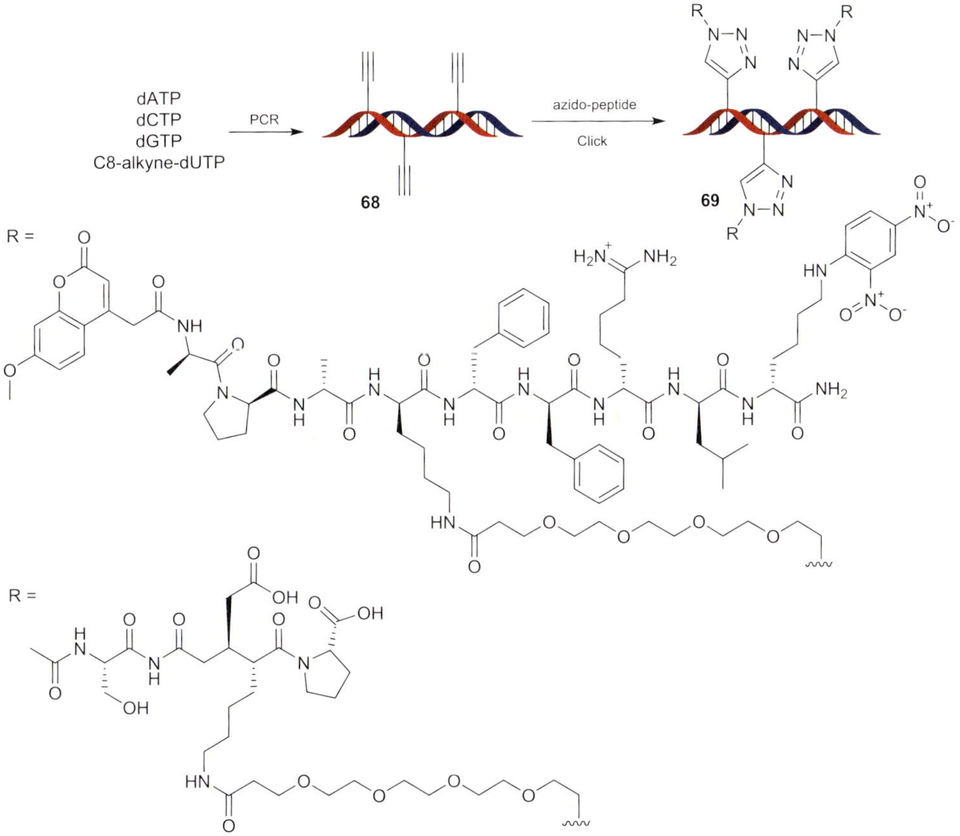


**Figure 10.20** Preparation of DNA–peptide hybrids via CuAAC.

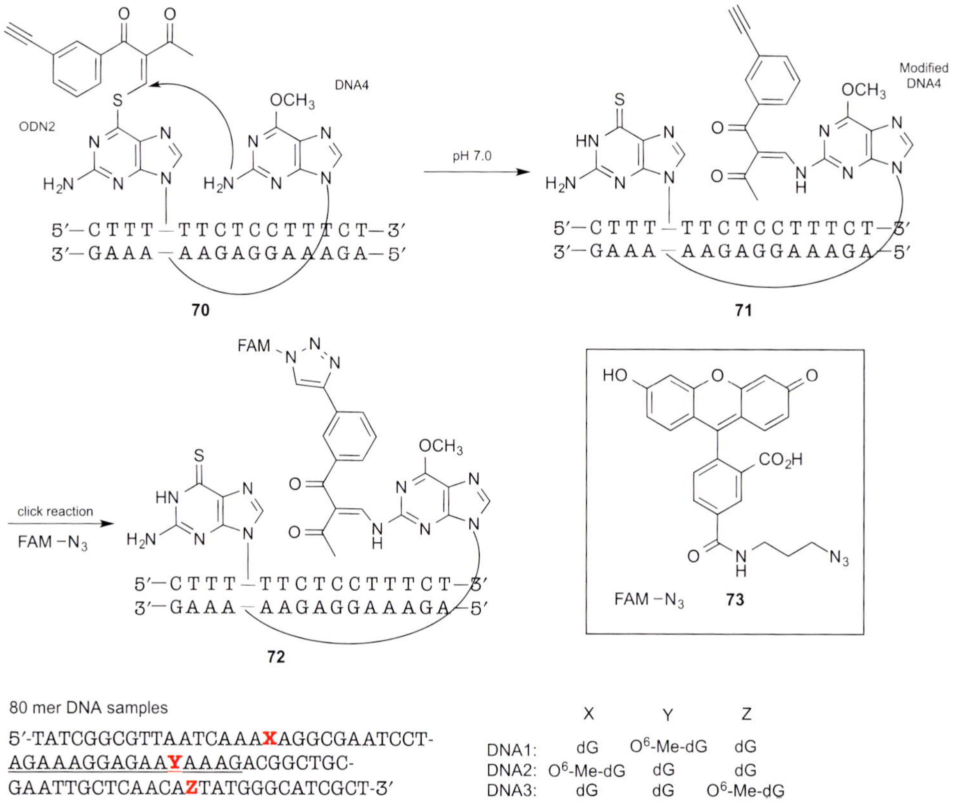


**Figure 10.21** Selective functionality transfer and click reaction on $O^6$-Me-dG with FAM-$N_3$.

in the carcinogenesis of various type of tissue [80]. Thus, in this work, an alkyne-containing transfer group 6-thiodG was carefully incorporated into the FT-ODN sequence so as to target $O^6$-Me-dG on the DNA strand. Then, the alkyne-containing transfer group of 6-thiodG was transferred to $O^6$-Me-dG at neutral pH. (Interestingly, while the transfer was very fast to the 2-amino group of $O^6$-Me-dG, no transfer was observed to the 2-amino group of the natural dG-containing template.) Then, in a second step, a CuAAC was performed between the transferred alkyne and the fluorescent azide derivative FAM-$N_3$. This protocol was then successfully applied to various 80-mer DNA strands containing an $O^6$-Me-dG at the targeted complementary site Y as well as nontarget sites X or Z for the transfer group of ODN1. After the transfer and CuAAC were completed, the mixtures were treated with 1% $NH_4OH$ to hydrolyze the unreacted vinyl sulfide derivative while the FAM adduct of DNA1 containing the modified $O^6$-Me-dG remained intact, therefore being the only fluorescent-labeled DNA in the mixture. Moreover, no detectable fluorescence was observed for DNA2 and DNA3 (no functionality group transfer) proving the high $O^6$-Me-dG and sequence selectivity of this procedure.

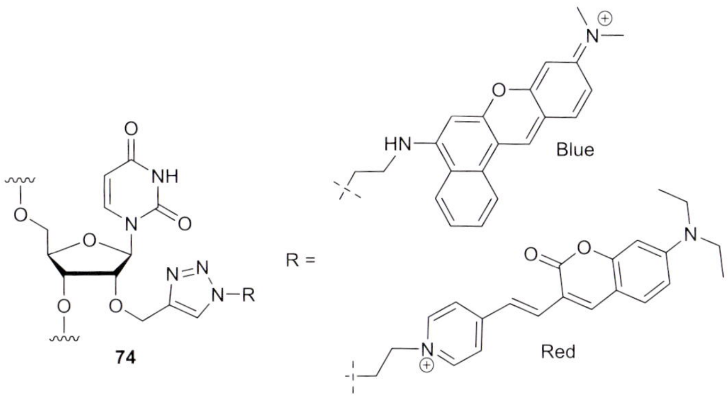


**Figure 10.22** Dye labeling at the 2-position of ribose in ODNs using CuAAC.

The 2′-position can also be a tagging site used in a postsynthetic derivation approach. Wagenknecht *et al.* successfully derivatized ODNs with phenoxazinium- (blue) and coumarin-modified (red) derivatives (Figure 10.22) [81]. Interestingly, the 2′-chromophore-containing uridines still recognize adenine as its counterbase and do not affect the thermal stability of the duplex when compared to the unmodified duplexes.

Wengel *et al.* also reported the derivation of a 2′-*O*-propargyl-modified uridine-containing ODN with a series of xanthene and cyanine azide derivatives (Figure 10.23) [82]. The first CuAAC was performed on the

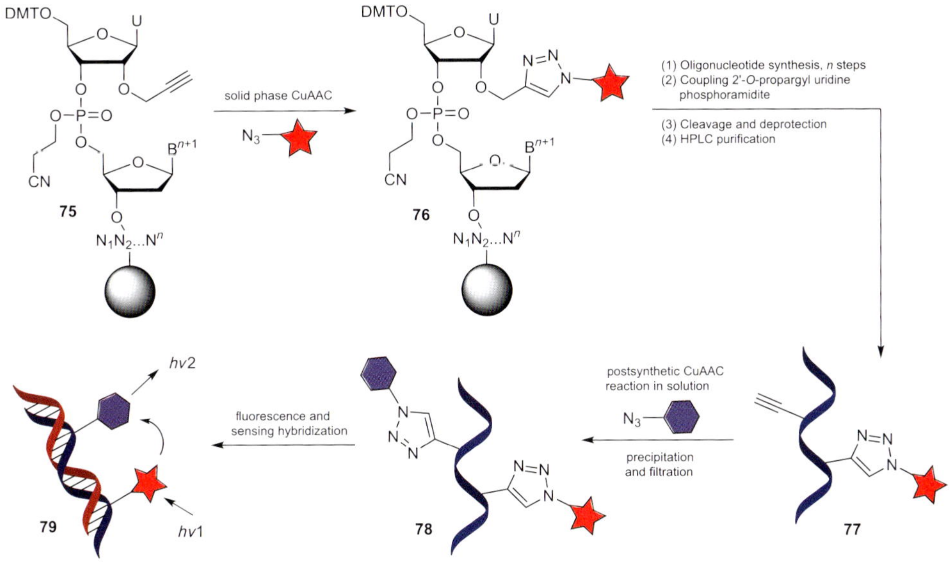


**Figure 10.23** Xanthene- and cyanine-labeled probe prepared by postsynthetic CuAAC reactions. (Reproduced from Serdjukow *et al.* [64], with permission of Royal Society of Chemistry.)

2′-*O*-propargyl-modified uridine analog during the automated solid-phase oligonucleotide synthesis. Subsequently, an additional 2′-*O*-propargyl-modified uridine derivative was incorporated before cleavage of the ODN from the solid support. The purified ODN was then subjected to a second click reaction in solution with selected dyes without requiring any presynthetic protection and deprotection steps. Authors have then studied in detail the thermal denaturation and fluorescence properties of the obtained labeled ODNs. As a result, efficient hybridization with complementary DNA and RNA was observed for several probes. Given its UV/Vis absorbance and fluorescence properties, those new internally labeled xanthene and cyanine ODN could be used for future application in fluorescence bioanalysis both *in vitro* and *in vivo*.

The preparation of dual emitting DNA probes in which the energy donor dye was inserted by postsynthetic CuAAC was also reported by Holzhauser *et al.* [83] Thus, a thiazol donor orange azide (thiazol orange (TO)) was clicked at the 2′-position of a uridine nucleotide to afford the thiazol-orange-labeled uridine nucleotide (U-TO) (Figure 10.24). The chosen site of attachment brought the TO into close proximity to the thiazole red (TR), which is a requirement for an efficient and noncollisional energy transfer. In this configuration, the base pairing of the modified uridine is maintained with its complementary adenosine nucleotide. Therefore, the uridine–adenosine pair was able to block the undesired excitonic interaction, which was observed in the past between TO and TR in certain orientation, when a base surrogate was used, affecting the efficiency of the energy transfer. Consequently, excitation at 490 nm provided a fluorescence, which was only the result of the energy transfer between U-TO and bs-TR since no fluorescence of the excitonically interacting TO TR pair was observed.

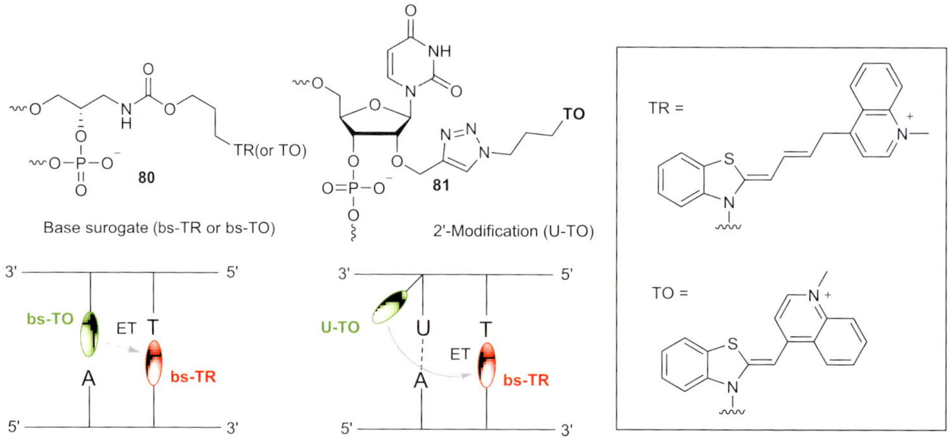


**Figure 10.24** (a) bs-TO and bs-TR as two base surrogates allow excitionic interactions that interfere with energy transfer (ET). (b) U-TO clicked at 2′-position of uridine, U-A base pairing block undesired excitonic interactions. (Reproduced from Astakhova and Wengel [82], with permission of Wiley.)

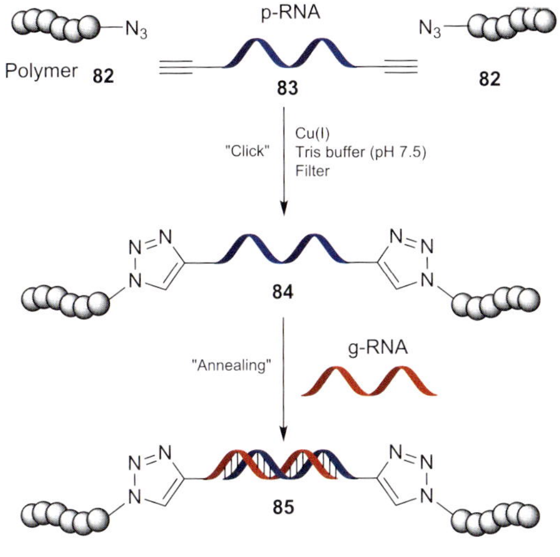


**Figure 10.25** Synthesis of the polymer-escorted siRNA.

In 2013, Averick and coworkers reported the CuAAC-assisted preparation of a siRNA polymer construct for the efficient delivery of RNAi agent for gene knockdown (Figure 10.25) [84]. Click reaction between bis-alkyne-terminated p-RNAs **83** with three different azide-containing polymers provided the desired bis-conjugation products **84,** which underwent final annealation with the complementary 21-mer guide strand (g-RNA) to afford the polymer escort duplex PEp-siRNA **85**. The polymer escort architecture appeared to be completely stable, remaining intact after 2-hour exposure to RNaseA, and offered an enhanced cell penetration to the RNA. As a result, this polymer escort architecture can be used to protect not only the bisconjugated p-RNA strand but also the annealed g-RNA sequence from the nuclease activity. A proof-of-concept experiment in which Drosophila S2 cells were transfected with Firefly luciferase (FLuc) and Renilla luciferase (RLuc) permitted the evaluation of RNAi-mediated knockdown in a dual luciferase assay. As expected, all three PEp-siRNA resulted in effective knockdown of the RLuc signal without requiring a transfection reagent. The concept was later validated for the knockdown of an endogenous gene in human embryonic kidney cells (HECK293) coding for a lymphocyte-specific protein tyrosine kinase (LcK), which is particularly involved in signal transductions in T-cells. When treated with Lck-PNEp-siRNA, specific and reproducible knockdown of Lck protein was observed without adding any transfection agent.

The CuAAC has also been used to identify the cellular target of platinum-based chemotherapeutics [85]. DeRose *et al.* have used Picazoplatin **90**, an azide-modified version of the anticancer drug of picoplatin **89**, which readily binds to DNA and RNA oligonucleotides, offering a site for selective click-mediated labeling with fluorophores such as the dansyl alkyne **87** (Figure 10.26) [86].

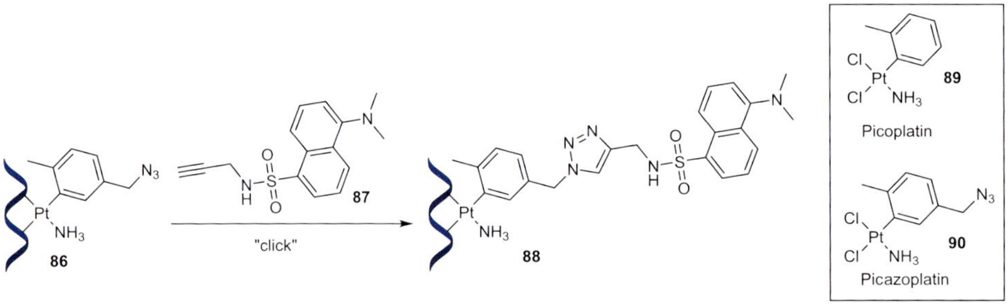


**Figure 10.26** Click reaction between DNA-bounded picazoplatin and a dansyl fluorophore.

The CuAAC reaction has been also used for the preparation of cyclic and catenated DNA. In 2007, Brown *et al.* reported the CuAAC-mediated labeling of an Single-Stranded Oligodeoxynucleotide (ssODN), in PEp-siRNA, bearing both an alkyne and an azide, respectively, at 5′- and 3′-extremities [87]. Under high dilution, those alkyne–azide-modified DNA strands can easily be cyclized under CuAAC conditions to produce cyclic ssODN **92** (Figure 10.27). Hybridization of the obtained cyclic DNA with its alkyne–azide-modified complementary sequence led to the formation of the covalently closed double-stranded catenane **94,** which, interestingly, exhibited a high resistance to exonuclease degradation. Similarly, Matsuda and coworkers synthesized a cyclic Double-Stranded Oligodeoxynucleotide (dsODN), capable of binding to the kB transcription factor and inhibiting *in vitro* and *ex-vivo* transcription [88].

Seela and coworkers have applied the CuACC to the construction and assembly of cross-linked [89] or branched Y-shaped [90] DNA (Figure 10.28). To prepare the Y-shaped DNA, a dendronized oligonucleotide containing a 7-tripropargylamine substituted 7-deaza-8-azaguanine was prepared (DNA-1), and subsequently, the two terminal alkynes were clicked with the bis-azide **96** to produce the bis-azido-DNA **97**. This intermediate was finally reacted in a second click reaction with DNA-2 containing a 7-ethynylated 7-deaza-8-azaguanine base to afford the desired Y-shaped three-armed DNA **99**. Annealing of the hyperbranched single strands with their complementary sequences resulted in supramolecular assemblies.

Seela also developed a new type of unsymmetrical diazido derivative that can, through its chelating properties, allow for the chemoselective tagging and

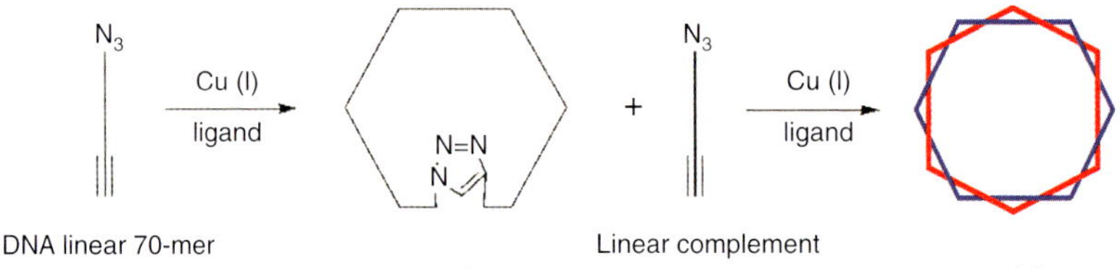


**Figure 10.27** Synthesis of catenated DNA by click chemistry.

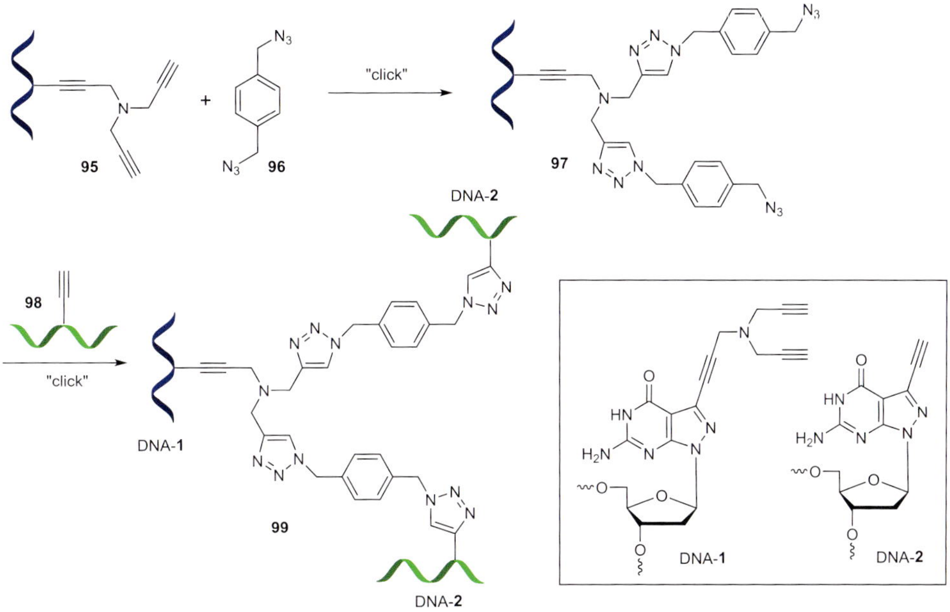


**Figure 10.28** Construction of branched DNA structures using CuAAC.

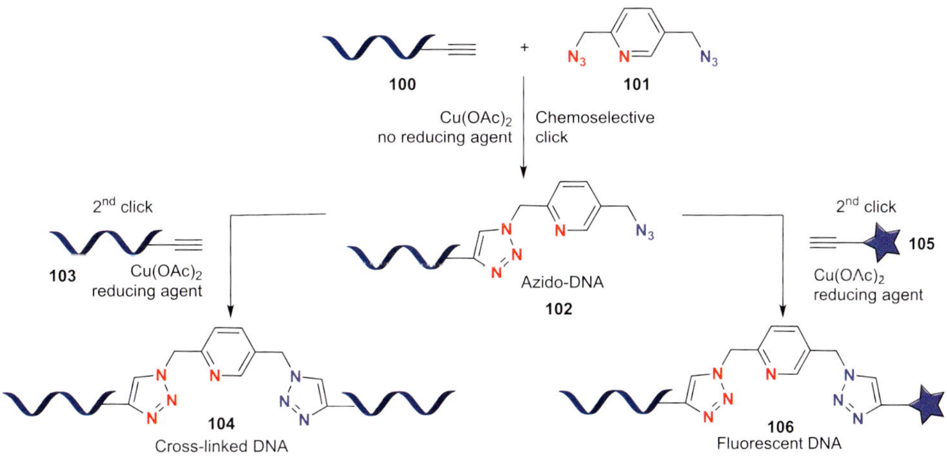


**Figure 10.29** Stepwise click reactions on DNA using a chelating and a nonchelating azido group containing linker.

cross-linking of DNA (Figure 10.29) [91]. Thus, 2,5-bis-(azidomethyl)pyridine **101** can selectively react with an alkynyl-modified DNA in the presence of copper(II) acetate, but in the absence of a reducing agent, to solely form azido DNA **102**. A second click reaction can then be performed with either an alkynyl DNA or a fluorescent tag to form the desired fluorescent DNA and/or cross-linked DNA.

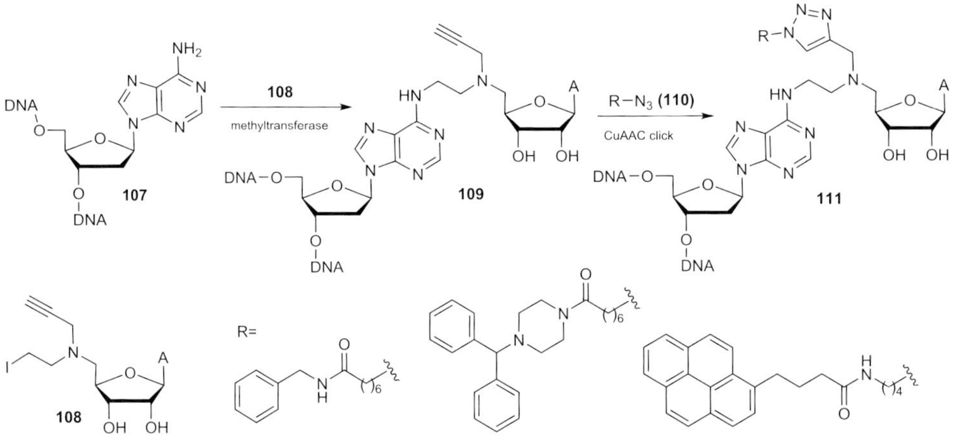


**Figure 10.30** Site-specific click reaction.

Weller and coworkers reported the discovery of an adenosine-derived *N*-mustard analog **108** that can be used to selectively introduce an alkynyl group onto ODNs via methyltransferases (Mtases) (Figure 10.30) [92]. Subsequently, the propargyl derivative formed was easily clicked with different azides to form various derivatized DNA species.

Alternatively, alkyne-containing phosphoramidite derivatives can be used for the labeling of the 5′-end of ODNs [93]. Thus, Morvan *et al.* reported the synthesis of bis-alkyne phosphoramidite **113,** which was incorporated to an ODN and subsequently reacted with different azide derivatives to afford thiohexyl, ferrocenyl, or carbohydrate moieties conjugated ODNs **115** (Figure 10.31) [94].

CuACC has also found important applications in the field of DNA nanotechnology [95]. For instance, Fishler *et al.* have reported the immobilization of azide-terminated gold nanoparticles to an alkyne-modified DNA, which led to the formation of a chain-like DNA–nanoparticle conjugates (Figure 10.32) [96].

During the past decade, the development of copper-free click reactions has facilitated the use of click reaction in the context of complex biological medium by negating the use of copper ions known to be toxic at concentrations employed for labeling [97]. Even though the catalyst-free SPAAC reaction between azido and cyclooctyne derivatives for the preparation of bioconjugates has been around since, at least, 2004 [98, 99], it is not until 2011 that it will be used for the bioconjugation of DNA [100]. Thus, a dibenzocyclooctyne (DIBO) phosphoramidite **122** was successfully introduced at the end of the SPOS to produce a 5′-DIBO-modified ODN, which was clicked via a SPAAC with different bioconjugates (Figure 10.33). Alternatively, the SPAAC step could also be performed at the end of SPOS before cleavage from solid support. In addition, a 500 bp sequence from bacteriophage $\lambda$ DNA was amplified via PCR using the DIBO-modified ODN as a primer, which confirmed the stability of the 5-DIBO group even under PCR conditions.

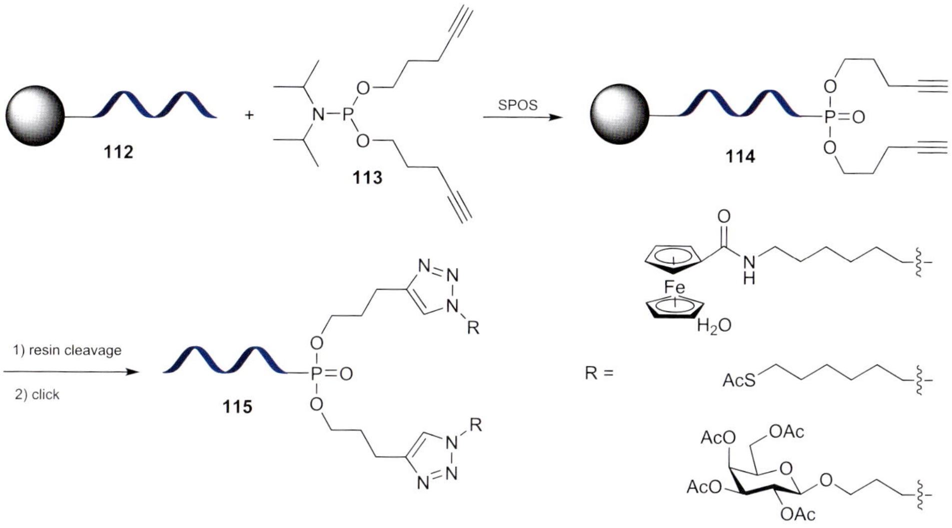


**Figure 10.31** Postsynthetic CuAAC click reactions with a 5′-bisalkyne ODN.

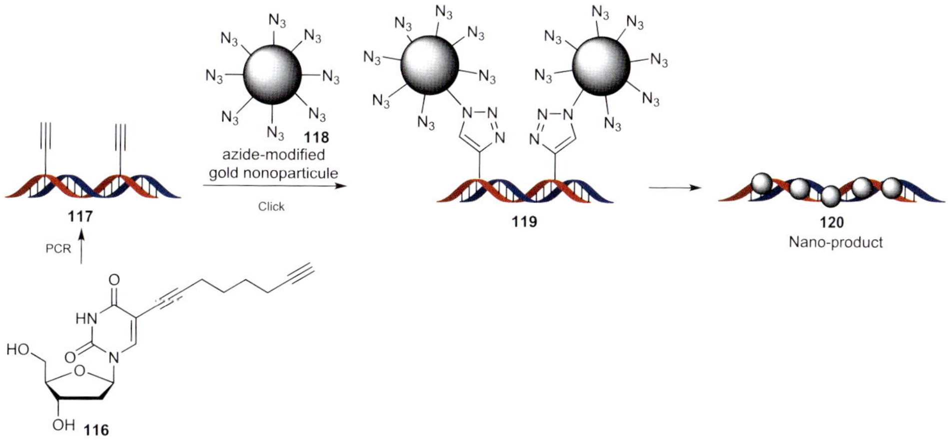


**Figure 10.32** Immobilization of azide-modified gold nanoparticles to alkyne-modified DNA.

Later, Brown and coworkers reported a fast and efficient methodology for DNA cross-linking using SPAAC click reaction [101]. A 14-mer ODN containing a DIBO attached to either 2′- or 5-position of a thymine along with several complementary azide-containing ODNs was synthesized and successfully clicked together (Figure 10.34).

Brown *et al.* also reported the use of SPAAC for the labeling of a DNA strand containing an azidomethyl dU with a fluorescent tag attached to a BCN (bicyclo[6.1.0]non-4-yne) derivative (Figure 10.35) [102].

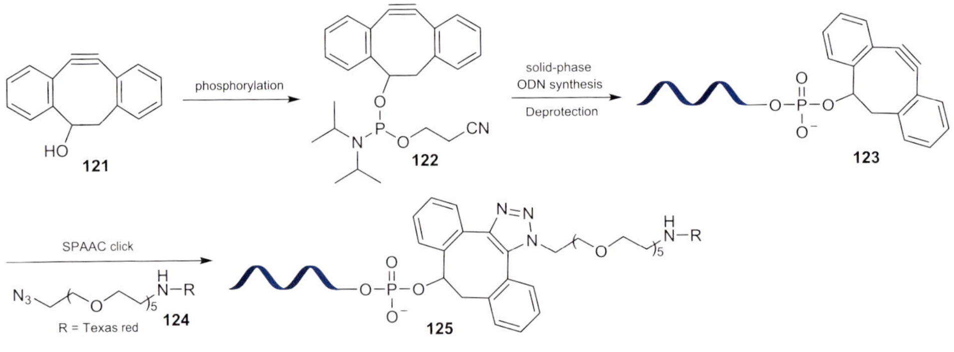


**Figure 10.33** SPAAC click between DIBO and Texas red tag for terminal labeling of DNA.

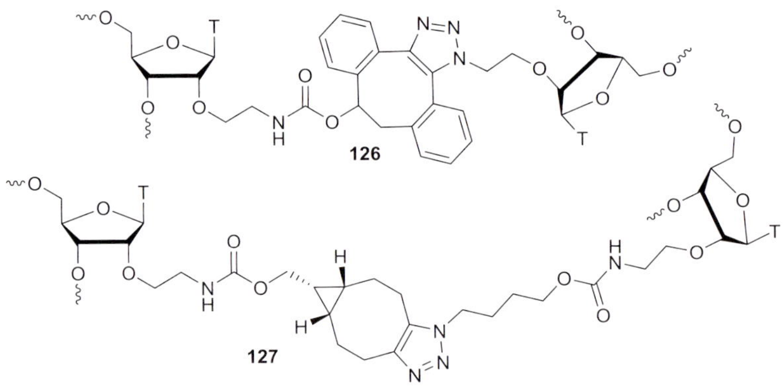


**Figure 10.34** DNA cross-linking using copper-free SPAAC click reactions with DIBO and BCN (R = DNA).

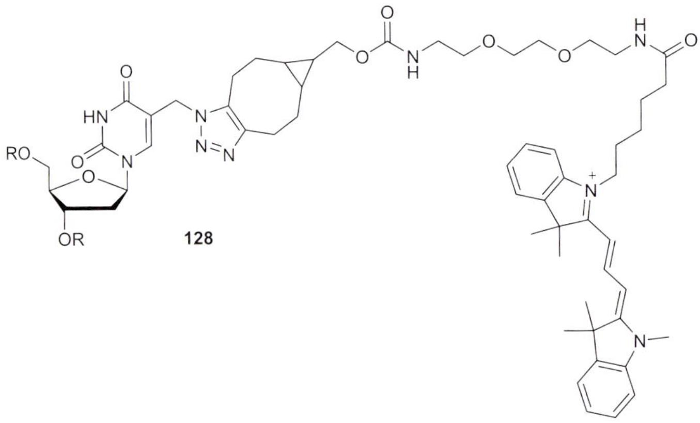


**Figure 10.35** SPAAC click product obtained by reaction between 5-azidomethyl dU and BCN-labeled fluorophore.

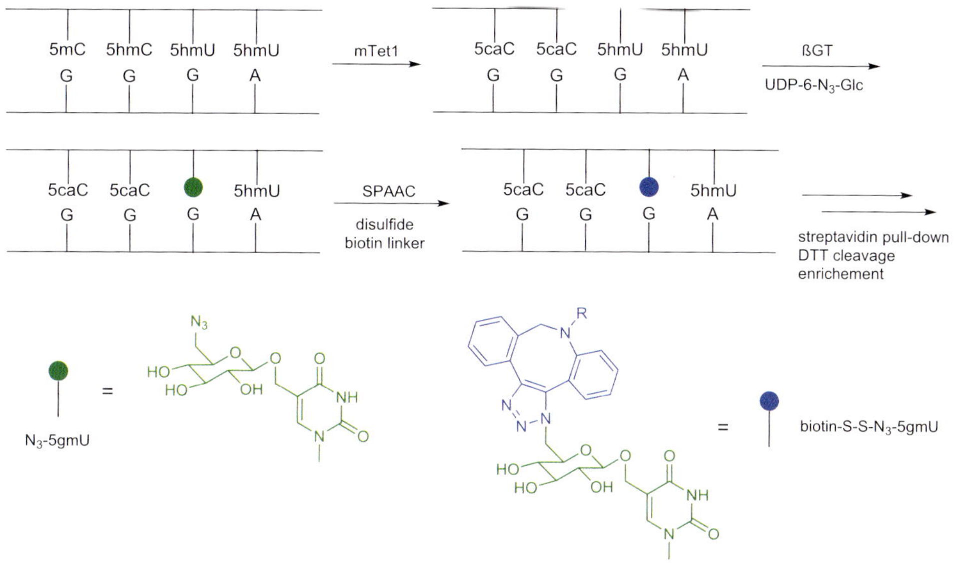


**Figure 10.36** Selective SPAAC-mediated biotin labeling of mismatched 5hmU.

Recently, Miao and coworkers developed a methodology for the detection of mismatched 5-hydroxymethyluracil (5hmU) in DNA by selective SPAAC-mediated labeling (Figure 10.36) [103]. It is assumed that 5-hmU is formed by deamination of 5-hydroxymethylcytosine (5-hmC), which is involved in various biological functions such as genomic imprinting, X chromosome inactivation, and cancer development [104]. First, a recombinant mouse Tet1 (mTet1) was used to oxidize all 5-methylcytosine (5mC) and 5-hydromethylcytosine (5hmC) present in the genomic DNA into 5-carboxylcytosine (5caC). Subsequently, β-glucosyltransferase (βGT) was used to selectively attach a modified $N_3$-glucose onto the hydroxyl group of 5hmU. This step was followed by a SPAAC-mediated labeling with a cyclic alkyne bearing a biotin. Furthermore, after capture of the mismatched-5hmU-containing fragments with streptavidin-coupled beads, the bound DNA fragments were released by cleavage of the disulfide bond. Deep sequencing of the enriched fragments gave an overview of the distribution of mismatched 5mhU.

## 10.3 Conclusion

Because of its modularity, high yields, simple reaction conditions, and purification procedures, the CuAAC has been especially suitable for the synthesis of a large number of modified nucleosides, nucleotides, and nucleic acids with a wide range of applications. Its use in biological systems was, for a while, limited by the inherent toxicity of $Cu^I$ in cells. The use of Cu(I)-stabilizing ligand, but also the recent development of highly strained alkynes capable of reacting in an uncatalyzed version of the azide–alkyne cycloaddition, may provide novel innovative solutions and be the key to new uncharted territories. Pandora's box is now opened!

## Acknowledgments

This work was supported in part by NIH CFAR grant 2P30-AI-050409 (to RFS).

## References


1. (a) Chu, C.K. and Baker, D.C. (eds) (1993) , Plenum Press, New York; (b) De Clercq, E. (2004) *J. Clin. Virol.*, **30**, 115–133; (c) Liu-Young, G. and Kozal, M.J. (2008) *AIDS Patient Care STDS*, **22**, 449–457; (d) Ayoub, W.S. and Keefe, E.B. (2008) *Aliment. Pharmacol. Ther.*, **28**, 167–177; (e) Jochmans, D. (2008) *Virus Res.*, **134**, 157–170; (f) Blagosklonny, M.V. (2004) *Cell Cycle*, **3**, 1033–1040; (g) Robak, T., Korycka, A., Kasznicki, M., Wrzesien-Kus, A., and Smolewski, P. (2005) *Curr. Cancer Drug Targets*, **5**, 421–444; (h) Miura, S. and Izuta, S. (2004) *Curr. Drug Targets*, **5**, 191–195.
2. Lee, Y.S., Park, S.M., Kim, H.M., Park, S.K., Lee, K., Lee, C.W., and Kim, B.H. (2009) *Bioorg. Med. Chem. Lett.*, **19**, 4688–4691.
3. Lakshman, M.K., Kumar, A., Balachandran, R., Day, B.W., Andrei, G., Snoeck, R., and Balzarini, J. (2012) *J. Org. Chem.*, **77**, 5870–5883.
4. Lakshman, M.K., Singh, M.K., Parrish, D., Balachandran, R., and Day, B.W. (2010) *J. Org. Chem.*, **75**, 2461–2473.
5. Cosyn, L., Gao, Z.G., Van Rompaey, P., Lu, C., Jacobson, K.A., and Van Calenbergh, S. (2006) *Bioorg. Med. Chem.*, **14**, 1403–1412.
6. Montagu, A., Roy, V., Balzarini, J., Snoeck, R., Andrei, G., and Agrofoglio, L.A. (2011) *Eur. J. Med. Chem.*, **4**, 778–786.
7. Lau, J.Y., Tam, R.C., Liang, T.J., and Hong, Z. (2002) *Hepatology*, **35**, 1002–1009.
8. Lin, C.C., Yeh, L.T., Vitarella, D., and Hong, Z. (2003) *Antiviral Chem. Chemother.*, **14**, 145–152.
9. Alvarez, R., Velazquez, S., San-Felix, A., Aquaro, S., Clercq, E.D., Perno, C.-F., Karlsson, A., Balzarini, J., and Camarasa, M.J. (1994) *J. Med. Chem.*, **37**, 4185–4194.
10. (a) Perez-Castro, I., Caamano, O., Fernandez, F., Garcia, M.D., Lopez, C., and De Clercq, E. (2007) *Org. Biomol. Chem.*, **5**, 3805–3813; (b) Cho, J.H., Bernard, D.L., Sidwell, R.W., Kern, E.R., and Chu, C.K. (2006) *J. Med. Chem.*, **49**, 1140–1148; (c) Goeminne, A., McNaughton, M., Bal, G., Surpateanu, G., Van der Veken, P., De Prol, S., Versees, W., Steyaert, J., Apers, S., Haemers, A., and Augustyns, K. (2007) *Bioorg. Med. Chem. Lett.*, **17**, 2523–2526; (d) Broggi, J., Diez-Gonzalez, S., Petersen, J.L., Berteina-Raboin, S., Nolan, S.P., and Agrofoglio, L.A. (2008) *Synthesis*, **1**, 141–148; (e) Broggi, J., Joubert, N., Aucagne, V., Zevaco, T., Berteina-Raboin, S., Nolan, S.P., and Agrofoglio, L.A. (2007) *Nucleosides Nucleotides Nucleic Acids*, **26**, 779–783; (f) Pradere, U., Roy, V., McBrayer, T.R., Schinazi, R.F., and Agrofoglio, L.A. (2008) *Tetrahedron*, **64**, 9044–9051.
11. Driowya, M., Puissant, A., Robert, G., Auberger, P., Benhida, R., and Bougrin, K. (2012) *Ultrason. Sonochem.*, **19**, 1132–1138.

12. Guezguez, R., Bougrin, K., El Akri, K., and Benhida, R. (2006) *Tetrahedron Lett.*, **47**, 4807–4811.
13. Ermolat'ev, D.S., Mehta, V.P., and Van der Eycken, E.V. (2007) *QSAR Comb. Sci.*, **26**, 1266–1273.
14. (a) Broggi, J., Joubert, N., Aucagne, V., Berteina-Raboin, S., Diez-Gonzalez, S., Nolan, S.P., Topalis, D., Deville-Bonne, D., Balzarini, J., Neyts, J., Andrei, G., Snoeck, R., and Agrofoglio, L. (2007) *Nucleosides Nucleotides Nucleic Acids*, **26**, 1391–1394; (b) Broggi, J., Joubert, N., Diez-Gonzalez, S., Berteina-Raboin, S., Zevaco, T., Nolan, S.P., and Agrofoglio, L.A. (2009) *Tetrahedron*, **65**, 1162–1170.
15. Dell'Isola, A., McLachlan, M.M., Neuman, B.W., Al-Mullah, H.M., Binks, A.W., Elvidge, W., Shankland, K., and Cobb, A.J. (2014) *Chem. Eur. J.*, **20**, 11685–11689.
16. Li, W., Xia, Y., Fan, Z., Qu, F., Wu, Q., and Peng, L. (2008) *Tetrahedron Lett.*, **49**, 2804.
17. Wang, M., Zhu, R., Fan, Z., Fu, Y., Feng, L., Yao, J., Maggiani, A., Xia, Y., Qu, F., and Peng, L. (2011) *Bioorg. Med. Chem. Lett.*, **21**, 354–357.
18. Xia, Y., Wang, M., Demaria, O., Tang, J., Rocchi, P., Qu, F., Iovanna, J.L., Alexopoulou, L., and Peng, L. (2012) *J. Med. Chem.*, **55**, 5642–5646.
19. (a) Youcef, R.A., Santos, M.D., Roussel, S., Baltaze, J.-P., Lubin-Germain, N., and Uziel, J. (2009) *J. Org. Chem.*, **74**, 4318–4323; (b) Kitamura, Y., Edayoshi, K., and Kitade, Y. (2012) *Tetrahedron Lett.*, **53**, 6987–6989.
20. Seley, K.L., Zhang, L., and Hagos, A. (2001) *Org. Lett.*, **3**, 3209–3210.
21. Seley, K.L., Zhang, L., Hagos, A., and Quirk, S. (2002) *J. Org. Chem.*, **67**, 3365–3373.
22. St Amant, A.H., Bean, L.A., Guthrie, J.P., and Hudson, R.H. (2012) *Org. Biomol. Chem.*, **10**, 6521–6525.
23. Parmenopoulou, V., Chatzileontiadou, D.S., Manta, S., Bougiatioti, S., Maragozidis, P., Gkaragkouni, D.N., Kaffesaki, E., Kantsadi, A.L., Skamnaki, V.T., Zographos, S.E., Zounpoulakis, P., Balatsos, N.A., Komiotis, D., and Leonidas, D.D. (2012) *Bioorg. Med. Chem.*, **20**, 7184–7193.
24. Kumar, S. (2015) *Nucleosides Nucleotides Nucleic Acids*, **34**, 371–378.
25. O'Mahony, G., Svensson, S., Sundgren, A., and Grotli, M. (2008) *Nucleosides Nucleotides Nucleic Acids*, **27**, 449–459.
26. Roy, V., Obikhod, A., Zhang, H.W., Coats, S.J., Herman, B.D., Sluis-Cremer, N., Agrofoglio, L.A., and Schinazi, R.F. (2011) *Nucleosides Nucleotides Nucleic Acids*, **30**, 264–270.
27. Zhou, L., Amer, A., Korn, M., Burda, R., Balzarini, J., De Clercq, E., Kern, E.R., and Torrence, P.F. (2005) *Antiviral Chem. Chemother.*, **16**, 375–383.
28. Habich, D. and Barth, W. (1989) *Heterocycles*, **11**, 2083–2088.
29. Hirota, K., Hosono, H., Kitade, Y., Maki, Y., Chu, C.K., Schinazi, R.F., Nakane, H., and Ono, K. (1990) *Chem. Pharm. Bull.*, **38**, 2597–2601.
30. Wu, J., Yu, W., Fu, L., He, W., Wang, Y., Chai, B., Song, C., and Chang, J. (2013) *Eur. J. Med. Chem.*, **63**, 739–745.
31. Sirivolu, V.R., Vernekar, S.K., Ilina, T., Myshakina, N.S., Parniak, M.A., and Wang, Z. (2013) *J. Med. Chem.*, **56**, 8765–8780.
32. Somu, R.V., Boshoff, H., Qiao, C., Bennett, E.M., Barry, C.E., and Aldrich, C.C. (2006) *J. Med. Chem.*, **49**, 31–34.
33. Lee, L., Chang, K.-H., Valiyev, F., Liu, H.-J., and Li, W.-S. (2006) *J. Chin. Chem. Soc.*, **53**, 1547–1555.
34. Glowacka, I.E., Balzarini, J., and Wroblewski, A.E. (2013) *Eur. J. Med. Chem.*, **70**, 703–722.
35. Elayadi, H., Smietana, M., Pannecouque, C., Leyssen, P., Neyts, J., Vasseur, J.J., and Lazrek, H.B. (2010) *Bioorg. Med. Chem. Lett.*, **20**, 7365–7368.
36. Krim, J., Taourirte, M., and Engels, J.W. (2012) *Molecules*, **17**, 179–190.
37. Diab, S.A., De Schutter, C., Muzard, M., Plantier-Royon, R., Pfund, E., and Lequeux, T. (2012) *J. Med. Chem.*, **55**, 2758–2768.
38. Glowacka, I.E., Balzarini, J., and Wroblewski, A.E. (2014) *Arch. Pharm.*, **347**, 506–514.

39. (a) Olejniczak, A., Wojtczak, B., and Lesnikowski, Z.J. (2007) *Nucleosides Nucleotides Nucleic Acids*, **26**, 1611–1613; (b) Wojtczak, B.A., Andrysiak, A., Gruner, B., and Lesnikowski, Z.J. (2008) *Chemistry*, **14**, 10675–10682; (c) Semioshkin, A., Laskova, J., Wojtczak, B., Andrysiak, A., Godovikov, I., Bregadze, V., and Lesnikowski, Z.J. (2009) *J. Organomet. Chem.*, **694**, 1375–1379.
40. Kosiova, I., Kovackova, S., and Kois, P. (2007) *Tetrahedron*, **63**, 312–320.
41. Arya, A., Mathur, D., Tyagi, A., Kumar, R., Kumar, V., Olsen, C.E., Saxena, R.K., and Prasad, A.K. (2013) *Nucleosides Nucleotides Nucleic Acids*, **32**, 646–659.
42. Jawalekar, A.M., Meeuwenoord, N., Cremers, J.G.O., Overkleeft, H.S., van der Marel, G.A., Rutjes, F.P.J.T., and van Delft, F.L. (2008) *J. Org. Chem.*, **73**, 287–290.
43. Baraniak, D., Kacprzak, K., and Celewicz, L. (2011) *Bioorg. Med. Chem. Lett.*, **21**, 723–726.
44. Schulz, J., Vimont, D., Bordenave, T., James, D., Escudier, J.M., Allard, M., Szlosek-Pinaud, M., and Fouquet, E. (2011) *Chemistry*, **17**, 3096–3100.
45. Zayas, J., Annoual, M.D., Das, J.K., Felty, Q., Gonzalez, W.G., Miksovska, J., Sharifai, N., Chiba, A., and Wnuk, S.F. (2015) *Bioconjugate Chem.*, ASAP.
46. Lee, L.V., Mitchell, M.L., Huang, S.J., Fokin, V.V., Sharpless, K.B., and Wong, C.H. (2003) *J. Am. Chem. Soc.*, **125**, 9588–9589.
47. Danel, K., Larsen, L.M., Pedersen, E.B., Sanna, G., La Colla, P., and Loddo, R. (2008) *Bioorg. Med. Chem.*, **16**, 511–517.
48. Bori, I.D., Hung, H.Y., Qian, K., Chen, C.H., Morris-Natschke, S.L., and Lee, K.H. (2012) *Tetrahedron Lett.*, **53**, 1987–1989.
49. Dang Thi, T.A., Kim Tuyet, N.T., Pham The, C., Thanh Nguyen, H., Ba Thi, C., Thi Phuong, H., Van Boi, L., Van Nguyen, T., and D'hooghe, M. (2015) *Tetrahedron Lett.*, **56**, 218–224.
50. Dobkowski, M., Szychowska, A., Pieszko, M., Miszka, A., Wojciechowska, M., Alenowicz, M., Ruczynski, J., Rekowski, P., Celewicz, L., Barciszewski, J., and Mucha, P. (2014) *J. Pept. Sci.*, **20**, 696–703.
51. Lolk, L., Pøhlsgaard, J., Jepsen, A.S., Hansen, L.H., Nielsen, H., Steffansen, S.I., Sparving, L., Nielsen, A.B., Vester, B., and Nielsen, P. (2008) *J. Med. Chem.*, **51**, 4957–4967.
52. (a) Jin, X., Yang, R., Jin, P., Xiao, Q., and Ju, Y. (2007) *Synthesis*, **19**, 2967–2972; (b) Jin, X., Ding, H., Yang, R., Xiao, Q., and Ju, Y. (2008) *Synthesis*, **6**, 865–870.
53. Mindt, T.L., Struthers, H., Brans, L., Anguelov, T., Schweinsberg, C., Maes, V., Tourwe, D., and Schibli, R. (2006) *J. Am. Chem. Soc.*, **128**, 15096–15097.
54. Nikan, M., Bare, G.A.L., and Sherman, J.C. (2011) *Tetrahedron Lett.*, **52**, 1791–1793.
55. El-Sagheer, A.H. and Brown, T. (2009) *J. Am. Chem. Soc.*, **131**, 3958–3964.
56. (a) El-Sagheer, A.H., Sanzone, A.P., Gao, R., Tavassoli, A., and Brown, T. (2011) *Proc. Natl. Acad. Sci. U.S.A.*, **108**, 11338–11343; (b) Tsaftaris, A., Pasentzis, K., and Argiriou, A. (2010) *Biotechnol. Lett*, **32**, 157–161.
57. Birts, C.N., Sanzone, A.P., El-Sagheer, A.H., Blaydes, J.P., Brown, T., and Tavassoli, A. (2014) *Angew. Chem. Int. Ed.*, **53**, 2362–2365.
58. (a) Isobe, H. and Fujino, T. (2014) *Chem. Rec.*, **14**, 41–51; (b) Isobe, H., Fujino, T., Yamazaki, N., Guillot-Nieckowski, M., and Nakamura, E. (2008) *Org. Lett.*, **10**, 3729–3732.
59. Chittepu, P., Sirivolu, V.R., and Seela, F. (2008) *Bioorg. Med. Chem.*, **16**, 8427–8439.
60. Ingale, S.A. and Seela, F. (2014) *Tetrahedron*, **70**, 380–391.
61. (a) Lin, N., Yan, Z., Altier, C., Li, M., Carrasco, N., Suyemoto, M., Johnston, L., Wang, S., Wang, Q., Fang, H., Caton-Williams, J., and Wang, B. (2007) *Nucleic Acid Res.*, **35**, 1222–1229; (b) Li, M., Lin, N., Huang, Z., Du, L., Altier, C., Fang, H., and Wang, B. (2008) *J. Am. Chem. Soc.*, **130**, 12636–12638.
62. Grijalvo, S., Ocampo, S.M., Perales, J.C., and Eritija, R. (2010) *J. Org. Chem.*, **75**, 6806–6813.

63. Godeau, G., Staedel, C., and Barthelemy, P. (2008) *J. Med. Chem.*, **51**, 4374–4376.
64. Serdjukow, S., Kink, F., Steigenberger, B., Tomas-Gamasa, M., and Carell, T. (2014) *Chem. Commun.*, **50**, 1861–1863.
65. (a) Gramlich, P.M.E., Wirges, C.T., Manetto, A., and Carell, T. (2008) *Angew. Chem. Int. Ed.*, **47**, 8350–8358; (b) El-Sagheer, A.H. and Brown, T. (2010) *Chem. Soc. Rev.*, **39**, 1388–1405.
66. (a) Burrows, C.J. and Muller, J.G. (1998) *Chem. Rev.*, **98**, 1109–1152; (b) Thyagarajan, S., Murthy, N.N., Narducci Sarjeant, A.A., Karlin, K.D., and Rokita, S.E. (2006) *J. Am. Chem. Soc.*, **128**, 7003–7008.
67. Seo, T.S., Li, Z., Ruparel, H., and Ju, J. (2003) *J. Org. Chem.*, **68**, 609–612.
68. Debets, M.F., Van Berkel, S.S., Dommerholt, J., Dirks, A.J., Rutjes, F.P.J.T., and Van Delft, F.L. (2011) *Acc. Chem. Res.*, **44**, 805–815.
69. Beyer, C. and Wagenknecht, H.A. (2010) *Chem. Commun.*, **46**, 2230–2231.
70. Gierlich, J., Burley, G.A., Gramlich, P.M.E., Hammond, D.M., and Carell, T. (2006) *Org. Lett.*, **8**, 3639–3642.
71. Seela, F. and Pujari, S.S. (2010) *Bioconjugate Chem.*, **21**, 1629–1641.
72. Gierlich, J., Gutsmiedl, K., Gramlich, P.M.E., Schmidt, A., Burley, G.A., and Carell, T. (2007) *Chem. Eur. J.*, **13**, 9486–9494.
73. Sirivolu, V.R., Chittepu, P., and Seela, F. (2008) *ChemBioChem*, **9**, 2305–2316.
74. Gramlich, P.M.E., Warncke, S., Gierlich, J., and Carell, T. (2008) *Angew. Chem. Int. Ed.*, **47**, 3442–3444.
75. Kovacic, S., Samii, L., Lamour, G., Li, H.B., Linke, H., Bromley, E.H.C., Woolfson, D.N., Curmi, P.M.G., and Forde, N.R. (2014) *Biomacromolecules*, **15**, 4065–4072.
76. Onizuka, K., Nishioka, T., Li, Z., Jitsuzaki, D., Taniguchi, Y., and Sasaki, S. (2012) *Chem. Commun.*, **48**, 3969–3971.
77. (a) Beranek, D.T. (1990) *Mutat. Res.*, **231**, 11–30; (b) Shrivastav, N., Li, D., and Essigmann, J.M. (2010) *Carcinogenesis*, **31**, 59–70.
78. (a) Loechler, E.L., Green, C.L., and Essigmann, J.M. (1984) *Proc. Natl. Acad. Sci. U.S.A.*, **81**, 6271–6275; (b) Saffhill, R., Margison, G.P., and O'Connor, P.J. (1985) *Biochim. Biophys. Acta*, **823**, 111–145.
79. Altshuler, K.B., Hodes, C.S., and Essigmann, J.M. (1996) *Chem. Res. Toxicol.*, **6**, 980–987.
80. Margison, G.P., Koref, M.F.S., and Povey, A.C. (2002) *Mutagenesis*, **17**, 483–487.
81. Berndl, S., Herzig, N., Kele, P., Lachmann, D., Li, X.H., Wolfbeis, O.S., and Wagenknecht, H.A. (2009) *Bioconjugate Chem.*, **20**, 558–564.
82. Astakhova, I.K. and Wengel, J. (2013) *Chem. Eur. J.*, **19**, 1112–1122.
83. Holzhauser, C., Rubner, M.M., and Wagenknecht, H.A. (2013) *Photochem. Photobiol. Sci.*, **12**, 722–724.
84. Averick, S.E., Paredes, E., Dey, S.K., Snyder, K.M., Tapinos, N., Matyjaszewski, K., and Das, S.R. (2013) *J. Am. Chem. Soc.*, **135**, 12508–12511.
85. Ding, S., Qiao, X., Kucera, G.L., and Bierbach, U. (2012) *J. Med. Chem.*, **55**, 10198–10203.
86. (a) White, J.D., Osborn, M.F., Moghaddam, A.D., Guzman, L.E., Haley, M.M., and DeRose, V.J. (2013) *J. Am. Chem. Soc.*, **135**, 11680–11683; (b) Moghaddam, A.D., White, J.D., Cunningham, R.M., Loes, A.N., Haley, M.M., and DeRose, V.J. (2015) *J. Chem. Soc., Dalton Trans.*, **44**, 3536–3539.
87. Kumar, R., El-Sagheer, A., Tumpane, J., Lincoln, P., Wilhelmsson, L.M., and Brown, T. (2007) *J. Am. Chem. Soc.*, **129**, 6859–6864.
88. Nakane, M., Ichikawa, S., and Matsuda, A. (2008) *J. Org. Chem.*, **73**, 1842–1851.
89. Pujari, S.S., Ingale, S.A., and Seela, F. (2014) *Bioconjugate Chem.*, **25**, 1855–1870.
90. Xiong, H., Leonard, P., and Seela, F. (2012) *Bioconjugate Chem.*, **23**, 856–870.
91. Ingale, S.A. and Seela, F. (2013) *J. Org. Chem.*, **78**, 3394–3399.
92. Weller, R.L. and Rajski, S.R. (2005) *Org. Lett.*, **7**, 2141–2144.

93. (a) Morvan, F., Meyer, A., Jochum, A., Sabin, C., Chevolot, Y., Imberty, A., Praly, J.P., Vasseur, J.J., Souteyrand, E., and Vidal, S. (2007) *Bioconjugate Chem.*, **18**, 1637–1643; (b) Moni, L., Pourceau, G., Zhang, J., Meyer, A., Vidal, S., Souteyrand, E., Dondoni, A., Morvan, F., Chevolot, Y., Vasseur, J.J., and Marra, A. (2009) *ChemBioChem*, **10**, 1369–1378; (c) Seela, F., Xiong, H., and Budow, S. (2010) *Tetrahedron*, **66**, 3930–3943; (d) Seela, F. and Ingale, S.A. (2010) *J. Org. Chem.*, **75**, 284–295.
94. Ligeour, C., Meyer, A., Vasseur, J.J., and Morvan, F. (2012) *Eur. J. Org. Chem.*, **9**, 1851–1856.
95. (a) Gerrard, S.R., Hardiman, C., Shelbourne, M., Nandhakumar, I., Norden, B., and Brown, T. (2012) *ACS Nano*, **6**, 9221–9228; (b) Qing, G., Xiong, H., Seela, F., and Sun, T. (2010) *J. Am. Chem. Soc.*, **132**, 15228–15232; (c) Cutler, J.I., Zheng, D., Xu, X., Giljohann, D.A., and Mirkin, C.A. (2010) *Nano Lett.*, **10**, 1477–1480; (d) Lundberg, E.P., El-Sagheer, A.H., Kocalka, P., Wilhelmsson, L.M., Brown, T., and Norden, B. (2010) *Chem. Commun.*, **46**, 3714–3716; (e) Shi, L., Jing, C., Ma, M., Li, D., Halls, J.E., and Marken, F. (2013) *Angew. Chem. Int Ed.*, **52**, 6011–6014; (f) Xu, X., Daniel, W.L., Wei, W., and Mirkin, C.A. (2010) *Small*, **6**, 623–626; (g) Choi, J.Y., Kim, Y.T., and Seo, T.S. (2013) *ACS Nano*, **7**, 2627–2633.
96. Fischler, M., Sologubenko, A., Mayer, J., Clever, G., Burley, G., Gierlich, J., Carell, T., and Simon, U. (2008) *Chem. Commun.*, **2**, 169–171.
97. Soares, E.V., Hebbelinck, K., and Soares, H.M.V.M. (2003) *Can. J. Microbiol.*, **49**, 336–343.
98. (a) Agard, N.J., Prescher, J.A., and Bertozzi, C.R. (2004) *J. Am. Chem. Soc.*, **126**, 15046–15047; (b) Codelli, J.A., Baskin, J.M., Agard, N.J., and Berozzi, C.R. (2008) *J. Am. Chem. Soc.*, **130**, 11486–11493; (c) Sletten, E.M. and Bertozzi, C.R. (2008) *Org. Lett.*, **10**, 3097–3099; (d) Baskin, J.M., Prescher, J.A., Laughlin, S.T., Agard, N.J., Chang, P.V., Miller, I.A., Lo, A., Codelli, J.A., and Bertozzi, C.R. (2007) *Proc. Natl. Acad. Sci. U.S.A.*, **104**, 16793–16797.
99. (a) Ning, X.H., Guo, J., Wolfert, M.A., and Boons, G.J. (2008) *Angew. Chem. Int. Ed.*, **47**, 2253–2255; (b) Debets, M.F., van Berkel, S.S., Schoffelen, S., Rutjes, F.P.J.T., van Hest, J.C.M., and van Delft, F.L. (2010) *Chem. Commun.*, **46**, 97–99.
100. Marks, I.S., Kang, J.S., Jones, B.T., Landmark, K.J., Cleland, A.J., and Taton, T.A. (2011) *Bioconjugate Chem.*, **22**, 1259–1263.
101. Shelbourne, M., Brown, T., El-Sagheer, A.H., and Brown, T. (2012) *Chem. Commun.*, **48**, 11184–11186.
102. Ren, X.M., El-Sagheer, A.H., and Brown, T. (2015) *Analyst*, **140**, 2671–2678.
103. Yu, M., Song, C.-X., and He, C. (2015) *Methods*, **72**, 16–20.
104. (a) Gal-Yam, E.N., Saito, Y., Egger, G., and Jones, P.A. (2008) *Annu. Rev. Med.*, **59**, 267–280; (b) Jones, G.E. (2010) *Nat. Rev. Mol. Cell Biol.*, **11**, 231.

# Index



*Click Reactions in Organic Synthesis*, First Edition. Edited by Srinivasan Chandrasekaran.